Lecture Notes in Computer Science 13422

More information about this series at https://link.springer.com/bookseries/558

Bohan Li · Lin Yue · Chuanqi Tao · Xuming Han ·
Diego Calvanese · Toshiyuki Amagasa (Eds.)

Web and Big Data

6th International Joint Conference, APWeb-WAIM 2022
Nanjing, China, November 25–27, 2022
Proceedings, Part II

 Springer

Editors
Bohan Li
Nanjing University of Aeronautics
and Astronautics
Nanjing, China

Chuanqi Tao
Nanjing University of Aeronautics
and Astronautics
Nanjing, China

Diego Calvanese ⓘ
Free University of Bozen-Bolzano
Bolzano, Italy

Lin Yue ⓘ
Newcastle University
Callaghan, NSW, Australia

Xuming Han
Jinan University
Guangzhou, China

Toshiyuki Amagasa ⓘ
University of Tsukuba
Tsukuba, Japan

ISSN 0302-9743   ISSN 1611-3349 (electronic)
Lecture Notes in Computer Science
ISBN 978-3-031-25197-9   ISBN 978-3-031-25198-6 (eBook)
https://doi.org/10.1007/978-3-031-25198-6

This Springer imprint is published by the registered company Springer Nature Switzerland AG
The registered company address is: Gewerbestrasse 11, 6330 Cham, Switzerland

Preface

These volumes (LNCS 13421–13423) contain the proceedings of the 6th Asia-Pacific Web (APWeb) and Web-Age Information Management (WAIM) Joint Conference on Web and Big Data (APWeb-WAIM). Researchers and practitioners from around the world came together at this leading international forum to share innovative ideas, original research findings, case study results, and experienced insights in the areas of the World Wide Web and big data, thus covering web technologies, database systems, information management, software engineering, knowledge graphs, recommender system and big data.

The 6th APWeb-WAIM conference was held in Nanjing during 25–27 November 2022. As an Asia-Pacific flagship conference focusing on research, development, and applications in relation to Web information management, APWeb-WAIM builds on the successes of APWeb and WAIM. Previous APWeb events were held in Beijing (1998), Hong Kong (1999), Xi'an (2000), Changsha (2001), Xi'an (2003), Hangzhou (2004), Shanghai (2005), Harbin (2006), Huangshan (2007), Shenyang (2008), Suzhou (2009), Busan (2010), Beijing (2011), Kunming (2012), Sydney (2013), Changsha (2014), Guangzhou (2015), and Suzhou (2016). And previous WAIM events were held in Shanghai (2000), Xi'an (2001), Beijing (2002), Chengdu (2003), Dalian (2004), Hangzhou (2005), Hong Kong (2006), Huangshan (2007), Zhangjiajie (2008), Suzhou (2009), Jiuzhaigou (2010), Wuhan (2011), Harbin (2012), Beidaihe (2013), Macau (2014), Qingdao (2015), and Nanchang (2016). The combined APWeb-WAIM conferences have been held in Beijing (2017), Macau (2018), Chengdu (2019), Tianjin (02020), and Guangzhou (2021). With the ever-growing importance of appropriate methods in these data-rich times and the fast development of web-related technologies, we believe APWeb-WAIM will become a flagship conference in this field.

The high-quality program documented in these proceedings would not have been possible without the authors who chose APWeb-WAIM for disseminating their findings. APWeb-WAIM 2022 received a total of 297 submissions and, after the double-blind review process (each paper received at least three review reports), the conference accepted 75 regular papers (including research and industry track) (acceptance rate 25.25%), 45 short research papers, and 5 demonstrations. The contributed papers address a wide range of topics, such as big data analytics, advanced database and web applications, data mining and applications, graph data and social networks, information extraction and retrieval, knowledge graphs, machine learning, recommender systems, security, privacy and trust, and spatial and multimedia data. The technical program also included keynotes by Ihab F. Ilyas Kaldas, Aamir Cheema, Chengzhong Xu, Lei Chen, and Haofen Wang. We are grateful to these distinguished scientists for their invaluable contributions to the conference program.

We would like to express our gratitude to all individuals, institutions, and sponsors that supported APWeb-WAIM 2022. We are deeply thankful to the Program Committee members for lending their time and expertise to the conference. We also would like to acknowledge the support of the other members of the organizing committee. All of them helped to make APWeb-WAIM 2022 a success. We are grateful for the guidance of the Honorary Chairs (Zhiqiu Huang), Steering Committee representative (Yanchun Zhang) and the General Co-chairs (Aoying Zhou, Wojciech Cellary and Bing Chen) for their guidance and support. Thanks also go to the Workshop Co-chairs (Shiyu Yang and Saiful Islam), Tutorial Co-chairs (Xiang Zhao, Wenqi Fan and Ji Zhang), Demo Co-chairs (Jianqiu Xu and Travers Nicolas), Industry Co-chairs (Chen Zhang Hosung Park), Publication Co-chairs (Chuanqi Tao, Lin Yue and Xuming Han), and Publicity Co-chairs (Yi Cai, Siqiang Luo and Weitong Chen).

We hope the attendees enjoyed the exciting program of APWeb-WAIM 2022 as documented in these proceedings.

November 2022

Toshiyuki Amagasa
Diego Calvanese
Xuming Han
Bohan Li
Chuanqi Tao
Lin Yue

Organization

General Chairs

Aoying Zhou	East China Normal University, China
Bing Chen	Nanjing University of Aeronautics and Astronautics, China
Wojciech Cellary	WSB University, Poland

Steering Committee

Aoying Zhou	East China Normal University, China
Divesh Srivastava	AT&T Research Institute, USA
Jian Pei	Simon Fraser University, Canada
Lei Chen	Hong Kong University of Science and Technology, China
Lizhu Zhou	Tsinghua University, China
Masaru Kitsuregawa	University of Tokyo, Japan
Mingjun Huang	National University of Singapore, Singapore
Tamor Özsu	University of Waterloo, Canada
Xiaofang Zhou	University of Queensland, Australia
Yanchun Zhang	Victoria University, Australia

Program Committee Chairs

Diego Calvanese	Free University of Bozen-Bolzano, Italy
Toshiyuki Amagasa	University of Tsukuba, Japan
Bohan Li	Nanjing University of Aeronautics and Astronautics, China

Publication Chairs

Chuanqi Tao	Nanjing University of Aeronautics and Astronautics, China
Lin Yue	University of Newcastle, Australia
Xuming Han	Jinan University, China

Program Committee

Alex Delis	University of Athens, Greece
An Liu	Soochow University, China
Aviv Segev	University of South Alabama, USA
Bangbang Ren	National University of Defense Technology, China
Baokang Zhao	NUDTCS, China
Baoning Niu	Taiyuan University of Technology, China
Bin Guo	Northwestern Polytechnical University, China
Bin Cui	Peking University, China
Bin Xia	Nanjing University of Posts and Telecommunications, China
Bin Zhao	Nanjing Normal University, China
Bolong Zheng	Huazhong University of Science and Technology, China
Byung Suk Lee	University of Vermont, USA
Carson Leung	University of Manitoba, Canada
Cheng Long	Nanyang Technological University, Singapore
Chengliang Chai	Tsinghua University, China
Cheqing Jin	East China Normal University, China
Chuanqi Tao	Nanjing University of Aeronautics and Astronautics, China
Cuiping Li	Renmin University of China, China
Dechang Pi	Nanjing University of Aeronautics and Astronautics, China
Dejun Teng	Shandong University, China
Demetrios Zeinalipour-Yazti	University of Cyprus, Cyprus
Derong Shen	Northeastern University, China
Dong Li	Liaoning University, China
Donghai Guan	Nanjing University of Aeronautics and Astronautics, China
Fabio Valdés	FernUniversität in Hagen, Germany
Fei Chen	Shenzhen University, China
Genoveva Vargas-Solar	CNRS, France
Giovanna Guerrini	University of Genoa, Italy
Guanfeng Liu	Macquarie University, Australia
Guodong Long	University of Technology Sydney, Australia
Guoqiong Liao	Jiangxi University of Finance & Economics, China
Haibo Hu	Hong Kong Polytechnic University, China
Hailong Liu	Northwestern Polytechnical University, China
Haipeng Dai	Nanjing University, China

Haiwei Pan	Harbin Engineering University, China
Haiwei Zhang	Nankai University, China
Hancheng Wang	Nanjing University, China
Hantao Zhao	Southeast University, China
Harry Kai-Ho Chan	University of Sheffield, UK
Hiroaki Ohshima	University of Hyogo, Japan
Hong Chen	Renmin University, China
Hongzhi Wang	Harbin Institute of Technology, China
Hongzhi Yin	University of Queensland, Australia
Hua Dai	Nanjing University of Posts and Telecommunications, China
Hua Wang	Victoria University, Australia
Hui Li	Xiamen University, China
Javier A. Espinosa-Oviedo	University of Lyon, France
Ji Zhang	University of Southern Queensland, Australia
Jia Yu	Washington State University, USA
Jiajie Xu	Soochow University, China
Jiali Mao	East China Normal University, China
Jian Yin	Sun Yat-Sen University, China
Jiangtao Cui	Xidian University, China
Jianqiu Xu	Nanjing University of Aeronautics and Astronautics, China
Jianxin Li	Deakin University, Australia
Jilin Hu	Aalborg University, Denmark
Jing Jiang	University of Technology Sydney, Australia
Jizhou Luo	Harbin Institute of Technology, China
Jun Gao	Peking University, China
Junhu Wang	Griffith University, Australia
Junjie Yao	East China Normal University, China
K. Selçuk Candan	Arizona State University, USA
Kai Zeng	Microsoft, USA
Kai Zheng	University of Electronic Science and Technology of China, China
Kazutoshi Umemoto	University of Tokyo, Japan
Krishna Reddy P.	International Institute of Information Technology, India
Ladjel Bellatreche	ISAE-ENSMA, France
Le Sun	Nanjing University of Information Science and Technology, China
Lei Duan	Sichuan University, China
Lei Zou	Peking University, China
Leong Hou U.	University of Macau, China

Liang Hong	Wuhan University, China
Lin Li	Wuhan University of Technology, China
Lin Yue	University of Queensland, Australia
Liyan Zhang	Yanshan University, China
Lizhen Cui	Shandong University, China
Lu Chen	Zhejiang University, China
Luyi Bai	Northeastern University, China
Makoto Onizuka	Osaka University, Japan
Maria Damiani	University of Milan, Italy
Maria Luisa Damiani	University of Milan, Italy
Markus Endres	University of Augsburg, Germany
Meng Wang	Southeast University, China
Miao Xu	University of Queensland, Australia
Mingzhe Zhang	University of Queensland, Australia
Min-Ling Zhang	Southeast University, China
Mirco Nanni	ISTI-CNR Pisa, Italy
Mizuho Iwaihara	Waseda University, Japan
My T. Thai	University of Florida, USA
Nicolas Travers	Léonard de Vinci Pôle Universitaire, France
Peer Kroger	Christian-Albrechts-Universität Kiel, Germany
Peiquan Jin	University of Science and Technology of China, China
Peisen Yuan	Nanjing Agricultural University, China
Peng Peng	Hunan University, China
Peng Wang	Fudan University, China
Qian Zhou	Nanjing University of Posts and Communications, China
Qilong Han	Harbin Engineering University, China
Qing Meng	Southeast University, China
Qing Xie	Wuhan University of Technology, China
Qiuyan Yan	China University of Mining and Technology, China
Qun Chen	Northwestern Polytechnical University, China
Quoc Viet Hung Nguyen	Griffith University, Australia
Reynold Cheng	University of Hong Kong, China
Rong-Hua Li	Beijing Institute of Technology, China
Saiful Islam	Griffith University, Australia
Sanghyun Park	Yonsei University, Korea
Sanjay Madria	Missouri University of Science & Technology, USA
Sara Comai	Politecnico di Milano
Sebastian Link	University of Auckland, New Zealand

Sen Wang	University of Queensland, Australia
Senzhang Wang	Central South University, China
Shanshan Li	Yanshan University, China
Shaofei Shen	University of Queensland, Australia
Shaojie Qiao	Chengdu University of Information Technology, China
Shaoxu Song	Tsinghua University, China
Sheng Wang	Wuhan University, China
Shengli Wu	Jiangsu University, China
Shiyu Yang	Guangzhou University, China
Shuai Xu	Nanjing University of Aeronautics and Astronautics, China
Shuigeng Zhou	Fudan University, China
Siqiang Luo	Nanyang Technological University, Singapore
Tianrui Li	Southwest Jiaotong University, China
Tingjian Ge	University of Massachusetts, USA
Tung Kieu	Aalborg University, Denmark
Victor Sheng	Texas Tech University, USA
Wang Lizhen	Yunnan University, China
Wei Hu	Nanjing University, China
Wei Shen	Nankai University, China
Wei Song	Wuhan University, China
Weiguo Zheng	Fudan University, China
Weijun Wang	University of Goettingen, Germany
Weitong Chen	Adelaide University, Australia
Wen Zhang	Wuhan University, China
Wenqi Fan	Hong Kong Polytechnic University, China
Wolf-Tilo Balke	TU Braunschweig, Germany
Xiang Lian	Kent State University, USA
Xiang Zhao	National University of Defense Technology, China
Xiangfu Meng	Liaoning Technical University, China
Xiangguo Sun	Chinese University of Hong Kong, China
Xiangliang Zhang	University of Notre Dame, USA
Xiangmin Zhou	RMIT University, Australia
Xiao Pan	Shijiazhuang Tiedao University, China
Xiao Zhang	Shandong University, China
Xiaohui Yu	Shandong University, China
Xiaohui Tao	University of Southern Queensland, Australia
Xiaoli Wang	Xiamen University, China
Xiaowang Zhang	Tianjin University, China
Xie Xiaojun	Nanjing Agricultural University, China

Xin Cao   University of New South Wales, Australia
Xin Wang   Tianjin University, China
Xingquan Zhu   Florida Atlantic University, USA
Xuelin Zhu   Southeast University, China
Xujian Zhao   Southwest University of Science and Technology,
                           China
Xuming Han   Jinan University, China
Xuyun Zhang   Macquarie University, Australia
Yaokai Feng   Kyushu University, Japan
Yajun Yang   Tianjin University, China
Yali Yuan   University of Göttingen, Germany
Yanda Wang   University of Bristol, UK
Yanfeng Zhang   Northeastern University, China
Yanghua Xiao   Fudan University, China
Yang-Sae Moon   Kangwon National University, Korea
Yanhui Gu   Nanjing Normal University, China
Yanjun Zhang   Deakin University, Australia
Yasuhiko Morimoto   Hiroshima University, Japan
Ye Liu   Nanjing Agricultural University, China
Yi Cai   South China University of Technology, China
Yingxia Shao   BUPT, China
Yong Tang   South China Normal University, China
Yong Zhang   Tsinghua University, China
Yongpan Sheng   Chongqing University, China
Yongqing Zhang   Chengdu University of Information Technology,
                           China
Youwen Zhu   Nanjing University of Aeronautics and
                           Astronautics, China
Yu Gu   Northeastern University, China
Yu Liu   Huazhong University of Science and Technology,
                           China
Yu Hsuan Kuo   Amazon, USA
Yuanbo Xu   Jilin University, China
Yue Tan   University of Technology Sydney, Australia
Yunjun Gao   Zhejiang University, China
Yuwei Peng   Wuhan University, China
Yuxiang Zhang   Civil Aviation University of China, China
Zakaria Maamar   Zayed University, UAE
Zhaokang Wang   Nanjing University of Aeronautics and
                           Astronautics, China
Zheng Zhang   Harbin Institute of Technology, China
Zhi Cai   Beijing University of Technology, China

Zhiqiang Zhang	Zhejiang University of Finance and Economics, China
Zhixu Li	Soochow University, China
Zhuoming Xu	Hohai University, China
Zhuowei Wang	University of Technology Sydney, Australia
Ziqiang Yu	Yantai University, China
Zouhaier Brahmia	University of Sfax, Tunisia

Additional Reviewers

Bo Tang	Southern University of Science and Technology, China
Fang Wang	Hong Kong Polytechnic University, China
Genggeng Liu	Fuzhou University, China
Guan Yuan	China University of Mining and Technology, China
Jiahao Zhang	Hong Kong Polytechnic University, China
Jinguo You	Kunming University of Science and Technology, China
Long Yuan	Nanjing University of Science and Technology, China
Paul Bao	Nagoya University, Japan
Philippe Fournier-Viger	Shenzhen University, China
Ruiyuan Li	Chongqing University, China
Shanshan Yao	Taiyuan University of Technology, China
Xiaofeng Ding	Huazhong University of Science and Technology, China
Yaoshu Wang	Shenzhen University, China
Yunpeng Chai	Renmin University of China, China
Zhiwei Zhang	Beijing Institute of Technology, China

Contents – Part II

Machine Learning

Graph Data and Social Networks

Accelerating Hypergraph Motif Counting Based on Hyperedge Relations

Yuhang Su[1]($\boxtimes$), Yu Gu[1], Yang Song[2], and Ge Yu[1]

[1] College of Computer Science and Engineering, Northeastern University, Shenyang 110819, Liaoning, China
suyuhang_neu@163.com, {yuge,guyu}@mail.neu.edu.cn
[2] College of Information Science and Engineering, Northeastern University, Shenyang 110819, Liaoning, China

Abstract. Hypergraphs can naturally represent inter-group relations that are prevalent in many application domains by hyperedges. Hypergraph motifs can be described as the structural patterns of three connected hyperedges. As an effective tool, it plays an important role in the local structure analysis of hypergraphs. In this paper, we study exact hypergraph motif counting which is a fundamental problem of hypergraph motif research. Existing algorithms don't adequately consider hyperedge relations in real-world hypergraphs, which lead to a large number of redundant computations. This motivates us to improve performance by exploiting hyperedge relations. In our work, we classify hypergraph motifs with different hyperedge relations. For different types of motifs, we use set theory to demonstrate and propose different optimization methods to reduce the computation of excessive intersections. We also further reduce the cost of the proposed method by preserving hyperedge intersections when constructing the hyperdege projected graph. Extensive experiments on real datasets validate the superiority of our algorithm compared to existing methods.

Keywords: Hypergraph · Hypergraph motif · Hypergraph motif counting · Hyperedge relation

1 Introduction

A hypergraph consists of vertices and hyperedges that can connect multiple vertices, and can be seen as a general form of ordinary graphs. Since hypergraphs can effectively simulate complex intergroup relationships between entities, they have a wide range of applications such as bioinformatics [5] and social network analysis [9] . Specifically, complex analyses over hypergraphs have also been extensively explored for hypergraph motifs [6], classification [4] and hyperedge prediction [10]. Network motifs have achieved great success in exploring and discovering local structural features of real-world graphs [7]. However, due to the different structures of ordinary graphs between hypergraphs, it is difficult to directly apply related techniques to hypergraphs. In order to better explore

© The Author(s), under exclusive license to Springer Nature Switzerland AG 2023
B. Li et al. (Eds.): APWeb-WAIM 2022, LNCS 13422, pp. 3–11, 2023.
https://doi.org/10.1007/978-3-031-25198-6_1

the local structural patterns of real-world hypergraphs, Lee et al. [6] success-fully define hypergraph motifs for the first time. Existing methods demonstrate the importance of hypergraph motifs in revealing hypergraph local structural patterns. However, existing algorithms do not effectively explore hyperedge relations to improve the computational efficiency. This motivates us to fully explore hyperedge relations (intersection and containment) to achieve the acceleration of hypergraph motifs counting. The major contributions are concluded as follows.

- We explore the widely existing hyperedge relations in real-world hypergraphs and classify hypergraph motifs according to specific relations. For different types of motifs, by using set theory, we study and demonstrate different optimal calculation methods to reduce the cost of excessive intersections.
- For the remaining hypergraph motifs that cannot be optimized, we further reduce the cost of the algorithm by preserving the hyperedge intersection when constructing the hyperdege projected graph.
- We conduct extensive experiments to verify that our algorithm outperforms existing algorithms. In total processing time, our algorithm is more than two times faster than existing algorithms.

2   Related Work

We examine existing related work on network motif counting for ordinary graph. Most of them apply the following three techniques to speed up motif counting: 1) Combinatorics: In order to speed up exact network motif counting, the existing work [8] adopt combinatorial relations computation methods. 2) MCMC sampling: Most approximate network motif counting algorithms estimate the number of motif instances by sampling [2,3]. 3) Color coding: The approximate network motif counting algorithm [1] uses color coding to randomly color each vertex and use this coloring information to randomly sample. However, due to the different structures of ordinary graphs and hypergraphs, it is difficult to directly apply related techniques to hypergraphs. We also review existing related work on hypergraph motifs. Hypergraph motifs are the basic building blocks of hypergraphs as defined by [6]. Unlike network motifs, it is formed by three connected hyperedges with 26 different connection patterns. Hypergraph motifs differ from network motifs in that they do not limit the number of vertices. Extensive experiments verify that hypergraph motifs play an important role in revealing local structural patterns of real-world hypergraphs. The only existing exact hypergraph motif counting algorithm is proposed by [6]. Although the algorithm efficiently implements hypergraph motif counting, it performs a lot of redundant intersection computations.

3   Hypergraph Motif Classification and Computation Acceleration Based on Hyperedge Relations

3.1   Basic Definition

Definition 1 (Hypergraph). *A hypergraph is represented by $G = (V, E)$, where V is a finite set of vertices, $E = \bigcup_{i=1}^{|E|} e_i$ is a finite set of hyperedges. Each hyperedge $e_i \in E$ is a non-empty subset of V.*

Definition 2 (Hyperdege Projected Graph). *A hyperdege projected graph of $G = (V, E)$ is an ordinary graph $PG = (E, H)$, where $H = \{(e_i, e_j) \mid e_i \cap e_j \neq \varnothing\}$. We use $\overline{H}_{ij}$ to denote the intersection of e_i and e_j, that is, $\overline{H}_{ij} = \{v_i \in V \mid v_i \in e_i \cap e_j\}$.*

Definition 3 (Hypergraph Motif). *Given three connected hyperedges $\{e_i, e_j, e_k\}$, hypergraph motifs are used to describe the connectivity patterns of the three connected hyperedges. Formally, a hypergraph motif is a binary vector of size 7 whose elements represent the emptiness of the following seven sets: (1) $e_i \setminus e_j \setminus e_k$, (2) $e_j \setminus e_k \setminus e_i$, (3) $e_k \setminus e_i \setminus e_j$, (4) $e_i \cap e_j \setminus e_k$, (5) $e_j \cap e_k \setminus e_i$, (6) $e_k \cap e_i \setminus e_j$ and (7) $e_i \cap e_j \cap e_k$.*

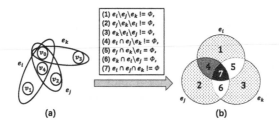

Fig. 1. Hypergraph motif and hypergraph motif instance

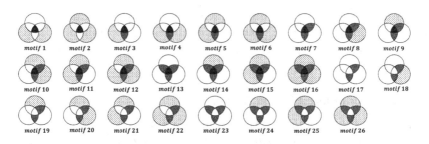

Fig. 2. The 26 hypergraph motifs

Example 1. As shown in Fig. 1(b), hypergraph motifs can be naturally represented in the Venn diagram. The three circles represent hyperedges e_i, e_j and e_k, respectively. The three circles are superimposed and divided into seven parts representing seven different sets. We usually use patterned parts to represent non-empty and white to represent empty. In fact, excluding symmetries and duplicated hyperedges, we can describe the pattern of all connected three hyperedges by means of 26 hypergraph motifs in Fig. 2. If the connectivity pattern of the three hyperedges corresponds to a particular hypergraph motif, we consider the three connected hyperedges as an instance of this hypergraph motif. As shown in Fig. 1, (a) is an instance of the hypergraph *motif* 6. It is worth noting that *motif* 17–22 are open motifs in Fig. 2. More intuitively, the open motif is the one that has two hyperedges which are not connected. Obviously, given three hyperedges e_i, e_j and e_k, if their connection pattern (motif) is a open motif, then $|e_i \cap e_j \cap e_k| = 0$.

Definition 4 (Hypergraph Motif Counting). *Hypergraph motif counting is to calculate the number of instances corresponding to 26 hypergraph motifs on a hypergraph.*

3.2   Double-Single-Inclusion Motifs

Definition 5 (Double-Single-Inclusion Motifs). *Given three hyperedges e_i, e_j and e_k, if their connection pattern (motif) satisfies any of the following three conditions (1) $|e_i \cap e_j| = |e_j \cap e_k| = |e_j|$; (2) $|e_j \cap e_k| = |e_i \cap e_k| = |e_k|$; (3) $|e_i \cap e_j| = |e_i \cap e_k| = |e_i|$, we call it a Double-Single-Inclusion Motif (DSI motif for short).*

Example 2. As shown in Fig. 2, *motif* 1 and *motif* 4 are DSI motifs. More intuitively, the DSI motif is the one that has one hyperedge contained by the other two hyperedges.

Theorem 1. *Given three hyperedges e_i, e_j and e_k, if their connection pattern (motif) is a DSI motif, there exist the following conclusions : (1) if $|e_i \cap e_j| = |e_j \cap e_k| = |e_j|$ then $|e_i \cap e_j \cap e_k| = |e_j|$; (2) if $|e_j \cap e_k| = |e_i \cap e_k| = |e_k|$ then $|e_i \cap e_j \cap e_k| = |e_k|$; (3) if $|e_i \cap e_j| = |e_i \cap e_k| = |e_i|$ then $|e_i \cap e_j \cap e_k| = |e_i|$.*

Proof. We first prove the conclusion (1). Given three hyperedges e_i, e_j and e_k, if $|e_i \cap e_j| = |e_j \cap e_k| = |e_j|$, then e_i contains e_j and e_k also contains e_j. Therefore, there is $|e_i \cap e_j \cap e_k| = |e_j \cap e_k| = |e_j|$. Similarly, conclusions (2) and (3) can be proved. Theorem 1 is proved.

3.3   Single-Double-Inclusion Motifs

Definition 6 (Single-Double-Inclusion Motifs). *Given three hyperedges e_i, e_j and e_k, if their connection pattern (motif) satisfies any of the following three conditions (1) $|e_i \cap e_j| = |e_j|$ and $|e_i \cap e_k| = |e_k|$; (2) $|e_j \cap e_k| = |e_k|$ and $|e_i \cap e_j| = |e_i|$; (3) $|e_j \cap e_k| = |e_j|$ and $|e_i \cap e_k| = |e_i|$, we call it a Single-Double-Inclusion Motif (SDI motif for short).*

Example 3. As shown in Fig. 2, *motif* 3, *motif* 7 and *motif* 8 are SDI motifs. More intuitively, the SDI motif is the one that has one hyperedge containing the other two hyperedges.

Theorem 2. *Given three hyperedges e_i, e_j and e_k, if their connection pattern (motif) is a DSI motif, there exist the following conclusions : (1) if $|e_i \cap e_j| = |e_j|$ and $|e_i \cap e_k| = |e_k|$ then $|e_i \cap e_j \cap e_k| = |e_j \cap e_k|$; (2) if $|e_j \cap e_k| = |e_k|$ and $|e_i \cap e_j| = |e_i|$ then $|e_i \cap e_j \cap e_k| = |e_j \cap e_k|$; (3) if $|e_j \cap e_k| = |e_j|$ and $|e_i \cap e_k| = |e_i|$ then $|e_i \cap e_j \cap e_k| = |e_i \cap e_j|$.*

Proof. We first prove the conclusion (1). Given three hyperedges e_i, e_j and e_k, if $|e_i \cap e_j| = |e_j|$ and $|e_i \cap e_k| = |e_k|$, then e_i contains e_j and e_k. Therefore, there is $|e_i \cap e_j \cap e_k| = |(e_i \cap e_j) \cap (e_i \cap e_k)| = |e_j \cap e_k|$. Similarly, conclusions (2) and (3) can be proved. Theorem 2 is proved.

3.4   Single-Single-Inclusion Motifs

Definition 7 (Single-Single-Inclusion Motifs). *Given three hyperedges e_i, e_j and e_k, if their connection pattern (motif) satisfies any of the following three conditions (1) $|e_i \cap e_j| = |e_j|$ and $|e_i \cap e_k| \neq |e_k|$; (2) $|e_j \cap e_k| = |e_k|$ and $|e_i \cap e_j| \neq |e_i|$; (3) $|e_j \cap e_k| = |e_j|$ and $|e_i \cap e_k| \neq |e_i|$, we call it a Single-Single-Inclusion Motif (SSI motif for short).*

Example 4. As shown in Fig. 2, *motif* 5, *motif* 9 and *motif* 10 are SSI motifs. More intuitively, the SSI motif is the one that has one hyperedge containing only one of other two hyperedges.

Theorem 3. *Given three hyperedges e_i, e_j and e_k, if their connection pattern (motif) is a SSI motif, there exist the following conclusions : (1) if $|e_i \cap e_j| = |e_j|$ and $|e_i \cap e_k| \neq |e_k|$ then $|e_i \cap e_j \cap e_k| = |e_j \cap e_k|$; (2) if $|e_j \cap e_k| = |e_k|$ and $|e_i \cap e_j| \neq |e_i|$ then $|e_i \cap e_j \cap e_k| = |e_j \cap e_k|$; (3) if $|e_j \cap e_k| = |e_j|$ and $|e_i \cap e_k| \neq |e_i|$ then $|e_i \cap e_j \cap e_k| = |e_i \cap e_j|$.*

Proof. We first prove the conclusion (1). Given three hyperedges e_i, e_j and e_k, if $|e_i \cap e_j| = |e_j|$ and $|e_i \cap e_k| \neq |e_k|$, then e_i contains e_j. Therefore, there is $|e_i \cap e_j \cap e_k| = |(e_i \cap e_j) \cap e_k| = |e_j \cap e_k|$. Similarly, conclusions (2) and (3) can be proved. Theorem 3 is proved.

    As described in Subsect. 3.2–3.4, we propose 3 different special motifs through set theory. We also exploit set theory to give and prove their respective special properties. By determining the type of motifs, we can speed up the computation for the corresponding motifs through Theorems 1–3.

4   Hypergraph Motif Counting Algorithm Framework Optimization

For the remaining hypergraph motifs that cannot be optimized, we further reduce the overall complexity of the algorithm by preserving the hyperedge pair intersections in the preprocessing stage. 1) **Constructing Projected Graph.** As a preprocessing step (*Lines* 1–7), Algorithm 1 builds a complete hyperedge projected graph for subsequent motif counting. It first clears H for recording hyperedge pairs (*Line* 1). Then it finds all neighbors of each hyperedge e_i (*Lines* 2–4). E_v is used to denote the set of all hyperedges containing the vertices v. Finally it stores the hyperedge pair in H (*Line* 6). At the same time, it pre-stores the intersection (set of vertices) of the corresponding hyperedge pairs in $\overline{H}$ for computing acceleration (*Line* 7). The time complexity of this preprocessing step is $O(\sum_{(e_i, e_j) \in H} |e_i \cap e_j|)$. In fact, it needs to compute $e_i \cap e_j$ to find the neighbor e_j of hyperedge e_i, hence it does not affect the time complexity of the algorithm by pre-storing $e_i \cap e_j$ in $\overline{H}$. **2) Motif Counting.** Algorithm 1 (*Lines* 8–12) first finds two neighbors of each hyperedge e_i to form a hyperedge triple (*Lines* 8–9). H_{e_i} is used to represent all neighbors of hyperedge e_i in PG. Then it determines whether the three hyperedges belong to a particular motif (*Line* 10). If the corresponding hyperedge triple belongs to DSI or SDI or SSI or open motif, we use the function $\overline{h}(\{e_i, e_j, e_k\})$ to determine which motif it belongs to and accumulate at the corresponding position of M (*Line* 11). Since $\overline{h}(\{e_i, e_j, e_k\})$ does not need to compute $e_i \cap e_j \cap e_k$, the time complexity of $\overline{h}(\{e_i, e_j, e_k\})$ is $O(1)$. For the remaining motifs, we use function $h(\{e_i, e_i, e_i\}, \overline{H})$ to calculate (*Line* 12). Since the algorithm pre-stores the hyperedge pair intersections in $\overline{H}$, the time complexity of $h(\{e_i, e_i, e_i\}, \overline{H})$ is $O(min(|e_i \cap e_j|, |e_j \cap e_k|, |e_i \cap e_k|))$. In conclusion, the time complexity of our algorithm is better than that of existing algorithm ($O(min(|e_i|, |e_j|, |e_i|))$) in [6].

Algorithm 1: The Framework For Hypergraph Motif Counting

    Input : Hypergraph $G = (V, E)$, hyperedge projected graph $PG = (E, H)$
    Output: Exact count of each hypergraph motif in M

1 **initialize:** Set $H := \emptyset$;
2 **foreach** *hyperedge* $e_i \in E$ **do**
3   **foreach** *vertex* $v \in e_i$ **do**
4   **foreach** *hyperedge* $e_j \in E_v$ **do**
5   **if** $j > i$ **then**
6                 $H \leftarrow (e_i, e_j)$;
7                 $\overline{H} \leftarrow e_i \cap e_j$;

8 **foreach** *hyperedge* $e_i \in E$ **do**
9   **foreach** *unordered hyperedge pair* $e_j \in H_{e_i}$ *and* $e_k \in H_{e_i}$ **do**
10   **if** $\{e_i, e_j, e_k\} \in DSI$ *or* SDI *or* SSI *or* $open\ motif$ **then**
11             $M[\overline{h}(\{e_i, e_j, e_k\})] + +$;
        else
12             $M[h(\{e_i, e_j, e_k\}, \overline{H})] + +$;

13 *return* M ;

---

5 Experimental Settings and Results Analysis

5.1 Experimental Settings

1) Competitive Algorithms. The first is a native algorithm for hypergraph motif counting. This algorithm does not employ any optimization techniques. We call this algorithm **HMC** for short, and we regard **HMC** as a basic method. The second algorithm **HMCO** can be seen as **HMC** with optimization techniques only for the open motif and it is actually the exact motif algorithm in [6]. The third algorithm **HMCA** can be seen as **HMC** with optimization techniques for DSI, SDI, SSI and open motifs. Our algorithm **HMCP** can be seen as **HMCA** adding preserving intersections techniques for remaining hypergraph motifs. **2) Experiment Environment.** We obtained the source code of **HMCO** from the authors of [6]. The compiler for compiling source code is $g++\ 4.9.3-O3\ flag$. We conduct all experiments on a PC machine with equipment of $Intel\ i5\ 3.20\,\mathrm{GHz}$ and $16\,\mathrm{GB}\ RAM$. **3) Metrics.** We measure the execution time in milliseconds (ms). **4) Datasets.** We use 8 real-world datasets (http://konect.cc/) to evaluate the algorithms. The specific information of all real-world datasets is given in Table 1.

Table 1. Real-World Datasets Statistics

Data	$|V|$	$|H|$	H_Avg	H_Max	V_Avg	Edge of BiGraph
unicodelang	254	614	2.04	141	4.94	1,225
edit-crwiki	1,188	2,071	10.63	248	19.11	22,700
filmtrust	1,508	30,087	17.14	1,044	23.54	35,494
escorts	10,106	6,624	7.64	615	5.01	50,632
wang-amazon	26,112	799	36.37	812	1.11	29,062
tripadvisor	145,316	1,759	99.92	2,138	1.21	175,765
bag-kos	3,430	6,906	67.73	2,123	136.36	467,714
flickr	395,979	103,631	82.46	34,989	21.58	8,545,307

5.2 Experimental Results Analysis

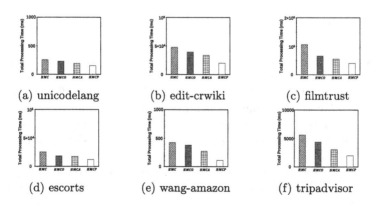

(a) unicodelang   (b) edit-crwiki   (c) filmtrust

(d) escorts   (e) wang-amazon   (f) tripadvisor

Fig. 3. Total Processing Time on Different Datasets (Vary Algorithm).

1) Total processing time. Figures 3(a)–(f) show the total time when processing the corresponding dataset. Based on the experimental results, we can obtain the following conclusions. 1) **HMC** performs the worst on all datasets, because it employs the brute force policy and lacks optimization method. 2) Simplifying the computation by considering only special motifs can also lead to better speedups. **HMCA** outperforms existing methods **HMCO**. This is because a large number of hyperedge inclusion relations are actually contained in the real-world hypergraph. Our optimization technique exploits these relationships to greatly reduce computational overhead. 3) **HMCP** always maintains the advantage on all datasets. The reason is twofold. One is to use Theorems 1–3 to reduce redundant intersection calculations. The second is that preserving the hyperedge pair intersections in the preprocessing stage provides speedup for computing the remaining hypergraph motifs. In general, **HMCP** is more than 2 times faster than existing method **HMCO**. In dataset wang-amazon, **HMCP** can bring a maximum speedup of four times. **2) Scalability.** To test the scalability of our algorithm, we use larger datasets. By varying the number of edges added to the hypergraph, we compare the performance of the four algorithms as shown in Figs. 4(a)–(b). The conclusion is that **HMCP** has better scalability than other algorithms. This is because our algorithm fully considers the hyperedge relationship to provide speedup. It is worth noting that the degree of the hyperedge increases as the number of edges increases. This will lead to more hyperedge inclusion relations, so the advantage of **HMCP** is more obvious.

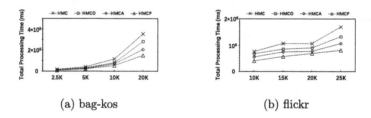

(a) bag-kos   (b) flickr

Fig. 4. Processing time on different datasets (vary number of edges)

6   Conclusion

In this paper, we propose effective techniques for accelerating hypergraph motif counting based on hyperedge relations. In our work, we classify hypergraph motifs with different hyperedge relations and demonstrate different optimization methods. For the remaining hypergraph motifs that cannot be optimized, we further reduce the overall complexity of the algorithm. Extensive experiments on real datasets show that our method is superior to the existing solutions.

Acknowledgements. This work is supported by the National Nature Science Foundation of China (62072083) and the Fundamental Research Funds of the Central Universities (N2216017).

References

1. Bressan, M., Leucci, S., Panconesi, A.: Motivo: fast motif counting via succinct color coding and adaptive sampling. Proc. VLDB Endow. **12**(11), 1651–1663 (2019)
2. Han, G., Sethu, H.: Waddling random walk: fast and accurate mining of motif statistics in large graphs. In: 2016 IEEE 16th International Conference on Data Mining (ICDM), pp. 181–190. IEEE (2016)
3. Huang, S., Li, Y., Bao, Z., Li, Z.: Towards efficient motif-based graph partitioning: an adaptive sampling approach. In: International Conference on Data Engineering (2021)
4. Jiang, J., Wei, Y., Feng, Y., Cao, J., Gao, Y.: Dynamic hypergraph neural networks. In: IJCAI, pp. 2635–2641 (2019)
5. Klamt, S., Haus, U.U., Theis, F.: Hypergraphs and cellular networks. PLoS Comput. Biol. **5**(5), e1000385 (2009)
6. Lee, G., Ko, J., Shin, K.: Hypergraph motifs: concepts, algorithms, and discoveries. Proc. VLDB Endow. **13**(11), 2256–2269 (2020)
7. Liu, J., Shao, Y., Su, S.: Multiple local community detection via high-quality seed identification over both static and dynamic networks. Data Sci. Eng. **6**(3), 249–264 (2021). https://doi.org/10.1007/s41019-021-00160-6
8. Pinar, A., Seshadhri, C., Vishal, V.: ESCAPE: efficiently counting all 5-vertex subgraphs. In: Proceedings of the 26th International Conference on World Wide Web, pp. 1431–1440 (2017)
9. Yang, D., Qu, B., Yang, J., Cudre-Mauroux, P.: Revisiting user mobility and social relationships in LBSNs: a hypergraph embedding approach. In: The World Wide Web Conference, pp. 2147–2157 (2019)
10. Yoon, S.E., Song, H., Shin, K., Yi, Y.: How much and when do we need higher-order information in hypergraphs? A case study on hyperedge prediction. In: The Web Conference 2020 (WWW 2020) (2020)

Next POI Recommendation Method Based on Category Preference and Attention Mechanism in LBSNs

Xueying Wang[1], Yanheng Liu[1,2], Xu Zhou[2,3]($\boxtimes$) (ib), Zhaoqi Leng[4], and Xican Wang[4]

[1] College of Computer Science and Technology, Jilin University, Changchun, China
[2] Key Laboratory of Symbolic Computation and Knowledge Engineering of Ministry of Education, Jilin University, Changchun, China
zhoux16@jlu.edu.cn
[3] Center for Computer Fundamental Education, Jilin University, Changchun, China
[4] College of Software, Jilin University, Changchun, China

Abstract. Focusing on learning the user's behavioral characteristics during check-in activities, the next point of interest (POI) recommendation is to predict user's destination to visit next. It is important for both the location-based service providers and users. Most of the existing studies have not taken full advantage of spatio-temporal information and user category preference, these are very important for analyzing user preference. Therefore, we propose a next POI recommendation algorithm named as CPAM that integrates category preference and attention mechanism to comprehensively structure user mobility patterns. We adopt the LSTM with multi-level attention mechanism to get user POI preference, which studies the weight of different contextual information of each check-in, and the different influence of each check-in the sequence to the next POI. In addition, we use LSTM to capture the user's category transition preference to further improve the accuracy of recommendation. The experiment results on two real-world Foursquare datasets demonstrate that CPAM has better performance than the state-of-the art methods in terms of two commonly used metrics.

Keywords: LSTM · Next POI recommendation · Contextual information · Location-based social networks · Attention mechanism

1 Introduction

With the rapid development of mobile networks, location-based social networks (LBSNs) are also widely used in recent years, such as Foursquare and Facebook [4]. Users can share their location and life by checking in locations. According to users' historical check-in information, it is convenient to construct users' movement trajectory and dig out their movement patterns. The next point of interest (POI) recommendation has become one of the most important tasks in

B. Li et al. (Eds.): APWeb-WAIM 2022, LNCS 13422, pp. 12–19, 2023.
https://doi.org/10.1007/978-3-031-25198-6_2

LBSNs and has a broad range of applications. Its primary objective is to predict the next POI that a user is likely to visit at a given time based on the user's check-in sequence [10]. The next POI recommendation plays a significant role in location-based services, and it can not only promotes customer experiences, but also helps improve the quality of location-related business services [2].

User's transition preference for POI category reflects user's mobility patterns at category level, in order to take full advantage of contextual information, we propose a next POI recommendation algorithm (CPAM) that combines category preference and attention mechanism. Experimental results on two real-world datasets demonstrate that CPAM algorithm is significantly better than other six comparative algorithms in terms of Recall and Map.

2   Related Work

Earlier approaches are to model the user's movement patterns through Markov chains to solve the sparse problem [6]. But existing Markov chain based methods are difficult to capture longer sequence contexts. In recent years, there has been a trend of methods applying deep learning for recommendation system. For example, Liu et al. proposed the ST-RNN which considers spatio-temporal information on the basis of RNN [5]. But RNN is not suitable for building long sequences. Subsequently, Zhang et al. proposed iMTL with multi-task learning framework based on LSTM [11], which comprehensively considered the category and space-time information in trajectory sequence. In addition, Some studies found that aggregating different contextual information (such as social relationship, time, location, etc.) into POI recommendation methods can alleviate data sparseness [12]. Attention mechanism can capture the degree of influence of different components [1]. It is also widely used for the next POI recommendation. Combining LSTM and attention mechanism can distinguish the differing degrees of influences that each time step may have on the next check-in. Huang et al. proposed ATST-LSTM, which adds attention mechanism on the basis of LSTM [3]. Li et al. proposed a codec framework, which could automatically learn the deep spatio-temporal representation of historical check-ins, but it did not consider the impact of spatio-temporal transition on check-in [14]. Wu et al. considered the long and short term preferences of users separately, and integrated attention mechanism, geographical location and category information of POI into the LSTM network [13]. The above studies all employ the attention mechanism to achieve better next POI recommendation performance.

3   Proposed Method

The model is mainly composed of three modules, as shown in Fig. 1. (1) Category module based on LSTM is to obtain the user's preference representation at category level; (2) POI module based on self-attention LSTM network to get user's preference representation at POI level; (3) Output layer is to generate a ranked list of next POIs.

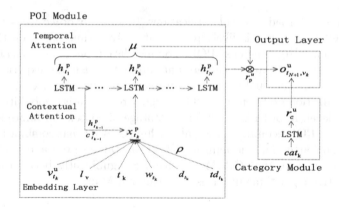

Fig. 1. The proposed CPAM framework

3.1   Category Module

Category sequence is the representation of check-in sequence at the semantic level. It reflects user's mobile preference at category level. Category module is designed to infer user category preference and participates in POI recommendation as an auxiliary function.

We learn the user's category preference $\mathbf{r}_c^u$ from category sequence $C_u = \{C_{t_1}^u, C_{t_2}^u, \cdots, C_{t_N}^u\}$, each element of C_u is denoted as $C_{t_k}^u = (u, cat_v)$. It indicates that the user u visits a POI v of category cat_v at time t_k. The latent vector of the category module is defined as follows.

$$\mathbf{x}_{t_k}^c = \mathbf{W}^C \mathbf{cat}_v + \mathbf{b}^C \tag{1}$$

where $\mathbf{W} \in \mathbb{R}^{E \times E}$ is the weight matrix, where E is the dimension of the hidden vector, $\mathbf{b} \in \mathbb{R}^E$ is bias. Then, $\mathbf{x}_{t_k}^c$ is input into the LSTM network to infer the hidden state $\mathbf{h}_{t_k}^c$ of user u.

$$\mathbf{h}_{t_k}^c = LSTM\left(\mathbf{x}_{t_k}^c, \mathbf{h}_{t_{k-1}}^c\right) \tag{2}$$

$$\mathbf{r}_c^u = \mathbf{h}_{t_N}^c \tag{3}$$

where $LSTM(\cdot)$ captures the sequential correlation of categories, $\mathbf{h}_{t_{k-1}}^c$ is the LSTM hidden state, which indicates the check-in category up to t_{k-1}.

3.2   POI Module

Embedding Layer. The historical check-in sequence of user u consists of the check-in tuple $A_{t_k}^u = \left(u, v_{t_k}^u, l_v, cat_v, t_k, w_{t_k}\right)$, we use it to learn the user's preference at the POI level. the latent vector of the embedding layer of the POI preference module is defined as follows:

$$\tilde{\mathbf{x}}_{t_k}^p = \mathbf{W}_v \mathbf{v}_{t_k}^u + \mathbf{W}_l l_v + \mathbf{W}_t t_k + \mathbf{W}_w \mathbf{w}_{t_k} + \mathbf{W}_d \mathbf{d}_{t_k} + \mathbf{W}_{td} \mathbf{td}_{t_k} + \mathbf{b} \tag{4}$$

where $\mathbf{v}_{t_k}^u$ is POI number, $\mathbf{l}_v$ is POI location, t_k is access timestamp, $\mathbf{w}_{t_k}$ is the day of the week, $\mathbf{d}_{t_k}$ is distance between $l_{t_k}^u$ and $l_{t_{k-1}}^u$, $\mathbf{td}_{t_k}$ is time difference between t_k and t_{k-1}. $\tilde{\mathbf{x}}_{t_k}^p$ is sent to the contextual attention layer.

Contextual Attention. Each feature of the embedded layer marks an attribute of the current check-in, and the extent to which these attributes affect the current check-in is different. Therefore, the proportion of different contextual information is studied with contextual attention mechanism in the current check-in.

$\tilde{\mathbf{x}}(i, t_k)$ represents the i-th attribute of the k-th historical check-in. $\rho(i, t_k)$ indicates the weight of the i-th feature in the k-th check-in. The softmax function is used for normalization.

$$\tilde{\rho}(i, t_k) = tanh\left(\mathbf{W}_i\left[\mathbf{h}_{t_{k-1}}^p, \mathbf{c}_{t_{k-1}}^p\right] + \mathbf{W}_i^{\tilde{x}}\tilde{\mathbf{x}}(i, t_k) + \mathbf{b}_i\right) \tag{5}$$

$$\rho(i, t_k) = \frac{exp\left(\tilde{\rho}(i, t_k)\right)}{\sum_{i=1}^{I} exp\left(\tilde{\rho}(i, t_k)\right)}, 1 \leq i \leq I \tag{6}$$

where I is the number of attributes, $\mathbf{h}_{t_{k-1}}^p$ is the LSTM hidden state, $\mathbf{c}_{t_{k-1}}^p$ is the LSTM cell state. Then, $\tilde{\mathbf{x}}(i, t_k)$ is multiplied by $\rho(i, t_k)$ to obtain the embedding vector, the updated attribute embedding vector is connected to obtain the aggregation $\mathbf{x}_{t_k}^p$ of the embedding layer based on contextual attention mechanism. $\mathbf{x}_{t_k}^p$ is sent to LSTM to infer the hidden state $\mathbf{h}_{t_k}^p$ at t_k.

$$\mathbf{x}(i, t_k) = \tilde{\mathbf{x}}(i, t_k) \times \rho(i, t_k) \tag{7}$$

$$\mathbf{x}_{t_k}^p = \sum_{i=1}^{I} \mathbf{W}(i)\mathbf{x}(i, t_k) + \mathbf{b} \tag{8}$$

$$\mathbf{h}_{t_k}^p = LSTM\left(\mathbf{x}_{t_k}^p, \mathbf{h}_{t_{k-1}}^p\right) \tag{9}$$

Temporal Attention. We use the temporal attention mechanism to adaptively select relevant historical check-ins activities to achieve a better recommendation of the next POI.

Let $\mathbf{H}^p$ be a matrix composed of all hidden vectors $\{\mathbf{h}_{t_1}^p, \mathbf{h}_{t_2}^p, \cdots \mathbf{h}_{t_N}^p\}$, where N is the length of the historical check-in sequence. The weight vector μ of historical check-in is generated through the temporal attention mechanism.

$$\mu = \frac{exp\left(g\left(\mathbf{h}_{t_k}^p, \mathbf{q}^u\right)\right)}{\sum_{i=1}^{N} exp\left(g\left(\mathbf{h}_{t_k}^p, \mathbf{q}^u\right)\right)} \tag{10}$$

the attention function $g\left(\mathbf{h}_{t_k}^p, \mathbf{q}^u\right)$ is as follows.

$$g\left(\mathbf{h}_{t_k}^p, \mathbf{q}^u\right) = \frac{\mathbf{h}_{t_k}^p(\mathbf{q}^u)^T}{\sqrt{E}} \tag{11}$$

where $\mathbf{q}^u$ is the query information. Then multiply the resulting weight vector μ by $\mathbf{H}^p$ to get user u's preference representation at the POI level.

$$\mathbf{r}_p^u = \sum_{k=1}^{N} \mu_k \mathbf{h}_{t_k}^p \tag{12}$$

3.3 Output Layer

We filter out a suitable POI for each user from all the accessed POIs, which must meet any of the following conditions: (1) the POI is the one that the user has visited before; (2) the POI is close to the POI that the user recently accessed to; (3) it is the POI that is visited most by all users, i.e., popular POI.

In the output layer, we calculate the POI preference obtained by the POI module and the category preference obtained by the category module with the selected POI v_k, and use the Softmax function to perform normalization, and the probability of all candidate POI is obtained as bellows.

$$o_{t_{N+1},v_k}^u = \frac{exp\left(\mathbf{r}_p^u \mathbf{v}_k \times \mathbf{r}_c^u \mathbf{cat}_v\right)}{\sum_{k=1}^{N} exp\left(\mathbf{r}_p^u \mathbf{v}_k \times \mathbf{r}_c^u \mathbf{cat}_v\right)} \tag{13}$$

3.4 Network Training

Bayesian Personalized Ranking (BPR) is used to define loss function for training the LSTM network in the category and POI modules [7], since BPR trains network models by learning pair-wise sorting and can effectively utilize information about POIs that the user does not visit. The data used for the category and POI modules consists of a set of triplets sampled from the original data, each triplet containing the user u and a pair of positive and negative samples.

The loss function of the category module is:

$$lc = \sum_{(c>c')\in\Omega_c} ln\left(1 + e^{-\left(o_t^c - o_t^{c'}\right)}\right) \tag{14}$$

where c' is the negative category of c, Ω_c is the training example, o_t^c is the predicted probability of user u visiting the POI of category c at time t, and $o_t^{c'}$ is the predicted probability of user u visiting the POI of category c'.

The loss function of the POI module is:

$$lp = \sum_{(v>v')\in\Omega_p} ln\left(1 + e^{-\left(o_t^v - o_t^{v'}\right)}\right) \tag{15}$$

By integrating the loss functions and regularization terms of the two modules, we strive to minimize the total loss function:

$$l = lc + lp + \frac{\varepsilon}{2}||\Theta^2|| \tag{16}$$

where ε is the regularization coefficient, Θ is the set of model parameters to learn. AdaGrad is employed to optimize network parameters since it can significantly improve the robustness of stochastic gradient descent.

4 Experiments

To verify the proposed method, we compare it with six baselines on two public real world check-in datasets named as Charlotte (CHA) and New York (NYC) from Foursquare. All the algorithms are coded in Python 3.8 and the framework is TensorFlow 2.3.1.

4.1 Datasets

The check-in data of CHA [11] is collected from January 2012 to December 2013 and the check-in data of NYC [9] is collected from April 2012 to February 2013. The CHA dataset includes 20,939 check-in records and NYC dataset includes 227,428 check-in records.

In this study, each check-in record consists of user, POI, the POI location, the check-in timestamp, the POI category, and the day of the week. Similar to the work of Zhang et al. [11], we use the first 90% of check-ins of each user as the training set and the last 10% as the test set.

4.2 Results and Analysis

We demonstrate the effectiveness of the CPAM method compared to the following six methods: PMF [8], ST-RNN [5], Time-LSTM [14], ATST-LSTM [3], LSPL [13], iMTL [11]. To investigate the effectiveness of CPAM, we focused on answering two research questions. **RQ1:** Can the performance of CPAM be improved by using attention mechanisms and category preference? **RQ2:** Can each component of CPAM help improve recommendation performance?

Table 1. The recommendation result of different methods on CHA and NYC dataset

Datasets	CHA				NYC			
Criteria	*Rec@5*	*Rec@10*	*MAP@5*	*MAP@10*	*Rec@5*	*Rec@10*	*MAP@5*	*MAP@10*
PMF	0.0868	0.1343	0.0181	0.0413	0.0322	0.125	0.0222	0.0263
ST-RNN	0.0890	0.1879	0.0333	0.061	0.0476	0.1964	0.025	0.0312
Time-LSTM	0.0943	0.2142	0.0625	0.0709	0.0794	0.2238	0.0372	0.0558
ATST-LSTM	0.1703	0.3083	0.0699	0.0819	0.1824	0.3269	0.0721	0.0821
iMTL	0.2138	0.3634	0.0833	0.0909	0.2184	0.3801	0.099	0.1057
LSPL	0.2539	0.3701	0.0909	0.1057	0.2702	0.3901	0.0925	0.1129
CPAM	**0.2785**	**0.4016**	**0.0921**	**0.1162**	**0.2777**	**0.4484**	**0.101**	**0.1234**

Answer to RQ1: Table 1 shows the performance of all methods, and the results of two evaluation indicators when k is set to 5 and 10 are listed. It is found that the recall and MAP value of Time-LSTM is higher than that of ST-RNN, which infers that LSTM has better performance than RNN in long sequence modeling. What's more, ATST-LSTM performs better compared with Time-LSTM, which indicates that adding the spatio information and attention mechanism of check-in sequence is beneficial to the modeling of POI check-in sequence. Compared to baseline methods, CPAM considers users' preferences for POI and category at the same time and it mines as much information contained in user check-in sequences as possible. So CPAM we proposed has a better recommendation performance.

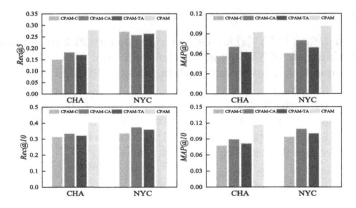

Fig. 2. The recommendation performance comparison of CPAM and its variants on CHA and NYC dataset

Answer to RQ2: In order to verify the performance brought by considering the contribution of category module, the contribution of contextual attention mechanism and the contribution of temporal attention mechanism, we design three different variants of CPAM: (1) CPAM-C removes the category module, that is, users' preferences at the category level are no longer considered. (2) CPAM-CA removes contextual attention from the POI module. (3) CPAM-TA removes the temporal attention mechanism from the POI module. Figure 2 illustrates the performance of CPAM and its variants. It is found that CPAM has better performance than its variants. The three components are indispensable, and they together improve the next POI recommendation performance.

5   Conclusion

A next POI recommendation algorithm based on category preference and attention mechanism is put forward in this paper. The proposed method CPAM considers the user's category preference and POI preference respectively, mines the user's movement behavior patterns through multi-level attention mechanism. The experimental results show CPAM performs better than the other six comparative methods. In the future, we further study the influence of user comment information for next POI recommendation.

Acknowledgements. This work is supported by the National Nature Science Foundation of China (62172186) and the Fundamental Research Funds for the Central Universities, JLU under Grant No.93K172021Z02.

References

1. Chen, W., Yue, L., Li, B., Wang, C., Sheng, Q.Z.: DAMTRNN: a delta attention-based multi-task RNN for intention recognition. In: Li, J., Qin, S., Li, X., Wang, S., Wang, S. (eds.) ADMA 2019. LNCS (LNAI), vol. 11888, pp. 373–388. Springer, Cham (2019). https://doi.org/10.1007/978-3-030-35231-8_27
2. Dai, S., Yu, Y., Fan, H., Dong, J.: Spatio-temporal representation learning with social tie for personalized poi recommendation. Data Sci. Eng. **7**(1), 44–56 (2022)
3. Huang, L., Ma, Y., Wang, S., Liu, Y.: An attention-based spatiotemporal LSTM network for next POI recommendation. IEEE Trans. Serv. Comput. **14**(6), 1585–1597 (2021)
4. Jiao, W., Fan, H., Midtbø, T.: A grid-based approach for measuring similarities of taxi trajectories. Sensors **20**(11), 3118 (2020)
5. Liu, Q., Wu, S., Wang, L., Tan, T.: Predicting the next location: a recurrent model with spatial and temporal contexts. In: Proceedings of the Thirtieth AAAI Conference on Artificial Intelligence, February 12–17, 2016, Phoenix, Arizona, USA, pp. 194–200. AAAI Press (2016)
6. Lv, Q., Qiao, Y., Ansari, N.: Big data driven hidden markov model based individual mobility prediction at points of interest. IEEE Trans. Veh. Technol. **66**(6), 5204–5216 (2017)
7. Rendle, S., Freudenthaler, C., Gantner, Z., Schmidt-Thieme, L.: BPR: bayesian personalized ranking from implicit feedback. CoRR abs/1205.2618 (2012)
8. Wu, Y., Li, K., Zhao, G., Qian, X.: Long- and short-term preference learning for next POI recommendation. In: Proceedings of the 28th ACM International Conference on Information and Knowledge Management, CIKM 2019, Beijing, China, November 3–7, 2019, pp. 2301–2304. ACM (2019)
9. Yang, D., Zhang, D., Zheng, V.W., Yu, Z.: Modeling user activity preference by leveraging user spatial temporal characteristics in lbsns. IEEE Trans. Syst. Man Cybern. Syst. **45**(1), 129–142 (2015)
10. Yu, D., Wanyan, W., Wang, D.: Leveraging contextual influence and user preferences for point-of-interest recommendation. Multimedia Tools Appl. **80**(1), 1487–1501 (2021)
11. Zhang, L., Sun, Z., Zhang, J., Lei, Y.: An interactive multi-task learning framework for next POI recommendation with uncertain check-ins. In: Proceedings of the Twenty-Ninth International Joint Conference on Artificial Intelligence, IJCAI 2020, pp. 3551–3557. ijcai.org (2020)
12. Zhao, P., Zhu, H., Liu, Y., Xu, J., Li, Z.: Where to go next: A spatio-temporal gated network for next POI recommendation. In: The Thirty-Third AAAI Conference on Artificial Intelligence, AAAI, pp. 5877–5884. AAAI Press (2019)
13. Zheng, C., Tao, D., Wang, J.: Memory augmented hierarchical attention network for next point-of-interest recommendation. IEEE Trans. Comput. Soc. Syst. **8**(2), 489–499 (2021)
14. Zhu, Y., Li, H., Liao, Y., Wang, B.: What to do next: Modeling user behaviors by time-lstm. In: Proceedings of the Twenty-Sixth International Joint Conference on Artificial Intelligence, IJCAI 2017, pp. 3602–3608. ijcai.org (2017)

Information Extraction and Retrieval

ToSA: A Top-Down Tree Structure Awareness Model for Hierarchical Text Classification

Deji Zhao[1], Bo Ning[1(✉)], Shuangyong Song[2], Chao Wang[2], Xiangyan Chen[2], Xiaoguang Yu[2], and Bo Zou[2]

[1] School of Information Science and Technology, Dalian Maritime University, Dalian, China
{dejizhao,ningbo}@dlmu.edu.cn
[2] JD AI Research, Beijing, China
{songshuangyong,wangchao208,chenxiangyan5,cdyuxiaoguang, cdzoubo}@jd.com

Abstract. Hierarchical text classification (HTC) is a challenging task that classifies textual descriptions with a taxonomic hierarchy. Existing methods have difficulties in modeling the hierarchical label structure. They focus on using the graph embedding methods to encode the hierarchical structure, ignoring that the HTC labels are based on a tree structure. There is a difference between tree and graph structure: in the graph structure, message passing is undirected, which will lead to the imbalance of message transmission between nodes when applied to HTC. As the nodes in different layers have inheritance relationships, the information transmission between nodes should be directional and hierarchical in the HTC task. In this paper, we propose a Top-Down Tree Structure Awareness Model to extract the hierarchical structure features, called ToSA. We regard HTC as a sequence generation task and introduce a priori hierarchical information in the decoding process. We block the information flow in one direction to ensure the graph embedding method is more suitable for the HTC task, then get the enhanced tree structure representation. Experiment results show that our model can achieve the best results on both the public WOS dataset and a collected E-commerce user intent classification dataset[3].

Keywords: Hierarchical multi-label text classification · Graph embedding · Text generation

1 Introduction

Hierarchical text classification (HTC) is a particular multi-label text classification task, where the classification results correspond to some nodes of a taxonomic hierarchy. It plays an important role in many real-world applications, such as webpage topic classification, product categorization and user feedback classification. Figure 1 is an example of the E-commerce user intent classification

B. Li et al. (Eds.): APWeb-WAIM 2022, LNCS 13422, pp. 23–37, 2023.
https://doi.org/10.1007/978-3-031-25198-6_3

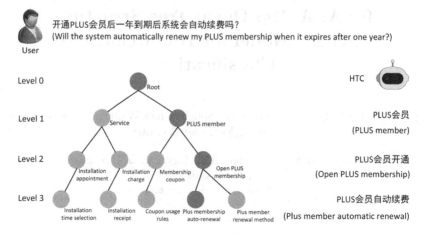

Fig. 1. An example of the E-commerce user intent classification dataset. In the HTC task, the user query corresponds to some nodes in the hierarchical prior tree. The blue nodes indicate the manually labeled results of the questions asked by the user in the HTC task. (Color figure online)

dataset[1], the goal of HTC is to use the structural relationship of labels to find the correct classification answers. The prior probability tree of HTC is constructed based on the labels that have been manually labeled, and the manually labeled labels have hierarchical structure information between them, so that the constructed prior probability tree can use the structural information of the labels between different levels. In the example of E-commerce user intent classification dataset in Fig. 1, level 1, level 2 and level 3 represent the three-level structure of all labels respectively. The role of the root node is to construct the labels in level 1 layer into a complete tree structure, so it is called level 0 and has no meaning at the level 0. In the HTC task, the user query corresponds to some nodes in the hierarchical prior tree. The blue nodes in each layer indicate the multi-label classification results of the user utterance, and the HTC task is dedicated to improving the accuracy of multi-label classification in the hierarchy. All nodes in the E-commerce user intent classification dataset are not fully listed in Fig. 1, and there are many more labels in the actual dataset.

Two kinds of methods are widely used for HTC, which are the local methods and the global methods. The local methods [2,9,12,18] focus on constructing multiple cascaded classifiers, and the number of classifiers depends on the number of label layers. Multiple classifiers built by local-based methods can learn features from different layers and then obtain multi-label classification results. The advantage of the local methods is that it can utilize more fine-grained hierarchical information. But those methods are easily affected by the parent clas-

---

[1] This E-commerce user intent classification dataset is collected from an intelligent service robot designed for creating an innovative online shopping experience in an E-commerce website.

sifiers. Each classifier's loss can easily affect the final model's performance. And the computational cost is very high.

In order to integrate the multiple cascaded classifiers' loss, researchers try to use the global methods. The global methods [19, 21] focus on building a whole classifier, which can utilize structural information of hierarchical prior tree. The global approach is rapidly gaining a lot of attention due to its low computational cost and high classification accuracy. So in this paper, we mainly focus on improving the accuracy of the global approach. The hierarchical prior tree is constructed from manually labeled labels in order to extract the structural relationship between labels. Recent global methods employ the graph convolutional neural network (GCN) [21] to utilize the structural features. GCN is a graph structure feature extractor, which can extract features from graph structure data to obtain an embedded representation of the graph. As the tree structure is a special form of graph structure, the method based on graph structure can also be used to extract tree structure features. It is not the first time to use GCN to model tree structure problems, Zhang et al. [20] employ GCN to fit the dependency tree in dependency parsing.

In this paper, we argue that it is inappropriate to use undirected graph structure approach to model directed tree structur data in the HTC task, and there are some fundamental differences between tree and graph structures. In graph embedding methods, node features are shared for the whole graph. The update direction and order of graph nodes is random. But in the HTC tree structure, the nodes should be updated from the root node instead of randomly selected nodes. Updating the features of the second layer nodes at first will not help much in decoding the first layer labels in the HTC task. And the information update direction is also very important, all nodes should be updated in the same direction as the prior tree, which is more suitable for the HTC task. Sequence to sequence (Seq2Seq) learning [15] is widely used in machine translation task and text generation task. The Seq2Seq method propose a encoder and decoder architecture, which has a large degree of freedom in its inputs and outputs, so it can be adapted to many tasks. Some researchers [13] use Seq2Seq method for multi-label classification before, but they focus on using external knowledge rather than hierarchical structure information. The Seq2Seq method can dynamically fuse other models into a holistic model without incurring hierarchical loss propagation. Based on the learning of the global method, we believe that the encoder and decoder structure is more suitable to integrate GCN in the decoding process.

To deal with the directional characteristic, we regard the hierarchical label extraction as a sequence generation task. Unlike the previous multi-label classification task, the decoder is formed as an auto-regressive structure, in which each time step can decode a corresponding label, and after time t, the decoding results of each time step form the final multi-label classification results together. Using encoder and decoder structure, the features of the parent layer labels can be effectively used to decode the current layer. Based on the framework of encoder-decoder, we employ a GCN to model the characteristics of a hierarchical tree. Then, aiming at the hierarchy of message transmission between tree nodes, we propose a method to suppress one directional data flow to ensure a complete hierarchical tree structure and get the enhance representation of hierarchical tree structure. We designed two unidirectional node update patterns, which are Top-Down and Down-Top. And we believe that using unidirectional node update patterns can enhance the representation of the tree structure in the HTC task.

The contributions of this paper are summarized as follows:

- We use autoregressive decoder for the HTC task and propose the ToSA model, which introduces GCN to get the representation of hierarchical structure, and the decoder can use hierarchical structure information effectively.
- In the process of modeling the hierarchical structure tree, we propose two unidirectional message passing methods in GCN, which are Top-Down and Down-Top to enhance the representation of tree structure.
- We compare ToSA with several baselines, and our model ToSA has achieved the best results on E-commerce user intent classification dataset Intent dataset and WOS public dataset.

2   Related Work

HTC is a particular multi-label text classification (MLC) problem, and the classification of MLC tasks results in multiple category labels. The labels of the HTC task have a hierarchical tree structure with the relationship between each level of labels, and the final classification results of the HTC task correspond to some nodes of the hierarchical tree. Existing methods for HTC could be categorized into two groups: the local method and the global method.

Researchers tend to construct different forms of multi-classifiers to fit the features in different level. Cesa-Bianchi et al. [5] propose a classification method using hierarchical SVM. Using this method is as efficient as training independent SVM-light classifiers for each node. Recently, Huang et al. [9] build a hierarchical attention-based recurrent layer, each recurrent layer can be considered as a local classifier. They classify the documents into the most relevant categories level by level via integrating texts and the hierarchical category structure. Kazuya et al. use text-cnn to extract parent labels' features, and then they use fine-tuning method to fit the children level labels' features. Banerjee et al. [2] propose a transfer learning method to train parent classifier and child classifier.

The global approach regards HTC as a flat MLC problem on the basis of making full use of structural information as much as possible. Early efforts for the HTC task focused on building a flat-based global classifier, and researchers often used Decision Tree and Naive Bayes [7] methods. These methods ignore the hierarchical structure information. There are many applications in industry nowadays, and with the emergence of pre-trained language models, which have also achieved good results in the HTC task. However, these methods ignore the structural information between the hierarchical labels, which means that the structural information is not utilized at all in the actual classification process. Moreover, the inference speed is slow in the actual classification process due to the large number of parameters. A Neural Network called MHC-CNN model is presented by Borges et al. [3] to predict all categories in the hierarchical structure. Harika et al. [1] present the first semi-supervised work for the multi-label classification. Cerri et al.[4] present a new hierarchical multi-label classification method based on multiple neural networks for the task of protein function prediction. In recent years, with the emergence of some new methods in deep learning, some researchers try to use meta-learning [19] to solve HTC problems. Mao et al. [14] use a reinforcement learning approach called HiLAP to transform the HTC task as a markov decision process. Zhou et al. [21] get the idea from the graph embedding method and use GCN to extract the structural features of the hierarchical tree, as a tree structure is a special graph structure. Du et al. [6] collect Multi-view data from different information sources or with distinct feature extraction approaches via clustering algorithm. It is not the first time to use GCN to model tree structure. Zhang et al. [20] employ GCN to fit the dependency tree in dependency parsing. Tree LSTM [16] can also model tree structure. Du et al. collected from different information sources or with distinct feature extraction approaches However, tree LSTM has a large number of parameters and is inferior to GCN in performance and training time. Sequence to Sequence learning [15] is widely used in mechine translation and text generation. Sequence to Sequence learning propose a encoder and decoder architecture, which has great flexibility for the input and output. Rojas et al. employ a encoder and decoder structure to fit the HTC task, as the architecture can fuse external knowledge.

In this paper, we believe that the difference between the two structures should be considered when using the undirected graph embedding method to solve the tree structure problem, so we control the message passing direction to make the graph embedding method more suitable for the tree structure.

3   Our Model-ToSA

In this section, we illustrate our model in detail, whose architecture is depicted in Fig. 2. The complete ToSA model consists of three parts: a user utterance information extraction **Encoder**, a hierarchical structure **Extractor**, and a label **Decoder**. Different colors represent different states of network, the yellow states represent the word embeddings of user utterance in the Encoder area, the green states represent the hidden vectors of the ToSA model. The outputs of the ToSA model are the three level of hierarchical labels. Since a generative model is used, we set a <BOS> tag at the beginning of the decoder and a <EOS> tag at the end of the decoder, which are not included in the hierarchical labels.

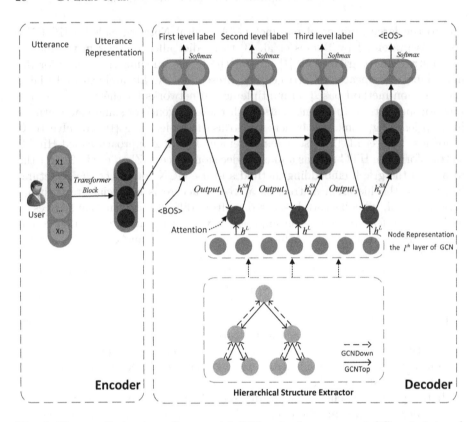

Fig. 2. The overall structure of our model. Different colors represent different states of the network. (Color figure online)

3.1 Encoder

Our encoder is composed of Transformer blocks [17]. Considering the length of sentences, this encoder can capture long-distance semantic dependency features. Transformer is a high-performance feature extractor, which consists of multi-head attention mechanism modules to extract features of sentences from different perspectives, and the attention mechanism is calculated as follows:

$$Attention\,(Q, K, V) = Softmax\left(\frac{QK^T}{\sqrt{d}}\right)V, \qquad (1)$$

where Q,K,V represent the query vectors, keys and values. The attention mechanism can calculate the similarity scores between query and keys, and the final vector is obtained by multiplying the calculated weights by values. Multi-head attention focuses on feature information from different perspectives, it randomly initializes some identical attention mechanisms, and the multi-head attention is calculated as follows:

$$M_i = Attention\,(Q_i, K_i, V_i), \qquad (2)$$

where i represents the ith Attention mechanism. The Q_i, K_i, V_i represent the matrix in the ith Attention mechanism respectively. To keep the latitude of the model constant after computing multiple attention mechanisms, a linear network is used to fuse the multi-headed attention information.

$$M = Concat\,(M_1, M_2, ..., M_i)\,,\qquad(3)$$

$$H = MW_l,\qquad(4)$$

where W_l represents linear weights, and the H represents the final sentence representation.

Given the input $\boldsymbol{X} = (x_1, x_2, ..., x_n)$, where x_i represents the word in the user utterance sentence. The encoder calculation formula is as follows, where H represents the sentence representation.

$$H = Transformer Block(\boldsymbol{X})\qquad(5)$$

In Transformer encoder block, the attention mechanism uses a self-attention mechanism in which the values of query, key, and values are equal, representing Q, K, and V in the formula, respectively. We use the Transformer block to encode the user's utterance into a fixed dimension.

3.2   Hierarchical Structure Extractor

Graph convolutional neural networks are widely used as structure extractors for aggregating node information in natural language processing. Graph neural networks are very effective for modeling structured information like knowledge graphs, and it can find the relationship between different nodes or infer the character of nodes after n hops. However, HTC is a typical tree structure, which is different from the undirected graph structure. We believe that GCN should fully learn the tree structure and balance the message transmission between tree nodes, so as to make GCN more suitable for HTC tasks. In this paper, we believe that controlling the GCN node message passing direction can improve the final effect of the model, so we design two hierarchical message passing patterns for the labeled prior tree, from down to top and from top to down respectively.

In order to ensure the balance between the root node of our hierarchical probability tree, we control the message passing direction in GCN, so as to model the hierarchical tree structure as a unidirectional tree and update the node information directly and hierarchically. As shown in Fig. 3 structure extractor, we propose two unidirectional tree patterns, which are GCN(Down) and GCN(Top).

We follow the previous method in the HTC task [9, 21] to build a hierarchical tree, each node in it represents a manually labeled label. We use the GCN method to extract the relationships between labels, the initialization of each node is random same as the original GCN. The role of the root node is to construct the first-level labels into a complete tree structure and the root node has no actual meaning.

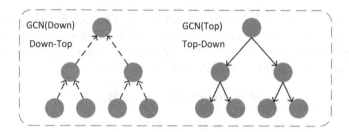

Fig. 3. Two unidirectional message passing patterns in the HTC task.

Given the hierarchical node set $Node = \{n_1, n_2, ..., n_i\}$, where n_i represents the $node_i$ in the hierarchical prior tree. The structure extractor is jointly trained with the whole model. The message passing paradigm between nodes is calculated as:

$$h^{(l+1)} = \sigma \left(b^{(l)} + \sum_{j \in N(i)} \frac{1}{c_{ji}} h_j^{(l)} W^{(l)} \right), \tag{6}$$

where $N_{(i)}$ is the set of neighbors of $Node_i$, c_{ji} is the product of the square root of node degrees, $b^{(l)}$ is bias, and σ is an activation function. We use the node vector features in the last layer of GCN, which have fused the structural information, as shown in Fig. 2 blue nodes.

3.3   Decoder

In the decoder stage, we use the hierarchical prior structure features while decoding the HTC labels in each time step t. The hierarchical structure features used for decoding are the steady state node vectors from the last layer of the GCN. Our model is jointly trained so that it can improve the accuracy of HTC labels and the representation of the tree structure.

Given the response sequence $Y = (y_1, y_2, ..., y_n)$, for each word y_t in Y, we employee the mask operation during the training process to avoid the model seeing the correct answer in advance. For each word y_t, we mask $\{y_{t+1}, ..., y_n\}$ and the model can only see $\{y_1, ..., y_t\}$ while decoding the hierarchical label in the time step t.

The formula for the decoder in the time step t can be written in the following equation:

$$Output_t = P(y_t | H, y_1, ..y_{t-1}), \tag{7}$$

where $Output_t$ is the label distribution at time t. Starting from the first $output$, we perform the attention operation on the $Output_t$ and the node features obtained by GCN. The obtained structure awareness representation is sent to the decoder in the next time step t. Every time we decode the current layer label, our model has inherited the information of the previous time step, which is the parent information of the current layer label. The structure awareness (h_t^{SA}) attention equation is as follows:

Algorithm 1. ToSA Algorithm

Input: $X = (x_1, x_2, ..., x_n)$, $Node = \{n_1, n_2, ..., n_i\}$

Output:$Labels$

1: **for** epoch in number of epochs **do**
2:     Encoder:
3:     $H = TransformerBlock(X)$
4:     Decoder:
5:     **for** label in number of labels **do**
6:         $h^{(l+1)} = Singel - directionGCN(Node)$
7:         $h_t^{SA} = Attention(Output_{t-1}, h^{(L)})$
8:         $Output_t = TransformerBlock(H, h_t^{SA})$
9:         $label_t = Softmax(Output_t)$
10:        $Labels = labels \cup label_t$
11:     **end for**
12: **end for**
13: **return** $Labels$

$$h_t^{SA} = Attention(Output_t, h^{(L)}), \tag{8}$$

where $h^{(L)}$ is the representation of the last layer of GCN. The last layer of the GCN indicates that the information transfer between nodes in the tree structure tends to a steady state. Using the attention mechanism can calculate the similarity between the output of this layer and the nodes of the hierarchical prior tree at different levels, so that correspondence relations can be established in order to improve the model's utilization of structural information.

In the decoding process, the overall probability of the label sequence generated by the model is calculated by the most likelihood estimation. The most likelihood of the response sequence can be calculated as follows:

$$P(Y|H;\theta) = \prod_{t=1}^{T} P\left(y_t|H, h_t^{SA}; \theta\right), \tag{9}$$

where h_t^{SA} is the attention fusion vector of label and hierarchical prior tree node features obtained by decoding at each time step t. The $output_t$ decoded by unidirectional message-passing GCN network finally use softmax to obtain the corresponding multiple labels.

3.4 ToSA Model Process

In order to summarize the encoder and decoder processes of the ToSA model more completely, we summarize Algorithm 1 to describe the model in detail. The ToSA model contains three parts, sentence encoder, structure extractor

Table 1. The statistics of datasets, NA represents no third layer.

Dataset	E-commerce	WOS
Number of sentences	80876	46985
Classes in level 1	10	7
Classes in level 2	140	143
Classes in level 3	1153	NA

and multi-label decoder, where the encoder uses the transformer block to encode the user's utterance and uses an attention mechanism to fuse the relationship between the labels captured by the unidirectional GCN during decoding. Since the decoder is an auto-regressive structure, each time step can decode a corresponding label, and after time t, the decoding results of each time step form the final multi-label classification results together. We use $Labels = labels \cup label_t$ in Algorithm 1 to represent the merging of each label result.

4   Experiment

4.1   Datasets and Evaluation Metrics

We used E-commerce user intent classification dataset and WOS dataset [21] for experiments. The E-commerce user intent classification dataset is more complex than WOS in quantity, as shown in Table 1.

We use standard evaluation metrics [8], including Micro F1 and Macro F1. Micro F1 takes the overall precision and recall of all the instances into account while Macro F1 equals the average F1-score of labels. So Micro F1 gives more weight to frequent labels, while Macro F1 equally weights all labels.

4.2   Baselines

We select several common baselines to test the model, such as TextCNN, BERT(Global), Seq2Seq(Att), HiAGM [21], HARNN [9]. The HiAGM and HARNN both are the most advanced HTC models in recent years.

- **TextCNN(Global)**: Text classification model based on Convolutional Neural Network [11]. TextCNN is widely used in industry application, as its small number of parameters and quickly response.
- **BERT(Global)**: BERT encoder for global classification directly [10], which is widely used in industry. We use it to evaluate global classification performance at the pre-trained level.
- **Seq2Seq(Att)**: Seq2Seq model with attention mechanism [15]. It's a classical encoder and decoder model, and many models use it as a baseline model. We use it to evaluate the effectiveness of generative method in the HTC task.

Table 2. Different components of ToSA variations.

Variations	Encoder	Decoder	Structure extractor
ToSA (Normal)	Transformer	Transformer	Without GCN
ToSA (BERT)	BERT(base)	Transformer	Without GCN
ToSA (GCN)	Transformer	Transformer	Undirected GCN
ToSA (GCNDown)	Transformer	Transformer	Down-Top GCN
ToSA (GCNTop)	Transformer	Transformer	Top-Down GCN
ToSA (GCNCat)	Transformer	Transformer	Bi-direction GCN

- **HARNN**: Using attention-based Recurrent Network Approach to model taxonomic hierarchy [9]. It is the latest local method in the HTC task and consider the each layer as a recurrent structure.
- **HiAGM**: Using Graph embedding method to model the hierarchy structure and it's the global method in the HTC task [21], which has achieved best results without external knowledge.

4.3   Variations of Our ToSA Model

Table 2 gives details about the different combinations of individual encoder, decoder and GCN for several variants of ToSA. In ToSA(BERT) model's encoder is composed of pre-trained levels of BERT, and the pre-trained BERT model uses the base version. The decoder is the normal transformer block with attention mechanism. The Undirected GCN is the normal GCN, which don't use the method proposed in this paper. We concatenate the node features in both Down-Top and Top-Down in Bi-direction structure extractor. We propose several ToSA variants of the model to verify that our proposed unidirectional messaging passing method is valid.

4.4   Implementation and Experiment Design

To keep the fairness across baseline models in the experiment, we used the same parameters on all groups of experimental models. The hidden layer size is set to 256, the batch size is set to 64. We use 8 heads attention, and the model parameters' optimizer is Adam. The learning rate is 0.001. We use Pytorch to run all models on four Tesla P40 GPU. The experiments will be designed to compare on several latitudes, the first set of experiments is to compare with the commonly used models, the second set of experiments is to compare with the recent state-of-the-art models, and the third set of experiments is to compare some variants of our proposed ToSA model.

The first set of experiments contains some basic models commonly used in industry. BERT pre-trained model is often used in industry as its ability to understand natural language, so we design experiments to compare our ToSA model with pre-trained models. TextCNN is a classical classification model that is also widely used in practical applications, and the Seq2Seq model is a classical encoder and decoder generation model. In addition, in the Seq2Seq model, we use the attention mechanism, and using the attention mechanism can significantly increase the performance of the Seq2Seq model. The first set of experiments was designed to verify the effectiveness of the model from different perspectives.

The second set of experiments used the most recent baseline models, where HARNN is a hierarchical label classification model based on a local approach and HiAGM model is a hierarchical label classification model based on a global approach. The second set of experiments is designed to compare the effect of the model proposed in this paper with the strong baseline model.

In the third set of experiments, we focus on which module in the ToSA model has the greatest impact on the ToSA model. First, we design ToSA(Normal) and ToSA(BERT) to improve the performance of the encoder and decoder structure. The purpose is to verify the effectiveness of the approach using generative encoder and decoder structure in multi-label classification. Then, we design a normal GCN model without using the methods in this paper, with the purpose of verifying that unidirectional propagation performs better than undirected propagation in GCN. Finally, we designed two models, called ToSA(GCNDown) and ToSA(GCNTop), based on the two different directional message passing patterns proposed in this paper. The purpose is to verify the effect of different directional messaging patterns on the ToSA model. ToSA(GCNCat) is used to verify whether fusing two different directions of information will be more effective than unidirectional message passing.

5   Experiment Results and Analysis

The experiment results are shown in Table 3. When performing global classification directly in the HTC task, the performance of using encoder-decoder architecture(Seq2Seq(Att)) is almost same as using BERT(base) global classification directly. And ToSA(Normal) is a common transformer structure and it performs better than Seq2Seq. The attention mechanism-based transformer module is much better than the RNN-based Seq2Seq model. Most importantly, the performance of ToSA(Normal) using the generative method to fit the HTC task is better than that of BERT(Global) method based on pre-training. As the Seq2Seq(Att) and ToSA(Normal) based on encoder and decoder architecture, the results of this set of experiments show that the encoder and decoder based architecture is more suitable for the HTC task.

In terms of the utilization of hierarchical information, the effect of ToSA(GCN) is significantly better than ToSA(Normal). And the ToSA(GCN) uses the undirected message passing method. This suggests that it is effective to focus on structural information in the encoder-decoder architecture through the unspecified design of the GCN. In terms of solving the HTC problem, the decoder in ToSA(GCN) uses the attention mechanism to consider the relationship between the current layer node and other layers in the hierarchy tree when

Table 3. Experiment results(%) of different models.

Dataset	E-commerce		WOS	
Model	Macro F1	Micro F1	Macro F1	Micro F1
TextCNN(Global)	67.98	88.57	76.18	77.31
BERT(Global)	74.45	90.92	78.33	82.06
Seq2Seq(Att)	72.69	91.46	79.22	80.27
HARNN	48.87	86.40	69.80	72.61
HiAGM	81.04	87.34	80.19	84.80
ToSA(Normal)	82.76	94.88	79.55	83.55
ToSA(BERT)	83.87	95.26	81.24	85.17
ToSA(GCN)	82.92	95.49	80.68	84.79
ToSA(GCNDown)	83.49	95.61	80.51	84.70
ToSA(GCNTop)	**84.47**	**95.75**	**81.25**	**85.92**
ToSA(GCNCat)	83.85	95.39	80.79	85.11

we decode the label of the current layer. Using attention mechanism can effectively integrate the hierarchical structure information with the decoder.

ToSA(GCNTop) and ToSA(GCNDown) use unidirectional information flow, and the results are significantly better than ToSA(GCN) embedded in the graph structure. The relationship features between layers are obtained by constraining the unidirectional information flow in GCN. This shows that the method of unidirectional message passing has more advantages than the traditional undirected ToSA(GCN). It also shows that it is unwise to employ the undirected graph methods to fit the task based on directed tree structure. And it's very effective to use the unidirectional message passing method to enhance the representation of tree structure. Moreover, the effect of using ToSA(GCNTop) is better than that of ToSA(GCNDown), and it also improves the model's performance compared with the recent models such as HiAGM [21], HARNN [9]. This is also in line with the cognition of Top-Down decoding in the encoder-decoder architecture, and the HTC is also decoded one by one from Top-Down.

In ToSA(BERT), we try to replace the encoder with the BERT [10] pre-trained encoder, and compared with our best ToSA(GCNTop). The ToSA(GCNTop) model achieves the best performance in terms of the performance and the number of parameters. In this set of experiments, it is well illustrate that using a GCN generative model with an enhanced tree structure representation achieves comparable results on the HTC task to a pre-trained model, which would have a large number of parameters and would be very time-consuming if applied to real-life situations.

We also try to integrate the bidirectional features in ToSA(GCNCat), we concatenate bidirectional node features, the results aren't as good as

ToSA(GCNTop), but it is still better than ToSA(GCN). We believe that compared with undirected message passing method, unidirectional information transmission proves its effectiveness again. And the performance of ToSA(GCNCat) is better than ToSA(GCNDown), which shows that in the process of concatenating Bi-directional information, ToSA(GCNDown) absorbs valuable information from ToSA(GCNTop), so ToSA(GCNTop) pattern is more suitable for the HTC task.

On both the two datasets, the values of Macro F1 are generally lower than those of Micro F1, because Macro F1 focuses more on the accuracy of the classification of each sample in the test set, while Micro F1 focuses on the weighted distribution of the number of samples in the test set. In general, any unidirectional ToSA model is better than the undirected ToSA model. The use of generative method is helpful for hierarchical HTC tasks, and the decoder can skillfully integrate structural information.

6   Conclusion

In this paper, we argued that using undirected graph structure to model the directed tree structure feature is not appropriate in the HTC task. We proposed a Top-Down hierarchical aware generative method for realizing the hierarchical text classification by controlling the direction of message passing in a graph embedding method and used generative method to fuse the relational features between tags during decoding. The experiments showed the superiority of our model. In the future, we will explore the application of more GCN variations in the HTC task and how to extract structural features more effectively.

References

1. Abburi, H., Parikh, P., Chhaya, N., Varma, V.: Fine-grained multi-label sexism classification using a semi-supervised multi-level neural approach. Data Sci. Eng. **6**(4), 359–379 (2021)
2. Banerjee, S., Akkaya, C., Perez-Sorrosal, F., Tsioutsiouliklis, K.: Hierarchical transfer learning for multi-label text classification. In: Proceedings of the 57th Annual Meeting of the Association for Computational Linguistics (2019)
3. Borges, H.B., Nievola, J.C.: Multi-label hierarchical classification using a competitive neural network for protein function prediction. In: The 2012 International Joint Conference on Neural Networks (IJCNN), pp. 1–8. IEEE (2012)
4. Cerri, R., Barros, R.C., PLF de Carvalho, A.C., Jin, Y.: Reduction strategies for hierarchical multi-label classification in protein function prediction. BMC Bioinform. **17**(1), 1–24 (2016)
5. Cesa-Bianchi, N., Gentile, C., Tironi, A., Zaniboni, L.: Incremental algorithms for hierarchical classification. In: Advances in Neural Information Processing Systems, pp. 233–240 (2004)
6. Du, G., Zhou, L., Yang, Y., Lü, K., Wang, L.: Deep multiple auto-encoder-based multi-view clustering. Data Sci. Eng. **6**(3), 323–338 (2021)

7. Fall, C.J., Törcsvári, A., Benzineb, K., Karetka, G.: Automated categorization in the international patent classification. In: ACM Sigir Forum, vol. 37, pp. 10–25. ACM, New York (2003)
8. Gopal, S., Yang, Y.: Recursive regularization for large-scale classification with hierarchical and graphical dependencies. In: Proceedings of the 19th ACM SIGKDD International Conference On Knowledge Discovery and Data Mining (2013)
9. Huang, W., et al.: Hierarchical multi-label text classification: An attention-based recurrent network approach. In: Proceedings of the 28th Acm International Conference On Information And Knowledge Management, pp. 1051–1060 (2019)
10. Kenton, J.D.M.W.C., Toutanova, L.K.: Bert: Pre-training of deep bidirectional transformers for language understanding. In: Proceedings of NAACL-HLT (2019)
11. Kim, Y.: Convolutional neural networks for sentence classification. In: Moschitti, A., Pang, B., Daelemans, W. (eds.) Proceedings of the 2014 Conference on Empirical Methods in Natural Language Processing, EMNLP 2014, 25–29 October 2014, Doha, Qatar, pp. 1746–1751. ACL (2014)
12. Li, B.H., Liu, Y., Zhang, A.M., Wang, W.H., Wan, S.: A survey on blocking technology of entity resolution. J. Comput. Sci. Technol. **35**(4), 769–793 (2020)
13. Li, B., et al.: Short text classification model combining knowledge aware and dual attention. J. Softw. **33**(10), 3565–3581 (2022)
14. Mao, Y., Tian, J., Han, J., Ren, X.: Hierarchical text classification with reinforced label assignment. In: Proceedings of the 2019 Conference on Empirical Methods in Natural Language Processing and the 9th International Joint Conference on Natural Language Processing (EMNLP-IJCNLP), pp. 445–455 (2019)
15. Sutskever, I., Vinyals, O., Le, Q.V.: Sequence to sequence learning with neural networks. In: Advances in Neural Information Processing Systems 27 (2014)
16. Tai, K.S., Socher, R., Manning, C.D.: Improved semantic representations from tree-structured long short-term memory networks. In: Proceedings of the 53rd Annual Meeting of the Association for Computational Linguistics and the 7th International Joint Conference on Natural Language Processing, pp. 1556–1566 (2015)
17. Vaswani, A., et al.: Attention is all you need. In: Advances in Neural Information Processing Systems (2017)
18. Wang, W., Feng, D., Li, B., Tian, J.: ATextCNN model: a new multi-classification method for police situation. In: Yang, X., Wang, C.-D., Islam, M.S., Zhang, Z. (eds.) ADMA 2020. LNCS (LNAI), vol. 12447, pp. 135–147. Springer, Cham (2020). https://doi.org/10.1007/978-3-030-65390-3_11
19. Wu, J., Xiong, W., Wang, W.Y.: Learning to learn and predict: A meta-learning approach for multi-label classification. In: Proceedings of the 2019 Conference on Empirical Methods in Natural Language Processing and the 9th International Joint Conference on Natural Language Processing (EMNLP-IJCNLP) (2019)
20. Zhang, Y., Qi, P., Manning, C.D.: Graph convolution over pruned dependency trees improves relation extraction. In: Proceedings of the 2018 Conference on Empirical Methods in Natural Language Processing, pp. 2205–2215 (2018)
21. Zhou, J., et al.: Hierarchy-aware global model for hierarchical text classification. In: Proceedings of the 58th Annual Meeting of the Association for Computational Linguistics (2020)

Clause Fusion-Based Emotion Embedding Model for Emotion-Cause Pair Extraction

Zhiwei Li[1], Guozheng Rao[1,3,5], Li Zhang[2], Xin Wang[1,5(✉)], Qing Cong[3],
and Zhiyong Feng[1,4,5]

[1] College of Intelligence and Computing, Tianjin University, Tianjin 300350, China
{chiwailee,rgz,wangx,zyfeng}@tju.edu.cn
[2] School of Economics and Management, Tianjin University of Science
and Technology, Tianjin 300457, China
zhangli2006@tust.edu.cn
[3] School of New Media and Communication, Tianjin University,
Tianjin 300072, China
chf@tju.edu.cn
[4] Shenzhen Research Institute of Tianjin University, Shenzhen 518000, China
[5] Tianjin Key Laboratory of Cognitive Computing and Applications,
Tianjin 300350, China

Abstract. The goal of emotion-cause pair extraction (ECPE) is to simultaneously extract all emotion clauses and their corresponding cause clauses in a document. In most existing methods, emotion clause representations and cause clause representations are usually obtained separately and are then fed into neural networks. However, the close relationship between emotion and cause is ignored, resulting in an insufficient representation of the emotion clause and the cause clause. To address this problem, we propose a new model, called the clause fusion-based emotion embedding model, to make full use of emotion-related knowledge by utilizing an emotion embedding method when obtaining the representation of the cause clause. First, the emotion word embedding is processed by the emotion clause encoder to get the emotion feature. Second, in clause fusion based emotion embedding network, the emotion clause-level feature in the sliding-window is fused to fused emotion features. The fused emotion features, cause word-level, and emotion word-level feature representation are embedded to get emotion embedding. Third, the emotion embedding is processed to the cause clause feature representation by a bidirectional long short-term memory. Forth, each emotion clause-level feature representation was paired with each cause clause-level feature representation to produce candidate pairs representation. Finally, in the clause pair encoder, a graph convolutional network is applied to model the pair-level context, and then contextual features are extracted for the candidate pairs. Experimental results show that our model achieves state-of-the-art performance on the Chinese benchmark ECPE corpus.

Keywords: Emotion-cause pair extraction · Emotion embedding ·
Graph convolutional network

© The Author(s), under exclusive license to Springer Nature Switzerland AG 2023
B. Li et al. (Eds.): APWeb-WAIM 2022, LNCS 13422, pp. 38–52, 2023.
https://doi.org/10.1007/978-3-031-25198-6_4

1   Introduction

Emotion cause extraction (ECE) [1, 2], which is a subtask of sentiment analysis, has attracted increasing research attention in the sentiment analysis community in recent years. Gui et al. [2] reformalized the ECE task as a clause-level extraction problem and released a new corpus for this task. The goal of the ECE task is to identify whether a clause in a document is the corresponding cause of the emotion annotation. The ECE task is well defined but has one serious drawback. Emotions in documents need to be manually annotated in advance. The high cost of annotation leads to limited practical application. To address this problem, Ding et al. [3] proposed a new task called emotion-cause pair extraction (ECPE). In the ECPE task, which is a clause-level extraction task, the goal is to extract all existing emotion-cause pairs (i.e., clause pairs consisting of emotion clauses and their corresponding cause clauses) in a given document as input. Fig. 1 shows an example of the ECPE task. The input in this example is a document consisting of five clauses. Clause c4 contains the "happy" emotion, and there are two corresponding causes: clause c2 ("a policeman visited the old man with the lost money") and clause c3 ("and told him that the thief was caught"). The final output is a set of valid emotion-cause pairs defined at the clause level: c4-c2, c4-c3. A two-step approach (ECPE-2Steps) was proposed by Ding et al. [3] to address this task. ECPE-2Steps is a two-step pipeline: first, the candidate emotion clauses and cause clauses are extracted separately. For example, in Fig. 1, the candidate emotion clause is c4, and the candidate cause clauses are c2 and c3. Second, candidate emotion clauses are matched with cause clauses pairwise to generate candidate clause pairs. Specifically, the candidate emotion-cause pairs are obtained by applying a Cartesian product to the candidate emotion clauses and cause clauses. Then, the candidate clause pairs are fed into a filter to obtain the valid pairs.

c1: Yesterday morning,

c2: a policeman visited the old man with the lost money,(cause)

c3: and told him that the thief was caught. (cause)

c4: The old man was very happy, (emotion)

c5: and deposited the money in the bank.

Fig. 1. An example of the ECPE task.

Although the ECPE-2Steps approach is reasonable and achieves good performance. However, it still has the following drawbacks: first, as a pipeline approach, step 1 of ECPE-2Steps needs to be completed before step 2. This means that errors produced in step 1 will be introduced into step 2. Second, the extraction of candidate emotions and cause clauses in step 1 are two independent sub-tasks. This means that the ECPE-2Steps approach does not use the association information between emotions and their corresponding causes. In fact, there is a causal relationship between emotions and their corresponding causes, and they are mutually indicative. But they are often underutilized in most existing approaches, and the extraction of emotion and cause feature is insufficient.

To address these shortcomings, we propose a novel end-to-end ECPE solution called the clause fusion-based emotion embedding model (CFEE). We first obtain the emotion word-level and clause-level feature representation through an emotion clause encoder and then feed both the emotion feature representations and the cause word-level feature representation into a clause fusion-based emotion embedding network to obtain the cause clause-level feature representation. Next, a Cartesian product is applied to the emotion and cause clause-level feature representations to generate emotion-cause pairs. Then, the emotion-cause pairs are fed into a graph convolutional network to obtain the final emotion-cause pair representation.

The main contributions of our work can be highlighted as follows:

- We propose a CFEE model, which is an end-to-end framework, for ECPE.
- We design a clause fusion-based emotion embedding network to fuse the emotion feature contexts and embed them into the cause word-level feature representation to facilitate the extraction of cause clause-level feature and improve the prediction of emotion-cause pairs.
- Experiments on the ECPE benchmark corpus demonstrate that our model achieves state-of-the-art performance. Furthermore, ablation experiments are performed to verify the effectiveness of the components in our model.

2   Related Work

2.1   Emotion-Cause Pair Extraction

The emotion-cause pair extraction (ECPE) task directly extracts potential emotion-cause pairs in documents without any emotion annotations. To address this problem, Xia and Ding et al. [3] proposed a two-step approach. In the first step, the emotion-cause pair extraction task is transformed into two separate subtasks, emotion extraction, and cause extraction. In the second step, emotions and causes are matched pairwise and then input into a filter. However, the approach has the following problems: 1) errors produced in step 1 will be introduced into step 2, which further increases the overall error; and 2) emotion extraction and cause extraction are executed independently, which means that the mutual indication between emotion and cause have not been exploited. To address these problems, Wei et al. [4] completed the task from a ranking point of view and proposed an end-to-end method that focuses on sentence internal modeling. The internal relationship between clauses in the document is modeled through graph attention and clause-pair representations and enhanced with kernel-based relative position embedding for effective ranking. In addition to the ranking method, Ding et al. [5] introduced a joint method to solve these problems, which can directly extract emotion-cause pairs. They adopted a two-dimensional representation scheme to represent emotion-cause pairs and integrated the two-dimensional representation, interaction, and prediction into a joint model. To capture the contextual information of pairs, they improved the standard 2D converter and proposed two kinds of converters: a constrained window 2D converter and a crisscross 2D converter. Ding et al. [6] and Chen et al. [7] also extracted emotion and cause clause representations separately and then concatenated them as pairs and fed them into a neural network. Considering that emotion clauses are usually close to their corresponding cause clauses, Cheng et al. [8] used a symmetric local search strategy for finding cause clauses based on emotion clauses or finding emotion clauses based on cause clauses.

In another way, some studies applied the sequence labeling method to handle this problem. Chen et al. [9] designed a special unified labeling scheme and changed the ECPE task into a unified sequence labeling task so that more than one emotion-cause pair could be extracted simultaneously. A new multilabel labeling scheme was proposed by Yuan et al. [10] to encode the distance between linked components into labels. The high computational cost caused by the Cartesian product of candidate clauses can be reduced by using this scheme.

However, except for the sequence labeling method, in a large number of the existing methods, emotion extraction and cause extraction are independent auxiliary subtasks, which means the fact that emotion and its corresponding cause have a causal relationship is ignored. In addition, these methods do not make full use of the association between emotion and cause.

2.2   Graph Convolutional Neural Network

The graph convolutional network (GCN) was first proposed by Kipf and Welling [11] for node classification, which was conducted on graph-structured data. Since then, GCN has shown powerful performance and impressive modeling capability. An increasing number of researchers in the NLP community utilize GCN in their tasks, such as relation extraction [23], emotion analysis, and text classification. Zhang et al. [12] used GCNs to effectively bring information together in parallel on any dependent structure and further applied a novel pruning strategy to the input trees. Sun et al. [13] presented a convolution over a dependency tree model for aspect-level sentiment analysis and further enhanced the embeddings with a graph convolutional network. In addition, Yao et al. [14] proposed using graph convolutional networks for text classification. They constructed a single text graph based on word co-occurrence and document word relations and then learned a text graph convolutional network. Ghosal et al. [15] used a graph convolutional neural network for emotion recognition in conversation. The method leveraged the self-and interspeaker dependency of the interlocutors to model the conversational context for emotion recognition.

3   CFEE Model

3.1   Task Definition

Given a document $D = \{c_1, c_2, \ldots, c_n\}$ with n clauses where $c_i, i \in \{1, 2, \ldots, n\}$ is the ith clause. Clause pairs C^p, which are candidate emotion-cause pairs, are formed by combining clauses in document D.

$$C^p = \{c^p_{1,1}, c^p_{1,2}, \ldots, c^p_{n,n}\} \tag{1}$$

$$c^p_{i,j} = \{c^e_i, c^c_j\} \tag{2}$$

where c^e_i represents the emotion clause in the document D and c^c_j represents the cause clause in the document D. The size of the set C^p is $n \times n$. The task of ECPE is to predict whether each candidate emotion-cause pair $(c^p_{i,j})$ is a true emotion-cause pair with a one-bit binary label (0 or 1), where '0' indicates that $c^p_{i,j}$ is not an emotional cause pair, that is, c^e_i is not an emotion clause or c^e_i is the emotion clause but c^c_j is not the corresponding cause clause, and '1' means that c^e_i is the emotion clause and c^c_j is the corresponding cause clause.

3.2   An Overview of CFEE

The overall structure of the clause fusion-based emotion embedding model for ECPE is shown in Fig. 2, which includes a Bi-LSTM-based emotion clause encoder network, a clause fusion-based emotion embedding network, and a graph convolutional network-based pair encoder network.

3.3   Emotion Clause Encoder

To better obtain the sequential information about the context in the clauses, we use the emotion clause encoder to obtain the emotion features of the clauses, namely, the clause-level emotion feature v^e. The encoder consists of two-layer Bi-LSTM networks, including a word-level Bi-LSTM network and a clause-level Bi-LSTM network. For an input text D with n clauses, we can obtain the emotion sequential representation $s^e = \{s^e_1, s^e_2, \ldots, s^e_n\}$ and the cause sequential representation $s^c = \{s^c_1, s^c_2, \ldots, s^c_n\}$ through the word embedding. The emotion word-level feature representation $r'^e = \{r'^e_1, r'^e_2, \ldots, r'^e_n\}$ is obtained through the first-layer word-level Bi-LSTM network (see Eq. 3) and r'^e is fed into the clause-level Bi-LSTM network to obtain the emotion clause-level feature representation $r^e = \{r^e_1, r^e_2, \ldots, r^e_n\}$ (see Eq. 4)

$$r'^e = BiLSTM^e_w(s^e) \tag{3}$$

$$r^e = BiLSTM^e_c(r'^e) \tag{4}$$

where $BiLSTM^e_w$ is a word-level Bi-LSTM that is used to extract an emotion word feature $r'^e \in \mathbb{R}^{2d_h}$ and $BiLSTM^e_c$ is a clause-level Bi-LSTM that is used to extract an emotion clause-level feature representation $r^e \in \mathbb{R}^{2d_h}$.

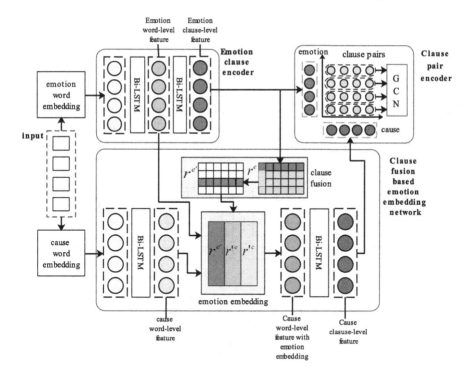

Fig. 2. An example of the ECPE task.

3.4   Clause Fusion-Based Emotion Embedding Network

To make better use of the enhanced effect of emotion features on the extraction of cause features, we design a clause fusion-based emotion embedding network to enhance the performance of extraction cause features. The network consists of four components, including a word-level Bi-LSTM network, a clause fusion component, an emotion embedding component, and a clause-level Bi-LSTM network.

The cause word-level feature representation $r'^c = \{r'^c_1, r'^c_2, \ldots, r'^c_n\}$ is obtained through the word-level Bi-LSTM network. Considering that most of the cause clauses are around the emotion clauses, we propose a sliding window-based emotion fusion method for the clause fusion component. We make a size-constrained window slide over the emotion clause-level feature representation r^e and then fuse all feature representations in the window as the feature of the central clause (see Fig. 3 (a)). Finally, the fused features are embedded into the cause word-level features. Inspired by the pooling operation, we design two fusion methods: mean fusion and sum fusion. Mean fusion refers to the average operation is performed on the feature representations within the window. Sum fusion means that the sum operation is performed on the feature representations within the window. In particular, when the size of the window is 0 means no feature representations need to be fused. In other words, the emotion and cause clause features are extracted independently. The calculation of the fused emotion feature representation $r^{e'}$ is as in Eq. 5 and Eq. 6:

$$r^{e'} = fusion_{mean}^{t}(r^e) \tag{5}$$

$$r^{e'} = fusion_{sum}^{t}(r^e) \tag{6}$$

where $fusion_{mean}$ represents the mean fusion, $fusion_{sum}$ represents the sum fusion, and $t \in \{0, 1, 3, 5\}$ denotes the size of the sliding window.

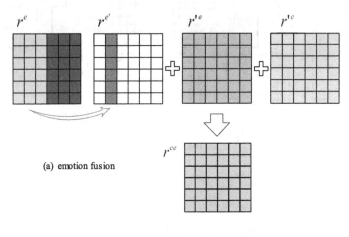

(a) emotion fusion

(b) emotion embedding

Fig. 3. (a) Clause fusion component with t = 3 and mean fusion, and (b) an emotion embedding component of clause fusion-based emotion embedding network.

As shown in Fig. 3 (b), in the emotion embedding component, we concatenate the cause word-level feature representation r'^c, the emotion word-level feature representation r'^e and the fused emotion clause-level feature representation fused $r^{(e')}$ to generate the cause word-level feature representation with emotion embedded r^{ce} (see Eq. 7). Then the cause word-level representation r^{ce} was fed into the cause clause-level Bi-LSTM network to obtain the cause clause-level feature representation, is as in Eq. 7:

$$r^{ce} = [r'^c, r'^e, r^{e'}] \tag{7}$$

$$r^c = BiLSTM_c^e(r^{ce}) \tag{8}$$

where [,] is the concatenating function and $BiLSTM_c^e$ is the clause-level Bi-LSTM that is used to extract a cause clause-level feature representation $r^c \in \mathbb{R}^{2d_h}$.

3.5   Clause Pair Encoder

To better obtain the contextual information at the clause-pair level, we construct a clause-pair graph to model candidate clause pairs, changing the transfer of clause-level contextual features into the transfer of clause pair-level contextual features.

First, we obtain the emotion-cause pair by pairing the emotion clause and the cause clause. An emotion-cause pair $c_{i,j}^p = (c_i^e, c_j^c)$ is represented as $v_{i,j}^p$, which includes the emotion clause-level feature representation r^e and the cause clause-level feature representation r^c:

$$v_{i,j}^p = [r_i^e, r_j^c] \quad (i, j \in 1, 2, \ldots, n) \tag{9}$$

Second, taking $v_{i,j}^p$ as a node of the graph, all nodes with the same emotion feature are used to build a simple graph, which is called a clause pair graph. A document with n clauses requires a total of n clause pair graphs to be constructed. As the example mentioned in Fig. 1, if there are 5 clauses, then a total of 5 clause-pair graphs need to be constructed. In addition, the distance between a cause clause and its corresponding emotion clause is relatively close. The statistical results of the open-source ECPE Chinese corpus of Xia and Ding [1] show that the distances between 95.80% of the emotion clauses and the corresponding cause clauses are less than 3. So, we only use the clause pairs whose distance from the central node is less 3 to construct the clause-pair graph. For an emotion clause c_i^e, the nodes in its corresponding pair graph no more than 5, as Eq. 10:

$$c_{i,[i-2:i+2]}^p = \{c_{i,i-2}^p, c_{i,i-1}^p, c_{i,i}^p, c_{i,i+1}^p, c_{i,i+2}^p\} \tag{10}$$

Since the influence of adjacent nodes with different distances is different, three different edges are designed. The first edge is the D0 edge, which is used to represent the self-loop edge of the self-migration of nodes. The second edge is the D1 edge connecting the nodes which have a distance of 1 between their candidate cause clauses. e.g., node $c_{i,i-1}^p$ and node $c_{i,i}^p$ need to be connected by the D1 edge. The third edge is the D2 edge connecting the nodes which have a distance of 2 between their candidate cause clauses. e.g., node $c_{i,i-1}^p$ and node $c_{i,i+1}^p$ need to be connected by the D2 edge.

For node $c_{i,j}^p$, its feature representation $g_{i,j}^p$ is obtained by transferring the feature from the clause-level encoding network. Specifically, the feature $g_{i,j}^p$ in the clause pair graph is obtained by integrating the nodes connected to it using different transformation parameters according to different edges they are linked to:

$$g_{i,j}^p = \sigma \left(\frac{1}{z} v_{i,j}^p W_{D0} + \frac{1}{z} \sum_{k \in D1} v_{i,k}^p W_{D1} + \frac{1}{z} \sum_{t \in D2} v_{i,t}^p W_{D2} \right) \tag{11}$$

where $W_{D0} \in \mathbb{R}^{d_{in} \times d_{out}}$, $W_{D1} \in \mathbb{R}^{d_{in} \times d_{out}}$ and $W_{D2} \in \mathbb{R}^{d_{in} \times d_{out}}$ are the weight matrices for the nodes lined to node $c_{i,j}^p$ with D0 edge, D1 edges, and D2 respectively, z is the normalization factor whose value is equal to the degree of the node, σ represents the nonlinear activation function, and ReLU [23] is used as the activation function.

3.6   Classification

Since most of the distances between emotions and causes are less than 3, it is helpful to add distance information when classifying emotion-cause pairs. Therefore, the classification representation of the node $c_{i,j}^p$ for the candidate clause is the concatenation of $g_{i,j}^p$ and $d_{i,j}$. Where $d_{i,j} \in \mathbb{R}^{d_{dis}}$ is the distance embedding. The softmax function is as follow:

$$\hat{y}_{i,j} = softmax\left(W_p^T[g_{i,j}^p, d_{i,j}] + b_p\right) \tag{12}$$

where $W_p \in \mathbb{R}^{(d_{out}+d_{dis}) \times d_p}$ is the weight matrix and $b_p \in \mathbb{R}^{d_{p_{out}}}$ is the bias vector. Finally, the probability distribution $\hat{y}_{i,j}$ and the corresponding predicted label $\widehat{EC}_{i,j}$ of the prediction of the candidate clause to node $c_{i,j}^p$ are obtained.

Model training uses cross-entropy loss as the loss function.

4   Experiment

4.1   Dataset and Experimental Settings

We use the Chinese benchmark dataset for ECPE released by Xia and Ding [3], which was reconstructed from the benchmark Chinese ECE dataset [2]. In the ECPE dataset, there are 1,945 documents, of which 1746 documents have one emotion-cause pair, 177 documents have two emotion-cause pairs and 22 documents have more than two emotion-cause pairs, and the percentages of the dataset are 89.77%, 91.0% and 1.13%, respectively. In addition, in 95.8% of the documents in the dataset, the distance between the emotion and the corresponding cause is less than or equal to 2. The specific statistics are shown in Table 1 For fair comparisons with Chen et al. [7], we use the same ratio to split the data: 90% are randomly selected for training, and the remaining data are used for testing. The results reported in the following experiments are an average of 10-fold cross-validation. Similarly, we repeat the experiments 20 times and report the average result. We use the precision (P), recall (R), and F1-score as evaluation metrics for the ECPE task as well as two subtasks: emotion extraction and cause extraction. In our experiments, we use word embedding, which was pretrained on 1.1 million Chinese Weibo corpora with the word2vec toolkit [16], and the dimension of word embedding is 200. Moreover, BERT representations [17] are

Table 1. Statistics of the distances between emotion clauses and their cause clauses in the Chinese ECPE corpus (Xia and Ding 2019)

Distance	Number	Percentage	Distance	Number	Percentage
0	511	23.6%	≤0	511	23.6%
1	1342	61.9%	≤1	1853	85.5%
2	224	10.3%	≤2	2077	95.8%

also utilized, where we use the base Chinese model. For word2vec, the dimension of word embeddings is 200, and the hidden units of Bi-LSTM and GCN are both 100. For BERT, the dimension of word embeddings is 768, and the hidden units of Bi-LSTM and GCN are both 200. While training, we use the Adam optimizer [18] to update all parameters. The mini-batch size and the learning rate are set to 32 and 0.005, respectively. To reduce overfitting, dropout [19] is applied to all feature vectors, including word embeddings and hidden representations, and it is set to 0.5. The method of emotion embedding uses mean fusion with a sliding window size of 1.

4.2 Comparative Approaches

1. **Inter-CE** [3]: This is an enhanced version of Indep [3] that is capable of capturing the correlation between emotions and causes. While extracting emotion clauses and cause clauses, emotion-cause extraction is used to improve emotion extraction. The method failure to fully exploit the causal relationship between emotion and cause.
2. **Inter-EC** [3]: This is another enhanced version of Indep. It uses emotion extraction to improve emotion-cause extraction while extracting emotion clauses and cause clauses.
3. **PairGCN** [7]: This method constructs a graph using the pair nodes and a pair graph convolutional network to model the dependency relations among candidate pairs. To make a fair, we use the results reproduced locally based on the author's open-source code for comparison.
4. **Hier-Bi-LSTM:** This is an end-to-end model that extracts emotion features and cause features using two hierarchical Bi-LSTMs independently, and the concatenation of an emotion feature and a cause feature is used to represent a candidate pair. Specifically, the hierarchical Bi-LSTM is similar to the one used in our clause-level context encoder, except that the input to the clause-level Bi-LSTM in the cause encoder is only the word-level cause feature.
5. **MTNECP** [20]: MTNECP is a unified multitask learning framework. It shares useful features across tasks and utilizes position-aware emotion information for cause extraction.
6. **MAM-SD** [21]: It is a mutually auxiliary multitask model to promote the extraction of emotion and cause clauses by adding two auxiliary tasks which are identical to the original tasks. It is also a self-distillation method for pairwise tasks to train the proposed model.
7. **Inter-EC-WC-BERT** [5]: This is a joint model that adopts a two-dimensional representation scheme to represent emotion-cause pairs, integrating the two-dimensional representation, interaction, and prediction. Window-constrained 2D transformer is applied in the model.
8. **Inter-EC-CR-BERT** [5]: This is a joint model that adopts a two-dimensional representation scheme to represent emotion-cause pairs, integrating the two-dimensional representation, interaction, and prediction. Cross-road 2D transformer is applied in the model.
9. **Trans-ECPE** [22]: This is a transition-based method that transforms the task into a parsing-like directed graph construction procedure.

4.3   Experimental Results

Table 2. Performance of our model and baselines without BERT using precision, recall, and F1-score as metrics on the ECPE task as well as the two subtasks. The best performance is in bold, the second-best performance is underlined, and '*' indicates the results reproduced locally based on the author's open-source code.

Model	EC pair extraction			Emotion extraction			Cause extraction		
	P	R	F1	P	R	F1	P	R	F1
Inter-CE	69.02	51.35	59.01	84.94	81.22	83.00	68.09	56.34	61.51
Inter-EC	67.21	57.05	61.28	83.64	81.07	82.30	70.41	60.83	65.07
Hier-Bi-LSTM	69.25	53.71	60.30	86.16	66.29	74.80	72.27	55.32	62.48
MTNECP	68.28	58.94	63.21	86.62	83.39	85.20	74.00	63.78	68.44
PairGCN*	71.03	56.29	62.68	87.07	70.54	77.84	73.53	57.89	64.66
MAM-SD	69.63	57.99	63.20	85.86	81.41	83.39	72.02	63.75	67.51
CFEE	73.88	56.54	63.98	88.35	69.30	77.62	75.73	57.98	65.60

Table 3. Performance of our model and baselines with BERT using precision, recall, and F1-score as metrics on the ECPE task as well as the two subtasks. The best performance is in bold, the second-best performance is underlined, and '*' indicates the results reproduced locally based on the author's open-source code.

Model	EC pair extraction			Emotion extraction			Cause extraction		
	P	R	F1	P	R	F1	P	R	F1
Inter-EC-WC-BERT	72.92	65.44	68.89	86.27	92.21	89.10	73.36	69.34	71.23
Inter-EC-CR-BERT	69.35	67.85	68.37	85.48	92.44	88.78	69.35	67.85	68.37
Trans-ECPE	73.74	63.07	67.99	87.16	82.44	84.74	75.62	64.71	69.74
Hier-Bi-LSTM-BERT	75.37	64.34	69.26	88.80	74.70	81.00	78.03	65.35	70.96
PairGCN-BERT*	76.50	67.17	71.42	88.12	78.73	83.05	79.08	68.78	73.46
CFEE-BERT	79.19	67.35	72.63	91.50	79.23	84.82	80.93	68.49	74.06

Table 2 and Table 3 show the results of the emotion-cause pair extraction (ECPE) task and two subtasks: emotion clause extraction (EE) and cause clause extraction (CE). The CFEE model and the CFEE-BERT model achieved the best results in the ECPE task. Specifically, our CFEE and CFEE-BERT models achieve 1.22% and 1.69% F1-score improvements for ECPE compared to the previous best models, MTNECP and PairGCN-BERT, respectively. CFEE-BERT also achieves a 3.52% precision improvement for ECPE compared to the previous best model PairGCN-BERT. We argue that the clause fusion-based emotion embedding network plays an important role in this process. It enhances clause-pair feature capture by embedding emotion context information. Moreover, for the cause extraction task, CFEE-BERT outperforms the baseline model

PairGCN-BERT in all three metrics. This finding indicates that the clause fusion-based emotion embedding network strengthens the ability of cause clause prediction.

The reason why our model outperforms the current state-of-the-art models is mainly because we use an emotion embedding model that adopts an end-to-end form to directly predict emotion-cause pairs. However, in previous research, the relationship between emotion and cause was not well utilized. And the extraction of emotion and cause is executed as two separate subtasks.

4.4   Effect of Emotional Embedding Position

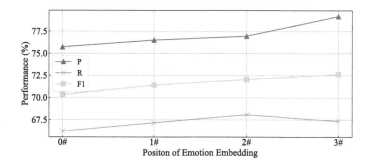

Fig. 4. Performance of our model with different positions of the emotion embedding on the ECPE task.

To study the effect of the position of emotion embedding in our full model (i.e., CFEE and CFEE-BERT), we design a set of ablation experiments for different embedding positions. To better test the influence of the embedding position, in the emotion fusion of the mode, the fusion window size is set to 1. The experimental results of ECPE are shown in Fig. 4

- **0#:** Clause-level feature representation extractions of emotion and cause are completely independent.
- **1#:** The cause word-level feature representation is directly concatenated with the emotion word-level feature representation and then fed into a Bi-LSTM network to obtain the cause clause-level feature representation.
- **2#:** The cause word-level feature representation is directly concatenated with the emotion clause-level feature representation and then fed into a Bi-LSTM network to obtain the cause clause-level feature representation.
- **3#:** Concatenate cause word-level feature representation, emotion word-level feature representation, and emotion clause-level feature representation and then fed into a Bi-LSTM network to obtain the cause clause-level encoding.

From the results in Fig. 4, we know that not only does the performance of the model change with the position of the emotion embedding, but the three metrics of the model also have the same trend. Different positions of the emotion embedding can produce a maximum performance gap of more than 2%. This means that the position of the emotion embedding is very important.

We also conduct experiments on the effect of the emotion fusion method and the size of the sliding window on the model effect. Based on embedding position 3#, we design multiple sets of experiments. The result of ECPE is shown in Fig. 5.

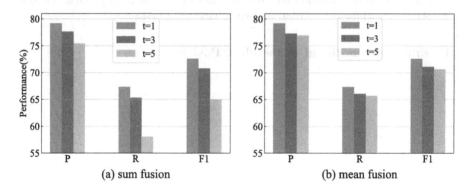

Fig. 5. Performance of our model with different positions of the emotion embedding on the ECPE task.

From Fig. 5, we can see that both fusion methods have trend performance decreases as the size of the sliding window increases. The reason is probably that a larger sliding window introduces more noise information. The results show that the size 1 sliding window is enough to capture the emotion feature we need.

5   Conclusion and Future Work

In this paper, we propose a novel clause fusion-based emotion embedding model (CFEE) to enhance the extraction of cause features, and further improve the extraction of emotion-cause pair. CFEE makes full use of the causal relationship between emotions and causes among local neighborhood clauses. The experiments on the Chinese benchmark corpus show that our model achieves state-of-the-art performance, and the ablation experiments demonstrate the effectiveness of our proposed modules. In the future, we would like to address the problem by increasing the mutual interaction between emotion and cause.

Acknowledgements. This work was supported by the National Natural Science Foundation of China (NSFC) under Grant 61832014, 61373165.

References

1. Chen, Y., Lee, S.Y.M., Li, S., Huang, C.-R.: Emotion cause detection with linguistic constructions. In: Proceedings of the 23rd International Conference on Computational Linguistics (Coling 2010), pp. 179–187 (2010)

2. Gui, L., Wu, D., Xu, R., Lu, Q., Zhou, Y.: Event-driven emotion cause extraction with corpus construction. In: 2016 Conference on Empirical Methods in Natural Language Processing, EMNLP 2016, pp. 1639–1649. Association for Computational Linguistics (ACL) (2016)

3. Xia, R., Ding, Z.: Emotion-cause pair extraction: a new task to emotion analysis in texts. In: Proceedings of the 57th Annual Meeting of the Association for Computational Linguistics, pp. 1003–1012 (2019)

4. Wei, P., Zhao, J., Mao, W.: Effective inter-clause modeling for end-to-end emotion-cause pair extraction. In: Proceedings of the 58th Annual Meeting of the Association for Computational Linguistics, pp. 3171–3181 (2020)

5. Ding, Z., Xia, R., Yu, J.: ECPE-2D: emotion-cause pair extraction based on joint two-dimensional representation, interaction and prediction. In: Proceedings of the 58th Annual Meeting of the Association for Computational Linguistics, pp. 3161–3170 (2020)

6. Ding, Z., Xia, R., Yu, J.: End-to-end emotion-cause pair extraction based on sliding window multi-label learning. In: Proceedings of the 2020 Conference on Empirical Methods in Natural Language Processing (EMNLP), pp. 3574–3583 (2020)

7. Chen, Y., Hou, W., Li, S., Wu, C., Zhang, X.: End-to-end emotion-cause pair extraction with graph convolutional network. In: Proceedings of the 28th International Conference on Computational Linguistics, pp. 198–207 (2020)

8. Cheng, Z., Jiang, Z., Yin, Y., Yu, H., Gu, Q.: A symmetric local search network for emotion-cause pair extraction. In: Proceedings of the 28th International Conference on Computational Linguistics, pp. 139–149 (2020)

9. Chen, X., Li, Q., Wang, J.: A unified sequence labeling model for emotion cause pair extraction. In: Proceedings of the 28th International Conference on Computational Linguistics, pp. 208–218 (2020)

10. Yuan, C., Fan, C., Bao, J., Xu, R.: Emotion-cause pair extraction as sequence labeling based on a novel tagging scheme. In: Proceedings of the 2020 Conference on Empirical Methods in Natural Language Processing (EMNLP), pp. 3568–3573 (2020)

11. Kipf, T.N., Welling, M.: Semi-supervised classification with graph convolutional networks. arXiv preprint arXiv:1609.02907 (2016)

12. Zhang, Y., Qi, P., Manning, C.D.: Graph convolution over pruned dependency trees improves relation extraction, pp. 2205–2215. Association for Computational Linguistics (2018)

13. Sun, K., Zhang, R., Mensah, S., Mao, Y., Liu, X.: Aspect-level sentiment analysis via convolution over dependency tree. In: Proceedings of the 2019 Conference on Empirical Methods in Natural Language Processing and the 9th International Joint Conference on Natural Language Processing (EMNLP-IJCNLP), pp. 5679–5688 (2019)

14. Yao, L., Mao, C., Luo, Y.: Graph convolutional networks for text classification. In: Proceedings of the AAAI Conference on Artificial Intelligence, pp. 7370–7377 (2019)

15. Ghosal, D., Majumder, N., Poria, S., Chhaya, N., Gelbukh, A.: DialogueGCN: a graph convolutional neural network for emotion recognition in conversation. arXiv preprint arXiv:1908.11540 (2019)

16. Mikolov, T., Sutskever, I., Chen, K., Corrado, G.S., Dean, J.: Distributed representations of words and phrases and their compositionality. In: Advances in Neural Information Processing Systems 26 (2013)

17. Devlin, J., Chang, M.-W., Lee, K., Toutanova, K.: BERT: pre-training of deep bidirectional transformers for language understanding. arXiv preprint arXiv:1810.04805 (2018)
18. Kingma, D.P., Ba, J.: Adam: a method for stochastic optimization. In: ICLR (Poster) (2015)
19. Srivastava, N., Hinton, G., Krizhevsky, A., Sutskever, I., Salakhutdinov, R.: Dropout: a simple way to prevent neural networks from overfitting. J. Mach. Learn. Res. **15**, 1929–1958 (2014)
20. Wu, S., Chen, F., Wu, F., Huang, Y., Li, X.: A multi-task learning neural net-work for emotion-cause pair extraction. In: ECAI 2020, pp. 2212–2219. IOS Press (2020)
21. Yu, J., Liu, W., He, Y., Zhang, C.: A mutually auxiliary multitask model with self-distillation for emotion-cause pair extraction. IEEE Access **9**, 26811–26821 (2021)
22. Fan, C., Yuan, C., Du, J., Gui, L., Yang, M., Xu, R.: Transition-based directed graph construction for emotion-cause pair extraction. In: Proceedings of the 58th Annual Meeting of the Association for Computational Linguistics, pp. 3707–3717 (2020)
23. Gu, G., Li, B., Gao, H., Wang, M.: Learning to answer complex questions with evidence graph. In: Wang, X., Zhang, R., Lee, Y.-K., Sun, L., Moon, Y.-S. (eds.) APWeb-WAIM 2020. LNCS, vol. 12317, pp. 257–269. Springer, Cham (2020). https://doi.org/10.1007/978-3-030-60259-8_20

Accelerated Algorithms for α-Happiness Query

Min Xie[⊠]

Shenzhen Institute of Computing Sciences, Shenzhen University, Shenzhen, China
xiemin@sics.ac.cn

Abstract. Extracting a good representative subset of tuples that meets the user's needs from a large database is an important problem in multi-criteria decision making. Many queries have been proposed for this purpose, including the top-k query and the skyline query. Unfortunately, these traditional queries either ask the user to specify their needs explicitly or overwhelm users with a large output size. Recently, an *α-happiness* query was proposed, which overcomes the deficiencies of existing queries: users do not need to specify any preference, while they can obtain a small set of tuples such that users are *happy* with the results, i.e., their favorite tuples in the returned subset is guaranteed to be not much worse than their favorite tuples in the whole database. In this paper, we study the α-happiness query. Inspired by the techniques of incremental convex hull computation, we develop two accelerated algorithms, which maintain useful information to avoid redundant computation, in both 2-dimensional and d-dimensional space ($d > 2$). We performed extensive experiments, comparing against the best-known method under various settings on both real and synthetic datasets. Our superiority is demonstrated: we can achieve up to two orders and 7 times of improvements in execution times in 2-dimensional and d-dimensional space, respectively.

Keywords: α-happiness · Incremental convex hull · Decision making

1 Introduction

Nowadays, a database system usually contains millions of tuples and end users may only want to find those tuples that fit their needs. This problem is known as *multi-criteria decision making* [5,18,19], and various queries were proposed to obtain a small representative subset of tuples without asking the user to scan the whole database. An example is the traditional top-k query [18,19], where a user has to provide her preference function, called the *utility function*. Based on the user's utility function, the *utility* of each tuple for this user can be computed, where a higher utility means that the tuple is more preferred. Finally, the best k tuples with the highest utilities are returned. Unfortunately, requiring the user to provide the exact utility function is too restrictive in many scenarios. In this case, the skyline query can be applied [5], which adopts the "dominance" concept. A tuple p is said to *dominate* another tuple q if p is not worse than q on each

B. Li et al. (Eds.): APWeb-WAIM 2022, LNCS 13422, pp. 53–68, 2023.
https://doi.org/10.1007/978-3-031-25198-6_5

attribute and p is better than q on at least one attribute. Intuitively, p will have a higher utility than q w.r.t. *all monotonic* utility functions. Tuples which are not dominated by any other tuples are returned in the skyline query. However, since there is no constraint on the output size of a skyline query, a skyline query can overwhelm the user with many results (at worst the whole database). Motivated by this, a query called α-*happiness* query was studied recently in [27] to overcome the deficiencies of both the top-k query (which requires the users to specify their utility functions) and the skyline query (which might have a large output size).

Informally, an α-happiness query computes a set of tuples, with size as small as possible, that makes the users *happy* where the degree of happiness is quantified as the *happiness ratio* of the user. Specifically, given a set of tuples, a user is $x\%$ happy with the set if the highest utility of tuples in the set is at least $x\%$ of the highest utility of all tuples in the whole database. In this case, we say that the happiness ratio of the user is $x\%$. Clearly, the happiness ratio is a value from 0 to 1. The larger the happiness ratio, the happier the user. The α-happiness query guarantees the happiness ratio of an end user is at least α. In practice, more tuples have to be returned to guarantee a higher happiness level, as expected. However, with more tuples returned, users have to spend more effort to examine the output, which is not desirable as they did in the skyline query. Hence, we want the solution to be as small as possible, to ensure the given happiness level.

Consider a car database application. Assume that Alice wants to buy a car from the car database where each car is described by two attributes, namely horse power (HP) and miles per gallon (MPG). To help Alice for finding her desired car, Alice can specify an α value, which represents the least happiness level she expects. In practice, Alice can set α to be 0.9, indicating that she wants a set of cars whose highest utility is at least 90% of the highest utility of all cars in the database. Then, we execute the α-happiness query, which returns a small set of cars from the database, hoping that Alice will be satisfied (since the happiness ratio of Alice is at least α, as specified). However, if Alice is not satisfied with those cars, she can increase the value of α, and execute the α-happiness query again to obtain more cars with better quality to ensure a higher α.

Although it is NP-hard to solve the α-happiness query [27], various practical algorithms were proposed in the literature. The best-known previous approach for the α-happiness query is CONE-GREEDY [27]. However, when we experimentally evaluated CONE-GREEDY, its execution time is unnecessarily long. This is because CONE-GREEDY did not keep sufficient information and thus, might perform redundant computation, resulting in a long query time. The situation is even worse when the user wants to perform multiple α-happiness queries with different values of α on the same database, which is common in reality since users might adjust the value of α to obtain more/less tuples to fit their needs. Motivated by this, we propose two novel approaches which accelerate CONE-GREEDY in both 2-dimensional and d-dimensional space ($d > 2$). Our algorithms are inspired by the incremental convex hull computation in computational geometry, and different from CONE-GREEDY, they effectively maintain the information needed during the computation and re-use them when necessary. Our experiments show that our algorithms substantially outperform CONE-GREEDY in execution time. Our major contributions are summarized as follows:

- To the best of our knowledge, we are the first one who connect the α-happiness query with the problem of incremental convex hull computation.
- We propose a 2-dimensional algorithm, called 2D-CH, for solving the α-happiness query exactly when each tuple is described by two attributes.
- In d-dimensional space, we propose a novel algorithm for the α-happiness query. In particular, our algorithm effectively maintain useful information, which can be re-used repeatedly, speeding up the overall query.
- We present extensive experiments on both synthetic and real datasets. Our evaluation shows that the proposed algorithms outperform the competitors substantially. Under some practical settings, our 2-dimensional algorithms achieve up to two orders improvement in running time, while our d-dimensional algorithms are around 7 times faster than the exiting ones.

Organization. The rest paper is organized as follows. We discuss the related works in Sect. 2. The α-happiness query and the solution overview are formally introduced in Sect. 3. In Sect. 4, we present the exact algorithm for the α-happiness query in 2-dimensional space and its d-dimensional extension. Finally, experiments are presented in Sect. 5 and Sect. 6 concludes this paper.

2   Related Work

Traditional queries for multi-criteria decision making include top-k queries and skyline queries. In top-k queries [10,13,19,21,28], a concrete utility function is given. Based on the function, the k tuples with the highest utilities are returned. However, it is sometimes difficult to obtain the exact utility function in practice. Alternatively, skyline queries [5] can be applied. However, it is found that the skyline query has a large output size, which is not desirable. Although there are some variants of skyline queries [9,15,20] which alleviate this drawback by introducing an integer k, which controls the output size, it is difficult for these queries to provide theoretical guarantee without knowing the exact utility function.

The α-happiness query was first considered in [2,12], called the *min-size regret query*, and it is later formalized by Xie et al. in [27]. Specifically, given a real number α, an α-happiness query minimizes the output size while keeping the users at least α happy (i.e., the minimum happiness ratio is at least α). The α-happiness query can be considered as a dual version of the well-known k-regret query [15,24,26], which, given an integer k, returns a set S of at most k tuples such that the "utility difference" between the maximum utility of S and the whole dataset D is minimized. See [25] for a recent survey. It has been shown that both the α-happiness query and the k-regret query are NP-hard problems [2,6,7,27].

Algorithms were proposed to get a solution for the α-happiness query, categorized as follows. (1) *ϵ-kernel based.* The first approach formulated it as the well-known ϵ-kernel problem [1] and several algorithms [2,6] were proposed to obtain a good approximation. (2) *Space partitioning based.* [2,3] discretized the function space and formulated the α-happiness query as a hitting set problem (or a set cover problem), which provides user-controlled approximations on happiness ratios and output sizes. (3) *Hybrid.* [12] proposed an algorithm which

Table 1. Car database and utilities

Car	HP	MPG	$f_{0.4,0.6}(p)$	$f_{0.2,0.8}(p)$	$f_{0.7,0.3}(p)$	Car	HP	MPG	$f_{0.4,0.6}(p)$	$f_{0.2,0.8}(p)$	$f_{0.7,0.3}(p)$
p_1	0.2	1	0.68	0.84	0.44	p_4	1	0.2	0.52	0.36	0.76
p_2	0.6	0.9	0.78	0.84	0.69	p_5	0.35	0.2	0.26	0.23	0.305
p_3	0.9	0.6	0.72	0.66	0.81	p_6	0.3	0.6	0.48	0.54	0.39

combines the ϵ-kernel and hitting set approaches, improving the efficiency of the existing algorithms. (4) *Geometric-based.* Xie et al. [27] provided a novel geometric interpretation of the α-happiness query, based on which they proposed the state-of-the-art algorithm, denoted by CONE-GREEDY for solving the problem. According to the experiments in [27], CONE-GREEDY outperforms the existing methods in both output sizes and execution times. We use it as the baseline in our experiments.

Compared with the existing studies [2,3,6,12,27], we utilize the techniques in incremental convex hull construction and propose accelerated algorithms. In particular, we maintain useful information so that intermediate results can be re-used repeatedly to avoid redundant computation. Our algorithms performs particularly well when the users execute multiple α-happiness queries on the same dataset. Our experimental superiority will be demonstrated in Sect. 5.

3   Problem and Overview

The input to our problem is a set D of n tuples (i.e., $|D| = n$) in a d-dimensional space (i.e., each tuple in D is described by d attributes). In this paper, we assume that d is a fixed constant. In the following, we first introduce the terminologies and the background. Then, we give an overview of our solution.

3.1   Preliminary

We use the same terminology as in [27]. We denote the i-th dimensional value of a tuple p in D by $p[i]$ where $i \in [1, d]$. In the rest paper, we also call each tuple as a point in a d-dimensional space. Without loss of generality, we assume that each dimension is normalized to $(0, 1]$, such that there exists a point p in D and $p[i] = 1$ for each $i \in [1, d]$ and a larger value on each dimension is more preferable to all users. Recall that in the car database, each car is associated with 2 attributes, HP and MPG; in the example shown in Table 1, the car database $D = \{p_1, p_2, p_3, p_4, p_5, p_6\}$ consists of 6 cars with normalized attribute values.

Following the assumption in existing studies [14,15,24,26,27], we assume that user's happiness is measured by an unknown *utility function*, which can be regarded as a mapping $f : \mathbb{R}^d_+ \to \mathbb{R}_+$. The *utility* of a point p w.r.t. f is denoted by $f(p)$. A user wants a point which maximizes the utility w.r.t. his/her utility function. Given a utility function f and $S \subseteq D$, we define the *maximum utility* of S w.r.t. f, denoted by $U_{max}(S, f)$, to be $\max_{p \in S} f(p)$.

In the following, we introduce two important terms, namely the *function-wise ratio* (*happiness ratio*) and the *minimum happiness ratio*.

Definition 1. *Given a set $S \subseteq D$ and a utility function f, the function-wise ratio of S w.r.t. f, denoted by* fRatio(S, f), *is defined to be* $\frac{U_{max}(S,f)}{U_{max}(D,f)}$.

Clearly, the value of a function-wise ratio ranges from 0 to 1 since $U_{max}(S, f) \leq U_{max}(D, f)$. Intuitively, when the maximum utility of S is closer to the maximum utility of D, the function-wise ratio of S w.r.t. the user's utility function becomes larger, which indicates that the user feels more satisfied with S. In this sense, the function-wise ratio is also called the *happiness ratio*.

As discussed in Sect. 1, it is difficult to know the user's exact utility function. Thus, we assume that all users' utility functions belong to a function class, denoted by $\mathcal{FC}$. A function class is defined to be a set of functions which share some common characteristics, e.g., the class of *linear utility functions* [15]. Given the function class $\mathcal{FC}$, the *minimum happiness ratio* of a set S can be regarded as the worst-case function-wise ratio w.r.t. a utility function in $\mathcal{FC}$.

Definition 2. *Given a set $S \subseteq D$ and a function class $\mathcal{FC}$, the minimum happiness ratio of S over $\mathcal{FC}$ is defined to be* $\inf_{f \in \mathcal{FC}}$ fRatio(S, f).

Example 1. To illustrate, assume that $\mathcal{FC}$ has 3 utility functions, namely $f_{0.4,0.6}$, $f_{0.2,0.8}$ and $f_{0.7,0.3}$ where $f_{a,b}(p) = a \times p[1] + b \times p[2]$. Consider p_1 in Table 1. Its utility w.r.t. $f_{0.4,0.6}$ is $f_{0.4,0.6}(p_1) = 0.4 \times 0.2 + 0.6 \times 1 = 0.68$. The utilities of other points in D are computed similarly. Given $S = \{p_1, p_4\}$, the maximum utility of S w.r.t. $f_{0.4,0.6}$ is $U_{max}(S, f_{0.4,0.6}) = \max_{p \in S} f_{0.4,0.6}(p) = f_{0.4,0.6}(p_1) = 0.68$. Similarly, $U_{max}(D, f_{0.4,0.6})$ is 0.78. Then, fRatio$(S, f_{0.4,0.6}) = \frac{U_{max}(S, f_{0.4,0.6})}{U_{max}(D, f_{0.4,0.6})} = \frac{0.68}{0.78} = 0.872$. Similarly, fRatio$(S, f_{0.2,0.8}) = 1$ and fRatio$(S, f_{0.7,0.3}) = 0.938$. The minimum happiness ratio of S over $\mathcal{FC}$ is $\min\{0.872, 1, 0.938\} = 0.872$. $\square$

Same as [2,12,14,15], we focus on the class of *linear utility functions*, denoted by $\mathcal{L}$, due to its popularity in modeling user preferences and assume each function in $\mathcal{L}$ is equally probable to be used by users. Other classes and distributions of utility functions are considered in [8,17,27] and are not our focus.

Specifically, we assume that each linear utility function f in $\mathcal{L}$ is associated with a d-dimensional non-negative *utility vector* u where $u[i]$ denotes the importance of the i-th dimension in user's preference. Mathematically, we can write: $f(p) = \sum_{i=1}^{d} u[i]p[i] = u \cdot p$. Without loss of generality, we assume $\sum_{i=1}^{d} u[i] = 1$. Thus, $\mathcal{L} = \{f| f(p) = u \cdot p$ where $u \in \mathbb{R}_+^d$ and $\sum_{i=1}^{d} u[i] = 1\}$. When it is clear, we refer each f in $\mathcal{L}$ by its utility vector u. Let minHap(S) be the minimum happiness ratio of S over $\mathcal{L}$. The α-happiness query is stated as follows.

Problem 1. Given a real number $\alpha \in [0, 1]$, the α-happiness query returns a set $S \subseteq D$ with minHap$(S) \geq \alpha$ such that the size of S, i.e., $|S|$, is minimized.

When there are multiple sets with the minimum size, an α-happiness query simply returns one of them. As stated in Sect. 1, the α-happiness query takes the advantages of both the top-k query and the skyline query: same as the skyline query, a user does not need to specify any preference and meanwhile, it returns a set with size as small as possible. Recall that minHap(S) is defined to be the worst-case happiness ratio w.r.t. *any* utility function in $\mathcal{L}$. If minHap(S) $\geq \alpha$, for *each* user, s/he will be at least α happy with S no matter which function s/he uses from $\mathcal{L}$. The α-happiness query is an NP-hard problem [2,6,7].

3.2   Geometric Interpretation

Note that $\mathcal{L}$ contains an *infinite* number of utility functions. Thus, it is not easy to compute minHap(S) for a given S directly according to Definition 2. To compute minHap(S) tractably, Xie et al. [27] interprets the problem geometrically.

We first introduce some geometric concepts. For each point $p \in D$, we define the *orthotope set* of p, denoted by Orth(p), to be a set of 2^d d-dimensional points constructed by $\{0, p[1]\} \times ... \times \{0, p[d]\}$. That is, for each $i \in [1, d]$, the i-dimensional value of a point in Orth(p) is equal to either 0 or $p[i]$. Given a set $S \subseteq D$, we define the orthotope set of S, denoted by Orth(S), to be $\bigcup_{p \in S}$ Orth(p). Given a set $S \subseteq D$, let Conv(S) be the *convex hull* (the smallest convex set) of the orthotope set of S [16]. Moreover, a point $p \in$ Conv(S) is said to be a *vertex* of Conv(S) if p is not in the convex hull of the other points in Orth(S). A *facet* of a convex hull is a bounded flat surface that forms a part of the boundary of the convex hull. We denote a facet by the set of points forming it.

Example 2. To illustrate, consider Table 1 where $D = \{p_1, p_2, p_3, p_4, p_5, p_6\}$. For the ease of presentation, we first visualize D in Fig. 1 where the X_1 and X_2 coordinate represent HP and MPG, respectively. The points in Orth(p_2) $= \{p_2, p_2', p_2'', O\}$ are shown in Fig. 1 where $p_2' = (0, p_2[2])$, $p_2'' = (p_2[1], 0)$ and O is the origin. Similarly, Orth(p_3) is shown in the same figure.

Given $S = \{p_2, p_3\}$, we define Orth(S) to be Orth(p_2) $\cup$ Orth(p_3). Then, the convex hull Conv(S) is shown in Fig. 2. There are 5 vertices in Conv(S), namely O, p_2', p_2, p_3 and p_3'' (labeled in Fig. 1), each of which is not in the convex hull of the other points in Orth(S). $\{p_2, p_3\}$ is a facet of Conv(S). □

Given a real value $\alpha \in [0, 1]$, we define the α-*shrunk set* of D, denoted by D_α', to be $\{p_\alpha' | p_\alpha' = \alpha p, \forall p \in D\}$ where p_α' is a proportionally shrunk point of p. When α is clear, we denote D_α' by D' and denote a point in D' by p'. Unless stated explicitly, we stick to the above notations in the rest of this paper.

Given two point sets, say S and T, if for each $p \in S$, p is inside Conv(T), we say Conv(T) *covers* Conv(S) since Conv(S) is totally contained inside Conv(T).

Example 3. Let $\alpha = 0.9$. The α-shrunk set D' (shown in white dots) of D (shown in black dots) is drawn in Fig. 2 where each point in D' is a proportionally scaled point in D. Given $S = \{p_2, p_3\}$, it is easy to observe from the figure that Conv(S) covers Conv(D') since Conv(D') is totally contained inside Conv(S). □

Xie et al. [27] shows that the α-happiness from the geometric perspective.

Lemma 1 ([27]). *Given $S \subseteq D$ and $\alpha \in [0, 1]$, S is a feasible set of the α-happiness query if Conv(S) covers Conv(D'), where D' is the α-shrunk set of D.*

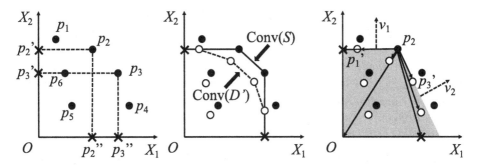

Fig. 1. Orthotope set   **Fig. 2.** Convex hull   **Fig. 3.** Conical hull

3.3 Solution Overview

According to Lemma 1, we can solve the α-happiness query by finding a minimum size set S such that $\mathsf{Conv}(S)$ covers $\mathsf{Conv}(D')$. To find such S, the CONE-GREEDY algorithm in [27] has the following two major steps (note that the correctness of the procedure below is proven in [27] and we omit it here for lack of space):

1. (Step 1) For each p in D, it computes a function set $\mathcal{F}_p$, whose utilities are maximized by p over points in D', i.e., $\mathcal{F}_p = \{f \in \mathcal{L} \mid f(p) \geq f(p') \ \forall p' \in D'\}$.
2. (Step 2) If finds a set S of tuples from D such that $\bigcup_{p \in S} \mathcal{F}_p = \mathcal{L}$.

Step 2 of CONE-GREEDY is reduced to the well-known set-cover problem in [27], where the greedy algorithm is adopted and it gives theoretical guarantees on the output size. We adopt the same approach for Step 2 in this paper. Interested readers can find more details in [27], and we focus on Step 1 next.

Note that when performing Step 1 in CONE-GREEDY, redundant operations might be done when computing $\mathcal{F}_p$ for distinct points in D. This is inefficient. In this paper, we adopt a novel strategy for computing $\mathcal{F}_p$, which maintains useful information for all points in D, so that we can re-use those information as much as possible. In the following, we briefly review the procedure in CONE-GREEDY and explain why it is inefficient. In Sect. 4, we give our advanced procedures.

Computing $\mathcal{F}_p$ in CONE-GREEDY. We first define "conical hull". Given a point p in D, let $V_p = \{t - p \mid$ for each vertex t of $\mathsf{Conv}(D')\}$. Then we define a *conical hull* of p, denoted by $\mathsf{Cone}(p)$, to be $\mathsf{Cone}(p) = \{q \in \mathbb{R}^d \mid q = p + \sum_{v \in V_p} wv$ where $w \geq 0\}$. Intuitively, $\mathsf{Cone}(p)$ can be regarded as a *convex cone* with apex at p. The boundaries of $\mathsf{Cone}(p)$ are *unbounded facets*, each of which is enclosed by some vectors in V_p and is a flat surface that forms the boundary of $\mathsf{Cone}(p)$.

In geometry, each facet of a conical hull is contained by a *unique* hyperplane (i.e., a subspace of dimensionality $d - 1$). Then, for each facet F of $\mathsf{Cone}(p)$, we define an *extreme vector* to be the unit vector (pointing out) perpendicular to the hyperplane containing F. Denote the set of *extreme vectors* of p by $\mathsf{Ext}(p)$.

Example 4. Consider the point p_2 in Fig. 3 as an example. We draw the vectors in $V_{p_2} = \{t - p_2|$ for each vertex t of $\mathsf{Conv}(D')\}$ in solid arrows. It is constructed by creating a vector for each vertex of $\mathsf{Conv}(D')$. The conical hull $\mathsf{Cone}(p_2)$ is showed in the shaded region in the figure, which is the set of all vectors with the form $p_2 + \sum_{v \in V_{p_2}} wv$ where $w \geq 0$. In this 2-dimensional example, the boundaries of $\mathsf{Cone}(p_2)$ are two unbounded facets, i.e., the rays shooting from p_2 to p_1' and from p_2 to p_3'. The extreme vectors of p_2 are dashed arrows $\mathsf{Ext}(p_2) = \{v_1, v_2\}$, each of which is perpendicular to a boundary facet of $\mathsf{Cone}(p_2)$. □

Based on the above concepts, Xie et al. [27] define the function set $\mathcal{F}_p$, which is a set of utility functions whose utilities are maximized by p, as follows:

Definition 3. *Given p in D and its $\mathsf{Ext}(p)$, the function set of p, denoted by $\mathcal{F}_p$, is defined to be $\{f \in \mathcal{FC}|\ f(p) = u \cdot p$ and $u = \sum_{v \in \mathsf{Ext}(p)} wv$ where $w \geq 0\}$.*

According to [27], $\mathcal{F}_p$ is uniquely defined by the extreme vectors in $\mathsf{Ext}(p)$. Thus, CONE-GREEDY obtain $\mathcal{F}_p$ by computing $\mathsf{Ext}(p)$ as follows:

1. It first computes the vertices in $\mathsf{Conv}(D')$;
2. For each p in D, it computes the set $V_p = \{t - p|$ for each vertex t of $\mathsf{Conv}(D')\}$ and the corresponding conical hull $\mathsf{Cone}(p)$; and
3. It obtains the extreme vectors $\mathsf{Ext}(p)$ based on boundary facets of $\mathsf{Cone}(p)$.

Note that in CONE-GREEDY, although the vertices in $\mathsf{Conv}(D')$ are used for all points in D, the vector set V_p is different for each distinct p in D. Therefore, the conical hull $\mathsf{Cone}(p)$ will be computed *independently* for distinct p in D, which might incur redundant computation, since the common information $\mathsf{Conv}(D')$ is not well utilized. In Sect. 4, we show our alternative ways for computing $\mathsf{Ext}(p)$, by maintaining useful information to avoid such redundant computation. Our algorithms are especially efficient when the user wants to execute multiple α-happiness queries on the same D with different values of α. Our experiments show that we are more efficient than the counterpart in CONE-GREEDY.

4   Algorithm

4.1   Conceptual Idea

Our algorithm is inspired by the incremental approach of convex hull computation [11]. Specifically, in incremental convex hull computation, a convex hull is built by inserting points iteratively. At the i-th iteration, we have the convex hull of the first i points, and we need to modify this convex hull to include the i-th point. For example in Fig. 4, if we are inserting p_2 to $\mathsf{Conv}(D')$ (shown in solid lines), the convex hull is updated and the vertices become $\{b_1, p_2, p_3', p_4', b_2, O\}$. To update $\mathsf{Conv}(D')$, new facets (e.g., $\{p_2, p_3'\}$, shown in dashed lines) are created, and old facets are removed (e.g., $\{p_1', p_2'\}$ and $\{p_2', p_3'\}$). It is not hard to observe that the newly created facets indeed give us the desired extreme vectors $\mathsf{Ext}(p)$, since each extreme vector is perpendicular to exactly one newly created facet (i.e., it is perpendicular to the unique hyperplane containing that facet). For example in Fig. 4, v_2, an extreme vector of p_2, is perpendicular to the newly created facet $\{p_2, p_3'\}$. Motivated by this, we can compute the desired $\mathsf{Ext}(p)$ for each p in D, by adapting the techniques of incremental convex hull computation, pretending that we are inserting p to the convex hull $\mathsf{Conv}(D')$.

4.2 Two-Dimensional Case: 2D-CH

In 2-dimensional space, the vertices (excluding the origin O) of the convex hull $\mathsf{Conv}(D')$ can be organized in a clockwise manner, say $t_1, t_2, \ldots, t_k$, where $\{t_i, t_{i+1}\}$ ($i \in [1, k-1]$) is a facet. For example, in Fig. 4, vertices of $\mathsf{Conv}(D')$ can be organized in order: $b_1, p_1', p_2', p_3', p_4', b_2$, where b_1 and b_2 are two orthotope points in $\mathsf{Orth}(D')$. $\{p_2', p_3'\}$ is facet of $\mathsf{Conv}(D')$. We store vertices of $\mathsf{Conv}(D')$ clockwise in a doubly-linked list so that we can create new facets efficiently.

Specifically, our 2-dimensional algorithm, called 2D-CH, is proposed by adopting the following strategy for computing the extreme vectors $\mathsf{Ext}(p)$ for p:

1. We first compute the convex hull $\mathsf{Conv}(D')$ and maintain its vertices in a doubly-linked list for efficient facet traversal for all points in D;
2. For each p in D that is not contained inside $\mathsf{Conv}(D')$, we compute the new facets, by pretending that we are inserting p to $\mathsf{Conv}(D')$ (see details below);
3. For each newly created facet, we obtain a desired extreme vector in $\mathsf{Ext}(p)$, which is the unique vector perpendicular to the new facet.

To insert a point p to $\mathsf{Conv}(D')$, we need to determine the correct positions for constructing the new facets. For example, in Fig. 4, p_3' is the desired position, and a new facet is created by connecting p_2 and p_3'. To determine such positions, we need the notion of "visibility". Formally, given a point p and a facet $\{t_i, t_{i+1}\}$ of $\mathsf{Conv}(D')$, $\{t_i, t_{i+1}\}$ is visible to p if p is *above* the unique hyperplane containing $\{t_i, t_{i+1}\}$. The following lemma (proof is intuitive and is omitted) tells us how to determine the correct positions with the notion of "visibility".

Lemma 2. *Given point p and two adjacent facets of $\mathsf{Conv}(D')$, say $F_1 = \{t_{i-1}, t_i\}$ and $F_2 = \{t_i, t_{i+1}\}$, when inserting p to $\mathsf{Conv}(D')$, we create a new facet by connecting p and t_i iff one facet in $\{F_1, F_2\}$ is visible to p and the other is not.*

For example in Fig. 4, $\{p_2', p_3'\}$ is visible to p_2, while $\{p_3', p_4'\}$ is not. To insert p_2 to $\mathsf{Conv}(D')$, we create a new facet by connecting p_2 and p_3' by Lemma 2. Since we maintain vertices of $\mathsf{Conv}(D')$ in a doubly-linked list, the correct position for creating facets can be found efficiently by binary search in the list.

After obtaining the new facets, the extreme vector set construction is straightforward. Note that in 2-dimensional space, there are exactly two extreme vectors for each p. Therefore, the corresponding function set $\mathcal{F}_p$ can be concisely represented by an *angle interval*. Specifically, we define the *angle of a vector v* in 2-dimensional spaces as the angle between the vector Ov and the y-axis, denoted by $\mathsf{Ang}(v)$. Given $\mathsf{Ext}(p) = \{v_1, v_2\}$ of a point p, we define the *angle interval of p*

to be $[\mathsf{Ang}(v_1), \mathsf{Ang}(v_2)]$. Then, it is easy to show that finding a set S such that $\bigcup_{p \in S} \mathcal{F}_p = \mathcal{L}$ is equivalent to finding a set S whose angle intervals covers $[0, \frac{\pi}{2}]$, where the latter problem is the *interval cover problem* [4]. We then employ the popular greedy strategy to solve the interval cover problem optimally.

Example 5. Consider p_2 in Fig. 4 where $\mathsf{Ext}(p_2) = \{v_1, v_2\}$. Since $\mathsf{Ang}(v_1) = 0$ and $\mathsf{Ang}(v_2) = 1.04$, we represent the function set $\mathcal{F}_{p_2}$ as an angle interval $[0, 1.04]$ (labeled in the figure). Similarly, we can compute the angle intervals for other points in D. By the greedy strategy, we find that the angle intervals of p_2 and p_3 covers the entire $[0, \frac{\pi}{2}]$, which gives us the desired set $S = \{p_2, p_3\}$. □

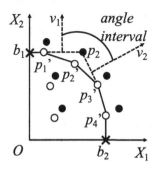

Fig. 4. 2D case

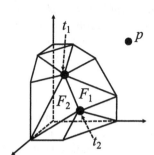

Fig. 5. 3D case

4.3   High-Dimensional Case: HD-CH

The problem is more complicated in the higher-dimensional case, since there is no natural order in the facets of a convex hull and each facet can have multiple adjacent facets (unlike exactly two adjacent facets in the 2-dimensional case).

To extend our algorithm to the high-dimensional case, we define the following notions in a high-dimensional convex hull. The boundaries of a facet are called *ridges*. Intuitively, the ridge signifies the adjacency of two neighbouring facets. For example, the ridges in a 2-dimensional space are points and the ridges in a 3-dimensional space are edges (i.e., the line segment jointed by two points). Given a point p, a ridge is called *horizon ridge* of p if it signifies the adjacency of a visible facet and an invisible facet of p. Intuitively, a horizon ridge indicates the maximum visible region from p to the convex hull. For example in Fig. 5, if F_1 is visible to p and F_2 is not visible to p, the ridge (i.e., edge in this case) $\{t_1, t_2\}$, which signifies the adjacency of F_1 and F_2, is a horizon ridge of p. For each horizon edge, we define an extreme vector of p to be the unit vector perpendicular to the unique hyperplane containing p and the horizon ridge.

With the above definitions, our high-dimensional algorithm, denoted as HD-CH, computes the extreme vector set $\mathsf{Ext}(p)$ as follows:

1. It first computes the convex hull $\mathsf{Conv}(D')$;
2. For each p in D, we maintain its visible facets in a queue $\mathcal{Q}$ and horizon ridges in a set $\mathcal{H}$. Initially, $\mathcal{H}$ is empty and we obtain the first facet F in $\mathcal{Q}$ by facet traversal on $\mathsf{Conv}(D')$. Neighboring facets of F is marked as "unchecked";
3. When there is a facet F in $\mathcal{Q}$ with unchecked neighboring facets, we pop F from $\mathcal{Q}$ and check its neighboring facets. Specifically, for each visible neighboring facet, we add it to $\mathcal{Q}$ for later processing; and for each invisible neighboring facet, we obtain a horizon ridge for p and it is inserted to $\mathcal{H}$;
4. Finally, for each horizon ridge in $\mathcal{H}$, we get an extreme vector (i.e., the unit vector perpendicular to the hyperplane containing p and the horizon ridge).

After obtaining the extreme vector set $\mathsf{Ext}(p)$, we adopt the same strategy as CONE-GREEDY for constructing the solution S. Note that HD-CH enjoys the same theoretical guarantee on the output size as CONE-GREEDY by similar analysis. Interested readers can find more details in [27] and we omit them here.

4.4 Discussion

Compared with the best-known previous approach CONE-GREEDY, our 2D-CH and HD-CH algorithms mainly differ in the procedure of constructing the extreme vector set $\mathsf{Ext}(p)$, by employing incremental computation on the convex hull $\mathsf{Conv}(D')$. Note that $\mathsf{Conv}(D')$ is a α-shrunk convex hull of $\mathsf{Conv}(D)$. Therefore, we can compute $\mathsf{Conv}(D)$ once and use it for α-happiness queries with different values of α, by properly scaling $\mathsf{Conv}(D)$. Moreover, given the convex hull $\mathsf{Conv}(D')$, we can use it for all points in D, for computing the desired function set $\mathcal{F}_p$ via facet traversal. In contrast, although CONE-GREEDY also computes the vertices $\mathsf{Conv}(D')$ for all points in D, it constructs the conical hull $\mathsf{Cone}(p)$ independently for each p in D, resulting in a large overall execution time. Even worse, when the user wants to execute an α-happiness query with a different value of α on the same dataset, the conical hull $\mathsf{Cone}(p)$ has to be re-computed from scratch for all points in D, since the vector set $V_p = \{t - p|$ for each vertex t of $\mathsf{Conv}(D')\}$ is radically different under different values of α.

5   Experimental Evaluation

We conducted experiments on a machine with 3.20 GHz CPU and 8 GB RAM. All programs were implemented in C/C++. Most experimental settings follow those in [2,12,27]. Both *synthetic* and *real datasets* were used in our experiments.

We generated the widely used anti-correlated datasets by a dataset generator [5]. Unless stated explicitly, for each synthetic dataset, the number of tuples is set to be 100,000 (i.e., $n = 100,000$), the dimensionality is set to be 3 (i.e., $d = 3$) and α is set to be 0.99. Following existing studies, we used three real datasets in our experiments: the *Island* dataset [15,27], the *Household* dataset [26] and the *El Nino* dataset [2,7,27]. Island is 2-dimensional, containing 63,383 points, which characterize geographic positions. Household consists of 1,048,576 family tuples in US in 2012 where each family is described by three economic attributes. El Nino contains 178,080 tuples with four oceanographic attributes taken at the Pacific Ocean. For all datasets, each attribute is normalized to (0, 1].

We implemented our algorithms, 2D-CH and HD-CH, and two variants 2D-CH$_{reuse}$ and HD-CH$_{reuse}$, which pre-compute the vertices and convex hulls and re-use them under different values of α. Our algorithms are compared against the state-of-the-art algorithm, CONE-GREEDY [27], for the α-happiness query. Note that although there are other algorithms proposed in the literature, [2,6, 12,15], they are shown to be worse than CONE-GREEDY in [27] and thus, we only compared CONE-GREEDY in the experiments for the ease of presentation. We used the same parameters reported in [27]. Unless specified explicitly, the performance of each algorithm is measured in terms of the *execution time*. Since 2D-CH and HD-CH only differ from CONE-GREEDY in the way of computing the function sets, their outputs are the same and we omit them for lack of space.

In the following, we show the experiments on the synthetic and real datasets in Sect. 5.1 and Sect. 5.2. We summarize our findings in Sect. 5.3.

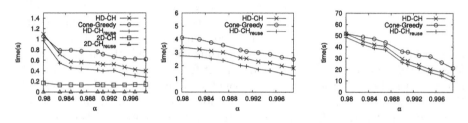

Fig. 6. 2D synthetic   Fig. 7. 3D synthetic   Fig. 8. 4D synthetic

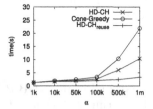
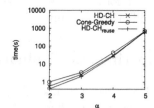

Fig. 9. Vary n Fig. 10. Vary d

5.1   Results on Synthetic Datasets

In Fig. 6, we evaluated our 2-dimensional algorithms, 2D-CH and 2D-CH$_{reuse}$ on a $2d$ anti-correlated dataset. For completeness, we also include the d-dimensional algorithm, HD-CH and HD-CH$_{reuse}$, in the figure (however, their performance will be analyzed in later experiments). As shown there, 2D-CH runs much faster than the other algorithms. In particular, it takes less than 0.2 s for all α and its running time is not sensitive to the value of α. This is because in a 2-dimensional dataset, there is an ordering on the vertices, and thus, constructing the functions sets on the α-shrunk convex hull $\mathsf{Conv}(D')$ can be done efficiently via binary search, which is not sensitive to α compared with the other methods. The performance of 2D-CH is further improved by 2D-CH$_{reuse}$, by pre-computing the vertices and re-using them for all points in D under different values of α. Note that CONE-GREEDY is the slowest in most cases, e.g., 2D-CH (resp. 2D-CH$_{reuse}$) achieves 5 times (resp. two orders) of improvements in execution times compared with CONE-GREEDY when $\alpha = 0.99$.

We proceed with the performance evaluation of our d-dimensional algorithms, HD-CH and HD-CH$_{reuse}$, on $3d$ and $4d$ anti-correlated datasets. The results are presented in Figs. 7 and 8. With the increasing value of α, all algorithms take less time to execute, in the cost of larger output sizes (not shown). This is because when α is large, the convex hull $\mathsf{Conv}(D')$ is "close" to $\mathsf{Conv}(D)$ and thus, each point in D can only "see" a small portion of $\mathsf{Conv}(D')$. Hence, it takes each point a shorter amount of time to construct the function set, which dominates the computational cost, but we need more points to cover the entire $\mathsf{Conv}(D')$. Nevertheless, CONE-GREEDY still has the largest execution time, e.g., it takes CONE-GREEDY 21 s on the 4-dimensional dataset when $\alpha = 0.999$, as opposed to 12 s by HD-CH. HD-CH$_{reuse}$ further improves the execution time of HD-CH by around 30%, This confirms our claim that our algorithms are especially efficient when the user wants to execute the α-happiness query on the same D with different values of α, since the convex hull can be efficiently pre-computed and used for different α-happiness queries.

We next evaluated the scalability of HD-CH and HD-CH$_{reuse}$, by varying the dimensionality d and the dataset size n in Figs. 9 and 10, where other parameters are fixed to the default setting stated at the beginning of this section. According to the results, we can see that our algorithm scales well w.r.t. both d and n. For example, on a large dataset with 1 million points, HD-CH$_{reuse}$ only takes 3 s to execute, 3 times and 7 times faster than HD-CH and CONE-GREEDY, respectively. When the dimensionality is 4, the execution time of CONE-GREEDY, HD-CH and HD-CH$_{reuse}$ is 42 s, 29 s and 26 s, respectively. In other words, HD-

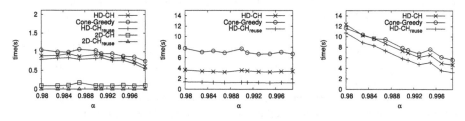

Fig. 11. Island   **Fig. 12.** Household   **Fig. 13.** El Nino

CH and HD-CH$_{reuse}$ outperform the state-of-the-art approach, w.r.t. both n and d, by accelerating the querying time.

5.2   Results on Real Datasets

In this section, we conducted experiments on three commonly used real datasets. The results are shown in Figs. 11, 12 and 13, respectively.

On the 2-dimensional Island dataset (Fig. 11), we plot the performance of both 2-dimensional and d-dimensional algorithms. Consistent to the performance on the synthetic datasets, the running times of our algorithms are much faster than the existing algorithms. For our d-dimensional algorithms HD-CH and HD-CH$_{reuse}$, they achieves 30% speedup against the state-of-the-art CONE-GREEDY algorithm. When considering our 2D-CH and 2D-CH$_{reuse}$ algorithms, which are designed for the 2-dimensional case, the improvement of execution time is significant, e.g., one order and two orders of improvement when $\alpha = 0.999$.

The result on the Household dataset are similar and it is shown in Fig. 12. Note that due to the small skyline size on Household, the execution times of all algorithms are not sensitive to the value of α. In this scenario, HD-CH still outperforms CONE-GREEDY, e.g., by reducing the average execution time from 7.2 s to 3.5 s. HD-CH$_{reuse}$ further improves the average execution time of HD-CH to 1.3 s, which clearly demonstrates that pre-computing the auxiliary structures is a promising way to speedup the query process. By maintaining intermediate information, we efficiently support the α-happiness queries for different values of α. Note that similar speedup cannot be achieved by the CONE-GREEDY algorithm. Although it also computes the vertices of $\mathsf{Conv}(D')$ for all points in D, it has to construct the conical hull independently for each point in D, resulting in a large overall execution time.

Finally, consider the experiments on the El Nino dataset in Fig. 13. Similar to the previous experiments, the performance of CONE-GREEDY is worse than that of HD-CH and HD-CH$_{reuse}$. When $\alpha = 0.999$, HD-CH$_{reuse}$ only spends half of the time compared with CONE-GREEDY to obtain the desired solution.

5.3   Summary

The experiments on both real and synthetic datasets demonstrated our superiority over the best-known previous approach. We observe the following. (1) On the 2-dimensional datasets, 2D-CH and 2D-CH$_{reuse}$ are the best algorithms, by achieving up to two orders of improvement in execution time, compared with the state-of-the-art algorithm. (2) On the d-dimensional datasets, HD-CH and HD-CH$_{reuse}$ runs much faster than the competitor, e.g., on the Household dataset, the average execution times of HD-CH, HD-CH$_{reuse}$ and CONE-GREEDY are 3.5 s, 1.3 s and 7.2 s, respectively. (3) Pre-computing the vertices and convex hulls is a promising way for reducing the query time, especially when the users want to execute multiple α-happiness queries on the same datasets. For example, when $n = 1,000,000$, it only takes HD-CH$_{reuse}$ 3 s to execute, 3 times faster than the HD-CH algorithm. (4) The scalability of our solutions is demonstrated, e.g., when varying the dimensionality or the dataset size, our algorithms are consistently faster than CONE-GREEDY.

6 Conclusions

This paper proposed two accelerated algorithms for the α-happiness query. Compared with the existing methods, we maintain useful information to avoid redundant computation, accelerating the query process. Our algorithms are particularly good at executing the α-happiness queries with different values of α on the same dataset D. We conducted comprehensive experiments to verify the speedup of our algorithms, which achieve up to two orders of improvement in execution time, compared with the best-known approach. As for future research, we consider introducing user interaction [22–24] in α-happiness queries, so that we can further reduce the solution set size while guaranteeing the happiness ratio.

Acknowledgements. This work was supported by Longhua Science and Technology Innovation Bureau LHKJCXJCYJ202003 and Guangdong Basic and Applied Basic Research Foundation 2022A1515010120.

References

1. Agarwal, P., Har-Peled, S., Varadarajan, K.: Approximating extent measures of points. JACM **51**, 606–635 (2004)
2. Agarwal, P.K., Kumar, N., Sintos, S., Suri, S.: Efficient algorithms for k-regret minimizing sets. In: SEA (2017)
3. Asudeh, A., Nazi, A., Zhang, N., Das, G.: Efficient computation of regret-ratio minimizing set: a compact maxima representative. In: SIGMOD (2017)
4. Bernhard, K., Vygen, J.: Combinatorial Optimization: Theory and Algorithms, 3rd edn. Springer, Heidelberg (2008). https://doi.org/10.1007/978-3-662-56039-6
5. Borzsony, S., Kossmann, D., Stocker, K.: The skyline operator. In: ICDE (2001)
6. Cao, W., et al.: k-regret minimizing: efficient algorithms and hardness. In: ICDT (2017)
7. Chester, S., Thomo, A., Venkatesh, S., Whitesides, S.: Computing k-regret minimizing sets. In: VLDB (2014)
8. Faulkner, T.K., Brackenbury, W., Lall, A.: k-regret queries with nonlinear utilities. In: VLDB (2015)
9. He, J., Han, X.: Efficient skyline computation on massive incomplete data. Data Sci. Eng. **7**(2), 102–119 (2022)
10. He, Z., Lo, E.: Answering why-not questions on top-k queries. In: TKDE (2014)
11. Kallay, M.: The complexity of incremental convex hull algorithms in RD. Inf. Process. Lett. **19**(4), 197 (1984)
12. Kumar, N., Sintos, S.: Faster approximation algorithm for the k-regret minimizing set and related problems. In: ALENEX (2018)
13. Liu, P., Wang, M., Cui, J., Li, H.: Top-k competitive location selection over moving objects. Data Sci. Eng. **6**(4), 392–401 (2021)
14. Nanongkai, D., Lall, A., Sarma, A., Makino, K.: Interactive regret minimization. In: SIGMOD (2012)

15. Nanongkai, D., Sarma, A., Lall, A., Lipton, R., Xu, J.: Regret-minimizing representative databases. In: VLDB (2010)
16. Peng, P., Wong, R.: Geometry approach for k-regret query. In: ICDE (2014)
17. Qi, J., Zuo, F., Yao, J.: K-regret queries: from additive to multiplicative utilities. CoRR (2016)
18. Qin, L., Yu, J., Chang, L.: Diversifying top-k results. In: VLDB (2012)
19. Soliman, M., Ilyas, I., Chang, C.: Top-k query processing in uncertain databases. In: ICDE (2007)
20. Tao, Y., Ding, L., Lin, X., Pei, J.: Distance-based representative skyline. In: ICDE (2009)
21. Tao, Y., Xiao, X., Pei, J.: Efficient skyline and top-k retrieval in subspaces. In: TKDE (2007)
22. Wang, W., Wong, R., Xie, M.: Interactive search for one of the top-k. In: SIGMOD (2021)
23. Xie, M., Chen, T., Wong, R.: Find your favorite: an interactive system for finding the user's favorite tuple in the database. In: SIGMOD (2019)
24. Xie, M., Wong, R., Lall, A.: Strongly truthful interactive regret minimization. In: SIGMOD (2019)
25. Xie, M., Wong, R., Lall, A.: An experimental survey of regret minimization query and variants: bridging the best worlds between top-k query and skyline query. VLDB J. **29**, 147–175 (2020)
26. Xie, M., Wong, R., Li, J., Long, C., Lall, A.: Efficient k-regret query algorithm with restriction-free bound for any dimensionality. In: SIGMOD (2018)
27. Xie, M., Wong, R., Peng, P., Tsotras, V.: Being happy with the least: achieving α-happiness with minimum number of tuples. In: ICDE (2020)
28. Xin, D., Han, J., Cheng, H., Li, X.: Answering top-k queries with multi-dimensional selections: the ranking cube approach. In: VLDB (2006)

SummScore: A Comprehensive Evaluation Metric for Summary Quality Based on Cross-Encoder

Wuhang Lin[1], Shasha Li[1], Chen Zhang[1], Bin Ji[1], Jie Yu[1(✉)], Jun Ma[1], and Zibo Yi[2]

[1] College of Computer, National University of Defense Technology, Changsha, China
{shashali,jibin,yj,majun}@nudt.edu.cn
[2] Information Research Center of Military Science PLA Academy of Military Science, Beijing 100142, China

Abstract. Text summarization models are often trained to produce summaries that meet human quality requirements. However, the existing evaluation metrics for summary text are only rough proxies for summary quality, suffering from low correlation with human scoring and inhibition of summary diversity. To solve these problems, we propose SummScore, a comprehensive metric for summary quality evaluation based on Cross-Encoder. Firstly, by adopting the original-summary measurement mode and comparing the semantics of the original text, SummScore gets rid of the inhibition of summary diversity. With the help of the text-matching pre-training Cross-Encoder, SummScore can effectively capture the subtle differences between the semantics of summaries. Secondly, to improve the comprehensiveness and interpretability, SummScore consists of four fine-grained submodels, which measure Coherence, Consistency, Fluency, and Relevance separately. We use semi-supervised multi-rounds of training to improve the performance of our model on extremely limited annotated data. Extensive experiments show that SummScore significantly outperforms existing evaluation metrics in the above four dimensions in correlation with human scoring. We also provide the quality evaluation results of SummScore on 16 mainstream summarization models for later research.

Keywords: SummScore · Comprehensive metric · Summary quality evaluation

1 Introduction

Automatic text summarization technology aims to compress a long document into a fluent short text, which is consistent with the key information of the original text and preserves the most salient information in the source document [6]. In

Supported by Hunan Provincial Natural Science Foundation Project (No. 2022JJ30668) and (No. 2022JJ30046).
W. Lin and S. Li—Contributed equally to this work.

B. Li et al. (Eds.): APWeb-WAIM 2022, LNCS 13422, pp. 69–84, 2023.
https://doi.org/10.1007/978-3-031-25198-6_6

recent years, automatic text summarization technologies have been significantly developed. However, the research on automatic summarization evaluation still fell behind [7]. Today, the mainstream evaluation metrics for automatic text summarization, such as ROUGE, BLEU, and Meteor, simply calculate n-gram overlap between candidates and references [1,11,14]. Studies [12,19] have shown that they are only rough proxies for summary quality evaluation. Some concerns of these metrics are shown as follows.

Firstly, the existing evaluation metrics strongly rely on expert-generated summaries as references, which are difficult to obtain. What's more, these metrics inhibit the diversity of summaries generated by the summarization model. Because the mainstream metrics only rely on the interaction between the reference summary. However, different summaries written by readers with different knowledge reserves and for different purposes are also correct. We cannot force different summaries to be evaluated simply by measuring the degree of alignment with a single reference summary. Such an evaluation metric will limit the diversity of summaries generated by the summarization model.

Secondly, some studies show that the mainstream evaluation metrics scoring do not correlate well with human scoring [3,19]. When humans evaluate the quality of summaries, they usually consider multiple fine-grained quality dimensions, such as rich information, non-redundancy, coherence, and well-structured. However, these metrics mainly focus on the similarity of literal and expressions, which cannot well evaluate semantic relevance and topic consistency. Moreover, they ignore the evaluation of language quality, such as logical consistency and language fluency. Many of the above-mentioned factors can affect the comprehensiveness and interpretability of the summary quality evaluation.

As illustrated in the examples in Fig. 1. Comparing the reference with the original text, when experts score the summary generated by model Bottom-Up, they find that the generated summary has factual errors (gray shaded fonts). The fact is that *Manuel Pellegrini (Manchester City)* wants to sign *Evangelos Patoulidis*. Therefore, except for Fluency, the experts give low scores for all quality dimensions. However, because of the large overlap of n-grams (blue fonts) between the summary and the reference, ROUGE scores high. For the BART model, because the generated summaries almost focus on the important information (orange fonts), and the text is of high quality and no redundant information. So, experts give it high marks. However, the wording is different from the reference summary, so ROUGE gives the summary a low rating. It can be seen from these two examples that ROUGE is a rough proxy that is unable to recognize semantic factual errors. Moreover, over-reliance on the literal matching of reference may lead to a suppression of the diversity of generated summaries. Therefore, a good summarization evaluation metric should be able to help identify: (i) semantically correct summaries with good word overlap with the original text or reference, and (ii) non-redundant and fluent summaries that contain enough correct facts, even if their wording is different from the reference.

To solve these problems, we propose SummScore, a comprehensive metric for summary quality evaluation based on Cross-Encoder. SummScore adopts the

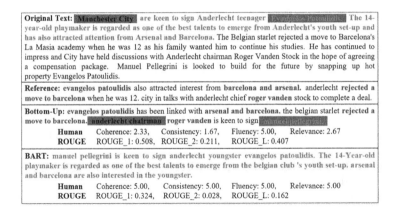

Fig. 1. A typical example showing ROUGE's problems.

original-summary paired measurement mode. The summaries are scored by comparing the semantics of the original text, avoiding the suppression of the diversity of the summaries caused by the forced alignment of a single reference summary. With the help of the text-matching pre-training Cross-Encoder, SummScore can effectively capture the subtle differences between the semantics of summaries. To improve the comprehensiveness and interpretability, SummScore consists of four fine-grained submodels, which measure Coherence, Consistency, Fluency, and Relevance separately.

We conduct our experiment in SummEval [7] dataset and measure the quality of our SummScore by calculating the Pearson correlation and Spearman correlation coefficient between SummScore scores and human annotation scores. We use semi-supervised multi-rounds of training to improve the performance of our model on extremely limited annotated data. Extensive experiments show that SummScore significantly outperforms existing evaluation metrics. In addition, we evaluate 16 mainstream summarization models with SummScore and publish the results for later research. Our contributions are summarized as follows:

- We propose SummScore, a novel evaluation metric for summary quality, which uses original text instead of the hard-to-obtain expert-generated gold summary as the reference to evaluate the quality of the generated summary.
- We trained four submodels of SummScore based on the Cross-Encoder framework to automatically evaluate the four fine-grained qualities of Relevance, Consistency, Fluency, and Coherence respectively. Experiments show SummScore has strong human relevance on all the four fine-grained dimensions.
- We evaluate 16 mainstream summarization models with SummScore and publish the results for later research.

2   Related Work

In this section, we will first introduce the common metrics on summarization and their main problems. Next, we will introduce the context-dependent metrics and the trained metrics in the evaluation of related natural language generation tasks. By borrowing the principles and advantages of the context-dependent metrics and the trained metrics, we design SummScore for summary quality evaluation.

Common Metrics in Summarization. The early common summary metrics are mainly represented by ROUGE [11], BLEU [14] and METEOR [1]. All of them obtain the summary quality score by calculating the token n-gram overlap between the summary and the reference. However, these lexical-based overlap metrics cannot capture the changes in semantics and grammar. Therefore, BERTscore [20] and MoverScore [21] use BERT to extract contextual embeddings and use embeddings matching to complete the similarity calculation between summary and references. However, these metrics, which rely on the alignment of single-reference abstracts, bring about suppression of abstract diversity.

Context-Dependent Metrics. To get rid of the constraints of reference summaries, ROUGE-C [9] improves ROUGE, which compares summaries with the original texts instead of reference summaries. ROUGE-C proves that using original text instead of reference can yield positive benefits, especially when the reference summary is not available. SUPERT [8] is an unsupervised referenceless summarization evaluation metric. SUPERT enables the quality assessment of the generated summaries with the help of pseudo-reference summaries created by salient sentences from the original text. Our model is also a context-dependent metric. Experiments show that our method not only gets rid of the comfort of reference summary but also supports the diversity of summary text generation.

Trained Metrics. There are training-based evaluation metrics in related natural language generation tasks. For machine translation, BLEND [13] and BEER [18] train a scoring model by combining a variety of existing untrained metrics, such as BLEU, METEOR, and ROUGE. As the pre-training models show promising performance, BERT for MTE [17] and BLEURT [16] are proposed for the machine translation system. By performing BERT fine-tuning training on a small amount of labeled data, they compute the similarity of the candidate and reference sentences. The difference is that BLEURT innovatively designs a set of pre-training signals and pre-trains BERT. We propose a trained-based summary evaluation metric SummScore, which consists of four submodels, corresponding to four quality dimensions. We believe that a single model may not be able to take into account the evaluation of various quality dimensions of the summary texts. At the same time, the independent scoring of multiple dimensions also helps to improve the interpretability of the summary quality score.

3   Our Methodology

3.1   Problem Definition

Our SummScore model is based on the **Cross-Encoder** [5] model in the field of information retrieval. In QA (Question answering) retrieval, when sorting

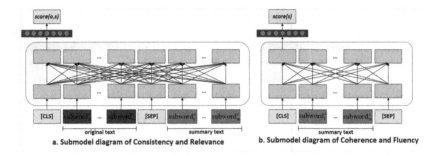

a. Submodel diagram of Consistency and Relevance

b. Submodel diagram of Coherence and Fluency

Fig. 2. Structural diagram of SummScore's submodels.

the candidate answers, the higher the similarity score between the answer and the question, the more accurate the answer is considered. The specific process can be realized by stitching the subword sequences of question text and answer text with [SEP] and inputting them into the Cross-Encoder model for training. Similarly, a summary can be regarded as a semantically similar text obtained after the original text is compressed. A heuristic idea is that the more similar the summary is to the original text, the higher the quality of the summary. The similarity here includes semantic similarity, content consistency, etc. Inspired by QA retrieval, we also regard the scoring of summary quality as a process of text similarity calculation between the original text and summary text.

As shown in Fig. 2, we formally define the summary quality evaluation problem as follows. Given the subword sequence O of the source document, where $O = \{o_1, ..., o_n\}$. Suppose that the subword sequence of the generated summary is S, where $S = \{s_1, ..., s_m\}$. The goal is to implement a function $score(O, S)$ and predict a score y to represent the similarity between document O and summary S. Given the training data with human annotation scores on summary quality, our goal is to train the function $score(O, S)$ so that it can regress to the human annotation score y'.

3.2 Structure of Model

The structure of SummScore's submodels is based on the Cross-Encoder. The Cross-Encoder [15] believes that the spliced sentence pair is a reasonable input mode, which is suitable for NSP(Next Sentence Prediction) [4] pre-training task and natural language inference task. Our SummScore is designed based on the principle of semantic similarity computation, and the used Cross-Encoder is pre-trained on related tasks. Hugging Face SentenceTransformers provides researchers with Cross-Encoder[1] after training on the semantic similarity benchmark dataset STS [2]. By inputting sentence pairs, the Cross-Encoder will predict a score between 0 and 5 representing the semantic similarity of the two texts.

Subsequently, we use the pre-trained Cross-Encoder to perform fine-tuning on fine-grained quality human-annotated data. Specifically, we add a regression

---

[1] https://www.sbert.net/examples/training/cross-encoder/README.html.

model based on MLP (Multilayer Perceptron) to Cross-Encoder to evaluate the score. The format of the input is a patchwork of sentence pairs. The first token of each sentence pair is always a special mark [CLS], and the sentences are separated by [SEP]. Finally, the final hidden state corresponding to the first special [CLS] token is taken as the sentence feature of the overall input. Feed the [CLS] embedding $V_{[CLS]}$ into the MLP to get the predicted score y:

$$V_{[CLS]} = Cross - Encoder([CLS], O, [SEP], S) \tag{1}$$
$$y = WV_{[CLS]} + b \tag{2}$$

where W and b are learnable parameters. The learning goal of the whole model is to fit the gold label y' with y. Our squared regression loss is:

$$\mathcal{L} = \frac{1}{N} \sum_{n=1}^{N} \|y - y'\|^2 \tag{3}$$

where N is the size of the sample.

3.3   Training Method of Submodels

Proxy metrics such as ROUGE and BERTscore usually return only a single value for the summary quality. It is difficult for people to clearly know how good or bad the current summary is from this score value. For example, does this summary capture the topic of the original text? How fluent is this summary? What are the main problems in this summary? That is, proxy metrics such as ROUGE and BERTscore are not well interpretable. Due to the poor interpretability of metrics scoring, it is also difficult for the summarization model to further improve the quality of the generated summary and the performance of the model.

The counterpart to machine scoring is human evaluation. It is a common fact that the human evaluation will first divide the quality of summaries into multiple fine-grained quality dimensions, and then score on the specific dimensions. A popular division is to divide the quality of the summary into four fine-grained quality dimensions (**Coherence, Consistency, Fluency**, and **Relevance**) [7,10]. Specifically, **Coherence:** the summary should be a coherent set of information about a topic, and whether the organizational structure between sentences is logical. **Consistency:** the summary should contain only the facts and themes of the original text. Both should be presented consistently and without hallucinatory facts. **Fluency:** the quality of the language. Whether there are grammatical errors that affect reading. **Relevance:** the summary should only contain important information from the source document, penalizing the summary that contains redundant information. Following this principle, our SummScore is also composed of four scoring submodels, and each corresponds to one of the above quality dimensions. Therefore, SummScore has the human-like ability to comprehensively evaluate the quality of summaries across multiple quality dimensions.

The model structures of the four submodels are consistent, but the mode of data input of the submodels of Coherence and Fluency is different. As shown in

Algorithm 1. The semi-supervised multi-round training

Input: Initial submodel M_0, annotated dataset $D_L(D_L^{train} \cup D_L^{val})$, unannotated data subset $D_U = \{D_1, ..., D_k\}$, epoch size for fine-tuning ep

Output: The best submodel M_{best}

1: /*Part1: the first round of supervised training with D_L^{train}.*/
2: Let $M_0^{best} = M_0$
3: **for** each $i \in \{0, 1, ..., ep - 1\}$ **do**
4:    Train M_0 on D_L^{train} an epoch agin and obtain M_{i+1}
5:    **if** $f(M_{i+1}, D_L^{val}) > f(M_{best}, D_L^{val})$ **then**
6:      $M_0^{best} = M_{i+1}$
7:    **end if**
8: **end for**

9: /*Part2: multiple rounds of semi-supervised training. */
10: Let $D = D_L^{train}, M^{best} = M_0^{best}$
11: **for** each $t \in \{1, 2, ..., k\}$ **do**
12:    Annotate D_t with M_{t-1}^{best} and obtain pseudo-annotated data D_t^{pseudo}
13:    $D = D \cup D_t^{pseudo}$
14:    Let $M_t^{best} = M_0$
15:    **for** each $i \in \{0, 1, ..., ep - 1\}$ **do**
16:      Train M_i on D_L^{train} an epoch agin and obtain M_{i+1}
17:      **if** $f(M_{i+1}, D_L^{val}) > f(M_t^{best}, D_L^{val})$ **then**
18:        $M_t^{best} = M_{i+1}$
19:      **end if**
20:    **end for**
21:    **if** $f(M_t^{best}, D_L^{val}) > f(M^{best}, D_L^{val})$ **then**
22:      $M^{best} = M_t^{best}$
23:    **end if**
24: **end for**

Fig. 2, among them, the scoring submodels for Fluency and Coherence no longer use the training mode of sentence pair. Because Fluency evaluates the linguistic quality of the summary itself. When experts annotate Fluency's scores, they can do it without referring to the original text. For the Coherence dimension, experts only focus on whether the summary text itself has a clear theme and rigorous sentence logic. In contrast, when experts score the quality dimensions of Consistency and Relevance, it is necessary to repeatedly compare the generated summary with the original text. Therefore, for the submodels of Coherence and Fluency dimension, we remove the original text information and change the formula (1) to the following form:

$$V_{[CLS]} = Cross - Encoder([CLS], S, [SEP]) \qquad (4)$$

Because the annotation data resources are very limited, we adopt a semi-supervised multi-round training method to maximize the correlation between SummScore and human ratings. The input of the algorithm includes the pre-trained Cross-Encoder M_0, which is used as the initial state of the SummScore's submodel. We have a small-scale manually annotated supervised dataset D_L. We divide D_L into the training set D_L^{train} and validation set D_L^{val}. In addition, we have a large amount of unsupervised data D_U generated by several mainstream summarization models. D_U is randomly divided into sub-datasets of the same size $\{D_1, ..., D_k\}$. Moreover, we also have a scoring function $f(\cdot)$ to judge whether the submodel is good or bad, which is achieved by comparing the correlation between the scores predicted by the submodel and the manually annotated scores on D_L^{val}. $f(\cdot)$ can be chosen from $max(Pearson)$, $max(Spearma)$ and $max(Pearson * Spearma)$. Our goal is to obtain the globally optimal submodel M_{best} with limited annotation data.

Our training is mainly divided into two parts, as shown in lines 1–8 and 9–24 of the Algorithm 1 respectively. In the first part of the algorithm, we first train the submodel on the small-scale supervised data D_L^{train}. In the beginning, we assume that the best submodel M_0^{best} in the first round of training is M_0 (line 2). Then, we perform fine-tuning training for ep times (line 3). After the i-th fine-tuning, the submodel is trained from M_i to M_{i+1} (line 4). After each fine-tuning, we compare the quality of M_0^{best} and M_{i+1}, and save the best model as M_0^{best} (line 5–7). After the first round of supervised training, we get the best model of the first round M_0^{best}.

In the second part of the algorithm, we will carry out multiple rounds of semi-supervised training to improve the performance of the submodel using unlabeled data. We assume that the initial global optimal model is M_0^{best}, and the current training available dataset D is D_L^{train} (line 10). Because the unsupervised dataset D_U is divided into k blocks, the algorithm will perform k rounds of semi-supervised training (Line 11). At the beginning of the t-th round of semi-supervised training, we will use the optimal model of the previous round M_{t-1}^{best} to label the sub-data D_t, and get the pseudo-labeled dataset D_t^{pseudo} (line 12). Then, the newly obtained pseudo-label data D_t^{pseudo} is extended to the available dataset D for the next round of semi-supervised training (line 13). After that, like the steps of Part1, start with the initial Cross-Encoder $M0$ and fine-tune the submodel ep epochs on data D (line 14–20). After the end of each epoch fine-tuning, the optimal submodel M_t^{best} of round t is retained (line 18). After each t-th round of semi-supervised training, we will also compare M^{best} and M_t^{best}, and keep the best model as the global optimal model M^{best}(line 22). After all t rounds of semi-supervised training, we finally obtain the globally optimal submodel M^{best} of SummScore for each fine-grained quality dimension.

4   Experiments Settings

We conduct experiments on SummEval [7] dataset, which contains 1600 manually annotated summaries. Each summary is evaluated on the four fine-grained quality dimensions according to criteria [10] and is scored by 5 independent crowdsource workers and 3 independent experts. Annotation scores span from 1 (worst) to 5 (best). We calculate the average of the annotation scores of the 3 experts as the final supervision score for each data and randomly divide the data into a training set D_L^{train} of 1000 pieces of data and a test set D_L^{test} of 600 pieces of data. At the beginning of each round of semi-supervised training, we randomly sample 100 pieces of data from the training set D_L^{train} as the validation set D_L^{val} for model selection and pass it to $f(\cdot)$ for model selection.

In addition to the above small-scale annotated data, we also use a large amount of unannotated data consisting of original texts and machine-generated summaries. These unannotated data will be randomly divided into k equal-sized parts in the experiment. Specifically, these divided data are mainly used in the semi-supervised training of the model to further help SummScore improve performance.

Our SummScore is based on Cross-Encoder[2] of Hugging Face SentenceTransformers. We expect that the scoring process of SummScore will be as fast as possible without taking up too much video memory of the machine. Therefore, we abandon the pre-training model with large-scale parameters, such as $RoBERTa_{LARGE}$ (24 layers), and only select the public $DistilRoBERTa_{BASE}$ (6 layers), and $RoBERTa_{BASE}$ (12 layers) for fine-tuning. So the GeForce GTX 1060 can meet all the experimental needs of SummScore. We set the epoch size for fine-tuning to be 6 and the batch size to be 4. When the amount of newly expanded pseudo-annotated data reaches 10,000 (about ten times the annotated data), the model can obtain satisfactory performance. At the beginning of each new round of semi-supervised training, SummScore will perform linner warmup training with $1/10$ of the single round steps. We use Adam as our optimizer with a learning rate of $2e-5$ and a weight decay of 0.01. Consistent with previous research works [17,20], we use Pearson and Spearman correlation coefficients to judge the correlation between the scoring metrics and manual scoring.

5   Experiments

5.1   Comparative Experiments

For the convenience of comparison, we conduct our comparative experiments in groups. First, we compare our model with several well-known training-free metrics. These metrics include BLEU [14], TF-IDF, ROUGE, BERTscore, and SUPERT. These metrics have their innovative principles and advantages, which have a profound impact on the development of the corresponding field. In particular, ROUGE and BERTscore are very popular and well-received in summary quality evaluation. Our SummScore is based on pre-training fine-tuning. Therefore, we also select two representative metrics based on the pre-trained model fine-tuning: BLEURT and BERT for MTE. For a fair comparison, we maintain the experimental design consistent with SummScore and conduct fine-tuning on the same data. Similarly, BLEURT and BERT for MTE also adopt multi-round semi-supervised training to eliminate the influence of training methods.

Table 1 shows our experimental results. Scores represent the Pearson correlation and Spearman correlation of each metric with respect to human annotations. It can be seen that compared with the training-free metrics, SummScore

---

[2] https://www.sbert.net/examples/training/cross-encoder/README.html.

Table 1. The results of the correlation experiment of the evaluation metrics on the test set of SummEval

Quality dimension		Coherence		Consistency		Fluency		Relevance	
Metric		Pearson	Spearma	Pearson	Spearma	Pearson	Spearma	Pearson	Spearma
Training-free	BLEU-1	0.0278	0.0272	0.2023	0.1552	0.1367	0.0696	0.2459	0.1992
	BLEU-2	0.0419	0.0384	0.1531	0.1456	0.1206	0.0810	0.2104	0.2002
	BLEU-3	0.0588	0.0668	0.1129	0.1367	0.1092	0.0841	0.1901	0.2240
	BLEU-4	0.0513	0.0764	0.0896	0.1271	0.1053	0.0898	0.1567	0.2193
	TF-IDF	0.0667	0.0689	0.0772	0.0930	0.0797	0.0727	0.0893	0.0472
	ROUGE-1	0.1757	0.1593	0.2242	0.1767	0.1651	0.1055	0.3264	0.2955
	ROUGE-2	0.1229	0.1237	0.1515	0.1489	0.1297	0.1062	0.2502	0.2509
	ROUGE-3	0.1219	0.1339	0.1184	0.1319	0.1182	0.1106	0.2160	0.2472
	ROUGE-L	0.1569	0.1457	0.1671	0.1486	0.1578	0.1499	0.2415	0.2284
	BERTscore-r	0.1414	0.1297	0.2456	0.1921	0.2142	0.1553	0.3474	0.2960
	BERTscore-p	0.1746	0.1428	0.1059	0.0821	0.2414	0.1683	0.1930	0.1513
	BERTscore-f	0.1792	0.1534	0.2043	0.1750	0.2614	0.1811	0.3135	0.2694
	SUPERT	0.2853	0.2599	0.3230	0.2931	0.3062	0.2280	0.3703	0.3256
Trained	BLEURT	0.4631	0.4410	0.3206	0.2233	0.4639	0.2193	0.5621	0.5286
	BERT for MTE	0.5532	0.5324	0.3721	0.3058	0.4601	0.2645	0.5638	0.5315
	BERT for MTE$_{DistilRobertaBase}$	0.6080	0.6036	0.4630	0.3512	0.4787	0.3509	0.5813	0.5500
Ours	SummScore$_{DistilRobertaBase}$	**0.6704**	**0.6684**	**0.4839**	**0.4080**	**0.7071**	**0.5586**	**0.6018**	**0.5538**
	SummScore$_{RobertaBase}$	**0.7061**	**0.7116**	**0.4852**	**0.4497**	**0.7348**	**0.5855**	**0.6746**	**0.6391**

far exceeds them. Moreover, except for SUPERT, these metrics have a low correlation with human annotations in all fine-grained dimensions. However, we find that they (e.g. BLEU, ROUGE, and SUPERT) tend to be more relevant to human judgments than Coherence and Fluency on Relevance and Consistency quality dimensions. The reason is that these metrics all need to compare the literal n-gram or semantic information of the reference summary when scoring. Because the reference summary is a compressed text that captures the central idea of the original text. Therefore, these metrics can achieve the purpose of preliminarily measuring the Relevance and Consistency of the original text topic semantics of the generated summary. Unfortunately, they are designed without considering the quality requirements of Coherence and Fluency. So these metrics tend to work poorly in the Coherence and Fluency dimensions.

Compared with the trained metric group, our model also outperforms all of them. However, we find that these metrics also perform well after multiple rounds of semi-supervised training on data. To eliminate the influence of the pre-trained language model, we also replace the pre-trained model of BERT for MTE with the same *DistilRobertaBase* trained on the STS dataset. We find that the performance of BERT for MTE model is more competitive. This proves that the idea of SummScore's quality evaluation design, which is inspired by the similarity matching principle of information retrieval, is reasonable.

5.2 Ablation Results

We believe that multi-round semi-supervised training is an important factor for improving SummScore. Because this training method brings about the rapid expansion of pseudo-annotated data and alleviates the problem of the small

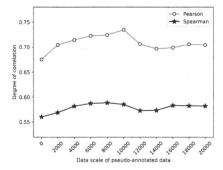

Fig. 3. Ablation experiments on the impact of pseudo-annotated data volume on the Fluency dimension of $SummScore_{RobertaBase}$.

amount of data. In order to explore the influence of the amount of pseudo-annotated data, we conduct corresponding ablation experiments. Only the ablation experiments of $SummScore_{RobertaBase}$ on the Fluency quality dimension are introduced here, and other quality dimensions have the same conclusion. In the ablation experiments, we expand the pseudo-annotation data with a span of 2000 pieces per round. The results of the ablation experiment are shown in Fig. 3. It can be clearly found that the performance of the model is significantly improved with the increase of pseudo-annotation data in the early stage. This indicates that the scale of data volume is an important factor to limit the model performance during this period. When the expanded pseudo-annotation data reaches about 10,000 (10 times the annotation data), the correlation of Fluency reaches its peak. This shows that at this time, the model has maximized the benefits from the increase in data volume. Subsequently, even with more data, the performance of the model does not improve and even begins to degrade. The ablation experiments show that reasonable multi-round semi-supervised training can effectively improve the performance of the SummScore in the case of scarce annotated data. This also provides a new training idea for later researchers to alleviate the limitation of small data volume in similar experimental scenes.

To explore the difference between original-summary pairing $[O||S]$ and common summary-reference $[S||ref]$ (adopted by BLEURT, BERT for MTE, and other models), we also conduct relevant analysis experiments. Table 2 shows the correlation of the two input methods on the $SummScore_{RobertaBase}$ model with human evaluation in the dimensions of Consistency and Relevance, respectively. We can find that $[O||S]$ can achieve better results than $[S||ref]$. Originally, we are also worried that the longer original information in $[O||S]$ may be more difficult for SummScore to learn than the short reference summary in $[S||ref]$. Analyzing the reasons for the better results of $[O||S]$, we believe that the reason is that the form of $[O||S]$ may be more consistent with the scoring process of humans in the dimensions of Consistency and Relevance. Because, normally, humans start to write a summary after reading the original text. In real life, few golden summaries can be repeatedly referred to write the new summaries.

In the scoring process, experts often score only after reading the original text. The input mode $[O||S]$ is also consistent with the expert scoring process.

Further experiments, we find that the original-summary mode $[O||S]$ can support the diversity of textual representations of summaries. The lower the ROUGE score, the more different the expression of the summary and the reference. However, those semantically correct summaries, which are expressed differently from the reference summaries, are also qualified summaries. Qualified summary metrics should be able to identify summaries that are diverse in expression but of acceptable quality. From the SummEval annotation data, we extract summary data with a low ROUGE score but a high human score. We plot the scatter plots of human scores and SummScore scores for these two input modes, respectively. Only the experiments in the Relevance dimension are listed here, and the results are shown in Fig. 4. We can find that $[O||S]$ is closer to the distribution of human scoring. However, for the summaries with high human scores, $[S||ref]$ is more likely to give low scores. Therefore, this can be illustrated that $[O||S]$ can recognize summaries with different literal expressions but qualified quality. This also shows that the $[O||S]$ mode can help to improve the diversity of summary generation.

Table 2. Ablation experiments on the influence of input modes $[S||ref]$ and $[O||S]$ on Consistency and Relevance.

	Pearson	Spearma
Consistency$_{[S||ref]}$	0.4291	0.3290
Consistency$_{[O||S]}$	**0.4852**	**0.4497**
Relevance$_{[S||ref]}$	0.6519	0.6172
Relevance$_{[O||S]}$	**0.6746**	**0.6391**

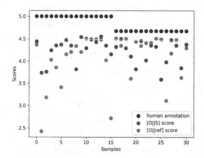

Fig. 4. Ablation experiments exploring the effect of input modes $[S||ref]$ and $[O||S]$ on the diversity of summaries.

5.3   Case Analysis

In Fig. 5, we show a typical example for case analysis. By reading the original text and the reference summary, we know that the original text is about

| | Coherence | Consistency | Fluency | Relevance |

Original Text: River Plate are keen to sign Manchester United striker Radamel Falcao but admit a deal is complicated. The Colombia forward spent eight years with the Argentine side before leaving for Porto in 2009 and River Plate are open to Falcao returning. During an interview with Esto es River program, vice president Matias Patanian said: 'We dream of Falcao Garcia. The doors are open.' River Plate are keen to sign former forward Radamel Falcao who has struggled on loan at Manchester United. River Plate vice president Matias Patanian admits the club 'dream of Falcao' and that 'the doors are open'. The 29-year-old has struggled during a season-long loan spell at Old Trafford this term - scoring just four Premier League goals - and it remains to be seen whether United will exercise the option to keep the frontman or whether he will return to parent club Monaco. However, **Falcao has been in good goalscoring form for his countrythis week,** finding the net three times in two games to equal Colombia's all-time goalscoring record with 24 goals. Joining River Plate at the age of 15 in 2001 before making his first-team debut four years later, Falcao went on to score 34 goals in 90 appearances for the Primera Division club. Falcao scored 34 goals in 90 appearances for the Argentine club during his four seasons in the first team.

Reference: river plate admit they 'dream' of manchester united striker radamel falcao. the colombia international spent eight years with the argentine club. falcao has managed just four goals in 19 premier league appearances.

Summary: radamel falcao has been in good goalscoring form for manchester united. the 29-year-old has struggled in a season-long loan at manchester united. the frontman has been scoring just four premier league goals.

	Coherence	Consistency	Fluency	Relevance
Human:	1.33	2.33	5.00	1.33
SummScore:	1.773	2.531	4.979	1.894
BERT for MTE:	2.031	4.896	4.928	3.052
ROUGE:	ROUGE_1:0.436	ROUGE_2:0.169	ROUGE_L:0.399	

Fig. 5. A classic example of performance comparison between SummScore and other metrics. (Color figure online)

River Plate are keen to sign Manchester United striker Radamel Falcao (orange fonts), and then some information about Radamel Falcao is introduced. However, we can find that the summary under test completely fails to capture the central idea of the original text. Therefore, in the Relevance dimension, both Summ-Score and experts give a low score of less than 2 points, but the baseline BERT for MTE scores a qualified score of 3 points. Further analysis, we find that there are hallucination errors (blue shaded text) in the summary under test. Radamel Falcao has good goalscoring form in *Colombia* rather than *Manchester United*. So both SummScore and experts give low marks for Consistency. Analyzing the structure between sentences, we find that the semantics of the summary to be tested is lack logic. For one moment, the summary tells us Radamel Falcao has good goalscoring form and another point that he struggles at Manchester United. Due to the lack of logic between sentences, it is difficult to read. So both Summ-Score and experts score low on the Coherence dimension, but BERT for MTE scores a high score close to 5. In terms of fine-grained quality dimension, it can be said that SummScore has better scoring ability than BERT for MTE, and the scoring effect is closer to human scoring. Because of the good n-gram overlap between the summary and the reference, ROUGE-1 gave this incomplete summary a high score of 0.436. As you can see, ROUGE is indeed a rough proxy indicator without explanatory power. ROUGE cannot tell us the specific quality of the summary, such as whether factual errors and grammatical errors exist.

5.4 Mainstream Summarization Models Evaluation

Finally, we use SummScore to evaluate the performance of 16 mainstream summarization models on the CNN/DailyMail dataset, and the scoring results are shown in Table 3. Please refer to the work SummEval [7] for a detailed introduction to these mainstream summarization models. We bold the top 5 scores of each quality dimension for further experimental analysis. We find that ROUGE

Table 3. Evaluation results of mainstream models on the SummScore indicator on the CNN/DailyMail.

Metrics	ROUGE-1	ROUGE-2	ROUGE-L	Coherence	Consistency	Fluency	Relevance
Extractive models							
LEAD-3	0.3994	0.1746	0.3606	**3.9146**	4.9766	4.9430	4.4264
NEUSUM	**0.4130**	**0.1893**	0.3768	3.1327	4.9712	4.9031	4.1876
BanditSum	**0.4137**	0.1868	0.3759	3.2399	4.9738	4.9139	4.2007
RNES	0.4088	**0.1878**	0.3719	**3.7673**	4.9771	4.9041	**4.4521**
Abstractive models							
Pointer Generator	0.3921	0.1723	0.3599	3.3892	4.9654	**4.9401**	4.1721
Fast-abs-rl	0.4057	0.1774	**0.3806**	2.2031	4.9255	4.6389	3.9024
Bottom-Up	0.4124	0.1870	**0.3815**	2.8551	4.9113	4.7716	3.8890
Improve-abs	0.3985	0.1720	0.3730	2.1961	4.8243	4.5633	3.6758
Unified-ext-abs	0.4038	0.1790	0.3675	3.4100	**4.9736**	4.8955	4.2684
ROUGESa	0.4016	0.1797	0.3679	3.2674	4.9700	4.8688	4.1793
Multi-task (Ent + QG)	0.3952	0.1758	0.3625	3.3573	4.9633	4.8870	4.1208
Closed book decoder	0.3976	0.1760	0.3636	3.3825	4.9688	4.8908	4.1866
T5	**0.4479**	**0.2205**	**0.4172**	3.6991	4.9126	4.8703	**4.3365**
GPT-2 (supervised)	0.3981	0.1758	0.3674	**3.7410**	3.9252	3.8176	3.6069
BART	**0.4416**	**0.2128**	**0.4100**	**4.2064**	4.9707	4.8545	**4.5683**
Pegasus	**0.4408**	**0.2147**	**0.4103**	**3.7148**	4.9176	4.8522	**4.3421**

favors the abstractive models, but SummScore seems to prefer the extractive models. In particular, the LEAD-3 model has achieved high SummScore scores on all four fine-grained qualities. The reason is that the first three sentences of the news are the most important part of the full text and the LEAD-3 is very suitable for news data. For Fluency and Consistency, extractive models tend to achieve higher scores. This is reasonable, because the summary of the abstractive model is generated according to the probability distribution of words, and the problems of fragment repetition and syntax errors can not be avoided. The generated summary may also have illusory facts that are inconsistent with the facts of the original text. However, the extractive model produces summaries by splicing sentences extracted from the original text. Because the sentences are written manually, this avoids grammatical errors and repetition. Moreover, the sentences are derived from the original, so there is no illusory fact. For Coherence and Relevance, there is a strong correlation between the two scores. Moreover, the ROUGE score is also correlated with the score of SummScore in these two quality dimensions. Almost models with high ROUGE scores also have high scores of Coherence and Relevance and vice versa.

6   Conclusion

In this paper, we propose SummScore based on the semantic matching principle of information retrieval, which is a trained scoring metric for summary quality evaluation based on Cross-Encoder. SummScore has good interpretability. It consists of four submodels, which measures the quality of the summary comprehensively from four fine-grained qualities of Coherence, Consistency, Fluency, and Relevance. We use semi-supervised multi-round training to improve model performance on limited annotated data. Extensive experiments show that SummScore significantly outperforms existing metrics in terms of human relevancy and helps improve the diversity of generated summaries. Finally, we also use SummScore to evaluate 16 mainstream summarization models and publish the results for later research.

References

1. Banerjee, S., Lavie, A.: Meteor: an automatic metric for MT evaluation with improved correlation with human judgments. In: Proceedings of the ACL Workshop on Intrinsic and Extrinsic Evaluation Measures for Machine Translation and/or Summarization, pp. 65–72 (2005)
2. Cer, D., Diab, M., Agirre, E., Specia, L.: Semeval-2017 task 1: semantic textual similarity multilingual and cross-lingual focused evaluation (2017)
3. Chen, W., Li, P., King, I.: A training-free and reference-free summarization evaluation metric via centrality-weighted relevance and self-referenced redundancy (2021)
4. Devlin, J., Chang, M.W., Lee, K., Toutanova, K.N.: BERT: pre-training of deep bidirectional transformers for language understanding. In: Proceedings of the 2019 Conference of the North American Chapter of the Association for Computational Linguistics: Human Language Technologies, Volume 1 (Long and Short Papers), pp. 4171–4186 (2018)
5. Ding, Y., Liu, J., Liu, K., Ren, R., Wang, H.: RocketQA: an optimized training approach to dense passage retrieval for open-domain question answering (2020)
6. Durmus, E., He, H., Diab, M.: FEQA: a question answering evaluation framework for faithfulness assessment in abstractive summarization (2020)
7. Fabbri, A.R., Kryściński, W., McCann, B., Xiong, C., Socher, R., Radev, D.: SummEval: re-evaluating summarization evaluation. Trans. Assoc. Comput. Linguist. **9**, 391–409 (2021)
8. Gao, Y., Zhao, W., Eger, S.: SUPERT: towards new frontiers in unsupervised evaluation metrics for multi-document summarization. arXiv preprint arXiv:2005.03724 (2020)
9. He, T., et al.: ROUGE-C: a fully automated evaluation method for multi-document summarization. In: 2008 IEEE International Conference on Granular Computing, pp. 269–274. IEEE (2008)
10. Kryściński, W., Keskar, N.S., Mccann, B., Xiong, C., Socher, R.: Neural text summarization: a critical evaluation (2019)
11. Lin, C.Y.: ROUGE: a package for automatic evaluation of summaries. In: Text Summarization Branches Out, pp. 74–81 (2004)
12. Lloret, E., Plaza, L., Aker, A.: The challenging task of summary evaluation: an overview. Lang. Resour. Eval. **52**(1), 101–148 (2018)
13. Ma, Q., Graham, Y., Wang, S., Liu, Q.: Blend: a novel combined MT metric based on direct assessment—CASICT-DCU submission to WMT17 metrics task. In: Proceedings of the Second Conference on Machine Translation, pp. 598–603 (2017)
14. Papineni, K., Roukos, S., Ward, T., Zhu, W.J.: BLEU: a method for automatic evaluation of machine translation. In: Proceedings of the 40th Annual Meeting of the Association for Computational Linguistics, pp. 311–318 (2002)

15. Reimers, N., Gurevych, I.: Sentence-BERT: sentence embeddings using Siamese BERT-networks (2019)
16. Sellam, T., Das, D., Parikh, A.P.: BLEURT: learning robust metrics for text generation. arXiv preprint arXiv:2004.04696 (2020)
17. Shimanaka, H., Kajiwara, T., Komachi, M.: Machine translation evaluation with BERT regressor. arXiv preprint arXiv:1907.12679 (2019)
18. Stanojević, M., Sima'an, K.: Fitting sentence level translation evaluation with many dense features. In: Proceedings of the 2014 Conference on Empirical Methods in Natural Language Processing (EMNLP), pp. 202–206 (2014)
19. Stiennon, N., Ouyang, L., Wu, J., Ziegler, D.M., Christiano, P.: Learning to summarize from human feedback (2020)
20. Zhang, T., Kishore, V., Wu, F., Weinberger, K.Q., Artzi, Y.: BERTScore: evaluating text generation with BERT. arXiv preprint arXiv:1904.09675 (2019)
21. Zhao, W., Peyrard, M., Liu, F., Gao, Y., Meyer, C.M., Eger, S.: MoverScore: text generation evaluating with contextualized embeddings and earth mover distance. arXiv preprint arXiv:1909.02622 (2019)

DASH: Data Aware Locality Sensitive Hashing

Zongyuan Tan[1,3], Hongya Wang[1,2,3(✉)], Ming Du[1], and Jie Zhang[1]

[1] School of Computer Science and Technology, Donghua University, Shanghai, China
tanzongyuan@mail.dhu.edu.cn, hywang@dhu.edu.cn
[2] State Key Laboratory of Computer Architecture, ICT, CAS, Beijing, China
[3] Shanghai Key Laboratory of Computer Software Evaluating and Testing, Shanghai, China

Abstract. Locality sensitive hashing (LSH) has been extensively employed to solve the problem of c-approximate nearest neighbor search (c-ANNS) in high-dimensional spaces. However, the search performance of LSH is degenerated with the number of data increasing. To this end, we propose an efficient method called *Data Aware Sensitive Hashing* (DASH) to deal with this drawback. DASH is the data-dependent hashing algorithm under considering the residual distance prior. DASH leverages this prior knowledge and provides theoretical guarantee for search results. Our experimental results with various datasets show that DASH achieves better search performance and the running time can reach up to about 4–40x speedups compared with other state-of-the-art methods.

Keywords: LSH · ANNS · High dimensions · Data-dependent hashing

1 Introduction

The nearest neighbor search (NNS) is the research focus all the time, which has been extensively applied to various fields, such as databases, machine learning and data mining. Given a query point q with dimension d in the Euclidean space, the problem of NNS is to return a point o^* in the dataset $\mathcal{D}$ with minimum distance to q. For low-dimensional NNS, the exact solutions have already been reported by based-tree methods. However, it has a great challenge to find the exact results for NNS in high-dimensional space due to the curse of dimensionality. Hence, an alternative scheme, i.e., the approximate nearest neighbor search (ANNS), has been extensively studied in recent two decades. Formally, the purpose of c-ANNS is to report a point $o \in \mathcal{D}$, whose distance with q is within $c \times r^*$, where r^* represents the distance between the query q and its exact nearest neighbor.

Locality sensitive hashing (LSH) is one of most effective techniques to solve high-dimensional c-ANNS problems, which is originally proposed for hamming space in [1], and later is extended to Euclidean space based on p-stable distribution [2]. Since LSH often needs to build hundreds of hash tables for achieving

B. Li et al. (Eds.): APWeb-WAIM 2022, LNCS 13422, pp. 85–100, 2023.
https://doi.org/10.1007/978-3-031-25198-6_7

good search results. This leads to prohibitively large space consumption. To deal with this problem, many LSH variants have been designed, e.g. [3–5]. However, LSH and its variants are data-independent hashing schemes.

As a matter of fact, many researchers turn to study the data-dependent hashing schemes to enhance the search performance for ANNS. In [7], the best LSH data structure is constructed by partitioning the original datasets into several subsets to form data-dependent hashing scheme. Based on [7], another data-dependent hashing scheme is proposed via spectral theorem [8]. Although the both methods have rigorous theoretical guarantee for search results returned, it is hard to put them to practice. Moreover, DSH [9] improves the hashing functions to address the problem that the elements of buckets for traditional LSH are unbalance. But DSH is lack of similar probability guarantee with traditional LSH. Our proposal DASH is the data-dependent hashing algorithm under considering the residual distance prior, and has the nature of probability guarantee for LSH. Similarly, BayesLSH [23] also puts forward a prior distribution and exploits Bayes inference to give the probability guarantee for the results returned, while its prior distribution is data-independent.

Motivations. Most of LSH and its variants have desirable theoretical guarantee for the search results, but often suffer from time inefficient, although they can achieve answering c-ANN queries with sub-linear query overhead. Also, the calculation cost on Euclidean distance between the query and its candidates is great because of finding a large amount of useless candidates. Product quantization (PQ) [10] provides effective means to estimate Euclidean distance for any high-dimensional points in Euclidean space. It constructs pre-calculation distance table by the manner of the asymmetric distance computation (ADC) or symmetric distance computation (SDC). This speeds up the distance computation for any two points, compared against computing Euclidean distance directly. By taking the merit of LSH and PQ into account, we are expected to design an algorithm that not only has probability guarantee for c-ANNS, but also is able to speed up query processing.

Contributions. The main contributions of the paper are concluded as follow:

- we propose an algorithm called Data Aware Sensitive Hashing (DASH) to answer the k-ANNS in high-dimensional Euclidean space. DASH is time efficient method and provides quality guarantee for the search results returned with preassigned success probability.
- We propose a novel prior (residual distance prior) – the key observation for DASH, which is based on the statistics of residual distance on data points to any random query. Equipped with this prior knowledge, DASH is able to address the k-NNS problem in more efficient manner.
- Extensive experiments demonstrate that DASH achieves desirable search performance for a variety of real datasets with different sizes. Compared against other state-of-the-art algorithms, DASH can obtain at least 4x speedup in the running time over different datasets.

Organization. The rest of this paper is organized as follows. Some preliminaries are reviewed in Sect. 2. A key observation is found in Sect. 3. Our method and probability analysis are presented in Sect. 4. Experimental results and analysis are reported in Sect. 5. Section 6 discusses the related work. Finally, a brief conclusion is drawn in Sect. 7.

2   Preliminaries

2.1   Problem Definitions

In this paper, we consider a dataset $\mathcal{D}$ with n points denoted as vectors in a d-dimensional Euclidean space R^d. For any point o and query q, let $d(o, q)$ represent the Euclidean distance that is defined as $d(o, q) = \sqrt{\sum_i^d (o[i] - q[i])^2}$, where $o[i]$ and $q[i]$ are the coordinate value of i-th dimension for o and q, respectively. Given a query q and the distance measure $d(\cdot, \cdot)$, the exact nearest neighbor (NN) o^* of q is the point in $\mathcal{D}$ with the minimum distance to q, namely $o^* = argmin_{o \in \mathcal{D}} d(o, q)$. Then, the c-approximate nearest neighbor search (c-ANNS) is defined as follows:

Definition 1. *Given an approximate ratio c ($c \geq 1$), any query $q \in R^d$ and the distance measure $d(\cdot, \cdot)$, the problem of c-ANNS is to establish a data structure, which retrieves a point $o \in \mathcal{D}$ satisfying $d(o, q) \leq c \times d(o^*, q)$, in which $o^* \in \mathcal{D}$ denotes the exact NN of q.*

The c-ANNS can be extended to more generalized form of c-k-ANNS. Similarly, the problem of c-k-ANNS is to establish a data structure, which for any query $q \in R^d$, retrieves a set of k ordered points $o_i \in \mathcal{D}$ ($1 \leq i \leq k$) satisfying $d(o_i, q) \leq c \times d(o_i^*, q)$, in which $o_i^* \in \mathcal{D}$ denotes the i-th exact NN of q.

2.2   Product Quantization

Product Quantization (PQ) [10] has been proposed to address the ANNS problem in high-dimensional space. It divides the d-dimensional original space into M subspaces equally, with the dimension of each subspace being $\frac{d}{M}$. Correspondingly, all the original vectors are divided into M subvectors and the dimension of each subvector is $\frac{d}{M}$. Then all the subvectors in each subspace are quantized to k^* different centroids, which are learned from a part of original data by k-means algorithm. That is, each subvector is denoted by the index of its nearest centroid, where the index is an integer over interval $[1, k^*]$. Thus, each original vector is denoted by a tuple of M integers, which are called as PQ-code. All the $M \cdot k^*$ centroids are compose of the codebook C of the product quantizer jointly.

With the codebook and PQ-codes, the Euclidean distance between any two vectors in the original space is estimated from their PQ-codes. There exists two manners to estimate the distance, i.e., the asymmetric distance calculation (ADC, the distance is calculated by a original vector and a PQ-code) and symmetric distance calculation (SDC, the distance is calculated by two PQ-codes). When a query vector arrives, a $M \cdot k^*$ pre-calculation distance table is built with ADC manner.

2.3   Query-Aware LSH Scheme

Assume that there is a random projection vector $\vec{a}$ of dimension d, namely $\vec{a} = [a_1, a_2, \ldots, a_d]$, whose each entry is an i.i.d random variable drawn from Gaussian distribution $\mathcal{N}(0,1)$, independently. Given a vector $\vec{o}$ of dimension d, the hash function between two vectors $\vec{a}$ and $\vec{o}$ represented as $h(o) = \langle \vec{a}, \vec{o} \rangle$ is the projection of o along $\vec{a}$, which is regarded as an LSH signature. According to the property of p-stable distribution for $p = 2$ [2], for any given points o_1, o_2, $h(o_1) - h(o_2)$ follows Gaussian distribution $\mathcal{N}(0, d^2(o_1, o_2))$.

Given a bucket width $2w$, the strategy of query-aware is that when query point q arrives, its LSH signature is located as the projected centre to identify the interval with bucket width $2w$, i.e., the interval $[h(q) - w, h(q) + w]$. For arbitrary point o, let $s = d(o,q)$, if the LSH signature of a point o falls in the hash bucket with width $2w$, i.e., $|h(o) - h(q)| \leq w$, then o collides with q under the hash function $h(\cdot)$. Accordingly, the collision probability is formalized as following form:

$$p(s) = P_r(\psi(o) \leq w) = \int_{-\frac{w}{s}}^{\frac{w}{s}} \varphi(x) dx \tag{1}$$

with $\psi(o) = |h(o) - h(q)|$ and $\varphi(x)$ being the probability density function of Gaussian Distribution $\mathcal{N}(0,1)$.

Suppose that the number of independent hash functions and the collision threshold are m and l, respectively, where $l \leq m$. According to the collision counting technique [4], it is easy to derive that the probability $\mathcal{P}(\dagger Col(o) \geq l)$ of point returned obeys Binomial distribution $\mathcal{B}(m, p(s))$, with

$$\mathcal{P}(\dagger Col(o) \geq l) = \sum_{i=l}^{m} \binom{m}{i} (p(s))^i (1 - p(s))^{m-i} \tag{2}$$

in which $\dagger Col(o)$ is the number of collision between q and o under m hash functions.

3   Residual Distance Prior

The residual distance prior is based on the following observation that the difference of approximate distance and Euclidean distance with respect to a random query point forms a specific distribution. To elaborate this observation formally, we first define the notion of residual distance. For a random query point q and an arbitrary point o, their residual distance (denoted by e) is given as following form:

$$e = d(o,q) - \widehat{d}(o,q) \tag{3}$$

where $d(o,q)$ denotes the Euclidean distance between the original vector o and query q, while $\widehat{d}(o,q)$ denotes the approximate distance of $d(o,q)$. Clearly, the residual distance is query-dependent.

Observation. We fit e over different datasets, as shown in Fig. 1 and Fig. 2. One important finding is that e fitted over many existing datasets asymptotically follow a common distribution family, i.e., Gaussian distribution.

Next, we will show how to obtain the universal residual distance distributions. This is similar to the previous works [11]. For a random query point q, we first compute the Euclidean distances $d(o, q)$ and $\widehat{d}(o, q)$ to obtain their difference e. Then the histogram on the difference e forms the residual distance distribution with respect to this query q. Actually, the residual distance distributions constructed for various queries could be discrepant. However, we observe that these residual distance distributions have similar shapes, but their scales are different. If a single residual distribution is only selected in one of the residual distance distributions as an universal distribution, errors will be resulted in. Therefore, in order to reduce error, we select the residual distance distributions for certain queries to approximate the universal residual distance distribution.

Concretely, we show how to extract statistical parameters from various datasets and conduct parameter estimation for the Gaussian distribution. According to the description above, we propose to fit the residual distance by the Gaussian distribution, whose probability density function is formalized as follows:

$$f(x|\mu, \sigma^2) = \frac{1}{\sqrt{2\pi}\sigma} exp\left(-\frac{(x-\mu)^2}{2\sigma^2}\right) \qquad (4)$$

with μ being the shape parameter; σ being the scale parameter. In practice, the purpose we select the Gaussian distribution is that it has distinct advantages.

First, the Gaussian distribution fits the residual distances with respect to various datasets, as shown in Fig. 1 and Fig. 2. Please see Subsect. 5.1 for more details about the datasets. One can be found that the residual distance distributions follow Gaussian distribution asymptotically. Since the accuracy of $\widehat{d}(o, q)$ depends on M, namely the larger M, the more accuracy $\widehat{d}(o, q)$ would be, the residual distance distributions vary with M. To verify the availability, we present different results with the variation of M. In Fig. 1, $\widehat{d}(o, q)$ over various datasets are computed with PQ under $M = 8$, except the dataset ImageNet with $M = 10$. While in Fig. 2, $M = 25$ for ImageNet, and other datasets are $M = 16$. Note that the residual distance distributions are also fitted to be the Gaussian distribution when increase the magnitude of M, where we only show part of the residual distance distributions due to the limitation of space.

Next, there exists an effective method to estimate the parameters of residual distance distribution, which can be estimated by Maximum Likelihood Estimation (MLE). For any sample with n i.i.d. Gaussian random variables, i.e., $\{x_1, x_2, \ldots, x_n\}$, the likelihood function is given by following form:

$$\mathcal{L}(\mu, \sigma) = \prod_{i=1}^{n} f(x_i|u, \sigma^2) \qquad (5)$$

then we maximize the log-likelihood function $ln(\mathcal{L}(\mu, \sigma))$. By respectively computing the partial derivatives of $ln(\mathcal{L}(\mu, \sigma))$ for both two parameters μ and σ,

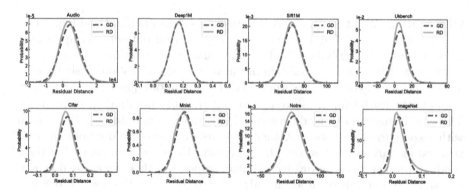

Fig. 1. The residual distance distribution (RD) asymptotically follows the Gaussian distribution (GD) for $M = 8$ and $M = 10$.

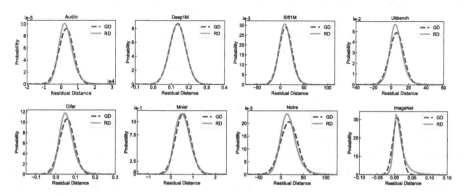

Fig. 2. The residual distance distribution (RD) asymptotically follows the Gaussian distribution (GD) for $M = 16$ and $M = 25$.

their estimation can be solved as $\hat{\mu} = \frac{1}{n}\sum_{i=1}^{n} x_i$ and $\hat{\sigma}^2 = \frac{1}{n}\sum_{i=1}^{n}(x_i - \hat{\mu})^2$. It can be found that the estimated parameters only hinge on arithmetic mean and variance of the sample, such that they can be easily obtained from given dataset.

4   Our Approach and Theoretical Analysis

4.1   Overview

Our algorithm is based on the search framework of QALSH [5]. It achieves that the exact distance calculation between the query and its candidates is converted to a look-up table operation, which greatly speeds up search time. Also, the similar probability guarantee with LSH is still obtained.

4.2 Algorithm Achievement

Our algorithm is based on the framework for memory version of QALSH [5], which can be divided into two parts: indexing construction and query processing.

Indexing Construction. We first select m LSH random projection vectors $\vec{a}$ generated from $\mathcal{N}(0,1)$. Then, each $o \in \mathcal{D}$ is randomly projected from d-dimensional space to m hash values $h_i(o)$, $i \in \{1, 2, \ldots, m\}$. For each projection vector $\vec{a}_i$, we construct a sorted hash table to store the key-value pair $(ID(o), h_i(o))$ for all points, where key $ID(o)$ is the identifier (ID) of each point o, and the hash table is sorted in the ascending order of $h_i(o)$. Finally, m hash tables reside on memory. Meanwhile, we employ PQ quantifying the dataset $\mathcal{D}$ to form PQ-codes and the pre-calculation distance tables for various queries are constructed with ADC manner.

Query Processing. When a query point $q \in R^d$ arrives, we also use m hash functions $h(\cdot)$ mapping it into corresponding hash values $h_i(q)$. Then for given bucket width $2w$, we will conduct a range search $[h_i(q) - w, h_i(q) + w]$ over each hash table. During this search processing, $\dagger Col(o)$ for each point o is updated dynamically. Recall that $\dagger Col(o)$ is the number of collision between o and q.

The pseudo-code of the probabilistic NNS on DASH is shown in Algorithm 1. DASH locates the hash values $h_i(q)$ via binary search, and the range search is conducted by gradually extending bucket width $2w$ by adding $\triangle w$ (Line 3), which is similar to the process of virtual rehashing [4,5] to access more points. However, the most significant difference of extending $2w$ is that it positively impacts the collision probability for supporting our probabilistic terminal condition, which will be discussed in the next subsection. When $|h_i(o) - h_i(q)| \leq w$, $\dagger Col(o)$ is updated (Line 5-6). If $\dagger Col(o)$ is not lower than the collision threshold l, o is regarded as the candidate of q (Line 7-8). Then, the calculation of exact distance $d(o, q)$ is converted to a look-up table operation, i.e., $d(o, q)$ is approximated with $\widehat{d}(o, q)$ calculated by the pre-calculation distance table constructed with PQ. Such an operation accelerates the query processing greatly.

Due to the fact that $\widehat{d}(o, q)$ is an estimation distance that has certain error with $d(o, q)$, DASH is unable to obtain desirable search accuracy under fixed bucket width compared with QALSH. To this end, we supplement the loss $\triangle$ based on $\widehat{d}(o, q)$ to obtain a more accuracy distance, where $\triangle$ is a constant value determined by Gaussian distribution $\mathcal{N}(\hat{\mu}, \hat{\sigma}^2)$, and we will present how to determine $\triangle$ in Subsect. 4.3. Finally, $d(o, q)$ is estimated as $\widehat{d}(o, q) + \triangle$. As $\widehat{d}(o, q) + \triangle$ is large enough, i.e., $\widehat{d}(o, q) + \triangle > d(o, q)$, it leads to DASH extending the bucket width to bring in more points in comparison with QALSH. This because for any given success probability P^*, the bucket width $2w$ is proportional to $\widehat{d}(o, q) + \triangle$. If $\widehat{d}(o, q) + \triangle > d(o, q)$, QALSH and DASH will be terminated early in $2w'$ and $2w''$ respectively, with $w' < w''$. Furthermore, another method is to heighten the success probability P^*. This condition is rigorous, which promotes DASH to return more candidates. Hence, DASH enhances the search accuracy effectively with the methods above.

Algorithm 1: NNS on DASH

Input: query point q, m hash tables, collision threshold l, bucket width $2w$, success probability P^*, approximate ratio c;

Output: o_{min} in the set of $\mathcal{C}$

1 $w = 0$; $\dagger Col(o) = 0$;
2 **while** *true* **do**
3     $2w = 2w + \triangle w$;
4   **for** *each* $i = 1$ *to* m **do**
5   **for** $|h_i(o) - h_i(q)| \leq w$ *for* o *in* i-*th hash table* **do**
6             $\dagger Col(o) = \dagger Col(o) + 1$;
7   **if** $\dagger Col(o) \geq l$ **then**
8                 $\mathcal{C} = \mathcal{C} \cup o$;
9   Calculate $(\widehat{d}(o,q) + \triangle)$ and $P_r(\widehat{d}(o,q) + \triangle)$;
10   **if** $P_r(\widehat{d}(o,q) + \triangle) \geq P^*$ **then**
11   break the while-loop;

12 **return** $o_{min} \in \mathcal{C}$;

If PQ is directly exploited to the acceleration, it will lead to the destruction on probability guarantee of LSH. Fortunately, we observe that the residual distance prior follows an universal distribution, as described in Sect. 3. According to the key observation, we determine the loss $\triangle$ as a constant value with certain probability. Then by taking this probability into account, we develop a new probability guarantee based on the framework of QALSH as the terminal condition of DASH. Actually the probability guarantee is similar to that in LSH, which is given in Subsect. 4.3. With the estimated distance $\widehat{d}(o,q) + \triangle$, it is easy to derive the collision probability $P_r(\widehat{d}(o,q) + \triangle)$ (Line 9). Assume the terminal condition has already been satisfied (Line 10), o_{min} in the set $\mathcal{C}$ is reported as the final result (Line 12).

k-NN Search. Our method can also be extended to perform k-NN search. It is sufficient to modify the terminal condition as $|\mathcal{C}| \geq k$ and $P_r(\widehat{d}(o,q) + \triangle) \geq P^*$ (Line 10 of Algorithm 1), and finally return k neighbors, i.e., $\{o^1_{min}, o^2_{min}, \ldots, o^k_{min}\}$. Hence, with this search framework, DASH can conduct probabilistic NNS and k-NNS.

4.3 Probability Analysis

Assume that the point o is the candidate of q. As discussed above, their estimated distance is $\widehat{d}(o,q) + \triangle$, and the loss $\triangle$ is determined based on $\mathcal{N}(\hat{\mu}, \hat{\sigma}^2)$. Next, we mainly focus how to obtain $\triangle$ with desirable probability.

According to the $j\sigma$ rule of Gaussian distribution, here $j \in \{1, 2, 3\}$, the probability $\mathcal{P}(\mu, j\sigma)$ for random variables drawn within $[\mu - j\sigma, \mu + j\sigma]$ is given as:

$$P(\mu, j\sigma) = F(\mu + j\sigma) - F(\mu - j\sigma) = \frac{2}{\sqrt{\pi}} \int_0^{\frac{j}{\sqrt{2}}} dx \tag{6}$$

where $F(\cdot)$ is the cumulative distribution function (CDF) of Gaussian distribution. From the Eq. (3), we have $d(o, q) = \widehat{d}(o, q) + e$, where $e \sim \mathcal{N}(\hat{\mu}, \hat{\sigma}^2)$. Then, one can derive based on Eq. (6) that $d(o, q)$ falls into the interval $[\widehat{d}(o, q) + \hat{\mu} - j\hat{\sigma}, \widehat{d}(o, q) + \hat{\mu} + j\hat{\sigma}]$ with probability $P(\hat{\mu}, j\hat{\sigma})$. Hence, the loss $\triangle$ is determined as $\triangle = \hat{\mu} + j\hat{\sigma}$, which is the worst-case to estimate $d(o, q)$ under $P(\hat{\mu}, j\hat{\sigma})$, with $d(o, q) = \widehat{d}(o, q) + \hat{\mu} + j\hat{\sigma}$.

Recall that $\psi(o) = |h(o) - h(q)|$. For given bucket width $2w$ and approximate ratio c, the collision probability $p(\widehat{d}(o, q) + \triangle)$ for both q and o is given as following form:

$$p(\widehat{d}(o, q) + \triangle) = P_r(\psi(o) \le cw) = \int_{-\frac{cw}{\widehat{d}(o,q)+\triangle}}^{\frac{cw}{\widehat{d}(o,q)+\triangle}} \varphi(x) dx \tag{7}$$

where $\varphi(x)$ is the probability density function of $\mathcal{N}(0, 1)$. A simple calculation for the Eq. (7) is $p(\widehat{d}(o, q) + \triangle) = 2norm(\frac{cw}{\widehat{d}(o,q)+\triangle}) - 1$, in which $norm(x) = \int_{-\infty}^x \varphi(t) dt$. Note that $norm(x)$ is the CDF of $\mathcal{N}(0, 1)$, which is the monotonically increasing function with respect to x. When c and $2w$ are fixed, $norm(\frac{cw}{\widehat{d}(o,q)+\triangle})$ is monotonically decreasing with $\widehat{d}(o, q) + \triangle$ increasing, so $p(\widehat{d}(o, q) + \triangle)$ monotonically decreases with $\widehat{d}(o, q) + \triangle$ increasing. We know that $\widehat{d}(o, q) + \triangle$ is the worse-case to estimate $d(o, q)$. Since $\triangle$ is a random variable drawn from $\mathcal{N}(\hat{\mu}, \hat{\sigma}^2)$ with probability $P(\mu, j\sigma)$, there is $\widehat{d}(o, q) + \triangle \le d(o, q)$ or $\widehat{d}(o, q) + \triangle \ge d(o, q)$.

Since $p(\widehat{d}(o, q) + \triangle)$ is known, $P(\dagger Col(o) \ge l)$ is obtained via the Eq. (2). To achieve the probability guarantee of the algorithm, we first make an assumption that the two events on the loss $\triangle$ determined and the nearest neighbor returned by collision counting are mutually independent under DASH, i.e., the both probability $P(\dagger Col(o) \ge l)$ and $P(\mu, j\sigma)$ obtained are independent. Then the overall probability $P_r(\widehat{d}(o, q) + \triangle)$ of finding the nearest neighbor for DASH can be expressed as:

$$P_r(\widehat{d}(o, q) + \triangle) = P(\dagger Col(o) \ge l) \cdot P(\mu, j\sigma) \tag{8}$$

This is regarded as the probabilistic terminal condition for DASH. Furthermore, if we could acquire $\widehat{d}(o, q) + \triangle$ in advance, then $p(\widehat{d}(o, q) + \triangle)$ is determined immediately by the Eq. (7). We only require to select suitable m, l and $P(\mu, j\sigma)$ to realize the success probability P^* specified beforehand, such that

$$P_r(\widehat{d}(o, q) + \triangle) = P^* \tag{9}$$

Example 1. Suppose the point o has been the candidate of q, and their estimated distance is $\widehat{d}(o, q) + \mu + 2\sigma$. This means $\triangle = \mu + 2\sigma$, with $P(\mu, 2\sigma) = 0.9544$.

If the collision probability $\mathcal{P}(\dagger Col(o) \geq l) = 0.9$, then the overall probability $P_r(\widehat{d}(o,q) + \triangle) = 0.9 \times 0.9544 \approx 0.86$. Similarly, when $d(o,q) = \widehat{d}(o,q) + \mu + 3\sigma$, we obtain $\mathcal{P}(\mu, 3\sigma) = 0.9974$, such that $P_r(\widehat{d}(o,q) + \triangle) = 0.9 \times 0.9974 \approx 0.9$.

To sum up, it is natural to yield the theorems below for the search results under DASH.

Theorem 1. *DASH returns the NN (o_{min}) of query q with the success probability at least P^*.*

Proof. First, we define the two events below:

E_1 : the loss $\triangle$ is determined based on residual distance distribution.

E_2 : the o_{min} is found by DASH.

Recall that we make an assumption on E_1 and E_2 being independent. As discussed earlier, if the points are contained under the fixed bucket width $2w$, then $P[E_1]$ and $P[E_2]$ can be obtained by the Eq. (6) and Eq. (2), respectively. Since the both events are mutually independent, $P[E_1 E_2] = P[E_1]P[E_2]$. However, DASH is guaranteed to answer the o_{min} with success probability P^*, then we have $P[E_1 E_2] \geq P^*$. Hence, this theorem is proved. □

Theorem 2. *DASH returns the k-NN ($\{o_{min}^i\}_{i=1}^k$) of query q with the success probability at least P^*.*

Proof. The proof of this theorem is similar to Theorem 1. □

5   Performance Evaluation

Our method is implemented in C++ and compiled with g++ 9.3 with -O3 optimization. The experiments for general-scale datasets were conducted on a laptop with six-cores Intel(R), i7-8750H @ 2.20GHz CPUs and 32 GB RAM, in Ubuntu 20.04. While others for large-scale datasets were conducted on a server with eight-cores Intel(R), E5-2620 v4 @2.1GHz CPUs and 256 GB RAM.

5.1   Datasets and Experiment Setting

We employ some publicly available real-life datasets in our experiments, whose data types cover audio, image and deep-learning data. Also, the 50 points are chose randomly from corresponding test sets as queries.

- **Cifar.** The Cifar dataset is a collection of 0.05 million 512-dimensional GIST feature vectors extracted from TinyImage.
- **Audio.** It is a 192-dimensional dastset that is composed of about 0.05 million audio feature vectors from DARPA TIMIT audio speed dataset.
- **Mnist.** The Mnist dataset contains about 0.07 million images of hand-written digits, which are represented as 784-dimensional vectors.
- **Notre.** It has about 0.3 million 128-dimensional features of a set of Flickr images and a reconstruction.

- **Sift.** The Sift dataset contains 1 million 128-dimensional SIFT vectors.
- **Deep.** The Deep dataset has 1 million data points with 256 dimensions that are deep neural codes of natural images obtained from the activations of a convolutional neural network.
- **Ukbench.** It is about 1 million 128-dimensional features of images.
- **ImageNet.** The ImageNet consists of about 2.4 million data points with 150-dimensional dense SIFT features.
- **Sift10M.** This dataset contains 10 million 128-dimensional SIFT vectors.
- **Sift100M.** This dataset contains 100 million 128-dimensional SIFT vectors.
- **Deep10M.** This dataset contains 10 million 96-dimensional DEEP vectors.
- **Deep100M.** This dataset contains 100 million 96-dimensional DEEP vectors.

The parameters have an important influence on the performance of our method. To this end, we empirically determine some near optimal parameters with respect to different datasets. For data compression, each vector is divided into $M = \frac{d}{2}$ subvectors and the centroids of each subspace are $k^* = 256$. The details for building the pre-calculation table can refer to [10]. In addition, we select the number of hash function $m = 60$ and the collision threshold $l = 50$ as the default for the experiments.

5.2 Evaluation Metrics

We employ the following metrics to evaluate the performance of our algorithm.

- **Recall.** We employ recall as a criterion to measure the accuracy for different algorithms. For the k-NNS, the recall is defined as the fraction on how many the k points answered by an algorithm are appeared in the true k nearest neighbors. Hence, it can be formalized as

$$Recall = \frac{|R' \cap R|}{|R|}$$

  where R' is a set of k points answered for a query and R is a set of true k nearest neighbors for the query. In our experiment, the Recall is computed as $R1@1, R10@10, \ldots, R100@100$.
- **Query Answering Time.** Another evaluation metrics is the query answering time, which is defined as the wall-clock time of an algorithm to answer k-NN.

In our experiments, we report the average recall and average running time as the final results, where both of them are the average over the queries.

5.3 Baseline Algorithms

There are many the state-of-the-art algorithms for approximate nearest neighbor search (ANNS), such as QALSH [5], VHP [6], PMLSH [18], R2LSH [19], SRS [20], HD-index [21], PQBF [22]. Since R2LSH performs better for ANNS compared with SRS, Hd-index and PQBF [19]. Also, we find that VHP, PMLSH and R2LSH are not compared with each other. Hence we select QALSH, VHP, PMLSH, R2LSH as baselines. Note that those algorithms work on memory. To implement the best performance of VHP, we use some parameter values presented in [6], in which hash functions $m = 60$, success probability $P^* = 0.9$ and the initial search window $t_0 = 1.4$. For QALSH, we employ the improved version, which can achieve higher accuracy and support $c = 1$. VHP and QALSH use identical hash functions m and collision threshold l. For PMLSH, we choose the parameters proposed in [18], with $m = 15$, $P^* = 1 - 1/e$ and the number of pivots $s = 5$. Also, the parameters of R2LSH are set as the default value suggested by the authors in [19], where $m = 40$ and $P^* = 0.9$. For the c-k-ANNS, we set $c = 1$ and $k \in \{1, 10, 20, 30, 40, 50, 60, 70, 80, 90, 100\}$.

5.4   Results and Analysis

Our method is based on the framework of QALSH. Due to the employment of PQ, our method requires to add extra consumption to construct indexing compared with QALSH, whereas the consumption is relatively small. For example, with respect to the large-scale dataset Deep100 M of size 36 GB used in the experiments, the additional time and space consumption are around 470 s and 4.5 GB, respectively, where QALSH needs about 46 GB memory space for constructing hash tables and the corresponding time is close to 820 s. It can be found that the additional time and memory space are only around a half and tenth of QALSH, respectively.

General-Scale Data. We study the performance mainly focusing on the average recall and query answering time. The distance for any two points is estimated with PQ, it is inevitable to result in certain estimated error and the destruction of probability guarantee. Nevertheless, our method can speed up the search processing with the pre-calculation distance table and obtain the similar probability guarantee with LSH based on residual prior distribution. Since the distance calculation with PQ negatively impacts the search accuracy, we use the original data to solve this issue. More explicitly, we consider that the real k nearest neighbors for any query are within the top-k' points returned, where $k' > k$ is a predefined constant. If we reorder the top-k' points with Euclidean distance, then it has higher possibility to find top-k exact nearest neighbors. Hence, we answer the best top-k in top-k' points as the final results for any query to achieve higher accuracy. The average recall $Rk@k$ by varying k from 1 to 100 under the success probability $P^* = 0.9$ is given in Fig. 3. One can be found that the average recall $Rk@k$ for our method is almost higher than other state-of-the-art methods with respect to different datasets.

Correspondingly, the running time curves for k-NNS are shown in Fig. 4. One can observe that the running time for R2LSH, PMLSH and VHP presents different over various datasets, while DASH is lower significantly than them. This is because when the candidates have been retrieved, other methods need to calculate the Euclidean distance, which cost a large amount of time; by contrast,

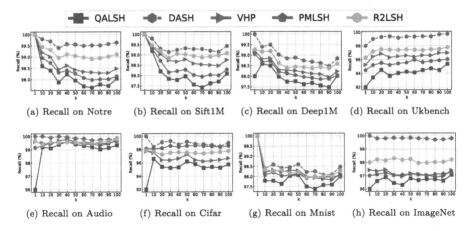

Fig. 3. The comparison on the accuracy among different methods.

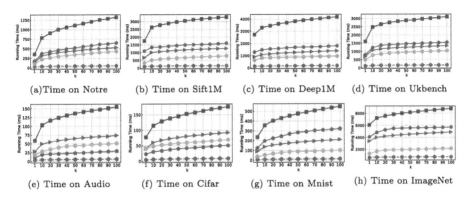

Fig. 4. The comparison on the running time among different methods.

DASH only needs to calculate the approximate distance by the pre-calculation distance table, so that the time consumption is relatively less. This means that DASH is more promising as k varies. For the dataset Mnist with high dimensionality, DASH performs better than those with low dimensionality, in which the speed can reach up to more than 40x in comparison with QALSH. Generally, DASH can achieve at least 5x speedup than other methods. Therefore, DASH is more superior pertaining to high dimensional datasets. Note that the cost of running time for finding k nearest neighbors is proportional to the increment of k, while the corresponding curve for DASH looks pretty stable than other algorithms because the magnitude of increment is relatively small.

Large-Scale Data. When DASH is applied to process more large-scale data, it also has significant superiority on accuracy and running time, as shown in Fig. 5. From the results, we can see that the accuracy obtain by DASH is higher than other algorithms with k increasing. The main reason is that DASH could access

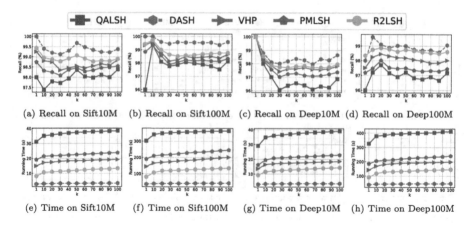

Fig. 5. The comparison on the search performance among different methods.

more points to obtain desired accuracy under the suitable success probability. In addition, DASH is able to achieve at least 4x speedups than other algorithms. It is benefited from the acceleration property of PQ. This indicates that the search performance of DASH has prominent superiority than other algorithms as the scale of datasets increases.

6   Related Work

Approximate nearest neighbor search (ANNS) has attracted extensive attention over decades. There exists a vast majority of works to solve the ANNS problem. For example, the space partitioning methods [12,13] perform well in the low-dimensional space, while their performance greatly decreases due to the "curse of dimensionality". The quantization-based methods play an important role in data compression at the cost of bringing the quantization error, e.g. PQ [10], such that query accuracy is relatively lower. Hence, many methods have been proposed to decrease the quantization error, such as OPQ [14] and TQ [15]. The graph-based methods [16,17] have favourable results for high-dimensional ANNS, which are benefited from effective indexing structure. Although it could reach up to high recall with few time, they are lack of quality guarantee. In addition, the hash-based methods employ a family of hash functions mapping the nearby points to the same bucket with high probability than the distant points. However, LSH needs to construct many hash tables to achieve desired accuracy. For this drawback, many LSH-based variants have been proposed, e.g., [18,19].

7   Conclusion

In this paper, we propose a time efficient data-dependent hashing sheme called Data Aware Sensitive Hashing (DASH) for approximate nearest neighbor search in high-dimensional space. DASH is based on the search framework of QALSH and takes the residual distance prior into account to evaluate a common distribution family for achieving probability guarantee. The extensive experiments are conducted to verify the efficiency and effectiveness of DASH by employing several real-life datasets. The results show that DASH obtains better search performance under the same reported quality compared against other methods.

Acknowledgements. The work reported in this paper is partially supported by NSF of Shanghai under grant number 22ZR1402000, the Fundamental Research Funds for the Central Universities under grant number 2232021A-08, State Key Laboratory of Computer Architecture (ICT,CAS) under Grant No. CARCHB 202118, Information Development Project of Shanghai Economic and Information Commission (202002009).

References

1. Indyk, P., Motwani, R.: Approximate nearest neighbors: towards removing the curse of dimensionality. In: Proceedings of ACM STOC, pp. 604–613 (1998)
2. Datar, M., Immorlica, N., Indyk, P., Mirrokni, V.: Locality-sensitive hashing scheme based on p-stable distributions. In: Proceedings of SoCG, pp. 253–262 (2004)
3. Ren, Z., Gu, Yu., Li, C., Li, F.F., Yu, G.: GPU-based dynamic hyperspace hash with full concurrency. Data Sci. Eng. **6**(3), 265–279 (2021). https://doi.org/10.1007/s41019-021-00161-5
4. Gan, J., Feng, J., Fang, Q., Ng, W.: Locality-sensitive hashing scheme based on dynamic collision counting. In: Proceedings of SIGMOD, pp. 541–552 (2012)
5. Huang, Q., Feng, J., Zhang, Y., et al.: Query-aware locality-sensitive hashing for approximate nearest neighbor search. In: Proceedings of VLDB, pp. 1–12 (2015)
6. Lu, K., Wang, H., Wang, W., Kudo, M.: VHP: approximate nearest neighbor search via virtual hypersphere partitioning. In: Proceedings of VLDB, pp. 1443–1455 (2020)
7. Andoni, A., Razenshteyn, I.: Optimal data-dependent hashing for approximate near neighbors. In: Proceedings of STOC, pp. 793–801 (2015)
8. Andoni, A., Naor, A., Nikolov, A., et al.: Data-dependent hashing via nonlinear spectral gaps. In: Proceedings of ACM SOTC, pp. 787–800 (2018)
9. Gao, J., Jagadish, H.V., et al.: DSH: data sensitive hashing for high-dimensional k-nnsearch. In: Proceedings of SIGMOD, pp. 1127–1138 (2014)
10. Jegou, H., Douze, M., Schmid, C.: Product quantization for nearest neighbor search. IEEE Trans. Pattern Anal. Mach. Intell. **33**(1), 117–128 (2010)
11. Dong, W., Wang, Z., Josephson, W., et al.: Modeling lsh for performance tuning. In: Proceedings of CIMK, pp. 669–678 (2008)
12. Guttman, A.: R-trees: A dynamic index structure for spatial searching. In: Proceedings of SIGMOD, pp. 47–57 (1984)
13. Bustos, B. Pedreira, O. Brisaboa, N.: A dynamic pivot selection technique for similarity search. In: Proceedings of SISAP, pp. 394–401 (2008)
14. Ge, T., He, K., Ke, Q., Sun, J.: Optimized product quantization for approximate nearest neighbor search. In: Proceedings of CVPR, pp. 2946–2953 (2013)
15. Babenko, A., Lempitsky, V.: Tree quantization for large-scale similarity search and classification. In: Proceedings of CVPR, pp. 4240–4248 (2015)

16. Yi, P., Li, J., Choi, B., Bhowmick, S.S., Xu, J.: FLAG: towards graph query auto-completion for large graphs. Data Sci. Eng. **7**(2), 175–191 (2022)
17. Malkov, Y.A., Yashunin, D.A.: Efficient and robust approximate nearest neighbor search using hierarchical navigable small world graphs. IEEE Trans. Pattern Anal. Mach. Intell. (2018)
18. Zheng, B., Xi, Z., Weng, L. et al.: PM-LSH: A fast and accurate LSH framework for high-dimensional approximate NN search. In: Proceedings of VLDB, pp. 643–655 (2020)
19. Lu, K. and Kudo, M.: R2LSH: A nearest neighbor search scheme based on two-dimensional projected spaces. In: Proceedings of ICDE, pp. 1045–1056 (2020)
20. Sun, Y., Wang, W., Qin, J., et al.: SRS: solving c-approximate nearest neighbor queries in high dimensional euclidean space with a tiny index. In: Proceedings of VLDB, pp. 1–12 (2014)
21. Arora, A., Sinha, S., Kumar, P., Bhattacharya, A.: Hd-index: Pushing the scalability-accuracy boundary for approximate knn search in highdimensional spaces. In: Proceedings of VLDB, pp. 906–919 (2018)
22. Liu, Y, Cheng, H, Cui, J.: PQBF: I/O-efficient approximate nearest neighbor search by product quantization. In: CIKM, pp. 667–676 (2017)
23. Satuluri, V., Parthasarathy, S.: Bayesian locality sensitive hashing for fast similarity search. In: Proceedings of VLDB, pp. 430–441 (2012)

AOPSS: A Joint Learning Framework for Aspect-Opinion Pair Extraction as Semantic Segmentation

Chengwei Wang[1,3], Tao Peng[1,2,3(✉)], Yue Zhang[1,3], Lin Yue[4], and Lu Liu[1,2,3]

[1] College of Computer Science and Technology, Jilin University, Changchun, China
{cwwang20,yue_zhang19}@mails.jlu.edu.cn, {tpeng,liulu}@jlu.edu.cn
[2] College of Software, Jilin University, Changchun, China
[3] Key Laboratory of Symbol Computation and Knowledge Engineer of the Ministry of Education, Changchun, China
[4] School of Information and Physical Sciences, The University of Newcastle, Newcastle, Australia
lin.yue@newcastle.edu.au

Abstract. Aspect-opinion pair extraction (AOPE) task, aiming at extracting aspect terms and their corresponding opinion terms in pairs, has caused widespread attention in recent years. Most studies focus on incorporating external knowledge, such as syntactic information. However, they are limited by the inadequate ability to capture long-distance information, and the utilization of external knowledge is more costly. In this paper, we propose AOPSS, a joint learning framework, to explore the AOPE task as semantic segmentation. As in most prior studies, we divide the AOPE task into two subtasks: entity recognition and relation detection. Specifically, AOPSS can synchronously capture task-invariant and task-specific features for the two subtasks without integrating any additional knowledge. Furthermore, we consider the interaction between entity and relation feature representations, which can improve the mutual heuristic effect for the two subtasks. Experimental results illustrate that our method achieves state-of-the-art performance on four public datasets, and we take further analysis to demonstrate the effectiveness of our approach.

Keywords: Sentiment analysis · Relation extraction · Entity recognition · Semantic segmentation · Joint learning

1 Introduction

In recent studies, scholars have focused on fine-grained aspect-based sentiment analysis (ABSA), which helps people to obtain sentiment polarity from the review sentences of a product or service. The subtasks of ABSA, aspect terms extraction (ATE) and opinion terms extraction (OTE), are usually treated as two independent tasks in former studies. ATE is to extract entities or phrases

[22], and OTE is to extract the expressions with sentiment polarity [3]. To obtain more fine-grained results, some efforts attempt to explore the aspect-opinion pair extraction (AOPE) task, aiming at extracting aspect terms and their corresponding opinion terms in pairs. To better clarify the difference between ATE, OTE, and AOPE tasks, we provide an example in Fig. 1.

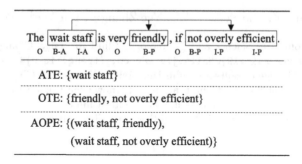

Fig. 1. A case explains the difference between ATE, OTE, and AOPE tasks. Below the review sentence is the sequence tags, where *B-A/B-P* represents the beginning of an aspect/opinion term, *I-A/I-P* represents the inside of an aspect/opinion term, and *O* represents it doesn't belong to any aspect or opinion term.

The AOPE task can be divided into two subtasks: entity recognition and relation detection. According to the statistics provided by [17], the overlap problem is about 24.42 % of the dataset. The case shown in Fig. 1 contains an overlap problem. For entity recognition, we expect to extract an aspect term *"wait staff"* by assigning *B-A* and *I-A* to *"wait"* and *"staff"*. Simultaneously, we expect to extract two opinion terms *"friendly"* and *"not overly efficient"* by assigning *B-P* to *"friendly"* and *"not"*, and *I-P* to *"overly"* and *"efficient"*. For relation detection, we expect to obtain two aspect-opinion pairs {(*"wait staff"*, *"friendly"*), (*"wait staff"*, *"not overly efficient"*)}.

Conventional methods utilize pipeline manner to extract aspect and opinion terms first and then perform pairing, but these extract-then-pairing models have the error propagation problem. The existing joint learning methods have difficulty in capturing long-distance information, which leads to the problem of local short-term feature combinations [1,19,29]. In this paper, we propose **A**spect-**O**pinion **P**air extraction as **S**emantic **S**egmentation (AOPSS), a joint learning framework, which can achieve good performance without introducing external knowledge and can capture long-distance information to extract aspect-opinion pairs more precisely. Specifically, we first calculate the interaction matrix of entity and relation feature representations in the encoding module. Subsequently, in the dual-channel semantic segmentation module, AOPSS can capture both local context and global interdependency, and obtain task-invariant and task-specific features synchronously for the two subtasks. Finally, we exploit the tagging and classification module for entity recognition and relation detection, respectively.

We conduct a series of experiments and analyses, demonstrating that AOPSS brings a significant outperformance over several current SOTA baselines, which verifies the effectiveness of our approach. In summary, the main contributions of our paper are threefold:

- We propose a joint learning framework to solve the AOPE task as semantic segmentation, which takes entity recognition and relation detection subtasks simultaneously without relying on external knowledge and not be affected by error propagation.
- Benefiting from the dual-channel semantic segmentation module, our approach can better capture both local context and global interdependency for the two subtasks, and it can leverage long-distance information sufficiently to solve the problem of local short-term feature combinations.
- Experimental results on four benchmark datasets show that our approach achieves state-of-the-art performance compared with baselines.

2   Related Work

Aspect-Opinion Pair Extraction. As a fine-grained sentiment analysis task, the AOPE task has attracted extensive attention in early research. [5,12] propose rule-based pipeline methods, which lay the foundation for the AOPE task. However, these methods are limited by template rules, and the pipeline-based methods are usually affected by error propagation. Subsequently, [6,21] propose joint learning methods, which are based on traditional machine learning and hand-craft feature. Although they provide a good direction for solving the issue of error propagation, the patterns that are not contained in the rules are also unrecognizable. Then, [3,20] utilize neural networks [23,26] to capture features automatically. However, their methods are limited by insufficient feature representation, and their extract-then-pairing models have error propagation problems. Recently, joint learning-based models [19,29] has drawn much attention, which effectively relieve the impact of error propagation, and achieve advanced performance.

Semantic Segmentation. At present, deep learning are widely applied in the field of medical data analysis [15,16,24,25]. Semantic segmentation is a fundamental task in computer vision, which has achieved impressive achievements in performing diagnosis and treatment of diseases. U-Net [13] is a frequently-used backbone for semantic segmentation, which can obtain a pixel-level segmentation matrix by fusing multi-scale features. [8,27] first introduce semantic segmentation into NLP tasks, achieving competitive results in their fields. Specifically, they utilize a U-shaped network, to extract high-dimensional features. Meanwhile, high-level semantic information and low-level surface information are combined through a skip-connection mechanism to complement the information lost.

3   Methodology

We will describe our proposed AOPSS, as shown in Fig. 2. It mainly consists of three modules: encoding module (Sect. 3.2), dual-channel semantic segmentation module (Sect. 3.3), and tagging and classification module (Sect. 3.4).

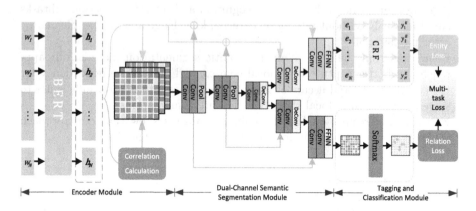

Fig. 2. The overview of our proposed AOPSS framework, which mainly consists of encoding module (**left**), dual-channel semantic segmentation module (**middle**), and tagging and classification module (**right**).

3.1   Problem Definition

Given a review sentence with N tokens: $S = \{w_1, \ldots, w_N\}$, where w_i denotes the $i-th$ token in sequence S. The relation detection subtask is to assign a label $y_{m,n}^R \in \{1, 0\}$ to identify whether a token pair is related. Thus, we regard it as a binary classification task, utilizing the relation segmentation matrix to obtain all possible aspect-opinion pairs. The entity recognition subtask is to assign a tag to each token w_i. Thus, we regard it as a sequence tagging task, converting the entity segmentation matrix into sequence form and then assigning a tag $y_i^E \in \{B - A, I - A, B - P, I - P, O\}$ to each token with CRF [7].

3.2   Encoding Module

BERT Encoder. We adopt pretrained BERT language model to obtain the initial representation for each token. We expect to convert an input review sentence $S = \{w_1, \cdots, w_N\}$ containing N tokens into a sequence of representations:

$$\mathbf{H^e} = \mathbf{BERT}\left(\{w_1, \cdots, w_N\}\right), \tag{1}$$

where the sentence embedding $\mathbf{H^e} = \{h_1, \cdots, h_N\} \in \mathbb{R}^{N \times d}$, and d is the embedding hidden dimension.

Interaction Matrix Calculation. We first utilize the entity feature representation to calculate the token-level relation feature representation, in which we adopt the similarity-based strategy mentioned by [27]. And then, we take the interaction between entity and relation feature representations to obtain the interaction matrix. The similarity-based strategy is calculated by concatenating element-wise similarity, cosine similarity, and bilinear similarity between h_m and h_n defined as follows:

$$\mathbf{F}(h_m, h_n) = [h_m \odot h_n; \cos(h_m, h_n); h_m W h_n] + \mathbf{H^e}, \tag{2}$$

where $\mathbf{F}(h_m, h_n) \in \mathbb{R}^{N \times N \times D}$ is the interaction matrix, token embedding $h_m, h_n \in \mathbf{H^e}$, D is the number of channels, and W is the learnable weight.

3.3 Dual-Channel Semantic Segmentation Module

Semantic segmentation applied in NLP tasks [8,27] have achieved excellent performance. They adopt a serial structure to perform feature extraction by a modified Unet. Inspired by their works, we design the dual-channel semantic segmentation module, a parallel structure, to perform both entity recognition and relation detection subtasks concurrently. It contains one down-sampling process and two up-sampling processes. We show the components of the module in the middle portion of Fig. 2. The down-sampling process consists of two down-sampling blocks, where each down-sampling block contains two separate convolution layers and a max-pooling layer. In addition, each up-sampling process is composed of two down-sampling blocks with skip-connection, where each up-sampling block contains two separate convolution layers and a deconvolution layer.

We first take the interaction matrix $\mathbf{F}(h_m, h_n)$ obtained from the encoding module (Sect. 3.2) as the input and convert it into a high-dimensional matrix to capture the task-invariant features by the down-sampling process. The down-sampling process doubles the D-channel matrix, which expands the receptive field to obtain the contextual semantic information of the segmentation target in the high-dimensional matrix to obtain rich global information. Then, we adopt the dual-channel strategy to perform entity recognition and relation detection subtasks simultaneously with the small-size high-dimensional matrix obtained from the down-sampling process. The two up-sampling processes halve the D-channel matrix, and the small-size high-dimensional matrix is restored to the original size, which can capture task-specific features for both two subtasks.

Since the down-sampling process leads to the loss of information, we adopt the skip-connection mechanism to provide supplementary information. For entity recognition, we introduce entity feature representation, which provides complete local semantic information. Then, we take the interaction between entity feature representation and low-level convolutional features in order to obtain the shared feature representation of entities and relations. For relation detection, we directly concatenate the cropped low-level feature representation. With the help of the skip-connection mechanism, the feature representation can be enhanced and gradient disappearance and network degradation problems can be reduced as well.

3.4 Tagging and Classification Module

After passing the dual-channel semantic segmentation module (Sect. 3.3), we can obtain the entity segmentation matrix and the relation segmentation matrix with $N \times N$ dimensions. Subsequently, the entity segmentation matrix is converted to a token-level entity feature sequence. And then, we take the sequence tagging and the relation classification subtasks, respectively.

Tagging Strategy. For entity recognition, it can be seen as a sequence tagging task. More specifically, we convert the entity segmentation matrix into a token-level entity feature sequence $E = \{e_1, \ldots, e_N\} \in \mathbb{R}^N$, and adopt CRF as our sequence tagging model to calculate the joint probability distribution of the sequence E. It finds the global optimal solution by calculating the correlation of adjacent labels. Formally, given a label sequence $y^E = \{y_1^E, \ldots, y_N^E\}$, CRF aims to calculate the conditional probability as follows:

$$P\left(y^E \mid E\right) = \frac{\exp\left(s\left(E, y^E\right)\right)}{\sum_{y^{E'} \in Y} \exp\left(s\left(E, y^{E'}\right)\right)},$$

$$s\left(E, y^E\right) = \sum_{i=1}^{N}\left(W_{y_{i-1}^E, y_i^E} \cdot e_i + b_{y_{i-1}^E, y_i^E}\right),$$

$$(3)$$

where $W_{y_{i-1}^E, y_i^E}$ and $b_{y_{i-1}^E, y_i^E}$ are learnable weight and bias corresponding to the neighboring labels $\left(y_{i-1}^E, y_i^E\right)$, and Y is the set of all possible tags.

Then, we take the negative log-likelihood function as the loss function of entity recognition subtask as follows:

$$L_{entity} = \log \sum_{y^{E*} \in Y}\left(\exp(s(E, y^{E*})) - s(E, y^E)\right),$$

$$(4)$$

where y^{E*} is the gold label sequence of the entity feature sequence E.

Classification Strategy. For relation detection, it can be seen as a binary classification task. More specifically, we adopt BCELoss[1] to identify a token pair is related or not. Formally, with the relation segmentation matrix $R \in \mathbb{R}^{N \times N}$, we can obtain the predicted relation distribution $y_{m,n}^R \in \{1, 0\}$ by calculating the conditional probability distribution $P\left(y_{m,n}^R \mid (h_m, h_n)\right)$ of each token pair. The loss function of relation detection subtask is constructed by calculating the BCELoss as follows:

$$L_{relation} = \sum_{y_{m,n}^{R*} \in P} \text{BCELoss}\left(y_{m,n}^R, y_{m,n}^{R*}\right),$$

$$(5)$$

where $y_{m,n}^{R*} = P\left(y_{m,n}^{R*} \mid (w_m, w_n)\right)$ denotes the gold relation distribution of relation segmentation matrix, and P is the set of all possible relations.

---

[1] $\text{BCELoss}(x, y) = -(y\log x + (1 - y)\log(1 - x))$.

Training. The loss function L is to guide the model during training, which consists of two parts: L_{entity} and $L_{relation}$. Our training object is to minimize the loss function L as follows:

$$L = \lambda L_{entity} + (1 - \lambda) L_{relation}, \tag{6}$$

where λ is the balance weight for joint learning strategy.

4 Experiments

4.1 Dataset

We evaluate our model on four public datasets annotated by [3], including 14Lap, 14Res, 15Res, and 16Res, which annotate aspect and opinion terms in pairs based on the original SemEval datasets derived from the SemEval challenge [9–11]. The statistics of datasets are shown in Table 1.

Table 1. Statistics of the experimental datasets, where #Sentences, #Aspects, #Opinions, and #Pairs denote the number of sentences, aspect terms, opinion terms, and aspect-opinion pairs.

Datasets	#Type	#Sentences	#Aspects	#Opinions	#Pairs
	Train	1259	2064	2098	2356
Res14	Dev	315	487	506	580
	Test	493	851	866	1008
	Train	899	1257	1270	1452
Lap14	Dev	225	332	313	383
	Test	332	467	478	547
	Train	603	871	966	1038
Res15	Dev	151	205	226	239
	Test	325	436	469	493
	Train	863	1213	1329	1421
Res16	Dev	216	298	331	348
	Test	328	456	485	525

4.2 Experimental Settings

We fine-tune the uncased BERT-base[2] pretrained model and set the hidden dimension d to 768. We adopt AdamW as the parameter optimizer with 0.1 warmup rate. The fine-tuning learning rate is set to $2e - 5$, and the training learning rate is set to $1e - 4$. In addition, the maximum sequence length is set to 100, and the batch size is set to 12. The balance weight λ is set to 0.1. We adjust the above hyper-parameters with the cross-validation method and evaluate our model with F1-score. Finally, we perform five experiments with random initialization and report the average experimental results. Our model is trained on the GeForce GTX 1080 Ti GPU.

---

[2] https://github.com/google-research/bert.

4.3   Baselines

We compare AOPSS with both pipeline-based models and joint learning-based models. We select two representative pipeline models as the baselines:

- **CMLA+CGCN** first employs **CMLA** [14] to jointly co-extract aspects and opinions, and then introduces **CGCN** [30] to perform pairing.
- **RINANTE+IOG** first utilizes **RINANTE** [2], a semi-supervised model, to extract aspects and opinions, and then it adopts **IOG** [3] to perform pairing.

Since joint-learning can solve the error propagation problem well, it achieves advanced performance in the AOPE. We select seven models as our baselines:

- **SpanMlt** [29] proposes a multi-task learning model, which extracts all candidate spans first and then performs entity recognition and relation detection.
- **GTS** [19] designs the Grid Tagging Scheme, composed of an arbitrary encoder and a designed inference strategy, to assign entity and relation tags.
- **SDRN** [1] proposes a synchronous double-channel recurrent network, which utilizes two channels for entity recognition and relation detection subtasks.
- **STER** [28] designs a group of gated RNNs networks to track all entities of a sentence in parallel.
- **ESGCN** [18] proposes an edge-enhanced syntactic graph convolutional network for enhancing the extraction and pairing of aspect and opinion terms.
- **SynFue+LAGCN** [17] incorporates rich syntactic features and adopts a high-order scoring method to calculate potential aspect-opinion pairs.
- **QDSL** [4] first conducts aspect term extraction by a question generation model and then obtains aspect-opinion pairs by conducting aspect-specified opinion terms extraction.

4.4   Main Results

The main results are shown in Table 2. We observe that our proposed AOPSS achieves the best results on eleven of twelve metrics.

For pipeline-based models, they adopt an extract-then-pairing manner, which can cause error propagation and lead to significant performance degradation. For joint learning-based models, SpanMlt, GTS, and SDRN employ the first order scoring mechanism that considers only one potential aspect-opinion pair when calculating the pairing score. For the four datasets, the RF, AF, and OF of AOPSS is 5.53%, 3.25%, and 2.86% higher than these methods on average. Meanwhile, ESGCN and SynFue+LAGCN utilize the higher order scoring mechanism to improve the calculation strategy of the pairing score. For the four datasets, the RF of AOPSS is 2.33% higher than these methods on average. Furthermore, STER provides a powerful baseline in entity recognition subtask. For the four datasets, the RF, AF, and OF of AOPSS is 2.81%, 0.85%, and 1.03% higher than STER on average. It takes advantage of the entity tracking network, which performs well in entity recognition, but has the limitation of its LSTM-based model to capture long-distance information. Simultaneously, QDSL establishes

Table 2. The experimental results on the annotation datasets of [3]. RF, AF, and OF represent the F1-score (%) of relation detection, aspect extraction, and opinion extraction, respectively. We conduct five experiments with random initialization on each model and then report the average experimental results.

Model	Lap14			Res14			Res15			Res16		
	RF	AF	OF	RF	AF	OF	RF	AF	OF	RF	AF	OF
CMLA+CGCN	53.03	-	-	63.17	81.22	80.48	55.76	76.03	74.67	62.70	-	-
RINANTE+IOG	57.10	-	-	67.74	81.34	83.33	59.16	73.38	75.40	-	-	-
SpanMlt	65.75	80.78	79.71	72.72	84.26	84.11	61.06	77.71	78.47	69.58	80.95	84.92
GTS	64.61	78.73	77.86	74.65	83.10	84.49	68.29	78.11	78.25	74.31	82.31	84.30
SDRN	63.63	78.28	76.48	73.08	84.70	84.01	68.00	76.30	79.71	75.02	78.93	86.22
STER	67.64	81.06	81.03	74.96	85.85	85.89	69.30	80.44	80.91	75.89	83.64	85.66
ESGCN	68.69	-	-	76.22	-	-	68.34	-	-	75.20	-	-
SynFue+LAGCN	68.88	-	-	76.62	-	-	68.91	-	-	76.59	-	-
QDSL	70.20	-	-	**78.05**	-	-	71.22	-	-	77.28	-	-
AOPSS	**70.84**	**82.22**	**82.49**	77.41	**86.44**	**86.73**	**72.66**	**81.14**	**81.60**	**78.13**	**84.59**	**86.77**

a strong baseline in relation detection subtask. For the four datasets, the RF of AOPSS is 0.57% higher than QDSL on average. It adopts a question generation model to capture aspect terms first and then match them with opinion terms. The results heavily depend on its question generation model.

To sum up, our proposed method achieves state-of-the-art performance on four datasets and takes advantage of its structural advantages to reduce gradient disappearance and network degradation problems effectively.

4.5 Ablation Study

We conduct ablation experiments to investigate the effectiveness of each module by comparing the RF-score. The results are shown in Table 3.

– **w/o BERT** means using GloVe[3] with Bi-LSTM to obtain the initialized embedding.
– **w/o Interaction Matrix** indicates that only the relation feature representation is used as input of the dual-channel semantic segmentation module without considering the interaction with the entity feature representation.

Table 3. The ablation study results on the four benchmark datasets, here we present RF-score (%) to illustrate the impact of each module.

	14Lap	14Res	15Res	16Res	Avg.
AOPSS	**70.84**	**77.41**	**72.66**	**78.13**	**74.76**
w/o BERT	64.81	73.69	66.05	73.28	69.46
w/o Interaction Matrix	70.41	76.52	72.02	77.24	73.79
w/o Entity Connection	67.63	75.97	68.45	76.50	72.14
w/o Dual-Channel	69.84	76.22	71.09	76.47	73.41

---

[3] https://nlp.stanford.edu/projects/glove.

- **w/o Entity Connection** implies removing the entity feature representation of the dual-channel semantic segmentation module.
- **w/o Dual-Channel** represents adopting a classify-then-extract strategy for semantic segmentation. It first obtains a token-level relation segmentation matrix for relation detection and then converts it into a sequence form and assigns a tag to each token with CRF for entity recognition.

In conclusion, our model has a performance degradation of 0.97 % – 5.3% for ablating each module, which demonstrates that each component of AOPSS is designed reasonably and effectively.

4.6   Closer Analysis

We conduct an in-depth analysis of the above experiments and list five representative cases, as shown in Table 4.

Table 4. Five representative cases selected from the annotated test dataset of [3].

Reviews	Ground-Truth	SDRN	STER	AOPSS
With the softwares supporting the use of other OS makes it much better.	(softwares, better) (OS, better)	(softwares, better)✓ (other OS, -)✗	(softwares, better)✓	(softwares, better)✓ (OS, better)✓
The food is great (big selection, reasonable prices) and the drinks are really good.	(food, great) (selection, big) (prices, reasonable) (drinks, good)	(food, great)✓ (food, big)✗ (prices, reasonable)✓ (drinks, good)✓	(food, great)✓ (prices, reasonable)✓ (drinks, good)✓	(food, great)✓ (selection, big)✓ (prices, reasonable)✓ (drinks, good)✓
The nicest part is the low heat output and ultra quiet operation.	(heat output, low) (heat output, nicest) (operation, quiet) (operation, nicest)	(heat output, low)✓ (operation, quiet)✓	(heat output, low)✓ (heat output, nicest)✓ (operation, quiet)✓	(heat output, low)✓ (heat output, nicest)✓ (operation, quiet)✓ (operation, nicest)✓
The Mini's body hasn't changed since late 2010- and for a good reason.	(body, hasn't changed) (body, good)	(-, -)✗	(-, hasn't changed)✗	(body, hasn't changed)✓ (body, good)✓
I can say that I am fully satisfied with the performance that the computer has supplied.	(performance, satisfied)	(performance, satisfied)✓ (performance, fully)✗	(performance, satisfied)✓ (performance, fully)✗	(performance, satisfied)✓

We compare the case study results of AOPSS with SDRN and STER to show the superior performance of our proposed method. From the experimental results, SDRN tends to give solutions in terms of syntactic structure, but the performance is limited by the first-order scoring method. In addition, STER is limited by its LSTM-based model, which has deficiencies in capturing long-distance information.

Our proposed method provides a good solution to the above-mentioned questions. The first review contains an overlap problem, where AOPSS can extract the aspect term *"OS"* and make pairings precisely. The results demonstrate that AOPSS is more adapted to annotated datasets. The second review includes a one-to-one pairing problem, but the relative positions of aspect-opinion pairs are not fixed in the sentence. The results prove that AOPSS can be more flexible to accommodate different syntactic structures. The third and fourth reviews contain two overlap problems involving complex entities and relations. AOPSS can better capture both local context and global interdependency for the two subtasks. Specifically, it can not only extract appropriate phrases, such as *"heat output"* in the third review and *"hasn't changed"* in the fourth review, but also capture long-distance relation pairs, such as (*"operation"*, *"nicest"*) in the third review. For the fifth review, both *"fully"* and *"satisfied"* can describe *"performance"* syntactically, but *"satisfied"* is more appropriate in semantics. The results illustrate that AOPSS can better understand the semantic information of the review sentence.

5   Conclusions

In this paper, we propose AOPSS, a joint learning framework, extracting aspect terms and their corresponding opinion terms in pairs. It provides a high capability, easy expansibility, and good interpretability scheme for solving the AOPE task. Concretely, we regard the AOPE task as two subtasks: entity recognition and relation detection, which can be solved by sequence tagging and binary classification, respectively. The two subtasks share the same down-sampling process to capture task-invariant features and use different up-sampling processes to obtain task-specific features. In addition, since the two subtasks are highly dependent, we take the interaction between entity and relation feature representation to enhance the mutual heuristic effect for the two subtasks. Experimental results on four benchmark datasets show that AOPSS outperforms other state-of-the-art baselines. Through closer analysis, we prove the effectiveness of our approach. In our future works, we hope to expand our work to other NLP fields to explore the application of semantic segmentation further.

Acknowledgements. This work is supported by the National Natural Science Foundation of China under grant No. 61872163 and 61806084, Jilin Province Key Scientific and Technological Research and Development Project under grant No. 20210201131GX, and Jilin Provincial Education Department project under grant No. JJKH20190160KJ.

References

1. Chen, S., Liu, J., Wang, Y., Zhang, W., Chi, Z.: Synchronous double-channel recurrent network for aspect-opinion pair extraction. In: Proceedings of the 58th Annual Meeting of the Association for Computational Linguistics, pp. 6515–6524 (2020)
2. Dai, H., Song, Y.: Neural aspect and opinion term extraction with mined rules as weak supervision. In: Proceedings of the 57th Annual Meeting of the Association for Computational Linguistics, pp. 5268–5277 (2019)
3. Fan, Z., Wu, Z., Dai, X.Y., Huang, S., Chen, J.: Target-oriented opinion words extraction with target-fused neural sequence labeling. In: Proceedings of the 2019 Conference of the North American Chapter of the Association for Computational Linguistics: Human Language Technologies, pp. 2509–2518 (2019)
4. Gao, L., Wang, Y., Liu, T., Wang, J., Zhang, L., Liao, J.: Question-driven span labeling model for aspect-opinion pair extraction. In: Proceedings of the Thirty-Fifth AAAI Conference on Artificial Intelligence, pp. 12875–12883 (2021)
5. Hu, M., Liu, B.: Mining and summarizing customer reviews. In: Proceedings of the Tenth ACM SIGKDD International Conference on Knowledge Discovery and Data Mining, pp. 168–177 (2004)
6. Klinger, R., Cimiano, P.: Bi-directional inter-dependencies of subjective expressions and targets and their value for a joint model. In: Proceedings of the 51st Annual Meeting of the Association for Computational Linguistics, pp. 848–854 (2013)
7. Lafferty, J.D., McCallum, A., Pereira, F.C.N.: Conditional random fields: Probabilistic models for segmenting and labeling sequence data. In: Proceedings of the Eighteenth International Conference on Machine Learning, pp. 282–289 (2001)
8. Liu, Q., Chen, B., Lou, J.G., Zhou, B., Zhang, D.: Incomplete utterance rewriting as semantic segmentation. In: Proceedings of the 2020 Conference on Empirical Methods in Natural Language Processing, pp. 2846–2857 (2020)
9. Pontiki, M., et al.: SemEval-2016 task 5: Aspect based sentiment analysis. In: Proceedings of the 10th International Workshop on Semantic Evaluation, pp. 19–30 (2016)
10. Pontiki, M., Galanis, D., Papageorgiou, H., Manandhar, S., Androutsopoulos, I.: SemEval-2015 task 12: Aspect based sentiment analysis. In: Proceedings of the 9th International Workshop on Semantic Evaluation, pp. 486–495 (2015)
11. Pontiki, M., Galanis, D., Pavlopoulos, J., Papageorgiou, H., Androutsopoulos, I., Manandhar, S.: SemEval-2014 task 4: Aspect based sentiment analysis. In: Proceedings of the 8th International Workshop on Semantic Evaluation, pp. 27–35 (2014)
12. Popescu, A.M., Etzioni, O.: Extracting product features and opinions from reviews. In: Proceedings of Human Language Technology Conference and Conference on Empirical Methods in Natural Language Processing, pp. 339–346 (2005)
13. Ronneberger, O., Fischer, P., Brox, T.: U-net: Convolutional networks for biomedical image segmentation. In: Medical Image Computing and Computer-Assisted Intervention, pp. 234–241 (2015)
14. Wang, W., Pan, S.J., Dahlmeier, D., Xiao, X.: Coupled multi-layer attentions for co-extraction of aspect and opinion terms. In: Proceedings of the Thirty-First AAAI Conference on Artificial Intelligence, pp. 3316–3322 (2017)
15. Wang, Y., Chen, W., Pi, D., Yue, L.: Adaptive multi-hop reading on memory neural network with selective coverage mechanism for medication recommendation. Acta Electron. Sin. **50**(4), 943–953 (2022)

16. Wang, Y., Chen, W., Pi, D., Yue, L., Xu, M., Li, X.: Multi-Hop Reading on Memory Neural Network with Selective Coverage for Medication Recommendation, In: Proceedings of the 30th ACM International Conference on Information & Knowledge Management, pp. 2020–2029 (2021)
17. Wu, S., Fei, H., Ren, Y., Ji, D., Li, J.: Learn from syntax: Improving pair-wise aspect and opinion terms extraction with rich syntactic knowledge. In: Proceedings of the Thirtieth International Joint Conference on Artificial Intelligence, pp. 3957–3963 (2021)
18. Wu, S., Fei, H., Ren, Y., Li, B., Li, F., Ji, D.: High-order pair-wise aspect and opinion terms extraction with edge-enhanced syntactic graph convolution. IEEE/ACM Trans. Audio Speech Lang. Process. **29**, 2396–2406 (2021)
19. Wu, Z., Ying, C., Zhao, F., Fan, Z., Dai, X., Xia, R.: Grid tagging scheme for aspect-oriented fine-grained opinion extraction. In: Findings of the Association for Computational Linguistics: EMNLP 2020, pp. 2576–2585 (2020)
20. Xu, H., Liu, B., Shu, L., Yu, P.S.: Double embeddings and CNN-based sequence labeling for aspect extraction. In: Proceedings of the 56th Annual Meeting of the Association for Computational Linguistics, pp. 592–598 (2018)
21. Yang, B., Cardie, C.: Joint inference for fine-grained opinion extraction. In: Proceedings of the 51st Annual Meeting of the Association for Computational Linguistics, pp. 1640–1649 (2013)
22. Yin, Y., Wei, F., Dong, L., Xu, K., Zhang, M., Zhou, M.: Unsupervised word and dependency path embeddings for aspect term extraction. In: Proceedings of the 25th International Joint Conference on Artificial Intelligence, pp. 2979–2985 (2016)
23. Yue, L., Shi, Z., Han, J., Wang, S., Chen, W., Zuo, W.: Multi-factors based sentence ordering for cross-document fusion from multimodal content. Neurocomputing **253**, 6–14 (2017)
24. Yue, L., Tian, D., Chen, W., Han, X., Yin, M.: Deep learning for heterogeneous medical data analysis. World Wide Web **23**(5), 2715–2737 (2020)
25. Yue, L., Zhao, H., Yang, Y., Tian, D., Zhao, X., Yin, M.: A mimic learning method for disease risk prediction with incomplete initial data. In: International Conference on Database Systems for Advanced Applications, pp. 392–396 (2019)
26. Zhang, C., et al.: Towards better generalization for neural network-based sat solvers. In: Pacific-Asia Conference on Knowledge Discovery and Data Mining, pp. 199–210 (2022)
27. Zhang, N., et al.: Document-level relation extraction as semantic segmentation. In: Proceedings of the Thirtieth International Joint Conference on Artificial Intelligence, pp. 3999–4006 (2021)
28. Zhang, Y., Peng, T., Han, R., Han, J., Yue, L., Liu, L.: Synchronously tracking entities and relations in a syntax-aware parallel architecture for aspect-opinion pair extraction. Appli. Intell. 1–16 (2022)
29. Zhao, H., Huang, L., Zhang, R., Lu, Q., Xue, H.: SpanMlt: A span-based multi-task learning framework for pair-wise aspect and opinion terms extraction. In: Proceedings of the 58th Annual Meeting of the Association for Computational Linguistics, pp. 3239–3248 (2020)
30. Zhou, Y., et al.: Graph convolutional networks for target-oriented opinion words extraction with adversarial training. In: 2020 International Joint Conference on Neural Networks, pp. 1–7 (2020)

Dual Graph Convolutional Networks for Document-Level Event Causality Identification

Yang Liu[1,2], Xiaoxia Jiang[2], Wenzheng Zhao[1], Weiyi Ge[2], and Wei Hu[1(✉)]

[1] State Key Laboratory for Novel Software Technology, Nanjing University,
Nanjing, Jiangsu, China
`whu@nju.edu.cn`
[2] Science and Technology on Information Systems Engineering Laboratory,
Nanjing, Jiangsu, China

Abstract. Event causalities organize events into a graph according to causal logics, which assists humans in decision making by causal reasoning among events. Despite many efforts to identify event causalities, most of them assume that only one causality exists in a sentence or causalities only occur in adjacent sentences, leading to the incapability of detecting multiple causalities or document-level causalities. In this paper, we propose a novel model for document-level event causality identification named DocECI. We define two heterogeneous document graphs, namely text structure graph and mention relation graph, and encode them with relational graph convolutional networks, which gradually aggregate the information of multi-granular nodes in a cascade manner and capture the causality patterns. Experiments on a benchmark dataset show that DocECI outperforms existing models by a significant margin. Moreover, a new experiment is conducted on causality direction identification, which is overlooked by existing models.

Keywords: Event causality · Document graph · R-GCN

1 Introduction

As an important semantic relation, event causalities organize events into an event graph according to causal logics, assisting humans to make better decisions in event prediction [21], public opinion monitoring [26] and many other scenarios. A causality is typically defined as a relation between two events if the occurrence of one leads to the other, which are called *cause* and *effect*, respectively. Causalities can be further divided into *implicit* and *explicit*, as well as *intra-sentence* and *cross-sentence*, depending on whether there is a causal connective and whether cause and effect appear within the same sentence.

Existing works on identifying event causalities can be classified into three categories [35]: pattern models, statistical models, and deep models. The former two suffer from the difficulty of feature engineering and limited expressiveness.

B. Li et al. (Eds.): APWeb-WAIM 2022, LNCS 13422, pp. 114–128, 2023.
https://doi.org/10.1007/978-3-031-25198-6_9

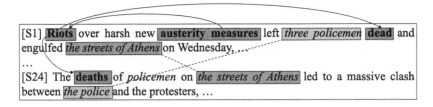

Fig. 1. A real example excerpted from the ESC dataset [2]. The document is annotated with event (bold font), participant and location (*italic font*) mentions, coreference relations (dashed lines), intra-sentence and cross-sentence causalities (solid lines with arrows, where arrows indicate the directions of causalities from causes to effects).

Deep models can capture patterns of implicit causalities, benefiting from the powerful expression capability of neural networks.

However, there are still many challenges such as multiple and document-level causality identification. Most existing works ignore the multiple causalities by making an impractical assumption that there is only one causality in a sentence [35]. However, as a real example shown in Fig. 1, there are two causalities in S1: **austerity measures** result in **riots**, which lead to three policemen **dead**. There are two common ways to handle multiple causalities [5]: The first method is to encode the sentence once for each causality, which requires huge computation. The second splits the sentence into several sub-instances with only one causality, but this makes the information in the sub-instances incomplete. Recently, some researchers [5,20] have attempted to address this issue by treating it as a sequence labeling task. However, their methods are infeasible for document-level causalities.

As for document-level causality identification, it is more challenging due to the long spans in documents. As shown in Fig. 1, there is a causality between S1 and S24: **riots** led to **deaths**. In fact, **dead** in S1 and **deaths** in S24 are coreferential due to the participant (*policemen*) and the location (*the streets of Athens*) are both coreferential. Although some works [8,27] combine statistical models with integer linear programming to identify document-level causalities, they face the problem of costly feature engineering and weak expressiveness.

Aiming at these two challenges, in this paper we propose a **Doc**ument-level **E**vent **C**ausality **I**dentification model named DocECI. Specifically, given a document, we first construct two heterogeneous document graphs called text structure graph and mention relation graph, and use relational graph convolutional networks (R-GCNs) [29] to model them. Then, we aggregate event and context representations through a context-aware aggregation layer, and identify all causalities in the document.

We evaluate our model on a benchmark dataset called ESC [2]. Our experimental results show that DocECI achieves the state-of-the-art performance. We also validate the indispensability of all components in DocECI through an ablation study. Additionally, we conduct a new experiment on causality direction identification, which is overlooked by the existing works.

To summarize, our main contributions in this paper are threefold:

- We study complex event causality identification. We find by our empirical study that a comprehensive understanding of event context and coreferential relations between event mentions can benefit this task.
- We propose a document-level event causality identification model named DocECI, which constructs a text structure graph and a mention relation graph to model global causality information at the document level.
- We conduct extensive experiments and show that DocECI sets up a new state-of-the-art for event causality identification. We also conduct a new experiment overlooked by the existing work on causality direction identification.

2   Related Work

2.1   Event Causality Identification

Early studies rely on manually-defined linguistic patterns to identify event causalities [9,15,16,28]. Their solutions are laborious and limited to small corpora, so they may not identify implicit causalities. Later works combine patterns with machine learning to improve the performance and reduce manual labor [1,10,38]. Gao et al. [8] propose a constraint-based method, which trains classifiers with linguistic features and improves the performance via integer linear programming. Other works extend the task of event causality identification to the joint identification task of event causalities and other relations, such as temporal relations [23,25] and event-arguments relations [27].

Intra-sentence Causality. Most existing works leverage deep models with external knowledge to identify intra-sentence causalities. De Silva et al. [6] combine linguistic features from WordNet with CNN. Li and Mao [19] propose a knowledge-oriented CNN to learn additional linguistic features from external knowledge bases. Liu et al. [22] present a knowledge enhanced model, which leverages ConceptNet to enrich event representations and uses a masking mechanism to mine event-agnostic, context-specific patterns. Zuo et al. [41] propose a knowledge enhanced data augmentation model KnowDis, which obtains unlabeled data via distant supervision and employs self-training to train models. Moreover, some works convert the event causality identification tasks into sequence labeling. They cope well with multiple causalities [5,20], but are difficult to identify cross-sentence causalities, especially causalities of long distance.

Cross-Sentence Causality. Kruengkrai et al. [18] propose a multi-column CNN, which takes dependency paths among causes and effects as background knowledge. Jin et al. [14] propose a cascade multi-structure neural network, which employs CNN to capture important features and causality patterns. These models are sentence-level. They focus on cross-sentence causalities between adjacent sentences, but pay less attention to document-level causalities.

2.2 Document-Level Relation Extraction

Most existing document-level relation extraction models aim to extract the relations among entities [12,24,30,32,40]. Specifically, the document graph-based methods [11,13] usually convert a document into a graph by taking entity mentions as nodes and relations as edges, and apply graph neural networks (GNNs) to model it. Subsequent works [4,37] improve these methods by modifying the graph and model structures. They construct multi-granular graph nodes, including words, mentions, entities, etc., and define different kinds of edges with heuristic rules.

3 Proposed Model

We define the document-level event causality identification task as follows. For an annotated document $\mathcal{D} = [w_1, \ldots, w_{n_w}]$ with event mentions $\mathcal{M} = [m_1, \ldots, m_{n_m}]$, where w_i $(1 \leq i \leq n_w)$ denotes the i-th word in the document, $m_j = [w_q, \ldots, w_p]$ $(1 \leq j \leq n_m)$ denotes the j-th event mention with $p - q + 1$ words, given any event pair (m_i, m_j) in $\mathcal{M}$, we aim to predict its relation $r \in \{CauseEffect, None\}$, i.e., whether there exists a causality between m_i and m_j or not.

The framework of our DocECI can be divided into five layers: (1) Encoding Layer transforms input words into dense vector representations; (2) **Text Structure Modeling Layer** (TSML) models word nodes and sentence nodes with the structural information of the document; (3) **Mention Relation Modeling Layer** (MRML) models mention nodes and sentence nodes with the relational information among different mentions to obtain potential patterns of causalities; (4) **Context-aware Aggregation Layer** (CAL) aggregates event representations with context representations from mentions and sentences. (5) Output Layer predicts a possible relation for each event mention pair.

3.1 Encoding Layer

In the encoding layer, we use BERT [7] and a BiLSTM network to obtain initial word representations. Given that the length of the document usually exceeds the maximum encoding length of BERT, we split a document into several segments and encode them separately:

$$[\overline{\mathbf{w}}_1, \ldots, \overline{\mathbf{w}}_{n_k}] = \text{BERT}([w_1, \ldots, w_{n_k}]), \tag{1}$$

where $\overline{\mathbf{w}}_i \in \mathbb{R}^{d_w}$, d_w denotes the size of word representations from BERT, and $[w_1, \ldots, w_{n_k}]$ denotes the k-th segment of the document.

This split operation results in the break of contextual information. Therefore, we use a BiLSTM network to integrate contextual information among different segments:

$$\mathbf{W} = [\mathbf{w}_1, \ldots, \mathbf{w}_{n_w}] = \mathcal{F}_1\Big(\text{BiLSTM}([\overline{\mathbf{w}}_1, \ldots, \overline{\mathbf{w}}_{n_w}])\Big), \tag{2}$$

where $\mathbf{w}_i \in \mathbb{R}^{d_w}$, $\mathcal{F}_1 : \mathbb{R}^{2 \times d_w} \rightarrow \mathbb{R}^{d_w}$ denotes to a linear function, and n_w denotes the length of the document.

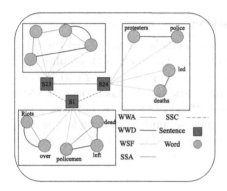 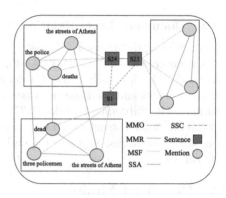

Fig. 2. Text structure graph.   **Fig. 3.** Mention relation graph.

3.2   Text Structure Modeling Layer

Inspired by Zhang et al. [37], we take words and sentences as nodes and define five types of edges to depict the relations among different nodes. Then, we construct an undirected text structure document graph $\mathcal{G} = (\mathcal{V}, \mathcal{E}, \mathcal{C})$ shown in Fig. 2, where $\mathcal{V}, \mathcal{E}, \mathcal{C}$ denotes the sets of nodes, edges and edge types, respectively. The five edge types are:

- WWA: Two word nodes are linked with an adjacency edge if they are adjacent within a sentence, which captures the word sequential information.
- WWD: A dependency edge is used to link two word nodes with an intra-sentence dependency relation, which captures the shallow syntactic structure. We use the spaCy tool[1] to obtain dependency relation.
- WSF: An affiliation edge between a word node and a sentence node if the word is in the sentence, depicting the hierarchical information of the document.
- SSA: Two sentence nodes are linked with an adjacency edge if they are adjacent, which captures the sentence sequential information.
- SSC: Two non-adjacent sentence nodes are linked with a complement edge to transmit the information of word nodes in distant sentences by 1-hop relation.

We use R-GCN to model this graph. For any word node $v_i \in \mathcal{V}$, we use the word representation $\mathbf{w}_i$ in $\mathbf{W}$ to initialize its representation $\mathbf{v}_i^{(0)}$. For sentence node $v_i = [w_q, \ldots, w_p]$, a max-pooling operation is applied over all corresponding word nodes to obtain its representation: $\mathbf{v}_i^{(0)} = \max([\mathbf{w}_q, \ldots, \mathbf{w}_p])$. At the $(l{+}1)$-th layer, R-GCN updates the representation of a node by aggregating its adjacent node representations through the message passing strategy. We also adopt a gating mechanism [37] to selectively remember the information contained in the node to prevent the over-smoothing problem of R-GCN [17]:

---

[1] https://spacy.io/.

$$\mathbf{u}_i^{(l)} = \sum_{c \in \mathcal{C}} \sum_{j \in \mathcal{N}_i^c} \frac{1}{|\mathcal{N}_i^c|} \mathbf{W}_c^{(l)} \mathbf{v}_j^{(l)} + \mathbf{W}_0^{(l)} \mathbf{v}_i^{(l)},$$

$$g_i^{(l)} = \text{sigmoid} \left(\mathcal{F}_2 ([\mathbf{u}_i^{(l)}; \sigma(\mathbf{u}_i^{(l)})]) \right), \tag{3}$$

$$\mathbf{v}_i^{(l+1)} = g_i^{(l)} \odot \tanh(\mathbf{u}_i^{(l)}) + (1 - g_i^{(l)}) \odot \sigma(\mathbf{u}_i^{(l)}),$$

where $\mathcal{N}_i^c$ is the set of nodes adjacent to v_i with edge type $c \in \mathcal{C}$, $\mathbf{W}_c^{(l)} \in \mathbb{R}^{d_w \times d_w}$ denotes an edge type-specific weight matrix and $\mathbf{W}_0^{(l)} \in \mathbb{R}^{d_w \times d_w}$ denotes the weight matrix for self-connection edge, $\mathcal{F}_2 : \mathbb{R}^{2 \times d_w} \to \mathbb{R}^{d_w}$ denotes to a linear function, $\sigma(\cdot)$ denotes the activation function, and $\odot$ stands for element-wise multiplication. Finally, we can get all node representations in $\mathcal{G}$: $\mathbf{V} = [\mathbf{v}_1^{(L)}, \dots, \mathbf{v}_{|\mathcal{V}|}^{(L)}]$, where L is the R-GCN layers.

3.3 Mention Relation Modeling Layer

Through the previous layer, the node representations contain the hierarchical, sequential, syntactic information, but miss the potential relational information among mentions. In this layer, we take sentences and mentions as nodes and define three types of edges in addition to SSA and SSC. Then, we construct an undirected mention relation document graph $\bar{\mathcal{G}} = (\bar{\mathcal{V}}, \bar{\mathcal{E}}, \bar{\mathcal{C}})$, where $\bar{\mathcal{V}}$ shown in Fig. 3, $\bar{\mathcal{E}}$, $\bar{\mathcal{C}}$ represents the sets of nodes, edges and edge types, respectively. The three new edge types are:

- MMO: Two mention nodes in the same sentence are linked with a co-occurrence edge, indicating the potential correlation between them.
- MMR: Two coreferential mention nodes are linked with a coreference edge, which helps model the relations among mentions across sentences. We assume that the coreference information is annotated beforehand.
- MSF: An affiliation edge is used to link a mention node and a sentence node if the mention is in the sentence. Unlike $\mathcal{G}$, affiliation edges in $\bar{\mathcal{G}}$ propagate the contextual information contained in the sentence node to the mention node.

We use R-GCN to model this graph as well. First, we leverage the idea of dense connection [34,37] and concatenate the representations of the 0-th and L-th layer in Sect. 3.2 with a linear transformation $\mathcal{F}_3 : \mathbb{R}^{2 \times d_w} \to \mathbb{R}^{d_w}$ to obtain the new representations:

$$\hat{\mathbf{V}} = [\hat{\mathbf{v}}_1, \hat{\mathbf{v}}_2, \dots, \hat{\mathbf{v}}_{|\mathcal{V}|}]$$
$$= \left[\mathcal{F}_3([\mathbf{v}_1^{(0)}; \mathbf{v}_1^{(L)}]), \mathcal{F}_3([\mathbf{v}_2^{(0)}; \mathbf{v}_2^{(L)}]), \dots, \mathcal{F}_3([\mathbf{v}_{|\mathcal{V}|}^{(0)}; \mathbf{v}_{|\mathcal{V}|}^{(L)}]) \right]. \tag{4}$$

We adopt $\hat{\mathbf{V}} = [\hat{\mathbf{v}}_1, \dots, \hat{\mathbf{v}}_{|\mathcal{V}|}]$ to initialize the node representations: (1) For mention node $\bar{v}_i = [w_q, \dots, w_p]$, a mean-pooling operation is applied; (2) For sentence node, we directly get its node representation from $\hat{\mathbf{V}}$. Then, we update the representations in the same way as previous layers. Finally, we obtain the node representations $\bar{\mathbf{V}} = [\bar{\mathbf{v}}_1^{\bar{L}}, \dots, \bar{\mathbf{v}}_{|\bar{\mathcal{V}}|}^{\bar{L}}]$, where $\bar{L}$ is the number of R-GCN layers.

3.4   Context-Aware Aggregation Layer

The representations of nodes in $\bar{\mathcal{G}}$ contain different types of information. Mention nodes contain the semantic information of the corresponding text mentions and the relational information with other mentions, while sentence nodes contain the contextual information, which is important for the relation reasoning among mentions. To further integrate the information in the representations of nodes in $\bar{\mathcal{G}}$, we propose an aggregation methods based on multi-head attention [31], which can capture different information in different subspaces.

The aggregation method organizes sentence representations from $\bar{\mathbf{V}}$ into the matrix $\mathbf{H}_s \in \mathbb{R}^{n_s \times d_w}$, where n_s denotes the number of sentences. Then, we take $\mathbf{H}_s$ as the query, key and value matrices to obtain the new sentence representation matrix $\hat{\mathbf{H}}_s$, which contains the global contextual information:

$$\hat{\mathbf{H}}_s = \text{MultiHead}(\mathbf{H}_s, \mathbf{H}_s, \mathbf{H}_s). \tag{5}$$

Finally, we concatenate the representation $\mathbf{h}_m$ of a mention from $\bar{\mathbf{V}}$ and the representation $\hat{\mathbf{h}}_s \in \hat{\mathbf{H}}_s$ of the sentence in which the mention lies to form the final representation of mention:

$$\hat{\mathbf{h}}_m = [\mathbf{h}_m; \hat{\mathbf{h}}_s]. \tag{6}$$

3.5   Output Layer

In this layer, we aim to predict the relation of an event pair. For each event mention pair (m_i, m_j), we first concatenate two event representations $\hat{\mathbf{h}}_i$, $\hat{\mathbf{h}}_j$ with their relative distance representation $\mathbf{d}_{(i,j)}$ as its ultimate representation, and then use a fully-connected layer with softmax to predict its relation. The relative distance of two events is the distance of their belonging sentences. Finally, we use a fully-connected layer with softmax to predict its relation:

$$\mathbf{y}_{(i,j)} = \text{softmax}\Big(\text{FC}([\hat{\mathbf{h}}_i; \hat{\mathbf{h}}_j; \mathbf{d}_{(i,j)}])\Big),$$
$$y_{(i,j)} = \text{argmax}(\mathbf{y}_{(i,j)}). \tag{7}$$

We use cross-entropy as the training loss:

$$J = - \sum_{\substack{m_i, m_j \in \mathcal{M} \\ i \neq j}} \Big[y_{(i,j)}^* \log(p_{(i,j)}) + (1 - y_{(i,j)}^*) \log(1 - p_{(i,j)}) \Big], \tag{8}$$

where $y_{(i,j)}^* \in \{0, 1\}$ denotes the true relation of (m_i, m_j), and $p_{(i,j)}$ denotes the probability of $y_{(i,j)} = 1$.

Table 1. Statistics of the ESC dataset.

	Intra-sentence		Cross-sentence	
	CauseEffect	None	CauseEffect	None
Training & Testing	1,859	4,662	3,818	34,475
Development	243	1,130	772	7,632

4 Experiments

4.1 Dataset and Experimental Settings

Our experiments are conducted on a benchmark dataset named EventStory-Line Corpus (abbr. ESC) [2]. ESC is the largest publicly available dataset for document-level event causality identification, which consists of 258 documents from 22 different topics, such as natural disasters and crimes. As suggested in [8], we use the last two topics as the development set, and conduct a 5-fold cross-validation on the remaining 20 topics (80% for training and 20% for testing). The statistics of the dataset are shown in Table 1. We employ Precision (P), Recall (R) and F1-score (F1) as the evaluation metrics, and report the average results on the five folds.

We leverage PyTorch to implement DocECI. We use BERT-base-cased as the default setting for BERT. The dimension size of LSTM hidden unit and relative distance representation are set to 256 and 50, respectively. The layer number of TSML and MRML are both set to 2. All edges in $\mathcal{G}$, $\bar{\mathcal{G}}$ are undirected and the default activation functions are ReLU. The model is trained with AdamW optimizer and the batch size is 12. The initial learning rate of BERT is 0.00004, while for the remaining modules is 0.002. We also use a negative sampling rate of 0.5 for training, owing to the sparseness of positive examples. Beyond that, we run spaCy tool to generate dependency parse trees. All experiments are conducted with an Intel Xeon 2.5 GHz CPU and a NVIDIA Tesla V100 GPU.

4.2 Comparative Models

Five sentence-level and document-level models are picked for comparison:

- *With external knowledge*: (1) **Tri-CNN** [6] is a knowledge-based model that constructs extra linguistic features via causal connectives and WordNet. (2) **KnowDis** [41] is for knowledge enhanced distant data augmentation. It obtains plenty of unlabeled data via distant supervision and adopts self-training to train model. (3) **MMG** [22] is a BERT-based model that leverages external knowledge from ConceptNet for reasoning and mines event-agnostic, context-specific patterns via mention masking generalization.
- *Without external knowledge*: (1) **SDP-LSTM** [3] is a dependency path-based sequential model that is first used for identifying event temporal relations. The work in [8] re-implements it as a baseline for event causality identification.

(2) **DCS** [8] is a constraint-based model, which identifies event causalities at the document level by modeling the global and fine-grained aspects of document-level causal structures and conducts optimization with integer linear programming.

Additionally, we use three sentence-level models for relation extraction, migrating them to the event causality identification task: **Att-LSTM** [39], **C-GCN** [36] and **R-BERT** [33]. Two methods are adopted to construct cross-sentence samples to make these models suitable for cross-sentence causality identification. Assume that two events are in S1 and S5, respectively. The first method combines these five sentences into one sentence as input, while the other only combines these two sentences into one sentence as input. Best results of the two methods are reported for comparison.

4.3   Score Replacement

The default training approach of DocECI is to train a classifier and predict intra-sentence and cross-sentence causalities uniformly. However, given that the underlying expression patterns of intra-sentence and cross-sentence causalities may be different, this approach may impair the performance of either intra-sentence causality identification or cross-sentence causality identification.

Following [8], we adopt an alternating training approach for DocECI: training three different classifiers for intra-sentence, cross-sentence and all pairs, respectively, and use the intra-sentence classifier to help predict cross-sentence pairs, which is called *score replacement*. For any cross-sentence pair, if it can be converted into an intra-sentence case through coreference relations and the score from intra-sentence classifier is higher, we use the prediction of intra-sentence classifier instead. A cross-sentence pair may have more than one intra-sentence case, we use the highest score produced by intra-sentence classifier as the final score. With score replacement, the learning rate of BERT is still 0.00004, while for the intra-sentence task, the cross-sentence task and the overall task, the learning rates are 0.002, 0.004 and 0.004, respectively.

4.4   Main Results

Table 2 shows the results of DocECI against the competitors on ESC. On the intra-sentence task: (1) DocECI achieves 53.8% on F1-score, which significantly outperforms other models and validates the effectiveness of our model. (2) Although Tri-CNN, KnowDis and MMG introduce external knowledge to improve the performance on the intra-sentence task, DocECI still outperforms these models, showing that DocECI can capture more information about event causalities. (3) Comparing DocECI with DCS, we find that our two document graphs are more helpful than the global causal structures of DCS on the intra-sentence task by a margin of 9.2% on F1-score. (4) R-BERT is a competitive model migrated from relation extraction, which obtains the second best precision among all competitors, just lags behind DocECI. This shows that DocECI

Table 2. Results on the ESC dataset. DocECI$_{sr}$ denotes DocECI with score replacement. The results with asterisk (*) are from original papers, the remainings are reproduced on the current dataset, based on the source code or our re-implementations.

	Intra-sentence			Cross-sentence			Overall		
	P	R	F1	P	R	F1	P	R	F1
Tri-CNN	27.0	44.9	33.7	12.6	37.8	18.9	15.6	49.8	23.8
KnowDis*	39.7	66.5	49.7	–	–	–	–	–	–
MMG	41.9	62.5	50.1	23.1	26.0	24.5	28.7	37.6	32.6
SDP-LSTM*	34.0	41.5	37.4	13.5	30.3	18.7	17.6	33.9	23.2
DCS*	38.8	52.4	44.6	35.1	48.2	40.6	36.2	49.5	41.9
Att-LSTM	34.7	44.6	39.0	17.5	44.7	25.1	22.5	42.4	29.4
C-GCN	32.1	57.5	41.2	19.3	40.3	26.1	22.3	50.9	31.0
R-BERT	42.8	54.5	48.0	36.2	30.2	33.0	41.4	34.9	37.9
DocECI	43.3	**71.2**	53.8	37.5	42.5	39.8	39.9	50.7	44.6
DocECI$_{sr}$	**45.8**	67.9	**54.7**	**44.3**	**59.5**	**50.8**	**45.3**	**62.5**	**52.5**

is more powerful to capture event causalities in the context than BERT. (5) With score replacement, DocECI achieves the highest F1-score (54.7%), showing that DocECI can capture more event causality patterns with the help of score replacement.

On the cross-sentence task, the performance of all models has varied degrees of decline, indicating that this task is more difficult. From the results, we have several findings: (1) DocECI achieves the best F1-score and significantly outperforms other models except DCS, which still shows the effectiveness of DocECI. (2) Compared with MMG, DocECI is clearly better by a margin of 15.3% on F1-score, which indicates that the event causalities patterns captured by DocECI are more helpful than the external knowledge from WordNet and ConceptNet. (3) Compared with DCS, DocECI slightly lags behind on F1-score(-0.8%). DCS achieves the second best F1-score (40.6%) due to the usage of score replacement. (4) Compared with the BERT-based models R-BERT and MMG, DocECI achieves substantial improvement of 6.8%–15.3%, which reflects the advantage of using document graphs to model documents. (5) With score replacement, DocECI obtains the best F1-score (50.8%), which indicates that score replacement really helps the cross-sentence task by converting it to the intra-sentence task.

In overall, DocECI achieves the best F1-score among all competitors, and score replacement brings a further improvement for it.

4.5 Causality Direction Identification

To study the sensitivity of models to causality directions, we conduct a new experiment overlooked by the existing works on causality direction identification.

Table 3. Results of causality direction identification on the ESC dataset.

	Intra-sentence			Cross-sentence			Overall		
	P	R	F1	P	R	F1	P	R	F1
Tri-CNN	15.2	22.4	18.1	6.1	19.5	9.3	8.0	25.1	12.1
MMG	**43.0**	**52.2**	**47.2**	14.4	16.3	15.3	29.3	31.6	30.4
Att-LSTM	21.9	16.7	19.0	10.1	25.5	14.5	11.9	28.0	16.7
C-GCN	21.2	47.8	29.4	14.1	28.2	18.8	14.6	**38.8**	21.2
R-BERT	42.4	41.1	41.8	30.2	18.6	23.0	37.9	25.6	28.9
DocECI	34.9	48.0	40.4	**35.5**	24.0	**28.7**	**34.7**	32.0	**33.3**
DocECI$_{sr}$	33.6	47.4	39.3	26.6	**29.9**	28.2	29.8	32.5	31.1

Specifically, we model it as a three-label classification problem, which is required to predict which event is *cause* and which is *effect* if there is a causality between this event pair. Table 3 depicts that the performance of all models on the task decreases. The main reason is that the models may need prior knowledge to identify subtle differences between two directions.

Our findings are: (1) DocECI still performs best on the causality direction identification task. (2) On the intra-sentence task, DocECI performs not well and lags slightly behind R-BERT and MMG. We argue that both R-BERT and MMG are sentence-level models, and they are better to capture the detailed context in a sentence rather than capture the document-level context, because their performance drops dramatically on the cross-sentence task, as shown in the table. (3) On the cross-sentence task, DocECI is superior to all baseline models, its F1-scores are 5.7% and 13.4% higher than R-BERT and MMG, respectively. (4) Score replacement slightly decreases the performance of DocECI. We argue that further distinguishing directions would lead to fewer positive samples, resulting in inadequate learning. In this case, instead of training three different classifiers, training a uniform classifier can make up for fewer positive samples.

4.6   Ablation Study

We conduct an ablation study to verify the influence of each module in DocECI. The results in Table 4 show that each module contributes to the final performance. We also conduct another ablation study about edge types, and the results are shown in Table 5. We can see that MMO and MMR contribute most to DocECI, showing that the relations among mentions (co-occurence and coreference) are more important for causality identification, especially for cross-sentence causality. WSF and MSF decrease the performance of DocECI on the intra-sentence task. We argue that the edges of these two types mainly capture information to improve cross-sentence causalities and they contribute most to the cross-sentence task.

Table 4. Results of ablation study on the ESC dataset.

	Intra-sentence			Cross-sentence			Overall		
	P	R	F1	P	R	F1	P	R	F1
DocECI$_{sr}$	**45.8**	**67.9**	**54.7**	44.3	59.5	50.8	**45.3**	**62.5**	**52.5**
w/o BERT	38.8	55.8	45.8	42.3	51.2	46.3	40.2	54.3	46.2
w/o TSML	43.9	64.2	52.2	**45.7**	52.5	48.9	44.5	58.3	50.5
w/o MRML	44.8	63.5	52.5	39.6	48.5	43.6	44.1	52.9	48.1
w/o CAL	44.5	60.8	51.4	41.1	58.7	48.3	43.7	59.8	50.5

Table 5. Results of ablation study of different types of edges on the ESC dataset.

	Intra-sentence			Cross-sentence			Overall		
	P	R	F1	P	R	F1	P	R	F1
DocECI$_{sr}$	45.8	67.9	54.7	**44.3**	**59.5**	**50.8**	45.3	**62.5**	**52.5**
w/o WWA & WWD	45.0	67.0	53.9	41.6	57.7	48.4	44.3	58.9	50.6
w/o MMO & MMR	44.3	61.4	51.5	29.5	26.5	27.9	39.4	34.7	36.9
w/o SSA & SSC	43.0	64.3	51.6	44.1	55.9	49.3	**45.6**	55.4	50.1
w/o WSF & MSF	**46.1**	**68.5**	**55.1**	41.8	57.1	48.3	45.0	56.9	50.2

4.7 Error Analysis

We illustrate two incorrect examples in Fig. 4, and the analysis is as follows:

- The first example is a false negative case. DocECI fails to identify the causality: **fire** causes **in**. The main reason is that DocECI cannot know the underlying semantics of such colloquial expressions using prepositions to express events. In the training, from the observation of existing data, DocECI is more inclined to think this is the expression of a preposition, so the representations learned by DocECI cannot cover the information at the event level.
- The second example is a false positive case. DocECI identifies a causality that is not in the references. We argue that there are two main reasons: (1) On one hand, DocECI fails to fully learn the deep-level causality patterns due to the small size of training data; (2) On the other hand, there may indeed have some causalities in the dataset that have not been labeled but captured by DocECI. In this example, O'Brien was **fired**, thus King needed to find a new coach and **checked** with Cheeks. In this sense, it is reasonable for DocECI to believe that there is a causality between these two events.

4.8 Hyper-parameter Sensitivity

We study the impact of the number of R-GCN layers used in TSML and MRML. We respectively set L and $\bar{L}$ from 1 to 3, and the results are shown in Table 6. It can be observed that DocECI performs differently on different tasks under

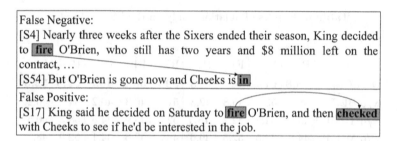

Fig. 4. Examples of incorrect prediction by DocECI.

Table 6. Results of DocECI w.r.t. different layer numbers on the ESC dataset.

		Intra-sentence			Cross-sentence			Overall		
		P	R	F1	P	R	F1	P	R	F1
$L = 1$	$\bar{L} = 1$	**49.4**	63.1	55.4	**48.0**	51.9	49.8	**48.3**	56.0	51.9
	$\bar{L} = 2$	45.7	61.7	52.5	42.7	57.6	49.0	46.4	58.7	51.8
	$\bar{L} = 3$	45.7	59.0	51.5	38.5	58.5	46.4	43.6	57.3	49.5
$L = 2$	$\bar{L} = 1$	47.6	67.4	**55.8**	44.9	57.2	50.3	45.7	60.3	52.0
	$\bar{L} = 2$	45.8	67.9	54.7	44.3	59.5	**50.8**	45.3	62.5	**52.5**
	$\bar{L} = 3$	37.7	**76.8**	50.6	39.5	**65.8**	49.4	40.2	**64.2**	49.4
$L = 3$	$\bar{L} = 1$	48.1	65.8	55.6	43.0	55.7	48.5	43.6	63.1	51.5
	$\bar{L} = 2$	44.4	68.0	53.7	42.1	56.6	48.3	45.3	61.1	52.0
	$\bar{L} = 3$	38.3	68.2	49.0	37.1	63.9	47.0	39.8	64.3	49.2

various settings. For example, DocECI achieves the best result (55.8%) on intra-sentence task when $L = 2$ and $\bar{L} = 1$, while it achieves the best result (50.8%) on the cross-sentence task when $L = 2$ and $\bar{L} = 2$. This shows that it is feasible to further improve the performance of model by training different classifiers for different tasks. Also from the results, we can find that choosing the proper number of R-GCN layers is important for modeling document. More layers may lead to the over-smoothing problem that all representations of nodes tend to be similar, while less layers may lead to inadequate expressiveness of model. Both of them have a negative impact on the model.

5   Conclusion and Future Work

In this paper, we present DocECI, a new model for document-level event causality identification with dual graph convolutional networks. We define text structure graph and mention relation graph to capture the structural information of text and the relational information among mentions, respectively, and use R-GCNs to encode them. Extensive experiments and ablation studies show that DocECI achieves the state-of-the-art performance on the ESC dataset. In future work, we will study causality direction identification, try to integrate external knowledge as prior knowledge, and explore more document modeling methods.

Acknowledgments. This work is supported by Science and Technology on Information Systems Engineering Laboratory (No. 05202006).

References

1. Blanco, E., Castell, N., Moldovan, D.I.: Causal relation extraction. In: LREC (2008)
2. Caselli, T., Vossen, P.: The event storyline corpus: a new benchmark for causal and temporal relation extraction. In: EventStory, pp. 77–86 (2017)
3. Cheng, F., Miyao, Y.: Classifying temporal relations by bidirectional LSTM over dependency paths. In: ACL, pp. 1–6 (2017)
4. Christopoulou, F., Miwa, M., Ananiadou, S.: Connecting the dots: document-level neural relation extraction with edge-oriented graphs. In: EMNLP-IJCNLP, pp. 4927–4938 (2019)
5. Dasgupta, T., Saha, R., Dey, L., Naskar, A.: Automatic extraction of causal relations from text using linguistically informed deep neural networks. In: SIGDIAL, pp. 306–316 (2018)
6. De Silva, T.N., Zhibo, X., Rui, Z., Kezhi, M.: Causal relation identification using convolutional neural networks and knowledge based features. Int. J. Comput. Syst. Eng. **11**(6), 696–701 (2017)
7. Devlin, J., Chang, M., Lee, K., Toutanova, K.: BERT: pre-training of deep bidirectional transformers for language understanding. In: NAACL, pp. 4171–4186 (2019)
8. Gao, L., Choubey, P.K., Huang, R.: Modeling document-level causal structures for event causal relation identification. In: NAACL, pp. 1808–1817 (2019)
9. Garcia, D.: COATIS, an NLP system to locate expressions of actions connected by causality links. In: EKAW, pp. 347–352 (1997)
10. Girju, R.: Automatic detection of causal relations for question answering. In: ACL, pp. 76–83 (2003)
11. Gupta, P., Rajaram, S., Schütze, H., Runkler, T.: Neural relation extraction within and across sentence boundaries. In: AAAI, pp. 6513–6520 (2019)
12. Ji, Y., Li, B., Liu, Y., Zhang, Y., Cai, K.: Multi-space knowledge enhanced question answering over knowledge graph. In: Asia-Pacific Web (APWeb) and Web-Age Information Management (WAIM) Joint International Conference on Web and Big Data, pp. 135–140 (2021)
13. Jia, R., Wong, C., Poon, H.: Document-Level n-ary relation extraction with multiscale representation learning. In: NAACL, pp. 3693–3704 (2019)
14. Jin, X., Wang, X., Luo, X., Huang, S., Gu, S.: Inter-sentence and implicit causality extraction from chinese corpus. In: PAKDD, pp. 739–751 (2020)
15. Khoo, C.S., Chan, S., Niu, Y.: Extracting causal knowledge from a medical database using graphical patterns. In: ACL, pp. 336–343 (2000)
16. Khoo, C.S., Kornfilt, J., Oddy, R.N., Myaeng, S.H.: Automatic extraction of cause-effect information from newspaper text without knowledge-based inferencing. Literary Linguist. Comput. **13**(4), 177–186 (1998)
17. Kipf, T.N., Welling, M.: Semi-supervised classification with graph convolutional networks. In: ICLR (2017)
18. Kruengkrai, C., Torisawa, K., Hashimoto, C., Kloetzer, J., Oh, J.H., Tanaka, M.: Improving event causality recognition with multiple background knowledge sources using multi-column convolutional neural networks. In: AAAI, pp. 3466–3473 (2017)
19. Li, P., Mao, K.: Knowledge-oriented convolutional neural network for causal relation extraction from natural language texts. Exp. Syst. Appl. **115**, 512–523 (2019)

20. Li, Z., Li, Q., Zou, X., Ren, J.: Causality extraction based on self-attentive BiLSTM-CRF with transferred embeddings. Neurocomputing **423**, 207–219 (2021)
21. Li, Z., Ding, X., Liu, T.: Constructing narrative event evolutionary graph for script event prediction. In: IJCAI, pp. 4201–4207 (2018)
22. Liu, J., Chen, Y., Zhao, J.: Knowledge enhanced event causality identification with mention masking generalizations. In: IJCAI, pp. 3608–3614 (2020)
23. Mirza, P., Tonelli, S.: CATENA: CAusal and TEmporal relation extraction from NAtural language texts. In: COLING, pp. 64–75 (2016)
24. Nan, G., Guo, Z., Sekulic, I., Lu, W.: Reasoning with latent structure refinement for document-level relation extraction. In: ACL, pp. 1546–1557 (2020)
25. Ning, Q., Feng, Z., Wu, H., Roth, D.: Joint reasoning for temporal and causal relations. In: ACL, pp. 2278–2288 (2018)
26. Qiu, J., Xu, L., Zhai, J., Luo, L.: Extracting causal relations from emergency cases based on conditional random fields. Procedia Comput. Sci. **112**, 1623–1632 (2017)
27. Qiu, S., Yu, B., Qian, L., Guo, Q., Hu, W.: Joint reasoning of events, participants and locations for plot relation recognition. In: APWeb-WAIM, pp. 700–715 (2020)
28. Radinsky, K., Davidovich, S., Markovitch, S.: Learning causality for news events prediction. In: WWW, pp. 909–918 (2012)
29. Schlichtkrull, M.S., Kipf, T.N., Bloem, P., van den Berg, R., Titov, I., Welling, M.: Modeling relational data with graph convolutional networks. In: ESWC, pp. 593–607 (2018)
30. Tang, H., et al.: HIN: hierarchical inference network for document-level relation extraction. In: PAKDD, pp. 197–209 (2020)
31. Vaswani, A., et al.: Attention is all you need. In: NIPS, pp. 5998–6008 (2017)
32. Wang, D., Hu, W., Cao, E., Sun, W.: Global-to-local neural networks for document-level relation extraction. In: EMNLP, pp. 3711–3721 (2020)
33. Wu, S., He, Y.: Enriching pre-trained language model with entity information for relation classification. In: CIKM, pp. 2361–2364 (2019)
34. Xu, K., Li, C., Tian, Y., Sonobe, T., Kawarabayashi, K., Jegelka, S.: Representation learning on graphs with jumping knowledge networks. In: ICML, pp. 5449–5458 (2018)
35. Yang, J., Han, S.C., Poon, J.: A survey on extraction of causal relations from natural language text. arXiv preprint arXiv:2101.06426 (2021)
36. Zhang, Y., Qi, P., Manning, C.D.: Graph convolution over pruned dependency trees improves relation extraction. In: EMNLP, pp. 2205–2215 (2018)
37. Zhang, Z., et al.: Document-level relation extraction with dual-tier heterogeneous graph. In: COLING, pp. 1630–1641 (2020)
38. Zhao, S., Liu, T., Zhao, S., Chen, Y., Nie, J.: Event causality extraction based on connectives analysis. Neurocomputing **173**, 1943–1950 (2016)
39. Zhou, P., et al.: Attention-based bidirectional long short-term memory networks for relation classification. In: ACL, pp. 207–212 (2016)
40. Zhou, W., Huang, K., Ma, T., Huang, J.: Document-level relation extraction with adaptive thresholding and localized context pooling. arXiv preprint arXiv:2010.11304 (2020)
41. Zuo, X., Chen, Y., Liu, K., Zhao, J.: KnowDis: knowledge enhanced data augmentation for event causality detection via distant supervision. In: COLING, pp. 1544–1550 (2020)

Self-supervised Label-Visual Correlation Hashing for Multi-label Image Retrieval

Yu Liu[1], Yanzhao Xie[1], Jingkuan Song[2(✉)], Rukai Wei[1], and Ke Zhou[1]

[1] Huazhong University of Science and Technology, Wuhan, China
{liu_yu,yzxie,weirukai,zhke}@hust.edu.cn
[2] University of Electronic Science and Technology of China, Chengdu, China
jingkuan.song@gmail.com

Abstract. Perceiving multiple objects within an image without the labels' supervision is the challenge of multi-label image hashing tasks. Existing unsupervised hashing approaches do reconstruction or contrastive learning for the representation of the object of interest but ignore the other objects in the image. We propose to use pseudo labels to provide candidate objects, making the image match the possible objects' features by the co-occurrence correlations between labels. As a result, we explore the co-occurrence correlations based on empirical models and design a data augmentation strategy in a self-supervised learning framework to learn label-level embeddings. We also build the image visual correlations and design a dual overlapping group sum-pooling (OGSP) component to fuse label-level and visual-level embeddings into image representations, alleviating noise from empirical models. Extensive experiments on public multi-label image datasets using pseudo labels demonstrate that our self-supervised label-visual correlation hashing framework outperforms state-of-the-art label-free hashing algorithms for retrieval. GitHub address: https://github.com/lzHZWZ/SS-LVH.git.

Keywords: Multi-label image hashing · Self-supervised learning · Co-occurrence correlations

1 Introduction

In label-free scenarios, image hashing algorithms [25] remain tricky for learning accurate hash codes for an image containing multiple objects. Existing unsupervised hashing methods ignore the existence of other objects and only perceive the object of interest in an image, resulting in limited performance. If features of all objects are extracted in advance using techniques like object detection [9], the computational cost will be a huge problem.

Recently, the study of co-occurrence correlation [4] has attracted our interest. This correlation reveals the probability of different objects appearing in an image. It can serve as potential supervisory information to aid in the perception of objects of interest. Meanwhile, since the co-occurrence correlation reflects a

B. Li et al. (Eds.): APWeb-WAIM 2022, LNCS 13422, pp. 129–143, 2023.
https://doi.org/10.1007/978-3-031-25198-6_10

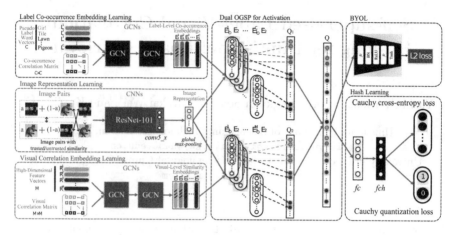

Fig. 1. The architecture of SS-LVH. (1) In the label co-occurrence embedding learning branch (blue frame), C denotes the number of labels, L^1 to L^C denote word vectors corresponding to pseudo labels, and E_1^1 to E_1^C denote label-level co-occurrence embeddings corresponding to L^1 to L^C. (2) In the image representation learning branch (red frame), the input is image pairs with trusted or untrusted similarity. *Conv5_x* is a layer generating the image feature. E_2 denotes the image representation calculated via *global max-pooling* layer with the feature. (3) In the visual correlation embedding learning branch (orange frame), M denotes the number of images sampled from the target dataset, R^1 to R^M denote high-dimensional feature vectors corresponding to sampled images, and E_3^1 to E_3^M denote visual-level similarity embeddings corresponding to R^1 to R^M. (4) The purple frame denotes the dual OGSP component. Based on *overlapping group (a dotted box) sum-pooling*, Q_1 (*i.e.*, semantic similarity representations) is fused by each E_1^i ($i \in \{1, 2, \ldots, C\}$) and E_2, and Q_2 (*i.e.*, visual similarity representations) is fused by each E_3^j ($j \in \{1, 2, \ldots, M\}$) and E_2, where ○ denotes Hadmard Product. The label-visual representation Q is acquired by concatenating Q_1 with Q_2. (5) The cyan frame completes self-supervised learning for Q in the way of BYOL. The *Tanh* function is added to improve the adaptation of hashing. (6) The golden frame achieves hash learning by the Cauchy distribution loss functions [25] consisting of Cauchy cross-entropy loss and Cauchy quantization loss. (Color figure online)

common phenomenon in the real world, it is reliable in label-free scenarios. As a result, the labels' co-occurrence correlations of the empirical model can provide relatively accurate priori information. Although some labels in the empirical model may not exist in the target dataset, co-occurrence correlation can ensure that the relevant objects (*e.g.*, basketball and players) in the image are simultaneously activated by Graph Convolutional Networks (GCNs) [25], partially solving the multi-object perception problem of unlabeled images.

Based on this motivation, we propose to incorporate co-occurrence correlations of pseudo labels [16] (*i.e.*, labels of the empirical model) into a self-supervised learning (SSL) framework [8] to design a multi-object hashing model. The architecture is shown in Fig. 1. We gather the co-occurrence probability of each pseudo label to build the adjacency matrix, and input the matrix into GCNs for the label-level embedding learning. To alleviate the noise caused by applying the empirical models to out-of-distribution (o.o.d) data, we introduce the adjacency matrix based on the visual correlations of all/sampling images (See Sect. 3), and input the matrix into the other GCN branch for the visual-level embedding learning. Since it is derived from images rather than labels, this visual-level embedding is more representative of the distribution of the target dataset [2]. We also use a feature extraction backbone to generate image representations. With the embeddings and image representations, we design a dual *overlapping group sum-pooling* (OGSP) component to fuse them. The embeddings and representations are fused into a vector by Hadmard Product and then is mapped to multiple cells by *group sum-pooling* with overlapped windows. Compared with the Multi-modal Factorized Bilinear (MFB) component used in LAH [25], the dual OGSP component preserves richer spatial information. As a result, the regions of interest will be highlighted through more representations. Furthermore, it can balance the activated representations based on two embeddings, improving generalization ability and accuracy. Finally, we employ the Cauchy distribution loss functions [1] to learn the activated representations into hash codes.

In this paper, we propose a self-supervised label-visual correlation hashing (SS-LVH) framework for multi-label image retrieval. In practice, we employ Bootstrap Your Own Latent (BYOL) [8] as the SSL framework in that we can learn compact representations without negative sampling. For this limitation, we design a data augmentation strategy that fuses the two images via different weights as the pretext task, used to enhance the learning for co-occurrence correlations. In addition, we incorporate the *Tanh* function into the BYOL framework to adapt hash learning. For the embeddings learning, we use BERT [5] to generate label embeddings and select the Classification Transformer (C-Tran) model pre-trained on Visual Genome 500 (VG-500) [13]) as the empirical model, the ResNet-101 [24] model as the representation backbone. Note that we will first use the BYOL framework to pre-train the model, and then access the hash loss functions for hashing. Extensive experiments on public multi-label image datasets using pseudo labels demonstrate that SS-LVH is conducive to retrieving images that share at least one label. Its performance is better than state-of-the-art label-free hashing methods. In addition, we demonstrate that all the components we introduced can improve retrieval performance.

The contributions of SS-LVH are summarized below.

(1) We proposed a novel SSL framework, *i.e.*, SS-LVH, for image hashing using co-occurrence correlations of pseudo labels. By perceiving multiple objects in an image via pseudo labels and their co-occurrence correlations, we achieve self-supervised hashing in a multi-label learning pattern.

(2) We designed a series of tricks to resist noise from o.o.d data, including the label-visual correlation learning scheme and the dual OGSP component, resulting in accurate multi-object activation.

(3) SS-LVH outperforms state-of-the-art unsupervised/self-supervised hashing methods in terms of multi-label image retrieval on three public datasets. We demonstrate that the co-occurrence correlations can benefit the label-free hash learning performance.

2   Related Work

The SS-LVH is designed via the co-occurrence information of objects in images and the SSL framework. Recent research in these fields is described below.

Co-occurrence Correlations. The object co-occurrence correlations in the images can represent the intrinsic logical relation of objects included by images. Wang *et al.* propose CNN-RNN [22], utilizing the semantic redundancy and the co-occurrence dependency to construct an end-to-end classification model. ML-GCN [4] is a novel trainable multi-object image recognition framework, which employs GCN to map the label representations (*i.e.*, word embeddings), including co-occurrence information and inter-dependency of objects in images. In SS-LVH, we also exploit this insight to construct a co-occurrence correlation matrix to delegate the object's inter-dependency.

Multi-label Image Hashing. The multi-label hashing methods can improve the accuracy of image retrieval. Lai *et al.* propose Instance-aware hashing (IAH) [12], which first conducts the instance-aware retrieval via learning-based hashing. Song *et al.* propose Deep Region Hashing (DRH) [20] with a cost-free hashing strategy, and can generate the hash codes for whole image as well as the object candidate regions. Xie *et al.* propose Label-Attended Hashing (LAH) [25] that combines the co-occurrence correlations of labels to learn hash codes.

Self-supervised Learning. We follow SSL for guiding our model to acquire the appropriate image representations without hand-crafted labels. In this field, contrastive methods [3,8] have shown impressive results, with the fundamental ideology pulling representations of different views transformed from the same sample closer together (*i.e.*, positive pairs) while spreading representations of different data views (*i.e.*, negative pairs). Chen *et al.* propose the method Sim-CLR [3] based on contrastive insight, which utilizes a learnable nonlinear transformation between data representations and the contrastive loss, thus improving the quality of representations. BYOL [8] utilizes the learnable target network as 'target' and weighted moving average to make target network learning smoother and efficiently.

3   Preliminary on SS-LVH

We introduce how to create correlations and training image pairs. Given the target image dataset $\mathcal{X}_D = \{x_i\}_{i=1}^N$ and a subset $\mathcal{X}_S = \{x_i\}_{i=1}^M$ of $\mathcal{X}_D$, where $x_i \in \mathbb{R}^D$ is the i-th image, C, N, and M are the number of labels, images, and sampled images, respectively.

Label Co-occurrence Correlation Matrix. As shown in Fig. 2, we employ $\mathcal{L}_L = \{L^i\}_{i=1}^C$ to calculate the correlation matrix $\mathcal{M}_L \in \mathbb{R}^{C \times C}$ based on the co-occurrence probability of each label, where $\mathcal{L}_L$ is a set of word vectors. For the image x_i, we gather the label probability $p_i \in \mathbb{R}^{C \times 1}$ from the last layer of C-Tran, where p_i denotes the probability of each object contained in the i-th

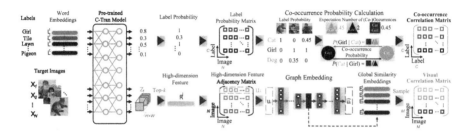

Fig. 2. The generation of the co-occurrence correlation matrix and visual correlation matrix. (Color figure online)

image and it is the i-th column of the label probability matrix $\mathcal{M}_P \in \mathbb{R}^{C \times N}$. Assume that $\mathcal{M}_P(i,j)$ denotes the element of the i-th row and the j-th column in $\mathcal{M}_P$. We change the values of $\mathcal{M}_P(i,j) \geq 0.5$ to 1, reserve the values of $0.5 > \mathcal{M}_P(i,j) \geq 0.3$ (expanding the range of candidates to correct bias), and assign the rest of elements to 0. Note that these settings were determined after we calculated the difference between the generated labels and the actual labels on VOC2007 [6]. To alleviate the sparsity issue caused by a large C, we generalize the method in LAH that regards the occurrence of a label as a discrete state (0 or 1) and calculate the co-occurrence probability $P_{j,i}$, $i.e.$, the probability of the j-th label's occurrence when the i-th label appears, as below.

$$P_{j,i} = \frac{T_{j,i}}{T_i} = \frac{\sum_{k=1}^{N} \mathcal{M}_P(i,k) \times \mathcal{M}_P(j,k)}{\sum_{k=1}^{N} \mathcal{M}_P(i,k)}, \tag{1}$$

where T_i denotes the expectation number of occurrences for the i-th label and $T_{j,i}$ denotes the expectation number of co-occurrences between the i-th label and the j-th one. Note that although $T_{i,j} = T_{j,i}$, $P_{j,i} \neq P_{i,j}$ when $T_i \neq T_j$. As shown in Fig. 2, only three images contain the $girl$ or cat, where $P_{cat,girl} \neq P_{girl,cat}$ because $T_{girl} = 2$ (purple triangle) and $T_{cat} = 1.45$ (red triangle). To promote the convergence efficiency and prevent over-fitting, we lower the long-tail effect by using the threshold μ to binarize $P_{j,i}$. Then, we fill $\mathcal{M}_L$ by $P_{j,i}$, which can be described as:

$$\mathcal{M}_L(i,j) = \begin{cases} 0, & \text{if } P_{j,i} \leq \mu, \\ 1, & \text{otherwise .} \end{cases}$$

To further overcome the problem of over-smooth caused by using the correlation matrix in GCNs, we employ the weighted scheme like LAH to determine $\mathcal{M}_L$. The $\mathcal{M}_L$ is described below.

$$\mathcal{M}_L(i,j) = \begin{cases} \frac{\alpha}{\sum_{j=1, i \neq j}^{C} \mathcal{M}_L(i,j)}, & \text{if } i \neq j, \\ 1 - \alpha, & \text{otherwise ,} \end{cases} \tag{2}$$

where $\alpha \in (0,1)$. We update a node feature with the effect from α. For instance, a node feature will be more determined by its neighbor nodes when $\alpha \to 1$.

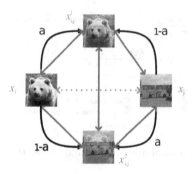 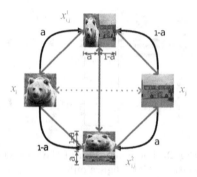

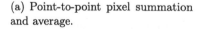

(a) Point-to-point pixel summation and average.

(b) Splicing along the horizontal and vertical.

Fig. 3. Data augmentation strategy. The black lines represent fusion weights. And the blue lines and orange dotted lines represent trusted and untrusted similarities, respectively. (Color figure online)

Image Visual Correlation Matrix. We employ $\mathcal{R}_V = \{R^i\}_{i=1}^{M}$ to calculate the correlation matrix $\mathcal{M}_V \in \mathbb{R}^{M \times M}$. Nevertheless, we are caught between employing $\mathcal{X}_S$, which results in information loss, and using $\mathcal{X}_D$, which results in a huge cost. Therefore, we learn the embeddings for $\mathcal{X}_D$ and sample the embeddings. As shown in Fig. 2, we collect $\mathcal{X}_D$ through the C-Tran model. Different from conventional features acquired from the convolution layer, our features consist of Z_1 to $Z_{H \times W}$, where each $Z_i \in \mathbb{R}^{1 \times 2048}$ consists of values at the same position for all feature maps, H and W represent the width and height of the feature map, respectively. Since features in C-Tran are generated by the relationship between pixels, we pick the top-k (See Sect. 5) values on each Z_i to construct R^i, *i.e.*, the high-dimension feature of the i-th image. Then, we form the adjacency matrix $\mathcal{M}_A$ by cosine distances between R^i and R^j for the graph embedding learning. Assume that $u_i = \mathcal{M}_A(i, \cdot)$ denotes the vector in the i-th row of $\mathcal{M}_A$. It also represents the similarity between the i-th image and others in $\mathcal{X}_D$. We employ SDNE [21] to encode u_i into the embedding E_0^i, and get the subset $\{E_0^i\}_{i=1}^{M}$ of $\{E_0^i\}_{i=1}^{N}$ to calculate $\mathcal{M}_V$, where $\{E_0^i\}_{i=1}^{M}$ is obtained by random sampling, but preferably in an amount equal to the number of pseudo labels and covering all categories in the target dataset. Each element of $\mathcal{M}_V$ is calculated by cosine distances between E_0^i and E_0^j.

Data Augmentation Strategy. We propose a label similarity transformation strategy (2 patterns) to fuse two images via different weights. As shown in Fig. 3, x_i and x_j are images in $\mathcal{X}_D$, while $x_{i,j}^1$ and $x_{i,j}^2$ are images composed by x_i and x_j with different weights, where weights are a and $1 - a$, and $a \in (0, 1)$ (See Sect. 5). We depict the method of point-to-point pixel summation and average in Fig. 3(a) and the splicing method along with the horizontal and vertical in Fig. 3(b). The label similarity transformation strategy produces more image pairs with the trusted similarity, alleviating the sparsity issue of similar pairs when

C is too large. Note that, since the composite images don't contaminate the correlation matrices, the co-occurrence correlations from the target dataset still are decisive.

4   SS-LVH

SS-LVH learns a nonlinear hash function $f_h : x \mapsto h \in \{-1,1\}^{\mathcal{K}}$ from input space to Hamming space using CNNs and two GCNs, encoding each image x into a $\mathcal{K}$-bit hash code $h = f_h(x)$. For the target images (untrusted pairs) or composite images (trusted pairs), $i.e.$, x_i and x_j, if their pseudo-multi-labels contain at least one same label, their similarity labels $s_{ij} = 1$. Otherwise, $s_{ij} = 0$. $f_h(x)$ should preserve the similarities, $i.e.$, $\mathcal{S} = \{s_{ij}\}$, in hash codes.

In the representation learning stage, we input $\mathcal{L}_L$ and $\mathcal{M}_L$, pairwise images $\{(x_i, x_j, s_{ij})\}$, and $\mathcal{R}_V$ and $\mathcal{M}_V$ into the label co-occurrence embedding learning branch, the image representation learning branch, and the visual correlation embedding learning branch, respectively. Then, $\{E_1^i\}_{i=1}^C$, E_2, and $\{E_3^i\}_{i=1}^M$ are generated and sent to the dual OGSP component. The fusion results Q_1 and Q_2 are concatenated to the label-visual representation Q. SS-LVH learns Q in the way of BYOL. In the hash learning stage, we fix the learned parameters and learn with the Cauchy distribution loss functions. Finally, SS-LVH transforms Q into a $\mathcal{K}$-dimensional continuous code $\mathcal{Z} \in \mathbb{R}^{\mathcal{K}}$ in the fc layer, and then transforms $\mathcal{Z}$ into a $\mathcal{K}$-dimensional hash code by $h = sgn(tanh(\mathcal{Z})) \in \{1, -1\}^{\mathcal{K}}$ in the fch layer. Finally, preserving similarity of pairwise images and lowering the quantization error, SS-LVH learns the non-linear hash function $f_h(x)$. The details of each part are described below.

Image Representation Learning. Following LAH, we employ ResNet-101 as the backbone to learn the image representation. For the image x that has been transformed to the dimension of $D = 448 \times 448 \times 3$, $i.e.$, $x \in \mathbb{R}^{448 \times 448 \times 3}$, we capture a $2048 \times 14 \times 14$-dimensional feature vector from the $conv5\_x$ layer. Then, we generate $E_2 \in \mathbb{R}^{2048 \times 1}$ through the $global\ max\text{-}pooling$ (GMP) layer.

GCN for Learning of Embeddings. GCN can smooth the features by the given correlation. More specifically, by the propagation of weights, it learns a function f_{gcn} on the graph to achieve feature extraction. For example, on the label co-occurrence embedding learning branch, we assume that $\mathcal{L}_L^{(i)}$ represents the input in the i-th layer and $\mathcal{L}_L^{(i+1)}$ denotes updated node features. The propagation function in each GCN layer is described below.

$$\mathcal{L}_L^{(i+1)} = f_{gcn}(\widehat{\mathcal{M}_L}\mathcal{L}_L^{(i)}\mathcal{W}_L^{(i)}), \qquad (3)$$

where $\mathcal{W}_L^{(i)}$ is the weight on the i-th graph layer, $\widehat{\mathcal{M}_L} = \tilde{D}^{-\frac{1}{2}}\widetilde{\mathcal{M}_L}\tilde{D}^{-\frac{1}{2}}$ with $\widetilde{\mathcal{M}_L} = \mathcal{M}_L + I_C$ and $\tilde{D}(i,i) = \sum_j \widetilde{\mathcal{M}_L}(i,j)$. In the implementation, we use two GCN layers with word vectors generated by BERT. The dimensions of the last layer of $\mathcal{L}_L$ and $\mathcal{R}_V$ are designed to match E_2.

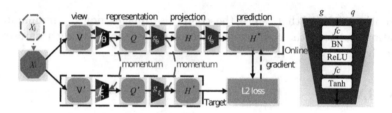

Fig. 4. Self-supervised learning process.

Dual OGSP for Activation. OGSP employs a one-dimensional overlapping window to perform *sum-pooling* over vectors and utilizes Hadmard Product (*i.e.*, ∘) to fuse embeddings and image representations. Each fusion result is mapped onto multiple values corresponding to multiple groups (*i.e.*, dotted boxes shown in Fig. 1), resulting in richer information highlighting regions of interest. For the i-th label activated representation Q_L^i, we define that the size of the group is $^i\mathcal{G}_L^g$ and the stride is $^i\mathcal{G}_L^s$, where $i \in \{1, 2, \ldots, C\}$. Meanwhile, $\mathcal{Q}_L^i = E_1^i \circ E_2$, where $E_1^i, E_2 \in \mathbb{R}^{2048 \times 1}$ and $\mathcal{Q}_L^i(k)$ denotes the k-th element of $\mathcal{Q}_L^i$. Thus, the j-th element of Q_L^i is described below.

$$Q_L^i(j) = \sum_{k=1+(j-1)\cdot {}^i\mathcal{G}_L^s}^{{}^i\mathcal{G}_L^g+(j-1)\cdot {}^i\mathcal{G}_L^s} \mathcal{Q}_L^i(k), \qquad (4)$$

where $j \in \{1, 2, \ldots, \lceil \frac{2048 - {}^i\mathcal{G}_L^g + {}^i\mathcal{G}_L^s}{{}^i\mathcal{G}_L^s} \rceil\}$. Note that when the number of elements is not enough, we fetch elements from the head of vector to fill. Based on this, $Q_L^i = [Q_L^i(1); Q_L^i(2); \ldots; Q_L^i(\lceil \frac{2048 - {}^i\mathcal{G}_L^g + {}^i\mathcal{G}_L^s}{{}^i\mathcal{G}_L^s} \rceil)]$ and the label semantic similarity representation Q_1 is described below.

$$Q_1 = [Q_L^1; Q_L^2; \ldots; Q_L^C], \qquad (5)$$

where $Q_1 \in \mathbb{R}^{\sum_{i=1}^{C} \lceil \frac{2048 - {}^i\mathcal{G}_L^g + {}^i\mathcal{G}_L^s}{{}^i\mathcal{G}_L^s} \rceil \times 1}$. In the same way, we define and calculate $^i\mathcal{G}_V^g$, $^i\mathcal{G}_V^s$, $\mathcal{Q}_V^i = E_3^i \circ E_2$, and Q_V^i. Then, the visual similarity representation Q_2 is described below.

$$Q_2 = [Q_V^1; Q_V^2; \ldots; Q_V^M], \qquad (6)$$

where $Q_2 \in \mathbb{R}^{\sum_{i=1}^{M} \lceil \frac{2048 - {}^i\mathcal{G}_V^g + {}^i\mathcal{G}_V^s}{{}^i\mathcal{G}_V^s} \rceil \times 1}$. Finally, the label-visual representation Q is described below.

$$Q = [Q_1; Q_2]. \qquad (7)$$

Generally, we recommend that $\forall i$, $^i\mathcal{G}_L^s = {}^i\mathcal{G}_V^s = \mathcal{G}^s$, $^i\mathcal{G}_L^g = {}^i\mathcal{G}_V^g = \mathcal{G}^g$ (See Sect. 5) because $Q \in \mathbb{R}^{\sum_{i=1}^{C+M} \lceil \frac{2048 - \mathcal{G}^g + \mathcal{G}^s}{\mathcal{G}^s} \rceil \times 1}$ conduces to the trade-off between two representations.

Self-supervised Learning. As shown in Fig. 4, the framework consists of the *online* (brown dotted frame) and *target* (green dotted frame) networks, whose parameters are θ and ξ respectively. The parameters ξ are an exponential moving average of θ. Given a target decay rate $\tau \in [0, 1]$, the update after each training step is described below.

$$\xi \leftarrow \tau\xi + (1 - \tau)\theta. \tag{8}$$

The *target* network has the same architecture as the *online* network except for the *prediction* function q. The two *views* V and V' come from image augmentations $\mathcal{T}$ and $\mathcal{T}'$ respectively. In our task, $\mathcal{T}$ and $\mathcal{T}'$ can be conventional methods for a single image, or our similarity transformation strategy for x_i and x_j. f denotes the representation extraction function corresponding to networks in blue, red, orange and purple frames shown in Fig. 1. *Representation* corresponds to Q shown in Fig. 1. $Q = f_\theta(V)$ and $Q' = f_\xi(V')$. g is a *projection* function consists of the BatchNorm layer (BN), ReLU, *fc* layer and *Tanh*, where *Tanh* is injected to adapt to hash task. $H = g_\theta(Q)$ and $H' = g_\xi(Q')$. q is a *prediction* function and $H'' = q_\theta(H)$, where q has the same architecture as g. Finally, we L_2-normalize $\widehat{H'} = \frac{H'}{\|H'\|_2}$ and $\widehat{H''} = \frac{H''}{\|H''\|_2}$. The loss between the normalized predictions and target projections is described below.

$$\mathcal{L}_{\theta,\xi} = \|\widehat{H'} - \widehat{H''}\|_2^2 = 2 - \frac{2\langle H', H''\rangle}{\|H'\|_2 \cdot \|H''\|_2}. \tag{9}$$

According to Eq. (9), we calculate $\overline{\mathcal{L}_{\theta,\xi}}$ by feeding V to the *target* network and V' to the *online* network. At each training step, the task is to minimize $\widehat{\mathcal{L}_{\theta,\xi}} = \mathcal{L}_{\theta,\xi} + \overline{\mathcal{L}_{\theta,\xi}}$ with respect to θ only. The optimizer of self-supervised learning are described below.

$$\theta \leftarrow Opt(\theta, \nabla_\theta\widehat{\mathcal{L}_{\theta,\xi}}, \eta), \tag{10}$$

where Opt is the stochastic gradient descent optimizer and η is a learning rate. When we use conventional augmentation strategies, we will initialize a low learning rate for label co-occurrence embedding learning branch that enhances activation ability for global visual similarity representations. Contrastively, when we adopt our similarity transformation strategy, we will initialize lower the learning rate for visual correlation embedding learning branch to enhance activation ability for local semantic similarity representations. Finally, we only keep f_θ involving in hash function learning.

Cauchy Loss for Hash Learning. To generate hash codes with high aggregation degree of similar samples within short Hamming distance, we employ Cauchy loss functions used in DCH [1], resulting in the best retrieval performance in Hamming radius ≤ 2.

The Cauchy loss functions consist of the Cauchy cross-entropy loss and the Cauchy quantization loss. For the h_i and h_j corresponding to $\{(x_i, x_j, s_{ij})\}$, the probability function based on the Cauchy distribution is written as:

$$\Gamma(\delta(h_i, h_j)) = \frac{\gamma}{\gamma + \delta(h_i, h_j)}, \tag{11}$$

where $\Gamma(*)$ is well-defined probability function, $\delta(h_i, h_j)$ denotes the Hamming distance between h_i and h_j, γ is the scale hyper-parameter of the Cauchy distribution and controls aggregation degree. Generally, $\gamma = 0.15$.

Assume that $h_i(j)$ is the j-th element of h_i. The sign function $sgn(h_i)$ is described below.

$$sgn(h_i(j)) = \begin{cases} -1, & \text{if } h_i(j) \leq 0, \\ 1, & \text{otherwise .} \end{cases} \tag{12}$$

For the quantization error $\| h - sgn(h) \|$, we combine γ and the Cauchy distribution to describe the prior for each hash code as:

$$\phi_{h_i} = \frac{\gamma}{\gamma + \delta(|h_i|, \mathbf{1})}, \tag{13}$$

where $\mathbf{1} \in \mathbb{R}^{\mathcal{K}}$. To cooperate with continuous relaxation, we set $\delta(h_i, h_j) = \frac{K}{2}(1 - \frac{\langle h_i, h_j \rangle}{\|h_i\|_2 \cdot \|h_j\|_2})$ to approximate the Hamming distance and to optimize the loss function.

Based on Eq. (11) and the logarithm Maximum a Posteriori estimation of the hash codes, the Cauchy cross-entropy loss function $\mathscr{L}_C$ is described below.

$$\mathscr{L}_C = \sum_{s_{ij}} \omega_{ij}(s_{ij} log \frac{\delta(h_i, h_j)}{\gamma} + log(1 + \frac{\gamma}{\delta(h_i, h_j)})), \tag{14}$$

where

$$\omega_{ij} = \begin{cases} |\mathcal{S}|/|\mathcal{S}_s|, & s_{ij} = 1, \\ |\mathcal{S}|/|\mathcal{S}_d|, & s_{ij} = 0, \end{cases}$$

where $\mathcal{S}_s = \{s_{ij} \in \mathcal{S} : s_{ij} = 1\}$ is the set of similar pairs, $\mathcal{S}_d = \{s_{ij} \in \mathcal{S} : s_{ij} = 0\}$ is the set of dissimilar pairs. For $\forall i, j$ and $i \neq j$, if $\exists \mathcal{M}_P(i, k) = \mathcal{M}_P(j, k) = 1$, we obtain $s_{ij} = 1$; otherwise, $s_{ij} = 0$. Meanwhile, according to Eq. (13), the Cauchy quantization loss function $\mathscr{L}_Q$ is described below.

$$\mathscr{L}_Q = \sum_{i=1}^{N} log(1 + \frac{\delta(|h_i|, \mathbf{1})}{\gamma}). \tag{15}$$

According to the deduction of Bayesian learning in DCH [1], the complete hash loss function is described below.

$$\mathscr{L} = \lambda \mathscr{L}_C + (1 - \lambda)\mathscr{L}_Q, \tag{16}$$

where λ is a hyper-parameter to balance two loss functions.

5   Experiment

Experimental Settings. We select three multi-label image datasets, including MS-COCO [15], VOC2007 [6], and MIRFLICKR-25K [10]. We randomly select 10,000, 4,000, and 5,000 images from three datasets respectively as target datasets to train models. Following parameters in LAH and BYOL, we train all datasets without using hand-crafted labels. Then, we randomly select 5000, 1000, and 1000 images from remaining images as the query set to test models respectively. The classification results in terms of Mean Average Precision

Table 1. MAP of re-ranking within Hamming radius 2 (MAP@H $\leq$ 2) at different bits on three public multi-label image datasets.

Method	MS-COCO					VOC2007					MIRFLICKR-25K				
	16 bits	32 bits	48 bits	64 bits	128 bits	16 bits	32 bits	48 bits	64 bits	128 bits	16 bits	32 bits	48 bits	64 bits	128 bits
DistillHash [26]	0.605	0.617	0.628	0.630	0.627	0.403	0.410	0.424	0.422	0.420	0.628	0.631	0.633	0.636	0.637
DU3H [27]	0.611	0.620	0.630	0.634	0.633	0.421	0.442	0.448	0.446	0.444	0.636	0.645	0.647	0.646	0.643
TBH [19]	0.607	0.613	0.615	0.618	0.617	0.423	0.441	0.447	0.451	0.448	0.638	0.639	0.642	0.646	0.645
DHNR [23]	0.606	0.609	0.611	0.613	0.611	0.434	0.438	0.439	0.438	0.437	0.624	0.631	0.637	0.647	0.645
Bi-half [14]	0.609	0.616	0.622	0.626	0.626	0.428	0.433	0.438	0.442	0.441	0.640	0.642	0.647	0.650	0.649
WDHT [7]	0.594	0.597	0.608	0.613	0.610	0.389	0.393	0.401	0.411	0.410	0.603	0.616	0.621	0.623	0.616
MGRN [11]	**0.618**	0.621	0.627	0.636	0.636	0.447	0.449	0.452	0.452	0.452	0.631	0.636	0.641	0.645	0.649
DATE [17]	0.611	0.621	0.633	0.639	0.638	0.481	0.488	0.493	0.505	0.507	0.641	0.650	0.656	0.657	0.657
CIBHash [18]	0.617	<u>0.623</u>	<u>0.638</u>	<u>0.641</u>	<u>0.641</u>	<u>0.489</u>	<u>0.504</u>	<u>0.517</u>	<u>0.519</u>	<u>0.518</u>	**0.644**	**0.655**	<u>0.659</u>	<u>0.660</u>	<u>0.659</u>
SS-LVH	0.617	**0.637**	**0.644**	**0.653**	**0.658**	**0.509**	**0.513**	**0.519**	**0.526**	**0.531**	0.639	**0.655**	**0.661**	**0.663**	**0.663**

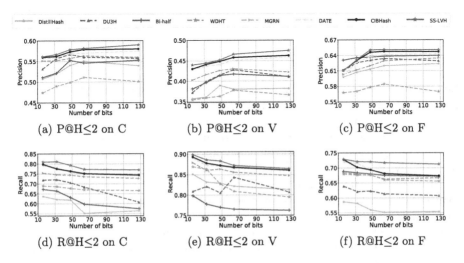

Fig. 5. P@H $\leq$ 2 and R@H $\leq$ 2 with different code lengths on the MS-COCO (C), VOC2007 (V) and MIRFLICKR-25K (F) datasets.

(MAP) on MS-COCO, VOC2007, and MIRFLICKR-25K are 0.518, 0.403, and 0.451, respectively when we test datasets by the empirical model. In addition, we employ the methods of the word vectors generation and evaluation metrics used in LAH, where LAH measures the quality of hash codes within Hamming radius 2: MAP within Hamming Radius 2 (MAP@H $\leq$ 2), Precision curves within Hamming Radius 2 (P@H $\leq$ 2), and Recall curves within Hamming Radius (R@H $\leq$ 2).

We compare SS-LVH with nine state-of-the-art label-free hashing methods, including five unsupervised methods (DistillHash [26], DU3H [27], TBH [19], DHNR [23], and Bi-half [14]), two label-embedding-based weakly-supervised methods (WDHT [7] and MGRN [11]), and two SSL methods, *i.e.*, contrastive learning methods (DATE [17] and CIBHash [18]).

Implementation Details. For the label-level co-occurrence embeddings learning, we employ labels of VG-500 and set $C = 500$. For the label-level visual similarity embeddings learning, we set $k = 10$, and $M = 500$ to equal C. For

Table 2. MAP within Hamming radius 2 (MAP@H $\leq$ 2) of SS-LVH and its variants on three public multi-label image datasets. VB denotes *Visual Correlation Embedding Learning Branch*. GSP denotes *Group Sum-Pooling*. OW denotes *Overlapping Window*. SSL denotes *Self-Supervised Learning*. STS denotes *Label Similarity Transformation Strategy*. ✓ means to enable the component, otherwise disable it.

Order	VB	MFB	GSP	OW	SSL	Tanh	STS	MS-COCO			VOC2007			MIRFLICKR-25K		
								32 bits	64 bits	128 bits	32 bits	64 bits	128 bits	32 bits	64 bits	128 bits
1	✓		✓	✓	✓	✓	✓	0.637	0.653	0.658	0.513	0.526	0.531	0.655	0.663	0.663
2			✓	✓	✓	✓	✓	0.620	0.624	0.623	0.485	0.487	0.481	0.626	0.633	0.618
3	✓	✓			✓	✓	✓	0.441	0.451	0.451	0.354	0.358	0.359	0.406	0.411	0.403
4	✓		✓		✓	✓	✓	0.632	0.648	0.652	0.510	0.519	0.524	0.648	0.650	0.652
5	✓		✓	✓				0.531	0.568	0.570	0.367	0.374	0.377	0.541	0.544	0.546
6	✓		✓	✓	✓			0.623	0.630	0.635	0.483	0.513	0.518	0.634	0.647	0.653
7	✓		✓	✓	✓	✓		0.633	0.639	0.653	0.510	0.522	0.529	0.650	0.660	0.661
8	✓		✓	✓	✓		✓	0.628	0.643	0.649	0.507	0.514	0.527	0.637	0.654	0.659

OGSP, we set $\mathcal{G}^g = 128$ and $\mathcal{G}^s = 32$. For BYOL, we adopt color transformation as the conventional data augmentation strategy. When we input untrusted pairs transformed by the conventional strategy, we initialize $\eta_1 = 0.05$, $\eta_2 = 0.05$, $\eta_3 = 0.1$ and $\eta_4 = 0.03$, where η_1, η_2, η_3 and η_4 denote learning rates of the label co-occurrence embedding learning branch, image representation learning branch, visual correlation embedding learning branch, and other components respectively. When we input trusted pairs using our strategy, we set $a = 0.35$, $\eta_1 = 0.1$, and $\eta_3 = 0.05$. With 1000 epochs, we set the batch sizes to 128 and 32 for the conventional data augmentation strategy and our one respectively, the weight decay to 10^{-6}, and the base target decay rate to $\tau = 0.99$. For the hash learning, we set $\eta_1 = \eta_2 = \eta_3 = 10^{-4}$ and $\eta_5 = 0.05$ with batch size 128, where η_5 is the learning rate of hash learning component. The momentum of optimization is 0.9 and the weight decay is 10^{-4}.

Comparisons with State-of-the-Arts. The MAP@H $\leq$ 2 of all comparison methods are listed in Table 1, where the underline and bold fonts represent the highest value in the comparison algorithms and all algorithms respectively. These results show that SS-LVH has a stable advantage over other algorithms. Especially at 128 bits, the improvements are 1.7%, 1.3% and 0.4% on MS-COCO, VOC2007 and MIRFLICKR-25K, respectively. Meanwhile, we find that except for contrastive learning methods, the performance of other algorithms will decline when the code length is beyond 32 or 64 bits. We think that the generalization ability derived from visual correlation improves the ability of hash code for carrying semantic information, while our incorporation pattern for the semantic and visual information can enhance this advantage.

To reflect the proportion of retrieved images related to the query image, we show the P@H $\leq$ 2 performance in Fig. 5(a), Fig. 5(b), and Fig. 5(c). SS-LVH achieves remarkable results on three datasets, and averagely exceeds the runner-up algorithm (*i.e.*, CIBHash) by 0.87%, 0.63% and 0.16% respectively. These results verify the superiority of SS-LVH in the perception of objects and semantic information. In addition, we show the aggregation degree of similar image and the R@H $\leq$ 2 performance in Fig. 5(d), Fig. 5(e), and Fig. 5(f). SS-LVH is dominant

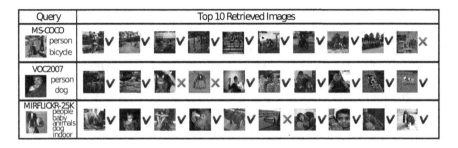

Fig. 6. The top 10 images returned by SS-LVH when we input query images.

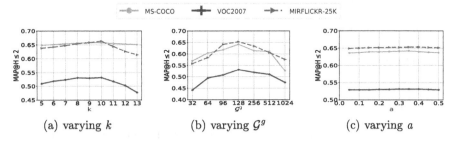

(a) varying k (b) varying $\mathcal{G}^g$ (c) varying a

Fig. 7. MAP@H ≤ 2 w.r.t. k, $\mathcal{G}^g$, and a with 128 bits hash codes on the MS-COCO, VOC2007 and MIRFLICKR-25K datasets.

on three datasets and averagely exceeds the runner-up algorithm (*i.e.*, CIBHash) by 0.86%, 1.78% and 3.01% respectively. These results verify the superiority of SS-LVH in the aggregating similar data and the perception of inter-dependency.

To further demonstrate the retrieval effect of SS-LVH, we visualize the top 10 returned images for three query images in Fig. 6.

Ablation Study. To verify contributions of components including the visual correlation embedding learning branch, OGSP, *Tanh*, and label similarity transformation strategy, we list the influence on MAP@H ≤ 2 at different code lengths using different combinations in Table 2. The 1st row denotes the performance of SS-LVH as a benchmark. The performance without the visual correlation embedding learning branch shows in the 2nd row, where the performance averagely decreases by 2.7%, 3.9%, and 3.47% on three datasets, respectively. These declines manifest the necessity of incorporating this visual branch and only employing SSL is not enough to alleviate the noise problem. In addition, the degradation of performance at 128 bits is remarkable. This result confirms that the visual branch helps the hash code carry more semantic information. In the 3rd and 4th rows, we verify the effect of OGSP. Obviously, MFB is not compatible with these SSL components. We believe that this is because the factorized matrices of MFB disturb the spatial information during activation. Furthermore, the overlapping window brings 1.33%, 0.57%, and 1.53% of increments of performance on three datasets. Finally, we notice the influence of SSL components in the 5th to 8th rows. The result in the 5th row means the performance using pseudo labels without SSL components. Compared with this result, on average, the performance in the 6th row improves **7.3%**, **13.2%**, and **10.1%**, respectively on three datasets. This result shows that the SSL method is important for improvement of hashing performance. Based on SSL, the incorporation of *Tanh* brings 1.23%, 1.57%, and 1.23% of benefits, respectively, and the label similarity transformation strategy also brings 1.07%, 1.13%, and 0.53% of benefits, respectively, on three datasets on average. All in all, our components contribute to performance improvement and the configuration of SS-LVH is optimal.

Hyper-Parameters Sensitivity Analysis. We fix the hyper-parameters that have been verified in other papers and investigate the sensitivity of the designed components' parameters including top-k, $\mathcal{G}^g(\mathcal{G}^s = 32)$, and a. We determine the best hyper-parameter by fixing others with the default values and performing the linear search in candidates. Figure 7 illustrates MAP@H ≤ 2 with 128 bits hash codes on three datasets. According to highest values, SS-LVH can achieve the best retrieval performance when $k = 10$, $\mathcal{G}^g = 128$ and $a = 0.35$.

6 Conclusion

This paper proposes an SS-LVH framework for multi-label image retrieval. Compared with existing methods, we preserve the advantage derived from label co-occurrence correlations and perceive image visual correlation to alleviate the noise problem. The results on three datasets demonstrate the generalization ability and superiority of SS-LVH. Our dual OGSP component, label similarity transformation strategy, and introduction of *Tanh* in BYOL can improve the retrieval performance.

Acknowledgments. This work is supported by the National Natural Science Foundation of China No. 61902135 and No. 62172180, and the Joint Founds of ShanDong Natural Science Funds (Grant No. ZR2019LZH003).

References

1. Cao, Y., Long, M., Liu, B., Wang, J.: Deep cauchy hashing for hamming space retrieval. In: CVPR, pp. 1229–1237 (2018)
2. Cao, Y., Long, M., Wang, J., Liu, S.: Deep visual-semantic quantization for efficient image retrieval. In: CVPR, pp. 916–925. IEEE Computer Society (2017)
3. Chen, T., Kornblith, S., Norouzi, M., Hinton, G.E.: A simple framework for contrastive learning of visual representations. In: ICML, vol. 119, pp. 1597–1607. PMLR (2020)
4. Chen, Z., Wei, X., Wang, P., Guo, Y.: Multi-label image recognition with graph convolutional networks. In: CVPR, pp. 5177–5186 (2019)
5. Devlin, J., Chang, M., Lee, K., Toutanova, K.: BERT: pre-training of deep bidirectional transformers for language understanding. In: Burstein, J., Doran, C., Solorio, T. (eds.) NAACL-HLT, pp. 4171–4186. Association for Computational Linguistics (2019)

6. Everingham, M., Van Gool, L., Williams, C.K.I., Winn, J., Zisserman, A.: The pascal visual object classes (VOC) challenge. Int. J. Comput. Vis. **88**(2), 303–338 (2010)

7. Gattupalli, V., Zhuo, Y., Li, B.: Weakly supervised deep image hashing through tag embeddings. In: CVPR, pp. 10375–10384 (2019)

8. Grill, J., et al.: Bootstrap your own latent - a new approach to self-supervised learning (2020)

9. Huang, C., Yang, S., Pan, Y., Lai, H.: Object-location-aware hashing for multi-label image retrieval via automatic mask learning. IEEE Trans. Image Process. **27**(9), 4490–4502 (2018)

10. Huiskes, M.J., Lew, M.S.: The MIR flickr retrieval evaluation. In: SIGMM, pp. 39–43 (2008)

11. Jin, L., Li, Z., Pan, Y., Tang, J.: Weakly-supervised image hashing through masked visual-semantic graph-based reasoning. In: MM, pp. 916–924 (2020)

12. Lai, H., Yan, P., Shu, X., Wei, Y., Yan, S.: Instance-aware hashing for multi-label image retrieval. IEEE Trans. Image Process. **25**(6), 2469–2479 (2016)

13. Lanchantin, J., Wang, T., Ordonez, V., Qi, Y.: General multi-label image classification with transformers, pp. 16478–16488 (2021)

14. Li, Y., van Gemert, J.: Deep unsupervised image hashing by maximizing bit entropy. In: EAAI, pp. 2002–2010 (2021)

15. Lin, T.-S., et al.: Microsoft COCO: common objects in context. In: Fleet, D., Pajdla, T., Schiele, B., Tuytelaars, T. (eds.) ECCV 2014. LNCS, vol. 8693, pp. 740–755. Springer, Cham (2014). https://doi.org/10.1007/978-3-319-10602-1_48

16. Liu, Y., et al.: Deep self-taught hashing for image retrieval. IEEE Trans. Cybern. **49**(6), 2229–2241 (2019)

17. Luo, X., et al.: A statistical approach to mining semantic similarity for deep unsupervised hashing. In: MM, pp. 4306–4314. ACM (2021)

18. Qiu, Z., Su, Q., Ou, Z., Yu, J., Chen, C.: Unsupervised hashing with contrastive information bottleneck. In: IJCAI, pp. 959–965. ijcai.org (2021)

19. Shen, Y., et al.: Auto-encoding twin-bottleneck hashing. In: CVPR, pp. 2815–2824 (2020)

20. Song, J., He, T., Gao, L., Xu, X., Shen, H.T.: Deep region hashing for efficient large-scale instance search from images. CoRR abs/1701.07901 (2017)

21. Wang, D., Cui, P., Zhu, W.: Structural deep network embedding. In: SIGKDD, pp. 1225–1234 (2016)

22. Wang, J., Yang, Y., Mao, J., Huang, Z., Huang, C., Xu, W.: CNN-RNN: a unified framework for multi-label image classification. In: 2016 IEEE Conference on Computer Vision and Pattern Recognition, CVPR 2016, Las Vegas, NV, USA, 27–30 June 2016, pp. 2285–2294. IEEE Computer Society (2016)

23. Wang, Y., Song, J., Zhou, K., Liu, Y.: Unsupervised deep hashing with node representation for image retrieval. Pattern Recogn. **112**, 107785 (2021)

24. Wu, Z., Shen, C., van den Hengel, A.: Wider or deeper: revisiting the resnet model for visual recognition. Pattern Recogn. **90**, 119–133 (2019)

25. Xie, Y., Liu, Y., Wang, Y., Gao, L., Wang, P., Zhou, K.: Label-attended hashing for multi-label image retrieval, pp. 955–962 (2020)

26. Yang, E., Liu, T., Deng, C., Liu, W., Tao, D.: DistillHash: unsupervised deep hashing by distilling data pairs. In: CVPR, pp. 2946–2955 (2019)

27. Zhang, W., Wu, D., Zhou, Y., Li, B., Wang, W., Meng, D.: Deep unsupervised hybrid-similarity hadamard hashing. In: MM, pp. 3274–3282 (2020)

Shallow Diffusion Motion Model
for Talking Face Generation from Speech

Xulong Zhang[1], Jianzong Wang[1(✉)], Ning Cheng[1], Edward Xiao[2], and Jing Xiao[1]

[1] Ping An Technology (Shenzhen) Co., Ltd., Shenzhen, China
jzwang@188.com
[2] Aquinas International Academy, La Palma, USA

Abstract. Talking face generation is synthesizing a lip synchronized talking face video by inputting an arbitrary face image and audio clips. People naturally conduct spontaneous head motions to enhance their speeches while giving talks. Head motion generation from the speech is inherently difficult due to the nondeterministic mapping from speech to head motions. Most existing works map speech to motion in a deterministic way by conditioning certain styles, leading to sub-optimal results. In this paper, we decompose the speech motion into two complementary parts: pose modes and rhythmic dynamics. Accordingly, we introduce a shallow diffusion motion model (SDM) by equipping a two-stream architecture, *i.e.*, a pose mode branch for primary posture generation, and a rhythmic motion branch for rhythmic dynamics synthesis. On one hand, diverse pose modes are generated by conditional sampling in a latent space, guided by speech semantics. On the other hand, rhythmic dynamics are synced with the speech prosody. Extensive experiments demonstrate the superior performance against several baselines, in terms of fidelity, similarity, and syncing with speech.

Keywords: Talking face · Shallow diffusion · Head motion generation · Speech

1 Introduction

Head motion generation from the speech is to synthesize spontaneous head motions synchronized with input speech audio. Professional speakers are experts in utilizing such motions to effectively deliver information. This task is essential for applications such as digital avatars and social robots [11]. Notably, with this technique, amateur speakers can also generate their own "professional" talking videos, by mimicking moves from professional speakers.

With the development of deep neural networks for generation-related tasks [29,31,38,43,44], talking face can be driven by audio speech. While generating lip motions has been extensively studied in talking face generation [23], synthesizing plausible speech head motions remains an open issue. Specifically, lip motions can be well matched with the input audio using a deterministic mapping, *i.e.*, one to one mapping from phonemes to lip shapes. However, such models can not be trivially extended to the head, due to the highly stochastic nature of head motions during a talk speech. Practically, the speech head motion is highly freedom. Even if the same person gives the same speech twice in a row, there is no guarantee that the speaker would exhibit the same head motions. Moreover, a person usually switches poses from time to time during a long talking speech. The same speech audio does not necessarily lead to a fixed form of motions, and different speeches may go well with the same motion sequence.

Most existing works treat head and lip motion generation in a similar way [14, 41, 46], *i.e.*, the head landmarks are directly inferred from the input audio via a deep network. To simplify the non-deterministic mapping, some methods [15, 18] rely on a set of pre-defined postures, or condition on person-specific styles and templates. These solutions can mimic motions of certain speakers/styles to some degree, but they are limited in terms of motion diversity and fidelity, especially for long talk speeches. Therefore, it is critical to developing algorithms that model the non-deterministic mapping between speech and head motions.

Based on studies in linguistics and psychology [39], speech motion helps the organization and presentation during speech delivery and contributes to both semantics and intonation. Semantically, head motions contribute to the utterance content. For example, some motions are conventionalized and attached to certain linguistic properties (e.g., "nod"). These motions are widely used to facilitate communication. In terms of intonation, the rhythmic movement that matches the prosody of audio could attract the attention of the audience, with the stressed syllable during speech. Moreover, proper rhythmic motions also reflect the progress of the speech and deliver a vivid listening experience. Such speech motion usually has no specific linguistic meaning and manifests as simple and fast hand dynamics related to prosody.

Motivated by these studies, we consider the structure of speech motions from a novel perspective. We introduce the concept of pose mode as the mode of the pose distribution that speakers have for fragments of speech. Considering the speaker's posture in a speech video as a random vector, it follows a multi-modal distribution in the high dimensional space. Modes in such distribution (values with local maximal density) correspond to the habitual postures of speakers. Our work focuses on motions in talk videos, where speakers organize a long speech around a certain topic. Under this setting, the pose modes are mostly habitual postures with no specific global meaning. Consequently, the structure of speech motions can be considered as the sequential transitions of pose modes with rhythmic dynamics under each pose mode. Therefore, the non-deterministic mapping from speech to head motion is decomposed into two parts: a stochastic mapping from speech semantics to pose modes, and the mapping from speech prosody to rhythmic motion dynamics. Our contributions are summarized as follows:

1. To address the non-deterministic mapping from speech to head motions, we propose to decompose the motion into pose modes and rhythmic motions.

The former is stochastically generated with a shallow diffusion model, and the latter is effectively inferred by speech prosody.
2. Extensive experiments demonstrate that our model generates plausible freedom motions well synced with the speech, outperforming other baselines in terms of the fidelity, similarity, and syncing with speech.

2   Related Work

Talking face generation is a cross-modal image synthesis task, Brand *et al.* [2] proposed *Voice Puppetry* for the generation of full facial animation from speech. With audio-driven facial animation, it can assist animation generation and film production. In the following paragraphs, we will overview the prior works about the audio-driven facial animation methods, which consist of facial landmarks, lip-sync animation, speaker-related animation, and image generation.

Facial Landmarks. A deep neural network-based facial landmarks generation is proposed by Eskimez *et al.* [10]. It was used in the talking face generation and improved speech intelligibility robust to noisy conditions. Chen *et al.* [5] proposed a cascade GAN-based method to generate a talking face, instead of learning a direct mapping between audio and image, a high-level structure of facial landmarks is used as a middle representation. First, transfer audio to landmarks and then generate the image conditioned on the landmarks. Greenwood *et al.* [13] jointly learn full-face animation and head pose, the landmarks were used as the image representation. In the image, each person had 62 landmarks distributed about the face, the landmarks along with lip edges and eyes. and translation combined.

Lip-Sync Animation. Given an arbitrary audio speech and one image of an arbitrary speaker, generating lip movement sync with the speech content is the lip-sync animation task. With the increased power of GPU computation, end-to-end learning [24,25,27,30,35] from audio to video frames have huge progress. Chen *et al.* [4] proposed to train an end-to-end model with a novel correlation loss to synchronize lip changes and speech changes, which is robust to view angels, lip shapes, and facial texture. Song *et al.* [26] propose a conditional recurrent generation network to build a temporal model for accurate lip synchronization, it considers the temporal dependency across video frames. To boost the accuracy of lip synchronization, a lip-reading discriminator is added. Vougioukas *et al.* [34] proposed an end-to-end method, using a static image of a speaker and an audio speech, without relying on handcrafted intermediate features. The model is based on a temporal GAN, that uses discriminators for the audio-visual synchronization, it generates lip movements sync with the speech. The speech styles like shouting or mumbling are related to the motion of face motion, Zhou *et al.* [47] proposed a three-stage LSTM network architecture to produce animator-centric speech motion curves, it is a real-time lip-sync from audio.

Speaker Related Animation. Given audio of a specific person, to synthesize a high-quality video of him speaking, replicate the sound and cadence of a person's voice. The speaker-related animation needs to model not only the speech content, but also requires to model the target style how it speaks, and how it expresses itself. Suwajanakorn *et al.* [28] used a recurrent neural network to learn the mapping between audio to mouth shapes conditioned on the same person of Obama. With the speaker-related model, it learns the texture of the lip. Cudeiro *et al.* [8] proposed a model that factors identity from facial motion, conditioning on speaker labels during training allows the model to learn different speaking styles. Thies *et al.* [32] proposed method with a latent variable to model the face of the target speaker, it learns temporal stability while rendering to generate video frames.

Image Generation. Fišer *et al.* [12] introduced a method of wrapping-based portrait video generation, with a controllable amount of landmarks to perform non-parametric texture synthesis. For the face image, image to image translation is popularly used to talking face synthesis. Thies *et al.* [33] proposed Face2Face to animate the facial expressions of the target speaker and re-render the output video in a photo-realistic fashion. It shows the robust appearance of face transfer between talking face videos. GAN-based method was proposed by Kim *et al.* [19], a recurrent GAN captures the Spatio-temporal features of talking face and could copy facial expressions from source to target speaker. A cycle-consistency loss [42] is added to the model for the facial expression styles transfer. Zakharov *et al.* [40] proposed few-shot talking face generation method, it performs meta-learning on a large dataset. The model embeds the face landmarks into embedding vectors, and the generator network maps the face landmarks into the output frames.

3   Method

We proposed a method called the shallow diffusion motion model (SDM) to generate a talking face sequence according to a given speech. To this end, a mapping from speech to face motion is required. We decomposed the talking face into pose motion and rhythmic motion. Additionally, to address the over-smoothing of generation of the talking face, a shallow diffusion mechanism was proposed for the generation of the motion sequence. Correspondingly, there are three modules for the proposed method, and the framework is shown in Fig. 1. For the input image and the video frames, we use a pre-trained face landmark detector [3] to do a preprocess and use the movement of the landmarks as the motion of the talking face.

Given a speech audio S and the corresponding video frame sequence contain the taking face F. The content encoder extracted feature on the input speech S, the content encoder is built up by four convolutional layers. The model of SDM is to learn the mapping between $S_{(i)}$ and $F_{(i)}$ of the ith frame. In this work, we used mel-spectrum as the feature representation of speech $S_{(i)}$ and keypoint landmarks of the human face as the representation of the visual face frame $F_{(i)}$.

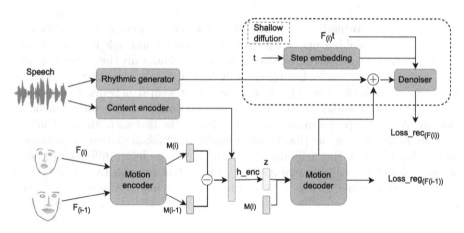

Fig. 1. The framework of talking face motion model with shallow diffusion model. The whole framework consists of two branches, a speech branch for the mapping of speech content and rhythm motions, a visual branch to learn the pose motion between frames.

To learn this mapping, we decompose the motion of talking face $F_{(i)}$ into two parts:

$$F_{(i)} = \overline{F}_{(i)} + \widetilde{F}_{(i)} \tag{1}$$

where $\overline{F}_{(i)}$ is content-related motion, and $\widetilde{F}_{(i)}$ is the rhythmic related motion of the talking face. The content-related motion can be regarded as the main motion of the head and the rhythmic-related motion could be the dynamics of the talking head. Finally, we use a shallow diffusion mechanism model for image generation.

3.1   Pose Motion Generation

The pose motion contains a motion encoder and a motion decoder. The content related pose motion conditioning on the content of the audio speech S_{cont} and the pose motion of the previous frame $F_{(i-1)}$.

$$\overline{F}_{(i)}^{*} = G_c(F_{(i-1)}, S_{cont}) \tag{2}$$

where the G_c represent the content related pose motion generator, and we use the superscript $*$ for the representation of the result of the generator.

The motion encoder encodes the Frame $F_{(i)}$ and $F_{(i-1)}$ into the latent vectors $M_{(i)}$ and $M_{(i-1)}$ separately. Conducting a subtraction between the neighbor frames could get the change the motion. With the condition on speech content to sync the motion change with the speech. The motion decoder does a reconstruction of the previous frame $F_{(i-1)}$ with the motion change variable and the latent vector $M_{(i)}$. The content-related motion can be formulated as a condition motion predictor. During the training phase, for each frame $F_{(i)}$ has fixed the previous frame of $F_{(i-1)}$, the module of the motion decoder conducts the reconstruction of the motion of frame $F_{(i-1)}$ as:

$$M^*_{(i-1)} = f_{dec}(z, M_{(i)}) = f_{dec}(M_{(i-1)}) \tag{3}$$

where f_{dec} is the motion decoder, and z is the latent variable of the layer h_enc. We use the the motion reconstruct to regularize the embedding space of the motion encoder and decoder:

$$\mathcal{L}_{reg} = ||M_{(i)} - f_{dec}(M_{(i)})|| + ||M_{(i-1)} - f_{dec}(M_{(i-1)})|| \tag{4}$$

This forces the motion decoder to use the information of the latent variable z.

3.2   Rhythmic Motion Generation

The rhythmic motion is changed according to the temporal domain, it is important for the talking face to control the motion with dynamics. We generate the rhythmic motion through the rhythmic dynamics of the prosodic information in speech. It can keep the sync of prosody between the visual and audio.

In the control of rhythmic motion generation, we use a rhythmic generator for the dynamics motion embedding. The rhythmic generator is mainly built up with a convolutional network. The rhythmic motion is independent of the motion learned from the contend related pose motion, and the loss is defined as:

$$\mathcal{L}_{ind} = ||\widetilde{M^*}_{(i)} - \overline{M^*}_{(i)}|| \tag{5}$$

The $\mathcal{L}_{ind}$ ensures the generated rhythmic motion pose $\widetilde{M^*}_{(i)}$ independent to the content related motion pose $\overline{M^*}_{(i)}$. It helps the dynamics of motion are not affected by the content of speech.

3.3   Shallow Diffusion Mechanism

The shallow diffusion mechanism is applied to the image animation generation. The main module of the shallow diffusion mechanism comes from the diffusion model [20–22]. The diffusion model contains two processes, a diffusion process to convert the image data into a Gaussian distribution step by step, and a reverse process to reconstruct the image data from Gaussian white noise. The pipeline of a diffusion model is shown in Fig. 2.

Diffusion Process. Let the distribution of data $F^0_{(i)}$ as $p(F^0_{(i)})$, the diffusion process converts the $F^0_{(i)}$ into $F^T_{(i)}$ step by step with a Markov chain with fixed parameters. The T steps conversion can be formulated as:

$$q(F^{1:T}_{(i)}|F^0_{(i)}) = \prod_{t=1}^{T} q(F^T_{(i)}|F^{t-1}_{(i)}) \tag{6}$$

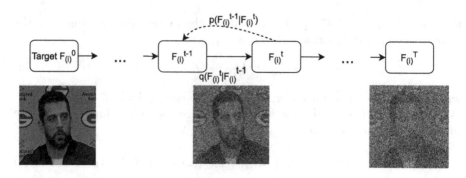

Fig. 2. The two processes of diffusion model. The diffusion process is from $F_{(i)}^0$ to $F_{(i)}^T$, the reverse process is from $F_{(i)}^T$ to $F_{(i)}^0$.

At each step $t \in (1, T)$, a Gaussian noise multiply with a variance of $\alpha \in [\alpha_1, \cdots, \alpha_T]$ is added to the $F_{(i)}^{t-1}$ to obtain $F_{(i)}^t$.

$$q(F_{(i)}^t | F_{(i)}^{t-1}) = \mathcal{N}(F_{(i)}^t; \sqrt{1-\alpha_t} F_{(i)}^{t-1}, \alpha_t \mathbf{I}) \tag{7}$$

If the parameters of α are well designed, and the step T is larger enough, the final $q(F_{(i)}^T)$ is equally an isotropic Gaussian distribution.

Reverse Process. The reverse process is from $F_{(i)}^T$ to $F_{(i)}^0$, which is follow the Markov chain with learnable parameters θ. The reverse process can be approximate it with the neural networks with the parameters θ. It can be formulated as:

$$p_\theta(F_{(i)}^{0:T}) = p(F_{(i)}^T) \prod_{t=1}^T p_\theta(F_{(i)}^{t-1} | F_{(i)}^t) \tag{8}$$

To learn the parameters θ, we optimizing the loss with stochastic gradient descent on:

$$\mathcal{L}_{diff} = D_{KL}(q(F_{(i)}^{t-1} | F_{(i)}^t, F_{(i)}^0) \| p_\theta(F_{(i)}^{t-1} | F_{(i)}^t)) \tag{9}$$

where $D_{KL}()$ is the Kullback-Leibler divergence. Finally, with the trained network, we can sample from $p(F_{(i)}^T) \sim \mathcal{N}(0, I)$ to generate the target data with the reverse process.

When the step of T is big enough, the trajectory from $F_{(i)}^0$ to Gaussian $F_{(i)}^T$ and the trajectory from $F_{(i)}^T$ to $F_{(i)}^0$ will meet in a step t. Inspired by this point, we can use an auxiliary predictor to predict the step of t. With the step of t to do a shallow diffusion process. And the reverse process could also start at the predicted step of t.

3.4   Training Losses

The reconstruct loss is applied to final reconstution of the image F_i^*:

$$\mathcal{L}_{rec} = ||F_i^* - F_{(i)}|| \tag{10}$$

Overall the total loss is defined as:

$$\mathcal{L}_{total} = \lambda_1 \mathcal{L}_{reg} + \lambda_2 \mathcal{L}_{ind} + \lambda_3 \mathcal{L}_{rec} \tag{11}$$

where the $\lambda_1 \sim \lambda_3$ is hyperparameters for balancing the different losses.

3.5   Testing Stage

The pipeline of talking face inference phase is followed as the function:

$$
\begin{aligned}
F_{(i)}^* &= G(F_{(i-1)}^*, S(i)) \\
&= f_{denoi}(f_{dec}(f_{con}(S(i)), f_{enc}(F_{(i-1)})), f_{rhy}(S(i)), t, F_{(i)^t})
\end{aligned}
\tag{12}
$$

where f_{denoi} is the reverse processes of diffusion, f_{rhy} is the rhythmic generator. The final results of talking face video are a stack of the frames $F_{(1)}^*, \cdots, F_{(n)}^*$ with the tool of ffmpeg.

4   Experiments and Results

4.1   Experimental Setup

Datasets. We used three datasets for the experimental evaluation, it contains VoxCeleb2 [6], LRW [7] and LRS3-TED [1]. The VoxCeleb2 contains more than 6000 celebrities and covers 1 million utterances in speech. The LRW is a large dataset containing 1000 speakers, and each speaker spoke 500 different words. The LRS3-TED includes face track over 400 h of videos from TED and TEDx, it has more challenges with head movements than others. We follow the raw split as the ratio of the dataset.

Training Details. We use the optimizer of ADAM with the learning rate of 2×10^{-4}, and the β_1 of 0.4, β_2 of 0.999. In the traning phase, we set the loss weight in Eq. 11 as λ_1 of 5, λ_2 of 2, and λ_3 of 1. The experiment was conducted on a single GPU of NVIDIA Tesla V100 with 16 GB memory.

Metrics of Evaluation. For the quantitative evaluation, we adopted several criteria, it includes Frchet Inception Distance (FID) [16], which was used to quantify the fidelity of the synthesized image, and structured similarity (SSIM) [36], it was used to compare the similarity of the synthesized image and real images. We use cosine similarity (CSIM) [40] to identify the speaker identity preserving ability, which computed the cosine distance between the embedding vectors of a face recognition network [9]. To check if the synthesized video contains sync movement of the lip to speech content, we use Landmarks Distance (LMD) [4] for evaluation.

4.2   Results and Analysis

Comparision with Talking Face Methods. We first compare the proposed method with the related works of talking face methods, we select the audio-driven method. With given a single images and an audio to generate the video of talking face, which has been studied in Zhou *et al.* [45], Song *et al.* [26], Chung *et al.* [17], Vougioukas *et al.* [34], Chen *et al.* [5], and Wiles *et al.* [37]. For a fair comparison, all the methods were input with the same image and speech from the test dataset. And we do a preprocess on the input image with the same cropping area. The quantitative evaluation results are shown in Table 1.

Table 1. Comparisions with different audio to video methods on the three public dataset of VoxCeleb2, LRW, and LRS3-TED. The score of FID and LMD smaller is better, while for SSIM and CSIM bigger is better. We bold each leading score.

Method	Datasets											
	VoxCeleb2				LRW				LRS3-TED			
	FID	SSIM	CSIM	LMD	FID	SSIM	CSIM	LMD	FID	SSIM	CSIM	LMD
Zhou *et al.* [45]	137	0.84	0.32	4.8	149	0.85	0.39	3.7	221	0.72	0.27	6.2
Song *et al.* [26]	163	0.78	0.27	5.6	134	**0.91**	0.45	**3.1**	204	0.62	0.28	6.5
Chung *et al.* [17]	159	0.79	0.29	5.4	132	**0.91**	0.44	**3.1**	212	0.58	0.32	6.7
Vougioukas *et al.* [34]	127	**0.85**	0.33	6.3	116	0.88	0.35	3.6	196	0.63	0.26	6.4
Chen *et al.* [5]	142	0.82	0.31	4.9	151	0.84	0.38	3.3	294	0.66	0.31	4.8
Wiles *et al.* [37]	117	0.65	0.31	4.8	107	0.69	0.31	3.2	172	0.57	0.28	5.6
Ours	**97**	0.74	**0.42**	**3.4**	**102**	0.76	**0.49**	**3.1**	**122**	**0.79**	**0.44**	**3.2**

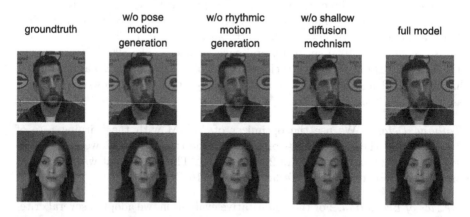

| groundtruth | w/o pose motion generation | w/o rhythmic motion generation | w/o shallow diffusion mechnism | full model |

Fig. 3. The ablation studies with visualization, three main modules of pose motion generation, rhythmic motion generation, and shallow diffusion mechnism are compaired with the full model.

Note that in the preprocess of our method, we did not include the affine transformation, which leads to a lower score in terms of SSIM. From the results shown in Table 1, we can see except for the SSIM score, our method could achieve

the best performance than other related audio to video methods in most evaluation metrics. As shown, the proposed method outperforms other baselines,
suggesting better generation ability in relating audio and motion. Our model
shows strong performances on the fidelity of the synthesized image by the low
FID scores, while other baselines fail to generate high fidelity images on some
speakers. Our model is more robust to lip motion syncing, leading to lower averaged LMD scores. Our model is more accurate keep the speaker identity in the
synthesized image, which leads to a high score of CSIM.

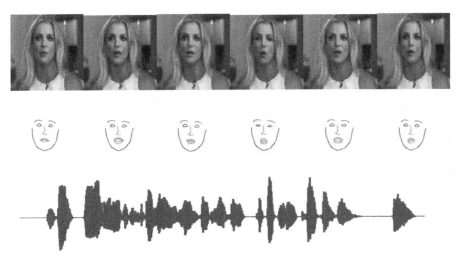

"Today, we are introducing our talking face method of ..."

Fig. 4. End to end video generation results. From bottom to top row are speech text,
speech audio, generation keypoint, and the final video sequence.

Ablation Studies. We compare the contributions of different modules in the ablation studies, the primary modules described in Sect. 3. We conduct the experiments on the dataset of VoxCeleb2. As shown in Fig. 3, we visualize the result
of each module compared with the full model.

From the results shown in Fig. 3, we can see the synthesized frames without
a shallow diffusion mechanism in the fourth column, the motion of the face
has a bigger distance from the groundtruth. We attribute this to the shallow
diffusion model, the diffusion-based module could synthesize the target image
more robustly, which could stabilize the generation and could lead to a faster
convergence during training. Another case we found in the ablation studies is
the pose motion generation module affects the lip part of the face in the second
column. Without the pose motion generation module, the synthesized image
could not control the mouth for the speech content.

Video Results. Further, we show the results based on our generated motion to the video frames in Fig. 4. The video can be generated end to end by inputting a speech and an image. We can apply the method to the arbitrary input image in the wild, it can generate any identity. It can be used for the recording of video presentations.

5 Conclusion

In this work, we propose an approach based on a shallow diffusion mechanism that synchronizes faces with speech content through rhythmic movements of the head. We solve the non-deterministic mapping problem by decomposing the difficult task into complementary parts. Given input speech audio, pose motion generation generates different pose patterns sequentially through conditional sampling, while rhythmic motion generation simultaneously enriches each pose pattern dynamically with audio-conditioned rhythms to achieve spontaneous movements. Our model generates highly diverse and visually plausible face images in a shallow diffusion mode, from the prediction time step to conducting the reverse process of diffusion.

Acknowledgement. This paper is supported by the Key Research and Development Program of Guangdong Province under grant No.2021B0101400003. Corresponding author is Jianzong Wang from Ping An Technology (Shenzhen) Co., Ltd (jzwang@188.com).

References

1. Afouras, T., Chung, J.S., Zisserman, A.: Lrs3-ted: a large-scale dataset for visual speech recognition. arXiv preprint arXiv:1809.00496 (2018)
2. Brand, M.: Voice puppetry. In: Proceedings of the 26th Annual Conference on Computer Graphics and Interactive Techniques, pp. 21–28 (1999)
3. Bulat, A., Tzimiropoulos, G.: How far are we from solving the 2d & 3d face alignment problem? (and a dataset of 230,000 3d facial landmarks). In: Proceedings of the IEEE International Conference on Computer Vision, pp. 1021–1030 (2017)
4. Chen, L., Li, Z., Maddox, R.K., Duan, Z., Xu, C.: Lip movements generation at a glance. In: Proceedings of the European Conference on Computer Vision (ECCV), pp. 520–535 (2018)
5. Chen, L., Maddox, R.K., Duan, Z., Xu, C.: Hierarchical cross-modal talking face generation with dynamic pixel-wise loss. In: Proceedings of the IEEE/CVF Conference on Computer Vision and Pattern Recognition, pp. 7832–7841 (2019)
6. Chung, J.S., Nagrani, A., Zisserman, A.: Voxceleb2: deep speaker recognition. arXiv preprint arXiv:1806.05622 (2018)
7. Chung, J.S., Zisserman, A.: Lip reading in the wild. In: Lai, S.-H., Lepetit, V., Nishino, K., Sato, Y. (eds.) ACCV 2016. LNCS, vol. 10112, pp. 87–103. Springer, Cham (2017). https://doi.org/10.1007/978-3-319-54184-6_6
8. Cudeiro, D., Bolkart, T., Laidlaw, C., Ranjan, A., Black, M.J.: Capture, learning, and synthesis of 3d speaking styles. In: Proceedings of the IEEE/CVF Conference on Computer Vision and Pattern Recognition, pp. 10101–10111 (2019)

9. Deng, J., Guo, J., Xue, N., Zafeiriou, S.: Arcface: additive angular margin loss for deep face recognition. In: Proceedings of the IEEE/CVF Conference on Computer Vision and Pattern Recognition, pp. 4690–4699 (2019)

10. Eskimez, S.E., Maddox, R.K., Xu, C., Duan, Z.: Generating talking face landmarks from speech. In: Deville, Y., Gannot, S., Mason, R., Plumbley, M.D., Ward, D. (eds.) LVA/ICA 2018. LNCS, vol. 10891, pp. 372–381. Springer, Cham (2018). https://doi.org/10.1007/978-3-319-93764-9_35

11. Eskimez, S.E., Zhang, Y., Duan, Z.: Speech driven talking face generation from a single image and an emotion condition. IEEE Trans. Multimedia **24**, 3480–3490 (2021)

12. Fišer, J., et al.: Example-based synthesis of stylized facial animations. ACM Trans. Graph. (TOG) **36**(4), 1–11 (2017)

13. Greenwood, D., Matthews, I., Laycock, S.: Joint learning of facial expression and head pose from speech. In: Interspeech (2018)

14. Guo, Y., Chen, K., Liang, S., Liu, Y.J., Bao, H., Zhang, J.: Ad-nerf: audio driven neural radiance fields for talking head synthesis. In: Proceedings of the IEEE/CVF International Conference on Computer Vision, pp. 5784–5794 (2021)

15. Gupta, A., Khan, F.F., Mukhopadhyay, R., Namboodiri, V.P., Jawahar, C.: Intelligent video editing: incorporating modern talking face generation algorithms in a video editor. In: Proceedings of the Twelfth Indian Conference on Computer Vision, Graphics and Image Processing, pp. 1–9 (2021)

16. Heusel, M., Ramsauer, H., Unterthiner, T., Nessler, B., Hochreiter, S.: Gans trained by a two time-scale update rule converge to a local nash equilibrium. Adv. Neural Inf. Processi. Syst. **30**, 1–12 (2017)

17. Jamaludin, A., Chung, J.S., Zisserman, A.: You said that?: synthesising talking faces from audio. Int. J. Comput. Vision **127**(11), 1767–1779 (2019)

18. Ji, X., Zhou, H., Wang, K., Wu, W., Loy, C.C., Cao, X., Xu, F.: Audio-driven emotional video portraits. In: Proceedings of the IEEE/CVF Conference on Computer Vision and Pattern Recognition, pp. 14080–14089 (2021)

19. Kim, H., et al.: Neural style-preserving visual dubbing. ACM Trans. Graph. (TOG) **38**(6), 1–13 (2019)

20. Kingma, D., Salimans, T., Poole, B., Ho, J.: Variational diffusion models. Adv. Neural Inf. Process. Syst. **34**, 21696–21707 (2021)

21. Kong, Z., Ping, W., Huang, J., Zhao, K., Catanzaro, B.: Diffwave: a versatile diffusion model for audio synthesis. In: 9th International Conference on Learning Representations, ICLR 2021 (2021)

22. Lam, M.W., Wang, J., Su, D., Yu, D.: Bddm: bilateral denoising diffusion models for fast and high-quality speech synthesis. In: International Conference on Learning Representations (2021)

23. Meshry, M., Suri, S., Davis, L.S., Shrivastava, A.: Learned spatial representations for few-shot talking-head synthesis. In: Proceedings of the IEEE/CVF International Conference on Computer Vision, pp. 13829–13838 (2021)

24. Qu, X., Wang, J., Xiao, J.: Enhancing data-free adversarial distillation with activation regularization and virtual interpolation. In: IEEE International Conference on Acoustics, Speech and Signal Processing, pp. 3340–3344. IEEE (2021)

25. Si, S., Wang, J., Peng, J., Xiao, J.: Towards speaker age estimation with label distribution learning. In: ICASSP 2022–2022 IEEE International Conference on Acoustics, Speech and Signal Processing (ICASSP), pp. 4618–4622 (2022). https://doi.org/10.1109/ICASSP43922.2022.9746378

26. Song, Y., Zhu, J., Li, D., Wang, A., Qi, H.: Talking face generation by conditional recurrent adversarial network. In: Proceedings of the Twenty-Eighth International Joint Conference on Artificial Intelligence, IJCAI 2019, pp. 919–925 (2019)

27. Sun, A., et al.: Reconstructing dual learning for neural voice conversion using relatively few samples. In: IEEE Automatic Speech Recognition and Understanding Workshop, pp. 946–953. IEEE (2021)

28. Suwajanakorn, S., Seitz, S.M., Kemelmacher-Shlizerman, I.: Synthesizing obama: learning lip sync from audio. ACM Trans. Graph. (ToG) **36**(4), 1–13 (2017)

29. Tang, H., Zhang, X., Wang, J., Cheng, N., Xiao, J.: Avqvc: one-shot voice conversion by vector quantization with applying contrastive learning. In: 2022 IEEE International Conference on Acoustics, Speech and Signal Processing (ICASSP2022), pp. 1–5. IEEE (2022)

30. Tang, H., et al.: TGAVC: Improving autoencoder voice conversion with text-guided and adversarial training. In: IEEE Automatic Speech Recognition and Understanding Workshop (ASRU2021), pp. 938–945. IEEE (2021)

31. Tang, J., Wu, Y., Li, M., Wang, Z.: Talking face generation based on information bottleneck and complementary representations. In: Proceedings of the 30th ACM International Conference on Information & Knowledge Management, pp. 3443–3447 (2021)

32. Thies, J., Elgharib, M., Tewari, A., Theobalt, C., Nießner, M.: Neural voice puppetry: audio-driven facial reenactment. In: Vedaldi, A., Bischof, H., Brox, T., Frahm, J.-M. (eds.) ECCV 2020. LNCS, vol. 12361, pp. 716–731. Springer, Cham (2020). https://doi.org/10.1007/978-3-030-58517-4_42

33. Thies, J., Zollhofer, M., Stamminger, M., Theobalt, C., Nießner, M.: Face2face: real-time face capture and reenactment of rgb videos. In: Proceedings of the IEEE Conference on Computer Vision and Pattern Recognition, pp. 2387–2395 (2016)

34. Vougioukas, K., Petridis, S., Pantic, M.: Realistic speech-driven facial animation with gans. Int. J. Comput. Vision **128**(5), 1398–1413 (2020)

35. Wang, Q., Zhang, X., Wang, J., Cheng, N., Xiao, J.: Drvc: a framework of any-to-any voice conversion with self-supervised learning. In: 2022 IEEE International Conference on Acoustics, Speech and Signal Processing (ICASSP2022), pp. 3184–3188. IEEE (2022)

36. Wang, Z., Bovik, A.C., Sheikh, H.R., Simoncelli, E.P.: Image quality assessment: from error visibility to structural similarity. IEEE Trans. Image Process. **13**(4), 600–612 (2004)

37. Wiles, O., Koepke, A., Zisserman, A.: X2face: a network for controlling face generation using images, audio, and pose codes. In: Proceedings of the European Conference on Computer Vision (ECCV), pp. 670–686 (2018)

38. Xu, L., Zhou, X.: A crowd-powered task generation method for study of struggling search. Data Sci. Eng. **6**(4), 472–484 (2021)

39. Yao, X., Fried, O., Fatahalian, K., Agrawala, M.: Iterative text-based editing of talking-heads using neural retargeting. ACM Trans. Graph. (TOG) **40**(3), 1–14 (2021)

40. Zakharov, E., Shysheya, A., Burkov, E., Lempitsky, V.: Few-shot adversarial learning of realistic neural talking head models. In: Proceedings of the IEEE/CVF International Conference on Computer Vision, pp. 9459–9468 (2019)

41. Zhang, C., Ni, S., Fan, Z., Li, H., Zeng, M., Budagavi, M., Guo, X.: 3d talking face with personalized pose dynamics. IEEE Trans. Visualization Comput. Graph. **29**, 1438–1449 (2021)

42. Zhang, X., Wang, J., Cheng, N., Xiao, E., Xiao, J.: CycleGEAN: cycle generative enhanced adversarial network for voice conversion. In: IEEE Automatic Speech Recognition and Understanding Workshop (ASRU2021), pp. 1–6. IEEE (2021)
43. Zhang, X., Wang, J., Cheng, N., Xiao, J.: Susing: su-net for singing voice synthesis. In: International Joint Conference on Neural Networks, IJCNN 2022. IEEE (2022)
44. Zhang, X., Wang, J., Cheng, N., Xiao, J.: Tdass: target domain adaptation speech synthesis framework for multi-speaker low-resource tts. In: International Joint Conference on Neural Networks, IJCNN 2022. IEEE (2022)
45. Zhou, H., Liu, Y., Liu, Z., Luo, P., Wang, X.: Talking face generation by adversarially disentangled audio-visual representation. In: Proceedings of the AAAI Conference on Artificial Intelligence, vol. 33, pp. 9299–9306 (2019)
46. Zhou, H., Sun, Y., Wu, W., Loy, C.C., Wang, X., Liu, Z.: Pose-controllable talking face generation by implicitly modularized audio-visual representation. In: Proceedings of the IEEE/CVF Conference on Computer Vision and Pattern Recognition, pp. 4176–4186 (2021)
47. Zhou, Y., Xu, Z., Landreth, C., Kalogerakis, E., Maji, S., Singh, K.: Visemenet: audio-driven animator-centric speech animation. ACM Trans. Graph. (TOG) **37**(4), 1–10 (2018)

Hierarchical Clustering and Measure for Tourism Profiling

Sonia Djebali(✉)🆔, Quentin Gabot(✉), and Guillaume Guerard(✉)🆔

Léonard De Vinci, Research Center, La Défense, 92 916 Paris, France
{sonia.djebali,quentin.gabot,guillaume.guerard}@devinci.fr

Abstract. Social network analysis has become widespread in recent years, especially in digital tourism. Indeed, the vast amount of data that tourists produce during their travels represents an effective source for interpreting their behaviors (geographics, demographics, psychographics, movement patterns). Since the classic measures unfit to those kind of information, this article presents a new measure to determine tourist profiles thanks to the digital traces left on social networks. This measure is based on geographic, demographic and pattern's behaviors of the tourists as the context and the content of their trips. The approach is simulated and evaluated experimentally with a hierarchical clustering on the traces left by tourists on TripAdvisor in the French capital *Paris*. Clusters found correspond to tourism segment determined by the Tourism Office of Paris.

Keywords: Tourism profiling · Machine learning · Distance measure

1 Introduction

The *World Tourism Organization* recorded 1.5 billion international tourist arrivals worldwide in 2019, an increase of 4% over 2018. Tourism is responsible for 10.3% of the world's gross domestic product and is considered one of the largest and fastest-growing industries. Tourism actors such as tourist offices, cultural and commercial services analyze the behavior of tourists to know their motivations as profiles, to adapt to their demands, and thus to help them make decisions [8]. Profiles are typically determined by surveys and polls. However, the emergence of social networks, such as Facebook, Flickr, TripAdvisor, and Booking, has created a new paradigm for the study of tourism profiling.

In the literature, to create tourist profiles, tourist experiences are processed and common characteristics of tourists with similar experiences are captured to extract knowledge. Profiling is mostly performed as in the case of recommendation systems by finding similar people as filtering methods. Those methods are biased because the profiles are selected in advance. In order not to induce bias, we consider that profiling should be unsupervised.

In this paper, we propose a new measure called `Tourists Profile Measure` (TPM), used by a hierarchical clustering, to determine tourist profiles considering

© The Author(s), under exclusive license to Springer Nature Switzerland AG 2023
B. Li et al. (Eds.): APWeb-WAIM 2022, LNCS 13422, pp. 158–165, 2023.
https://doi.org/10.1007/978-3-031-25198-6_12

geographic, demographic and behavioral information left by tourist on social medias. From the TPM measure, an hierarchical clustering algorithm determines groups of tourists' stays. They are examined to extract information, as a profile and perform various comparisons between them. This method can be applied to any dataset without the need for expertise.

The main contributions of our work can be summarized as follows:

- A summary of tourists' stays based on data shared via social networks.
- TPM, a new measure to qualify the proximity between two tourist stays.
- A knowledge extraction of profiles.

This article is organized as follows. In Sect. 2, we present related work on tourism profiling. In Sect. 3, we formalize and enrich our dataset. In Sect. 4, we present our new measure to compare the tourist experience and to generate the tourist profile using the classification method presented in Sect. 5. Our method is implemented and is the subject of a case study on a TripAdvisor dataset in Sect. 6. We finish with a conclusion about the presented works.

2   Literature Review

Our objective in this study is to establish tourism profiles that are not biased by this preliminary choice. We seek to create profiles using an unsupervised method to extract knowledge. To achieve this goal, we must address three major challenges. The first is how to define an experience in the context of tourism; the second is how to define tourist profiles and the third is how to extract knowledge from these profiles. The literature review presented below is structured along these three axes.

To Define Tourist's Experiences. The initial challenge of profiling tourists is to identify the key characteristics of tourist experiences. In the literature, some studies consider the demographic data of the tourist as a characteristic to achieve a classification [11]. Other studies explore other features such as interests, order of visits, semantic analysis of comments, or photo location [5]. Some studies consider stays with their context *i.e.* season, duration, weather, etc. [9]. The objective of our study is to determine tourist profiles, so we need all the information about tourists, the context of their stays, and their interests.

Define Profiles. Apart filtering methods and polls, most studies use machine learning approaches (supervised and unsupervised). Concerning supervised learning, many recent studies use polls and/or social network data to improve the profiling of tourists and enrich existing (already labeled) profiles [3]. Popular classification algorithms for profile enrichment include *K-Nearest Neighbors*, *Naive Bayes* and *Support Vector Machine* [4]. However, supervised learning methods have the same biases as filtering methods. In this case, an apriori choice of profiles on which to infer the rest of the data. About unsupervised learning, studies dealing with tourism recommendation systems consider a matrix composed of the set of tourism locations and implement methods such as *Latent Class Analysis* on it [6]. However, given the diversity of tourist places, it is often unlikely to find tourists with similar visiting experiences. Many other studies group tourists based on point-of-interest ratings to find tourist preferences [7]. However, the context of the stays or the social information of the tourists is often neglected.

Extracting Knowledge. Although unsupervised learning represents a popular and useful approach, it is more difficult to handle than supervised learning. One reason is the often opaque meaning or meaningless of the clusters discovered by unsupervised learning algorithms. It is a significant challenge to extract knowledge from them and analyze it against reality.

Many studies focus on a very precise piece of information deduced from tourists' stays and ignore essential elements such as the content of the stay (points of interest visited) or the context of the stay (duration and season). In the absence of a measure that can compare all of this information, the studies focus on either the content or the context. The main contribution of the paper is a new measure dedicated to the tourism profiling.

3   Touristic Data

We focus this on the study and analysis of tourist profiles based on the digital traces left by tourists on social networks. Digital traces refer to the digital data intentionally left by tourists on these networks. Data includes information about tourists, information about the places they visit, and their interactions.

Tourists' behaviors and decisions are influenced by a set of external parameters called contextual factors. They refer to the general background within which the tourist operates, like the season, weather conditions, length of the stay, social factors, etc. Contextual factors are not present during the extraction of digital traces. Therefore, we will enrich the data set.

Tourists make a series of stays consisting of visits to various places. A stay refers to a length of time beginning with the time the tourist leaves its usual place of residence and the time the tourist leave the destination area. Each stay is a chronological succession of places that the tourist has visited. To build this set of stays, we will rely on the comments left by tourists on the networks. The method was previously presented in a previous paper [1].

Contextual factors of a stay can be of two kinds, push factors and pull factors. Push factors cause tourists to go. These include natural motivations like the climate of the home country and institutionalized ones like school vacations. Pull factors attract tourists and relate to the destination area. They include the climate of the country visited, cultural events, or sports seasons. To study tourism profiling, we will focus on pull factors. We compute season and length of stay from the stay's building.

Determining the season of the tourist's country of origin is complex due to the lack of information of its departure. We will take into account only the season of the destination deduced from the dates of the beginning and the end of a stay

Table 1. Ontology of places.

Category	Subcategory
Heritage	Monuments, Parks and Gardens, Urbanism (neighborhoods, bridges, cemeteries, streets)
Cultural Buildings	Art galleries and Museums, Holy sites and Places of worship, Historic buildings, Theaters and Auditorium
Food and Services	Shops, Restaurants and Bars, Gastronomy, Hotels
Entertainment	Music buildings (concerts, discotheques), Cinemas, Amusement park, Sports
Viewpoints	(no sub-categories)
Nature	Woods, Watering place (river, lake), Beaches and Mountains

and the country visited. The duration of the stay is equal to the date difference between the first comment of the stay and the last comment of the same stay.

To study tourism content, we will classify tourist places based on an ontology. In the literature, many studies propose ontologies to categorize tourist places [2,10]. We compute a resume of these studies in Table 1. The first level will be composed of six key categories and the second level will be composed of several subcategories. Each place belongs to at least one category and one subcategory. Note that a place can belong to several categories and subcategories.

4   Tourism Profiling Measure

To use an unsupervised clustering algorithm, we propose a measure `Tourists Profile Measure` (TPM) that allows comparing stays. Our measure is used to compute the similarity between two stays by taking into account the context and the content of the stays. The TPM between two stays can be seen as the sum between the context distance and the content distance, both normalized. Given S_a and S_b two stays:

$$TPM(S_a, S_b) = distance_{context}(S_a, S_b) + distance_{content}(S_a, S_b) \quad (1)$$

The context distance is defined as an addition of the duration distance and the season distance. Let S_a and S_b be two stays with ΔS_a and ΔS_b their respective duration, p represents the normal distribution on the duration, the distance of duration between these two stays is defined as follows:

$$distance_{context_{duration}}(S_a, S_b) = |p(\Delta S_a) - p(\Delta S_b)| \quad (2)$$

We base our season distance on the seasonal calendar. Since the seasons are cyclical, we can represent them in a cyclic graph where the seasons are nodes. Let S_a and S_b be two stays with $Season_a$ and $Season_b$ their respective seasons:

$$distance_{context_{season}}(S_a, S_b) = \begin{cases} 0 \text{ if } Season_a \text{ and } Season_b \text{ are the same node} \\ 0.5 \text{ if } Season_a \text{ and } Season_b \text{ are adjacent} \\ 1 \text{ if } Season_a \text{ and } Season_b \text{ are distant nodes} \end{cases}$$

$$(3)$$

Content Distance. We recall that a stay contains a set of visited places. Our ontology allows us to know the number of visited places for each category and subcategory. It is composed of six subvectors corresponding of the main categories counting the monument of each subcategory. To calculate the distance between two content vectors, we sum the distance *cosinus* of each sub-vector. The *cosinus* compare the distribution of two vectors, not their magnitude which fit with a behaviours comparison. Note that we are computing a distance, so we are inverting the bounds of the cosine.

5 Creating Profiles

The unsupervised algorithm will work on the stays independently of the tourists who made them, which means that stays made by the same tourist can be in different groups. As a result, it is necessary to re-inject the tourist's demographic information into each of his or her stays. We generate the tourist profiles using a machine learning method that will consist of:

- To construct the distance matrix by calculating the distance based on the text between the stays in pairs. This matrix is symmetric.
- To use an unsupervised clustering algorithm that will take the distance matrix as input and derive groups. We use AGNES [12], a hierarchical algorithm with a Ward linkage and Elbow method for the number of clusters.
- To inject the tourists' demographic data into the groups containing at least one of his stays.

Each cluster is then analyzed to extract the tourist profile. The summary of a cluster consists in calculating: 1) the statistics on the length of the stay: average and standard variation; 2) the statistics on the cluster: average and standard variation of the numerical traces by stay and by group size; 3) the distribution of seasons; 4) the distribution of nationalities; 5) the distribution of categories and sub-categories of the content of the stays. In addition to a summary for each cluster, an overall summary of the data set is constructed. Finally, these summaries are analyzed to extract interesting information about tourist behavior to create typical tourist profiles.

6 Result and Discussion

To validate our tourism profiling method, we will apply it to data from the social network *TripAdvisor* over a period from 2015 to 2018. For our case study, we

Table 2. Statistics for each cluster.

Cluster	Duration		Places		Number of stays
	Mean	Standard deviation	Mean	Standard deviation	
Global	**1.92**	**2.139**	**4.935**	**2.923**	100%
1	1.3	0.747	3.736	2.021	5%
2	2.065	2.513	4.673	3.034	5.8%
3	1.624	1.615	3.621	2.039	7.6%
4	1.994	2.106	4.972	3.038	16.8%
5	1.332	0.761	5.435	2.353	6.3%
6	2.314	2.584	4.999	3.179	12.1%
7	4.191	3.458	6.987	3.681	4%
8	1.245	0.604	5.515	2.345	16.4%
9	1.269	0.664	5.827	2.434	4.4%
10	1.29	0.734	5.874	2.525	6.4%
11	1.565	1.422	3.532	2.093	15.2%

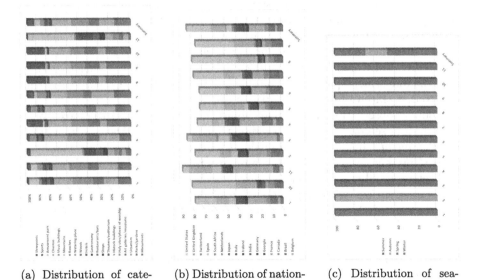

(a) Distribution of categories.

(b) Distribution of nationalities.

(c) Distribution of seasons.

Fig. 1. Profiles summaries.

have chosen the city of *Paris*, because it is one of the most attractive cities in the world, regularly ranking first among the most visited cities in the world.

Our database is composed of 4, 222, 838 comments distributed among 1, 571, 362 tourists for a ratio of approximately 2.7 comments per tourist (with the date of the comment and the concerned monument). We compute the stays and we obtain a set of 150, 306 stays. The Elbow method returns a total of 11 clusters, we summarise them and the whole data set in Table 2.

We can notice cluster 7 represents the biggest average of duration of stays of 4.19 days with an average of visited places the most important in the event 6.98 (two more than the average) but with the lowest density per day, the cluster represents 4% of the total. We can observe in Fig. 1a representing the percentage distribution of the visited subcategories in each cluster, the visits made in cluster 7 are very close to the global summary. We notice the presence of the 10 most represented nationalities in the global summary as seen in Fig. 1b. From the Fig. 1c, the entirety of the stays is realized during the Parisian summer period. We can conclude that cluster 7 contains tourists without any particular preferences on the visited places. These tourists tend to make/comment few visits to places during their long stay, which may imply a desire to take advantage of the summer sun and to enjoy the streets of Paris.

The analysis of cluster 7 is made without context, i.e. without comparisons to the results of sociological studies on tourist behaviors. For the remainder of the analysis, we will compare the profiles obtained with studies from tourist offices and sociological research on tourism. In the discussions, we will refer specifically to the public reports of the *Paris* regional tourism committee[1].

We notice, for example, that clusters 3 and 11 are mainly interested in *amusement parks* and the infrastructures that accompany them such as *hotels* and *restaurants*. Two nationalities are mainly present, *France* and *United Kingdom*. According to the Table and the Figure, both clusters come to *Paris* on average for one day and a half in winter (15.2% of all stays) and in spring (7.6%) to enjoy the amusement parks. This profile is confirmed by the reports from the Paris tourist office of French, and British tourists.

A similar observation can be made about clusters 5, 8, 9 and 10 (corresponding to the four seasons). The most represented categories of places are *Viewpoints* and *Monuments*. In terms of nationalities, countries from the anglosphere are the most present corresponding to the reports from the *Paris* tourist office.

Clusters 4 and 6 show a similar distribution of categories of places visited. In this case, an overwhelming proportion of places are related to the culture and urbanism of *Paris* for an average stay of two days. Nationalities far from *France* are more present showing the cultural appeal of *Paris* in the world. This tourism, having a particular attraction for indoor visits, is more dominant during the winter and spring seasons, with fewer outdoor attractions.

Clusters 1 and 2 represent a similar distribution of categories of places visited and nationalities with 5.0% and 5.8% of the total number of stays respectively with a majority of *parks/gardens*, *urbanism* and *amusement parks*. These clusters represent a summer tourism profile, privileging outdoor activities and summer attractions of *Paris* (fairs, amusement parks, music festivals).

The tourism profiles found by our method are very interesting in their accuracy with real-world data. Similar data set on the Hauts-de-France region and Nouvelle-Aquitaine region (popular region of France) have been studied in a similar way with equal relevant results.

---

[1] https://pro.visitparisregion.com/chiffres-du-tourisme/profil-clientele-tourisme.

7 Conclusion

In this article, we propose a method to discover tourist profiling. We have proposed a measure of distance based on both context and content data from tourist stays. We have shown that this measure highlights tourist profiles heretofore known in the literature, but with a finer knowledge. Our experiments demonstrate the validity of our results by comparing them to tourism management reports. Thus, the tourism industry can widely exploit our method in any geographical area without resorting to sociological studies of tourism, which are often complex to set up and must be spread over many years.

References

1. Ben Baccar, L., Djebali, S., Guérard, G.: Tourist's tour prediction by sequential data mining approach. In: Li, J., Wang, S., Qin, S., Li, X., Wang, S. (eds.) ADMA 2019. LNCS (LNAI), vol. 11888, pp. 681–695. Springer, Cham (2019). https://doi.org/10.1007/978-3-030-35231-8_50
2. Borràs, J., Moreno, A., Valls, A.: Intelligent tourism recommender systems: a survey. Exp. Syst. Appl. **41**(16), 7370–7389 (2014)
3. Bu, N.T., Pan, S., Kong, H., Fu, X., Lin, B.: Profiling literary tourists: a motivational perspective. J. Destination Mark. Manage. **22**, 100659 (2021)
4. Bzdok, D., Krzywinski, M., Altman, N.: Machine learning: supervised methods. Nat. Methods **15**(1), 5 (2018)
5. Gu, G., Li, B., Gao, H., Wang, M.: Learning to answer complex questions with evidence graph. In: Wang, X., Zhang, R., Lee, Y.-K., Sun, L., Moon, Y.-S. (eds.) APWeb-WAIM 2020. LNCS, vol. 12317, pp. 257–269. Springer, Cham (2020). https://doi.org/10.1007/978-3-030-60259-8_20
6. Jia, Z., Yang, Y., Gao, W., Chen, X.: User-based collaborative filtering for tourist attraction recommendations. In: 2015 IEEE International Conference on Computational Intelligence & Communication Technology, pp. 22–25. IEEE (2015)
7. Logesh, R., Subramaniyaswamy, V., Vijayakumar, V., Li, X.: Efficient user profiling based intelligent travel recommender system for individual and group of users. Mob. Netw. Appl. **24**(3), 1018–1033 (2019)
8. March, R., Woodside, A.G.: Tourism behaviortravelers' decisions and actions. No. G155. A1 M2655 2005, Ovid Technologies, Inc. (2005)
9. Massimo, D., Ricci, F.: Clustering users' POIs visit trajectories for next-POI recommendation. In: Pesonen, J., Neidhardt, J. (eds.) Information and Communication Technologies in Tourism 2019, pp. 3–14. Springer, Cham (2019). https://doi.org/10.1007/978-3-030-05940-8_1
10. Moreno, A., Valls, A., Isern, D., Marin, L., Borràs, J.: SigTur/E-Destination: ontology-based personalized recommendation of tourism and leisure activities. Eng. Appl. Artif. Intell. **26**(1), 633–651 (2013)
11. Refanidis, I., et al.: MYVISITPLANNER ^{GR}: personalized itinerary planning system for tourism. In: Likas, A., Blekas, K., Kalles, D. (eds.) SETN 2014. LNCS (LNAI), vol. 8445, pp. 615–629. Springer, Cham (2014). https://doi.org/10.1007/978-3-319-07064-3_53
12. Struyf, A., Hubert, M., Rousseeuw, P., et al.: Clustering in an object-oriented environment. J. Stat. Softw. **1**(4), 1–30 (1997)

Non-stationary Dueling Bandits for Online Learning to Rank

Shiyin Lu[1], Yuan Miao[2], Ping Yang[2], Yao Hu[2], and Lijun Zhang[1(✉)]

[1] National Key Laboratory for Novel Software Technology, Nanjing University,
Nanjing 210023, China
{lusy,zhanglj}@lamda.nju.edu.cn
[2] Alibaba Group, Hangzhou 311121, China
{miaoyuan.my,yangping.yangping,yaoohu}@alibaba-inc.com

Abstract. We study online learning to rank (OL2R), where a parameterized ranking model is optimized based on sequential feedback from users. A natural and popular approach for OL2R is to formulate it as a multi-armed dueling bandits problem, where each arm corresponds to a ranker, i.e., the ranking model with a specific parameter configuration. While the dueling bandits and its application to OL2R have been extensively studied in the literature, existing works focus on static environments where the preference order over rankers is assumed to be stationary. However, this assumption is often violated in real-world OL2R applications as user preference typically changes with time and so does the optimal ranker. To address this problem, we propose non-stationary dueling bandits where the preference order over rankers is modeled by a time-variant function. We develop an efficient and adaptive method for non-stationary dueling bandits with strong theoretical guarantees. The main idea of our method is to run multiple dueling bandits gradient descent (DBGD) algorithms with different step sizes in parallel and employ a meta algorithm to dynamically combine these DBGD algorithms according to their real-time performance. With straightforward extensions, our method can also apply to existing DBGD-type algorithms.

Keywords: Online learning to rank · Dueling bandits · Non-stationary environments

1 Introduction

As a powerful ranking optimization paradigm, learning to rank has found applications in a variety of information retrieval scenarios such as web search, online advertising, and recommendation systems [7,15]. In the classical offline learning to rank, a parameterized ranking model is first trained on collected queries and documents with relevance labels, and then deployed to respond to users' queries with predicted relevant documents. A drawback of offline learning to rank is that the process of collecting training data with relevance labels is highly time-consuming and expensive in large-scale applications [4]. Furthermore, as

B. Li et al. (Eds.): APWeb-WAIM 2022, LNCS 13422, pp. 166–174, 2023.
https://doi.org/10.1007/978-3-031-25198-6_13

the ranking model is fixed after being deployed, it cannot track the evolution of user needs [6].

To address these issues, recent advances in information retrieval have introduced online learning to rank (OL2R), where the ranking model is optimized based on its interactions with users on the fly [3]. Compared to its offline counterpart, OL2R has lighter computational overhead and higher updating frequency. At the heart of OL2R lies the trade-off between exploring new rankers and exploiting the seemingly optimal ranker. Thus, a natural and popular approach for OL2R is to formulate it as a dueling bandits problem [13,14], where each ranker is viewed as an arm and the ranking model is optimized through sequential noisy comparisons between rankers. While the dueling bandits based methods have been widely studied for OL2R, they are limited in that the preference order over rankers is assumed to follow stationary probability distributions. However, in real-world scenarios, user preference typically changes with time, making the stationary assumption invalid.

To better cope with real-world ranking tasks, we investigate dueling bandits with non-stationary preference probability distributions for OL2R. Specifically, let $\mathbf{w}$ and $\mathbf{w}'$ be two points in the parameter space of the ranking model. We model the probability that users prefer the ranking results produced by a ranker with parameter $\mathbf{w}$ over those of a ranker with parameter $\mathbf{w}'$ by a composite function $f_t(\mathbf{w}, \mathbf{w}') = \sigma(v_t(\mathbf{w}) - v_t(\mathbf{w}'))$, where σ is a static link function, and v_t denotes the utility function in round t. Compared to the existing works on dueling bandits, the novelty of our model is that the utility function can change with time t, capturing the non-stationarity of user preference. Since v_t and $v_{t'}$ can be different for $t \neq t'$, the optimal parameter $\mathbf{w}_t^*$ that maximizes v_t and hence the optimal ranker can change with time, making the non-stationary dueling bandits much harder to deal with than its stationary counterpart.

Nevertheless, by drawing inspiration from recent progress in dynamic online optimization [16,17], we develop an efficient and adaptive method for non-stationary dueling bandits. Our method follows the prediction with expert advice framework [1] and has a two layer hierarchical structure: multiple dueling bandits gradient descent (DBGD) [14] algorithms running parallel in the bottom and a meta algorithm aggregating the outputs of DBGDs in the top. Generally speaking, DBGDs aim at balancing the exploration-exploitation tradeoff, which also exists in the classical stationary dueling bandits, and the meta algorithm is responsible for tracking the change of utility functions, which is a new task arising only in our non-stationary setting. Under mild assumptions, we prove that our method guarantees no-regret learning, indicating that when the number of rounds goes infinity, the average performance of our method is the same as that of a clairvoyant who knows the optimal ranker in each round. Furthermore, we show that our method, while developed in the context of DBGD, can be also straightforwardly extended to existing variants of DBGD. Finally, we conduct extensive experiments on public datasets to demonstrate the effectiveness and efficiency of our method for OL2R in non-stationary environments.[1]

---

[1] Due to space limitation, proofs and experiments are postponed to the full version of this paper: www.lamda.nju.edu.cn/lusy/ns-ol2r.pdf.

2 Problem Setup

We study non-stationary dueling bandits for online learning to rank, which proceeds in a sequence of rounds. Let $\mathcal{W} \subseteq \mathbb{R}^d$ be the parameter space of a ranking model and T be the number of rounds. Following previous work [8,11,12], we refer to the ranking model with a specific parameter configuration as a ranker. In each round $t \in [T] = \{1, \ldots, T\}$, firstly a learner chooses two rankers with parameters $\mathbf{w}_t \in \mathcal{W}$ and $\mathbf{w}'_t \in \mathcal{W}$, respectively. Then, the ranking lists produced by the rankers are merged by an interleaving method [5,9]. The merged list is displayed to a user and a noisy preference order over the rankers is inferred from the user's click feedback. Specifically, the ranker whose ranking list receives more clicks is preferred. Finally, the learner updates the parameter of the ranking model based on the inferred preference order.

We denote by $\mathbf{w} \succ \mathbf{w}'$ the event that users prefer the ranking list produced by the ranker $\mathbf{w}$ than that of the ranker $\mathbf{w}'$. While the existing works only consider the setting where the probability of this event is fixed, we allow the probability to change with time so as to capture the non-stationary nature of user preference. Specifically, in round t, the probability of the event $\mathbf{w} \succ \mathbf{w}'$ is defined as

$$\Pr(\mathbf{w} \succ \mathbf{w}'|t) = f_t(\mathbf{w}, \mathbf{w}') = \sigma(v_t(\mathbf{w}) - v_t(\mathbf{w}')) \tag{1}$$

where σ is a static link function, and v_t denotes the utility function in round t. Following previous work [11,14], we make some standard assumptions as follows:

– The parameter space of the ranking model $\mathcal{W}$ is bounded

$$\max_{\mathbf{w} \in \mathcal{W}} \|\mathbf{w}\|_2 \le R. \tag{2}$$

– The link function σ is rotation-symmetric

$$\sigma(x) = 1 - \sigma(-x). \tag{3}$$

– The link function σ is monotonically increasing and satisfies

$$\sigma(-\infty) = 0, \quad \sigma(0) = 1/2, \quad \sigma(\infty) = 1.$$

– The link function σ is L_σ-Lipschitz, and all utility functions $v_t, t \in [T]$ are L_v-Lipschitz. Furthermore, the link function σ is also second order L_2-Lipschitz.[2]

Denoting $L = L_\sigma L_v$, the above assumptions directly imply the functions $f_t, t \in [T]$ are L-Lipschitz in both arguments.

Let $\mathbf{w}_t^* = \arg\max_{\mathbf{w} \in \mathcal{W}} v_t(\mathbf{w})$ denote the optimal ranker achieving the maximum utility in round t. We adopt dynamic regret as performance metric, defined as

$$\mathrm{DR}(T) = \sum_{t=1}^{T} \left(f_t(\mathbf{w}_t^*, \mathbf{w}_t) + f_t(\mathbf{w}_t^*, \mathbf{w}'_t) - 2f_t(\mathbf{w}_t^*, \mathbf{w}_t^*) \right).$$

Our goal is to design an online learning method for minimizing the above dynamic regret.

---

[2] In OL2R, a widely used link function is the sigmoid function $\sigma(x) = 1/(1+\exp(-x))$, which satisfies all of our assumptions.

3   Method

In this section, we first review the dueling bandits gradient descent (DBGD) algorithm and derive its dynamic regret bound, then present our method as well as its theoretical guarantee, and finally discuss the extensions of our method to existing DBGD-type algorithms.

3.1   Dueling Bandits Gradient Descent

As outlined in Algorithm 1, DBGD has two hyperparameters δ and γ, corresponding to the step sizes of exploration and exploitation, respectively. In each round t, DBGD first draws a vector $\mathbf{u}_t$ uniformly at random from the unit sphere $\mathbb{S} \triangleq \{\mathbf{x} \in \mathbb{R}^d : \|\mathbf{x}\|_2 = 1\}$ as an exploratory direction. Then, a candidate ranker is created with parameter

$$\mathbf{w}_t' = \Pi_{\mathcal{W}}[\mathbf{w}_t + \delta \mathbf{u}_t] \tag{4}$$

where $\mathbf{w}_t$ is the current parameter of the ranking model and $\Pi_{\mathcal{W}}[\cdot]$ denotes the operation of projecting a point to the parameter space $\mathcal{W}$. Next, the two rankers $\mathbf{w}_t$ and $\mathbf{w}_t'$ are compared by the probabilistic interleaving method [5], which can merge the ranking lists produced by the two rankers and infer a preference order over the two rankers from user clicks on the merged ranking list. Finally, based on the preference order, DBGD updates the parameter of the ranking model for the next round. Specifically, if $\mathbf{w}_t'$ wins, which reveals that the exploratory direction leads to better ranking performance, then the parameter of the ranking model moves along the exploratory direction with step size γ: $\mathbf{w}_{t+1} = \Pi_{\mathcal{W}}[\mathbf{w}_t + \gamma \mathbf{u}_t]$. Otherwise, the ranking model remains unchanged.

We rigorously analyze the learning properties of DBGD and derive a sublinear dynamic regret bound as follows.

Theorem 1. *Let C_T be the path length of the optimal rankers over T rounds, defined as*

$$C_T = \sum_{t=2}^{T} \|\mathbf{w}_t^* - \mathbf{w}_{t-1}^*\|_2. \tag{5}$$

By setting $\delta = \sqrt{\frac{2\lambda d}{(11+2\lambda)L\sqrt{T}}}$ and $\gamma = \sqrt{\frac{5R^2 + 2RC_T}{T}}$, the dynamic regret of DBGD satisfies

$$\mathbb{E}[DR(T)] \leq \sqrt{2(11 + 2\lambda)\lambda dL} \left(1 + \sqrt{5R^2 + 2RC_T}\right) T^{\frac{3}{4}}.$$

3.2   DBGD Meets Meta Learning

While DBGD can achieve a sub-linear dynamic regret bound for $C_T = o(\sqrt{T})$, it requires the value of the path-length C_T for tuning the step size γ, which is clearly impossible in practice since C_T depends on the unknown optimal rankers

---

Algorithm 1. DBGD

Require: step sizes of exploration δ and exploitation γ
1: Initialize a ranker $\mathbf{w}_1 \in \mathcal{W}$ arbitrarily
2: **for** $t = 1, 2, \ldots, T$ **do**
3:    Draw a vector $\mathbf{u}_t$ uniformly at random from $\mathbb{S}$
4:    Create an exploratory ranker $\mathbf{w}'_t = \Pi_{\mathcal{W}}[\mathbf{w}_t + \delta \mathbf{u}_t]$
5:    Compare $\mathbf{w}_t$ and $\mathbf{w}'_t$ by probabilistic interleaving
6:    **if** $\mathbf{w}'_t \succ \mathbf{w}_t$ **then**
7:        Set $\mathbf{w}_{t+1} = \Pi_{\mathcal{W}}[\mathbf{w}_t + \gamma \mathbf{u}_t]$
8:    **else**
9:        Set $\mathbf{w}_{t+1} = \mathbf{w}_t$
10:   **end if**
11: **end for**

$\mathbf{w}_1^*, \ldots, \mathbf{w}_T^*$. To address this issue, we employ the meta learning technique to automatically tune the step size γ, which has exhibited successes in online convex optimization [2,16,17]. The basic idea is to run multiple DBGDs in parallel, each of which is configured with a different step size γ and admits the sublinear dynamic regret bound for a class of path length. We develop our method in the prediction with expert advice framework, where each DBGD is viewed as an expert and the outputs of DBGDs are combined by an expert-tracking algorithm.

We now describe our method in detail, which is termed as DBGD Meets Meta Learning (DM²L) and consists of a meta algorithm and an expert algorithm.

Meta Algorithm As outlined in Algorithm 2, at the beginning of the meta algorithm, we invoke the expert algorithm with different step size γ. According to our theoretical analysis, we maintain

$$N = \left\lceil \log_2 \sqrt{1 + 4T/5} \right\rceil + 1 \tag{6}$$

experts and the step size γ of the i-th expert is configured as

$$\gamma_i = 2^{i-1} R \sqrt{5/T}, \quad i = 1, \ldots, N. \tag{7}$$

Each expert $i \in [N]$ is associated with a time-variant weight π_t^i, which is dynamically adjusted according to the real time performance of expert i. For deriving a tighter dynamic regret bound, we take a nonuniform initialization of weights:

$$\pi_1^i = \frac{N+1}{i(i+1)N}, \quad i = 1, \ldots, N. \tag{8}$$

In each round t, we first receive a ranker $\mathbf{w}_t^i$ from each expert $i \in [N]$ and aggregate these rankers according to the weights of experts $\pi_t^i, i \in [N]$ as $\mathbf{w}_t =$

$\sum_{i=1}^{N} \pi_t^i \mathbf{w}_t^i$. Then, we sample a vector $\mathbf{u}_t$ from the unit sphere $\mathbb{S}$ uniformly at random and compare $\mathbf{w}_t$ with $\mathbf{w}_t' = \Pi_{\mathcal{W}}[\mathbf{w}_t + \delta \mathbf{u}_t]$ by invoking the probabilistic interleaving method, which returns a noisy preference order $\mathbb{I}_{\{\mathbf{w}_t' \succ \mathbf{w}_t\}}$. Next, we update the weight of each expert using an exponential scheme

$$\pi_{t+1}^i = \frac{\pi_t^i \exp(-\alpha \ell_t(\mathbf{w}_t^i))}{\sum_{j=1}^{N} \pi_t^j \exp(-\alpha \ell_t(\mathbf{w}_t^j))}, \quad i = 1, \dots, N \tag{9}$$

where $\ell_t(\mathbf{w})$ is a surrogate loss function, defined as

$$\ell_t(\mathbf{w}) = -\frac{d}{\delta} \langle \mathbb{I}_{\{\mathbf{w}_t' \succ \mathbf{w}_t\}} \mathbf{u}_t, \mathbf{w} - \mathbf{w}_t \rangle$$

which approximately evaluates the real-time performance of the experts. Finally, both the preference order $\mathbb{I}_{\{\mathbf{w}_t' \succ \mathbf{w}_t\}}$ and the exploratory direction $\mathbf{u}_t$ are sent to each expert so that they can update their own rankers accordingly.

Expert Algorithm. As summarized in Algorithm 3, the expert algorithm is a variant of DBGD. In each round t, each expert $i \in [N]$ first sends its current ranker $\mathbf{w}_t^i$ to the meta algorithm. Then, each expert receives the same preference order $\mathbb{I}_{\{\mathbf{w}_t' \succ \mathbf{w}_t\}}$ and exploratory direction $\mathbf{u}_t$ from the meta algorithm. Finally, each expert updates its own ranker as

$$\mathbf{w}_{t+1}^i = \Pi_{\mathcal{W}}[\mathbf{w}_t^i + \gamma_i \mathbb{I}_{\{\mathbf{w}_t' \succ \mathbf{w}_t\}} \mathbf{u}_t], \quad i = 1, \dots, N. \tag{10}$$

Different from DBGD, we here take the same updating direction $\mathbb{I}_{\{\mathbf{w}_t' \succ \mathbf{w}_t\}} \mathbf{u}_t$ for all experts so that only two rankers $\mathbf{w}_t, \mathbf{w}_t'$ need to be compared in each round. While the updating direction is no longer opposite to the gradient of the smoothed function $\nabla h_t(\mathbf{w}_t^i)$, it is the inverse of the gradient of the surrogate loss function $\nabla \ell_t(\mathbf{w}_t^i)$. Thus, the updating rule of each expert can still be viewed as gradient descent and the dynamic regret of each expert can be analyzed following the proof of Theorem 1.

We present the theoretical guarantee of our method DM²L in the following theorem. Compared to DBGD, the main advantage of DM²L is that it can achieve the sub-linear dynamic regret bound without prior knowledge of the path length C_T and thus can adapt to unknown non-stationarity of environments.

Theorem 2. *By setting* $\delta = \sqrt{\frac{3\lambda d}{(11+2\lambda)L\sqrt{T}}}$ *and* $\alpha = 4/\sqrt{T}$ *and using the configurations in (6) and (7), DM²L achieves the following dynamic regret bound*

$$\mathbb{E}[\mathrm{DR}(T)] \leq \sqrt{3(11+2\lambda)\lambda dL} \left(1 + \sqrt{5R^2 + 2RC_T}\right) T^{\frac{3}{4}} + \lambda(1 + \ln(N+1))\sqrt{T}.$$

3.3   Extensions to DBGD-Type Algorithms

While our meta learning method is developed in the context of DBGD, it be also straightforwardly extended to existing DBGD-type algorithms such as MGD [10]

Algorithm 2. DM^{2}L: Meta Algorithm

Require: number of experts N, step sizes δ, $\gamma_1, \ldots, \gamma_N$, learning rate α
1: Invoke Algorithm 3 with γ_i for each expert $i \in [N]$
2: Initialize the weights of experts $\pi_1^i, i \in [N]$ by (8)
3: **for** $t = 1, 2, \ldots, T$ **do**
4:   Receive ranker $\mathbf{w}_t^i$ from each expert $i \in [N]$
5:   Aggregate the rankers as $\mathbf{w}_t = \sum_{i=1}^{N} \pi_t^i \mathbf{w}_t^i$
6:   Draw a vector $\mathbf{u}_t$ uniformly at random from $\mathbb{S}$
7:   Create an exploratory ranker $\mathbf{w}_t' = \Pi_{\mathcal{W}}[\mathbf{w}_t + \delta \mathbf{u}_t]$
8:   Compare $\mathbf{w}_t$ and $\mathbf{w}_t'$ by probabilistic interleaving
9:   Update the weight of each expert $\pi_t^i, i \in [N]$ by (9)
10: Send $\mathbb{I}_{\{\mathbf{w}_t' \succ \mathbf{w}_t\}}$ and $\mathbf{u}_t$ to each expert $i \in [N]$
11: **end for**

Algorithm 3. DM^{2}L: Expert Algorithm

Require: step size of exploitation γ_i
1: Initialize a ranker $\mathbf{w}_1^i \in \mathcal{W}$ arbitrarily
2: **for** $t = 1, 2, \ldots, T$ **do**
3:   Send ranker $\mathbf{w}_t^i$ to Algorithm 2
4:   Receive $\mathbb{I}_{\{\mathbf{w}_t' \succ \mathbf{w}_t\}}$ and $\mathbf{u}_t$ from Algorithm 2
5:   Update ranker $\mathbf{w}_{t+1}^i = \Pi_{\mathcal{W}}[\mathbf{w}_t^i + \gamma_i \mathbb{I}_{\{\mathbf{w}_t' \succ \mathbf{w}_t\}} \mathbf{u}_t]$
6: **end for**

---

and NSGD-DSP [11,12]. Note that the existing DBGD-type algorithms only differ in the exploratory direction and the updating direction. Thus, we can replace Steps 6–8 at Algorithm 2 with the corresponding exploration pseudocodes of the DBGD-type algorithm and set $\mathbf{u}_t$ used in Steps 9–10 at Algorithm 2 as the updating direction in the DBGD-type algorithm, while keeping Algorithm 3 and the other steps of Algorithm 2 unchanged. We termed the algorithms obtained by applying our meta learning method to MGD and NSGD-DSP as M^{3}L (MGD Meets Meta Learning) and NM^{2}L (NSGD-DSP Meets Meta Learning), respectively.

4   Conclusion

We have formulated a new bandits model for OL2R, termed as non-stationary dueling bandits, where the preference order over rankers can change with time. For this bandits model, we developed a meta learning method, which dynamically aggregates multiple DBGD algorithms with different step sizes. Theoretical analysis showed that under mild assumptions, our meta learning method enjoys a sub-linear dynamic regret bound. We also discuss the extensions of our meta learning method to existing DBGD-type algorithms. Extensive experiments on public datasets demonstrate the effectiveness and efficiency of our meta learning method for OL2R in non-stationary environments.

Acknowledgements. This work was partially supported by NSFC (61976112) and JiangsuSF (BK20200064). We thank the anonymous reviewers for their constructive suggestions.

References

1. Cesa-Bianchi, N., Lugosi, G.: Prediction, Learning, and Games. Cambridge University Press (2006)
2. van Erven, T., Koolen, W.M.: Metagrad: Multiple learning rates in online learning. In: Advances in Neural Information Processing Systems, vol. 29, pp. 3666–3674 (2016)
3. Grotov, A., Rijke, M.: Online learning to rank for information retrieval. In: Proceedings of the 39th International ACM SIGIR Conference on Research and Development in Information Retrieval, pp. 1215–1218 (2016)
4. Hofmann, K.: Fast and reliable online learning to rank for information retrieval. Ph.D. Dissertation (2013)
5. Hofmann, K., Whiteson, S., Rijke, M.: A probabilistic method for inferring preferences from clicks. In: Proceedings of the 20th ACM International Conference on Information and Knowledge Management, pp. 249–258 (2011)
6. Hofmann, K., Whiteson, S., Rijke, M.: Balancing exploration and exploitation in listwise and pairwise online learning to rank for information retrieval. Inform. Retriev. **16**(1), 63–90 (2013)
7. Liu, T.Y.: Learning to rank for information retrieval. Found. Trends Inf. Retriev. **3**(3), 225–331 (2009)
8. Oosterhuis, H., Rijke, M.: Differentiable unbiased online learning to rank. In: Proceedings of the 27th ACM International Conference on Information and Knowledge Management, pp. 1293–1302 (2018)
9. Radlinski, F., Kurup, M., Joachims, T.: How does clickthrough data reflect retrieval quality? In: Proceedings of the 17th ACM Conference on Information and Knowledge Management, pp. 43–52 (2008)
10. Schuth, A., Oosterhuis, H., Whiteson, S., Rijke, M.: Multileave gradient descent for fast online learning to rank. In: Proceedings of the 9th ACM International Conference on Web Search and Data Mining, pp. 457–466 (2016)
11. Wang, H., Kim, S., McCord-Snook, E., Wu, Q., Wang, H.: Variance reduction in gradient exploration for online learning to rank. In: Proceedings of the 42nd International ACM SIGIR Conference on Research and Development in Information Retrieval, pp. 835–844 (2019)
12. Wang, H., Langley, R., Kim, S., McCord-Snook, E., Wang, H.: Efficient exploration of gradient space for online learning to rank. In: Proceedings of the 41st International ACM SIGIR Conference on Research and Development in Information Retrieval, pp. 145–154 (2018)
13. Yue, Y., Broder, J., Kleinberg, R., Joachims, T.: The k-armed dueling bandits problem. J. Comput. Syst. Sci. **78**(5), 1538–1556 (2012)
14. Yue, Y., Joachims, T.: Interactively optimizing information retrieval systems as a dueling bandits problem. In: Proceedings of the 26th International Conference on Machine Learning, pp. 1201–1208 (2009)
15. Zang, Y., et al.: GISDCN: A graph-based interpolation sequential recommender with deformable convolutional network. In: International Conference on Database Systems for Advanced Applications, pp. 289–297. Springer, Cham (2022). https://doi.org/10.1007/978-3-031-00126-0_21

16. Zhang, L., Lu, S., Zhou, Z.H.: Adaptive online learning in dynamic environments. In: Advances in Neural Information Processing Systems, vol. 31, pp. 1323–1333 (2018)
17. Zhao, P., Wang, G., Zhang, L., Zhou, Z.H.: Bandit convex optimization in non-stationary environments. In: Proceedings of the 23rd International Conference on Artificial Intelligence and Statistics, pp. 1508–1518 (2020)

High-Order Correlation Embedding for Large-Scale Multi-modal Hashing

Junfeng An, Yingjian Li, Zheng Zhang$^{(\boxtimes)}$, Yongyong Chen$^{(\boxtimes)}$, and Guangming Lu

School of Computer Science and Technology, Harbin Institute of Technology, Shenzhen 518055, China

darrenzz219@gmail.com, YongyongChen.cn@gmail.com, luguangm@hit.edu.cn

Abstract. Benefitting from the superb storage and computational efficiency, hashing has received considerable research attention on large-scale multi-modal retrieval. However, most existing methods are mainly built based upon matrix optimization without high-order correlation and equally treat the training instances, which fail to fuse heterogeneous sources and ignore the heuristic information contained by the sampling order. To this end, we, for the first time, propose a novel tensor-based supervised discrete learning framework named Discrete Multi-modal Correlation Hashing (DMCH) to perform a high-order correlation preserved semantic hash learning. Specifically, DMCH stacks all the modality-private matrices into a third-order tensor to simultaneously exploit the high-order intrinsic correlations across heterogeneous sources, which explicitly enforces the consistent and private properties of different modalities. Moreover, DMCH selects the training samples from reliable to unreliable ones to extract heuristic information contained by the instance learning order, which increases the robustness of the model. Furthermore, the specific semantic labels are utilized as specific prior knowledge to preserve full-scale supervision instead of the widely-used pair-wise similarity. Finally, the jointly learning objective is formulated to concurrently preserve the modality-common information and modality-private semantics in the learned hash codes. Extensive experiments on four public datasets demonstrate the state-of-the-art performance of our proposed method.

Keywords: Multi-modal hashing · Tensor optimization · Similarity learning · Information retrieval

1 Introduction

With the explosion of digital social networks, large amounts of multimedia information are explosively generated and accumulated. How to efficiently retrieve the desired content from massive multimedia data is a hot research area. Typically, multi-modal hashing supporting fast similarity retrieval with the low-cost storage property [11], which collaborates heterogeneous features for discrete codes representation learning.

© The Author(s), under exclusive license to Springer Nature Switzerland AG 2023
B. Li et al. (Eds.): APWeb-WAIM 2022, LNCS 13422, pp. 175–182, 2023.
https://doi.org/10.1007/978-3-031-25198-6_14

Multi-modal hashing algorithm maps continuous heterogeneous multi-sources features into discrete hamming space for efficient retrieval. Existing multi-modal hashing methods can be roughly categorized into unsupervised and supervised ones. Unsupervised methods [4,6–8,12] model the training data based on their inter-modality and intra-modality relationship by multiple graphs or matrix factorization without considering any precise classified information. But they neglect the most descriptive element, semantic labels, resulting in sub-optimal representative hash codes. In contrast, the supervised ones [5,6,10] concentrate on facilitating the representation of the learned hash codes by integrating semantic labels to learn a more discriminative representation space, generating more representative discrete codes.

There are two deficiencies in the existing supervised hashing methods: 1) Inadequate modality fusion. Most existing hashing methods are based upon a matrix learning framework [1]. They naively utilize linear or non-linear projection to merge different modality-private projections as well as explore correlations between single modality features and discrete codes, without fully bridging the heterogeneous high-order correlation lying under multiple modalities, leading to insufficient modality exploration. 2) Equal importance of each instance. Existing methods impose equal weight on each item, which inevitably mixes noise and outliers into the model and weakens the model robustness.

In this work, we propose Discrete Multi-modal Correlation Hashing (DMCH) for multi-modal retrieval to address the two mentioned drawbacks. DMCH stacks all the projection matrices into a third-order tensor called projection tensor, which can depict both common semantic and private information of the different modalities. Moreover, we develop a heuristic strategy to progressively convert the continuous feature vectors into a discrete hash representation, by which the complementary semantics of different modalities can be adequately preserved into the learned discrete codes. Particularly, we leverage a modality-private projection matrix to convert high-dimensional features generated by the Gaussian kernel into the target discrete space. Finally, we apply label regression to enforce the specific category information into the target discrete space to improve the representation of the learned hash codes.

The main contributions are:

- To the best of our knowledge, DMCH is the first tensor-based multi-modal hashing retrieval framework to capture the high-order correlation by a unified projection tensor, bridging the heterogeneous gaps of all modalities and exploiting the private properties of each modality and boosting the retrieval performance.
- We develop a heuristic learning strategy by cognitively constructing the training sequence from the easy to complex samples to reduce the negative influence of outliers and increase the robustness of the learned hash method.
- Extensive experimental results show that our method can outperform these baselines.

2 Discrete Multi-modal Correlation Hashing

See Fig. 1.

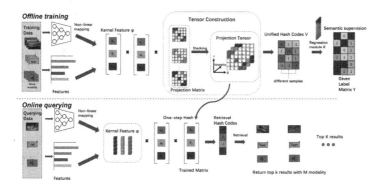

Fig. 1. The framework of DMCH. Firstly, it utilizes non-linear kernel to separate the high dimension original features. Then it stacks every modal-private matrix calculated with the cognitive learning strategy as the slice into a third-order projection tensor to capture the high-order coherent correlation. Finally, the learned projection tensor is applied in one-step hash method for the online multi-modal retrieval.

2.1 Preliminary Notation

Multi-modal hashing retrieval aims to learn the unified discrete codes $\mathbf{V} = [\mathbf{v}_1, \cdots, \mathbf{v}_N] \in \{-1, 1\}^{r \times N}$ for the heterogeneous data, where $\mathbf{v}_n \in \{-1, 1\}^{r \times 1}$ denotes the hash codes for the n-th sample. In this work, bold lower-case letters is the symbol of a column vector, e.g. $\mathbf{k}$, and the bold capital letters, e.g. $\mathbf{W}$, represents for a matrix, $\mathbf{W}^{\mathbf{T}}$ is the transpose of $\mathbf{W}$, $\mathbf{W}_{p,q}$ represents the (p, q)-element of $\mathbf{W}$. Frobenius norm of $\mathbf{W}$ is written as $\|\mathbf{W}\|_F^2 = tr(\mathbf{W}^{\mathbf{T}}\mathbf{W})$. Moreover, the calligraphy, e.g. $\mathcal{Z}$ denotes a third-order tensor. $\hat{\mathcal{Z}}$ represents the fast Fourier transform of $\mathcal{Z}$. $\|\mathcal{Z}\|_{\circledast}$ denotes the tensor nuclear norm of $\mathcal{Z}$, which is defined as the average of singular values of all the frontal slices of $\hat{\mathcal{Z}}$. Given multi-modal training set with M modalities and N instances as $\mathbf{T} = \{\mathbf{X}_1, \mathbf{X}_2, \cdots, \mathbf{X}_M\}$, where the representation of m-th modality is written as $\mathbf{X}_m = [\mathbf{x}_m^1, \mathbf{x}_m^2, \cdots, \mathbf{x}_m^N] \in \mathbb{R}^{d_m \times N}$ and the m-th modality has d_m dimensional features. We utilize the same label provided by $\mathbf{Y} \in \{0, 1\}^{c \times N}$ for N paired text-visual instances belonging to c categories, where $\mathbf{Y}_{fg} = 1$ suggests x_g belongs to f-th class, otherwise $\mathbf{Y}_{fg} = 0$.

2.2 Framework Construction

In this subsection, we introduce the construction process of our DMCH framework. We firstly construct the modality-private semantic extraction, then introduce the proposed cognitive strategy for the modality-private semantic extraction. In addition, the definition of the tensor-based high-order correlation exploration is given followed by the specific semantic supervision.

In order to separate the linear indivisibility heterogeneous features, all of the formulations in this section are based on the non-linear similarity-preserve mapping method with Gaussian kernel function. Specifically, for $\mathbf{X}_t = [\mathbf{x}_t^1, \mathbf{x}_t^2, \cdots, \mathbf{x}_t^N] \in \mathbb{R}^{d_t \times N}$, after the nonlinear conversion we have the following module $\varphi(\mathbf{X}_t) = [\varphi_t^1, \varphi_t^2, \cdots, \varphi_t^N] \in \mathbf{R}^{p \times n}$. Particularly, $\varphi_t^k = [exp(-\frac{\|\mathbf{x}_t^k - \mathbf{a}_t^1\|^2}{2\sigma^2}), \cdots, exp(-\frac{\|\mathbf{x}_t^k - \mathbf{a}_t^p\|^2}{2\sigma^2})]^T$ is the nonlinear feature vector for the k-th instance of the t-th modality, and $\{\mathbf{a}_t^j\}_{j=1}^p$ are the randomly sampled p anchor points selected from training samples, and σ is the Gaussian kernel width.

Modality-Private Semantic Extraction. To avoid the loss of information, we directly explore the relation between the modality feature space and the discrete space, which averts the construction of the intermediate continuous space that accumulates quantization errors. The formulation of this module is given by $\min_{\mathbf{P}_t} \sum_{t=1}^M \alpha^{(t)} \|\mathbf{V} - \mathbf{P}_t \varphi(\mathbf{X}_t)\|_F^2, s.t. \sum_{t=1}^M a^{(t)} = 1, 0 < a^{(t)} < 1$, where $\mathbf{V}$ is the hashing space, $\mathbf{P}_t$ is the private semantic projection of the t-th modality, and $\alpha^{(t)}$ is the modality fusion weight that indicates the contribution of the t-th modality for the continuous-discrete conversion.

Heuristic Training Strategy for Modality-Private Semantic Extraction. There are always inevitable outliers among the training data. In order to alleviate their negative influence and improve the model robustness, we introduce a cognitive learning strategy with the help of self-paced learning. It constructs the cognitive weighting vector that extracts heuristic knowledge from the reliable easy instances to the unreliable complex ones by calculating their different weights in each iteration. This module is given as follows: $\min_{\alpha^{(t)}, \mathbf{P}_t, \mathbf{V}} \sum_{t=1}^M \alpha^{(t)} \|(\mathbf{V} - \mathbf{P}_t \varphi(\mathbf{X}_t)) diag(\sqrt{\mathbf{r}_t^{spl}})\|_F^2 + f(\gamma, \mathbf{r}^{spl}), s.t. \sum_{i=1}^M a^{(i)} = 1, 0 < a^{(i)} < 1$, where $f(\gamma, \mathbf{r}^{spl})$ is the regulation term of cognitive vector $\mathbf{r}^{spl}$, $\mathbf{r}_t$ indicates the different importance of each instance from the t-th modality with the cognitive sample selecting strategy, and γ is the penalty parameter.

In order to jointly explore the high-order intra-modality relations of heterogeneous sources, we propose to construct a novel third-order projection tensor to capture the high-order correlation among the different modalities. Specifically, after learning the private information by $\mathbf{P}_t$, we stack all $\mathbf{P}_t$ into a projection tensor $\mathcal{P}$. $\mathcal{P}$ contains three dimensions, where the first two dimensions denote the slice of the high-order tensor, and the last represents the count of modalities, which induces modality-joint optimization. Considering the upper superiority and inspired by the huge success of the tensor nuclear norm in multi-modal clustering [3], and hyperspectral image restoration [2], we adopt the tensor singular value decomposition-based tensor nuclear norm to explore the low-rankess of the projection tensor. More details can be referred to in [9]. Then the object function is formulated as follows: $\min_{\mathbf{P}_t} \sum_{t=1}^M \alpha^{(t)} \|\mathbf{V} - \mathbf{P}_t \varphi(\mathbf{X}_t)\|_F^2 + \lambda \|\mathcal{P}\|_{\circledast}, s.t. \sum_{t=1}^M a^{(t)} =$

$1, 0 < a^{(t)} < 1$, where $\mathcal{P} = \phi(\mathbf{P}_1, \mathbf{P}_2, \cdots, \mathbf{P}_M), \phi(\mathbf{P}_1, \mathbf{P}_2, \cdots, \mathbf{P}_M)$ denotes the operation to form tensor $\mathcal{P} \in \mathbb{R}^{r \times p \times M}$, $\|\mathcal{P}\|_{\circledast} = \frac{1}{M} \sum_{t=1}^{M} \|\hat{\mathcal{P}}_t\|_*$, and $\hat{\mathcal{P}}$ represents the fast Fourier transform of $\mathcal{P}$ calculated by $\hat{\mathcal{P}} = \mathbf{fft}(\mathcal{P}, [\cdot], 3)$.

Specific Semantic Supervision. The specific category labels are the most discriminative elements in all of the learning sources. After heuristically extracting information from the real-data into discrete hashing space, we develop a maximizing semantic preservation strategy to enforce the specific classification knowledge into the learning process. In particular, considering both the target hashing space and semantic space are binary, we adopt linear regression to minimize the gap to strengthen the semantic supervision by the matrix $\mathbf{R}$. The model is formulated as follows: $\min_{V,R,P} \sum_{t=1}^{M} \alpha^{(t)} \|(\mathbf{V} - \mathbf{P}_t \varphi(\mathbf{X}_t)) diag(\sqrt{\mathbf{r}_t^{spl}})\|_F^2 +$

$f(\gamma, \mathbf{r}^{spl}) + \lambda \|\mathcal{P}\|_{\circledast} + \beta \|\mathbf{RV} - \mathbf{Y}\|_F^2 + \eta \|\mathbf{R}\|_F^2, s.t. \sum_{t=1}^{M} a^{(t)} = 1, 0 < a^{(t)} < 1, \mathbf{V} \in \{-1, 1\}^{r \times n}, \mathbf{Y} \in \{0, 1\}^{c \times n}$.

After generating the query discrete hash codes, our method returns multi-modal retrieval results by ranking the Hamming distance between the query codes and the learned hash codes of the database.

3   Experiments

In this paper, we use four public datasets to evaluate our methods, which are Wiki[1], MIRFlickr-25K[2], NUS-WIDE[3], and MS COCO[4].

3.1   Experiment Results

In this section, we report experimental comparisons and convergence efficiency.

Retrieval Accuracy Comparison. We report the experimental results comparison of 8, 16 and 32 bits hash codes with baseline methods including four unsupervised methods and two supervised ones. We use three commonly-used criterias i.e., mean Average Precision (mAP), top K-precision and precision-recall curve to measure the performance of our model and baselines.

Table 1 gives the mAP result of our method from 8bits to 32bits, compared with other methods. It shows that the mAP of our method is always better than the compared methods. And with different length of hash codes, our methods has stable performance. Other methods, on the contrary, show worse stability.

We report the 16 and 32 bits result on the MIR Flickr and NUS-WIDE in the figures. As shown in Fig. 2, with the increasing numbers of queries, the

---

[1] https://huggingface.co/datasets/wikipedia.

[2] https://press.liacs.nl/mirflickr/.

[3] https://lms.comp.nus.edu.sg/wp-content/uploads/2019/research/nuswide/NUS-WIDE.html.

[4] https://cocodataset.org/.

Table 1. mAP comparison of different methods under different length of hash codes.

Methods	Wiki			MIRFlickr-25k			NUS-WIDE			MS COCO		
	8 bits	16 bits	32 bits	8 bits	16 bits	32 bits	8 bits	16 bits	32 bits	8 bits	16 bits	32 bits
MFH [8]	0.2889	0.4364	0.4899	0.5841	0.5842	0.5844	0.3268	0.3272	0.3279	0.3957	0.3960	0.3963
MAH [4]	0.1111	0.1151	0.4118	0.5818	0.5818	0.5818	0.3225	0.3225	0.3225	0.3963	0.3963	0.3963
MVLH [7]	0.4064	0.4268	0.4671	0.6766	0.6286	0.6334	0.4504	0.4442	0.3638	0.4096	0.4200	0.4264
MvDH [6]	0.4548	0.4174	0.5356	0.6501	0.6593	0.7126	0.4821	0.4934	0.4963	0.3965	0.3972	0.3972
DMVH [10]	0.5274	0.5364	0.5335	0.5818	0.5818	0.5818	0.3225	0.3225	0.3225	0.3962	0.3962	0.3982
OMH-DQ [5]	0.5271	**0.5908**	0.5926	0.6406	0.6522	0.7372	0.4522	0.5471	0.5372	0.4176	0.4131	0.4418
Our method	**0.5702**	0.5518	**0.5930**	**0.8119**	**0.8520**	**0.8609**	**0.7236**	**0.7527**	**0.7468**	**0.5850**	**0.6112**	**0.6379**

topK-precision curves of 16, 32 bits hash codes DMCH returns more relevant results against baselines. As plotted in Fig. 3, our method has better balance of precision and the recall for the online retrieval task.

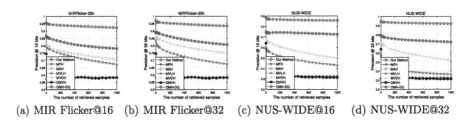

(a) MIR Flicker@16   (b) MIR Flicker@32   (c) NUS-WIDE@16   (d) NUS-WIDE@32

Fig. 2. Comparison of precision@TopK on MIR Flicker and NUS-WIDE of 16bits hash codes and 32bits hash codes, between baselines and our method.

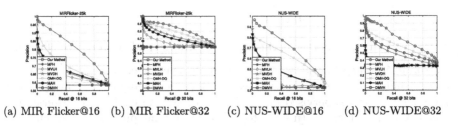

(a) MIR Flicker@16   (b) MIR Flicker@32   (c) NUS-WIDE@16   (d) NUS-WIDE@32

Fig. 3. Comparison of Precision-Recall curve on MIR Flicker and NUS-WIDE of 16bits hash codes and 32bits hash codes, between baselines and our method.

3.2 Ablation Study

In this section, we design a group of ablation experiments and give the analysis to prove the efficiency of our proposed modules. The comparison results of DMCH and its variations are reported in Table 2.

Ablation Study of High-Order Correlation. We propose a matrix-based optimization method without tensor construction called DMCH-matrix and other steps are the same with DMCH to prove the semantic preserved ability of the tensor structure in our proposed DMCH method. The $\mathbf{P}_t$ optimization is given as $\mathbf{P}_t = \frac{\alpha^{(t)}\mathbf{V}\varphi(\mathbf{X}_t)}{\alpha^{(t)}\varphi(\mathbf{X}_t)\varphi(\mathbf{X}_t)^T + \lambda\mathbf{I}}$, where $\mathbf{I}_{p \times p}$ is the identity matrix. From the map of DMCH-matrix in Table 2, we can find that the retrieval accuracy obviously declines compared with DMCH.

Ablation Study of the Cognitive Learning Strategy. We eliminate the sample weighting vector and design a variation of DMCH with equally weighted training sequence named DMCH-equal, and other optimization steps are the same with DMCH. The object function of $\mathbf{P}_t$ is $-2\alpha^{(t)}\mathbf{V}\varphi(\mathbf{X}_t)^T + 2\alpha^{(t)}\mathbf{P}_t\varphi(\mathbf{X}_t)\varphi(\mathbf{X}_t)^T + \rho\mathbf{P}_t - \rho\mathbf{Z}_t + \pi_t = 0$. Hence, we have $\mathbf{P}_t = \frac{2\alpha^{(t)}\mathbf{V}\varphi(\mathbf{X}_t)^T + \rho\mathbf{Z}_t - \pi_t}{2\alpha^{(t)}\varphi(\mathbf{X}_t)\varphi(\mathbf{X}_t)^T + \rho\mathbf{I}}$. From the comparison of DMCH-equal and DMCH in Table 2, we can obvious that our cognitive learning strategy can improve the robustness of the model and promote the retrieval accuracy.

Ablation Study of the One-Step Hashing Method. At the period of online retrieval, DMCH utilizes the learned projection tensor to preserve more semantic features into the discrete hash codes. To confirm its efficiency we design a two-step variation method called DMCH-twostep. The object function of the extract hash function $\mathbf{W}$ is $\min_{W} \|\mathbf{V} - \sum_{t=1}^{M}\alpha^{(t)}\mathbf{W}\varphi(\mathbf{X}_t)\|_F^2$, $s.t. \sum_{t=1}^{M} a^{(t)} = 1, 0 < a^{(t)} < 1, \mathbf{V} \in \{-1, 1\}^{r \times n}$. $\mathbf{W}$ is optimized as $\mathbf{W} = \frac{\mathbf{V}\sum_{t=1}^{M}\alpha^{(t)}\varphi(\mathbf{X}_t)}{(\sum_{t=1}^{M}\alpha^{(t)}\varphi(\mathbf{X}_t))(\sum_{t=1}^{M}\alpha^{(t)}\varphi(\mathbf{X}_t))^T + I}$. So, we can get out-of-sample discrete codes as $\mathbf{V}_{tst} = \sum_{t=1}^{M}\alpha_t\mathbf{W}\varphi(\mathbf{X}_t^{tst})$, where $\mathbf{X}_t^{tst}$ is the t-th modality of input sample, and $\mathbf{V}_{tst}$ is our retrieval hash codes. Compare the results of DMCH-twostep and DMCH in Table 2, we can easily find that the one-step learning method can preserve more representative information than the ablated compared method DMCH-twostep and improves the multimodal retrieval accuracy. From Table 2, we can find that every module in our DMCH performs obvious positive effect on the retrieval precision.

Table 2. mAP comparisons of DMCH and its variants with different bits of hash codes.

Methods	Wiki			MIRFlicker		
	8bits	16bits	32bits	8bits	16bits	32bits
DMCH-matrix	0.6067	0.5721	0.5552	0.8030	0.8359	0.8514
DMCH-equal	0.4917	0.6256	0.5374	0.8187	0.8405	0.8573
DMCH-twosteps	0.5123	0.5633	0.5107	0.8049	0.8256	0.8579
DMCH	**0.6249**	**0.6352**	**0.6134**	**0.8263**	**0.8426**	**0.8641**
Methods	NUS-WIDE			MS COCO		
	8bits	16bits	32bits	8bits	16bits	32bits
DMCH-matrix	0.6860	0.7532	0.7776	0.5779	0.6094	0.6486
DMCH-equal	0.7357	0.7627	0.7810	0.5733	0.6138	0.6509
DMCH-twosteps	**0.7494**	0.7782	0.7668	0.5754	0.6107	0.6487
DMCH	0.7392	**0.7791**	**0.7828**	**0.5770**	**0.6152**	**0.6527**

4   Conclusion

In this work, we developed the first cognitive tensor-based supervised multi-modal hashing framework named Discrete Multi-modal Correlation Hashing (DMCH). A cognitive third-order tensor based hash learning method is introduced to explore the inter-modality pair-wise relation between single-modality features and the learned hash bits, and it simultaneously captures the intra-modality high-order correlation among different modalities under the supervision of the specific label matrix. Then we develop a series of extensive experiments on various real-world datasets and achieve promising performance by comparing to some state-of-the-art methods.

Acknowledgements. This work was supported in part by Shenzhen Fundamental Research Fund under grants GXWD20201230155427003-20200824103320001 and JCYJ20210324132212030, and also supported by the Guangdong Natural Science Foundation under grant 2022A1515010819.

References

1. An, J., Luo, H., Zhang, Z., Zhu, L., Lu, G.: Cognitive multi-modal consistent hashing with flexible semantic transformation. IPM **59**(1), 102743 (2022)
2. Chang, Y., Yan, L., Zhao, X.L., Fang, H., Zhang, Z., Zhong, S.: Weighted low-rank tensor recovery for hyperspectral image restoration. IEEE TCYB **50**(11), 4558–4572 (2020)
3. Chen, Y., Wang, S., Peng, C., Hua, Z., Zhou, Y.: Generalized nonconvex low-rank tensor approximation for multi-view subspace clustering. IEEE TIP **PP(99)**, 1 (2021)
4. Liu, L., Yu, M., Shao, L.: Multiview alignment hashing for efficient image search. IEEE TIP **24**(3), 956–966 (2015)
5. Lu, X., Zhu, L., Cheng, Z., Nie, L., Zhang, H.: Online multi-modal hashing with dynamic query-adaption. In: Proceedings of ACM SIGIR, pp. 715–724 (2019)
6. Shen, X., Shen, F., Liu, L., Yuan, Y.H., Liu, W., Sun, Q.S.: Multiview discrete hashing for scalable multimedia search. ACM TIST **9**(5), 1–21 (2018)
7. Shen, X., Shen, F., Sun, Q.S., Yuan, Y.H.: Multi-view latent hashing for efficient multimedia search. In: Proceedings of the 23rd ACM MM, pp. 831–834 (2015)
8. Song, J., Yang, Y., Huang, Z., Shen, H.T., Luo, J.: Effective multiple feature hashing for large-scale near-duplicate video retrieval. IEEE TMM **15**(8), 1997–2008 (2013)
9. Xie, Y., Tao, D., Zhang, W., Liu, Y., Zhang, L., Qu, Y.: On unifying multi-view self-representations for clustering by tensor multi-rank minimization. IJCV **126**(11), 1157–1179 (2018)
10. Yang, R., Shi, Y., Xu, X.S.: Discrete multi-view hashing for effective image retrieval. In: Proceedings of the 2017 ACM ICMR, pp. 175–183 (2017)
11. Zhang, Z., Liu, L., Shen, F., Shen, H.T., Shao, L.: Binary multi-view clustering. IEEE TPAMI **41**(7), 1774–1782 (2018)
12. Zhang, Z., Luo, H., Zhu, L., Lu, G., Shen, H.T.: Modality-invariant asymmetric networks for cross-modal hashing. IEEE TKDE (2022)

A Self-training Approach for Few-Shot Named Entity Recognition

Yudong Qian and Weiguo Zheng$^{(\boxtimes)}$

School of Data Science, Fudan University, Shanghai, China
{20210980103,zhengweiguo}@fudan.edu.cn

Abstract. Named entity recognition (NER) is a basic task in natural language processing and can be used in a wide range of downstream tasks, such as question answering, text summarization, and machine translation. In recent years, deep-learning based methods achieve great performance in the NER task. It often demands a huge amount of data to train models. However, it is very expensive to collect sufficient training data in many real-world applications. Thus, it is important to develop NER systems for few-shot settings. In this paper, we propose a self-training approach for NER that employs the framework of the machine reading comprehension model when lacking training samples. Experimental results on NER benchmarks demonstrate that the proposed method in this paper outperforms the state-of-the-art methods.

Keywords: Named entity recognition · Few-shot learning · Semi supervised learning · Self-training

1 Introduction

Named entity recognition (NER) is an important task of natural language processing, it recognizes the predefined entity types from the input text. Early NER systems, e.g., NetOwl [1], relied on manually-defined rules. Some feature-based supervised learning methods regard the NER task as a multi-classification problem or sequence labeling problem, e.g., CRF [2]. However, traditional NER methods cannot capture the semantic information in the text, so it is difficult to improve the performance of these methods further. As deep learning methods, e.g., BiLSTM + CRF [3], have been widely applied in NER tasks, these methods can capture hidden features and exhibit better generalization ability than traditional methods.

Although deep-learning based methods have achieved great progress in NER tasks, many challenges remain to address, such as the lack of sufficient annotation data in some low-resource fields. Many NER systems have good results in general domain data sets, but they need a large amount of annotation data to train the model, and the acquisition of annotation data usually requires rich domain knowledge, as well as huge labor costs. However, high-quality annotation data is scarce in many practical scenarios. Therefore, it is of great significance to develop NER systems for few-shot settings.

B. Li et al. (Eds.): APWeb-WAIM 2022, LNCS 13422, pp. 183–191, 2023.
https://doi.org/10.1007/978-3-031-25198-6_15

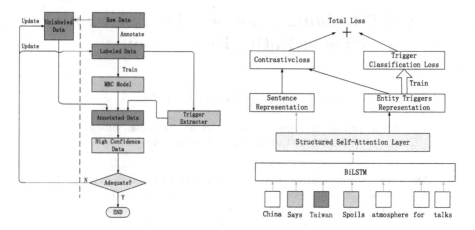

Fig. 1. Framework of our model   **Fig. 2.** Trigger representation learning

In this paper, we propose a few-shot NER model based on self-training, taking machine reading comprehension (MRC) as a built-in block. The overall structure of our model is shown in Fig. 1. Specifically, it is mainly composed of three steps: 1) The base model is trained first by using labeled data; 2) Compute the confidence of weak annotation data inferred by the trained model in the former step, and select high-confidence data to expand labeled data; 3) Iterate from step 1 to step 2 until the stop condition is achieved. The introduction of the MRC-based model can encode external knowledge about entities by setting appropriate queries, which benefits the application in few-shot settings. While the framework of self-training is adopted, we use entity triggers to compute the confidence of weak annotation data, which can mine information from different perspectives of labeled data and provide effective filtering rules to filter out noisy data. As self training has proved its effectiveness in few-shot settings, we apply a new confidence measure to the process of self-training and conduct experiments to show the effectiveness of our method.

In summary, the contributions of the paper can be summarized as follows:

- We propose a self-training based framework to recognize named entities in few-shot settings.
- We select machine reading comprehension model as the base model of our self-training framework, and the NER task is regarded as answering the corresponding queries. Besides, we compute confidence of weak labeled data based on entity triggers.
- Extensive experiments are conducted on two benchmarks to confirm the effectiveness of the proposed method.

2   Our Model

2.1   MRC-NER

We first transform the tagging-style annotation NER dataset into MRC-style. Specifically, we generate the query set $Q = \{q_{y_1}, \ldots, q_{y_k}\}$, where q_{y_i} denotes

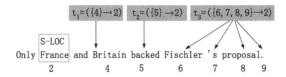

Fig. 3. Example of entity trigger

the query of entity type y_i. Then we can get corresponding answer set $A = \left\{ a_{start_1, end_1}, \ldots, a_{start_p, end_p} \right\}$ of input S, where $a_{start_p, end_p} = \{ w_{start}, \ldots, w_{end} \}$ denotes the corresponding entity mention. Therefore, we can get MRC-style annotation sample $(Question, Answer, Context)$. After transforming tagging-style dataset into MRC-style, we can extract the entity by answering the question of a certain type. Solving NER tasks by the MRC-based model has a key advantage against traditional methods: we can encode prior knowledge about entity categories through the query, and the specific description of similar entity categories can effectively eliminate ambiguity.

For few-shot learning, due to the limited annotation data, it is necessary to import external knowledge. Thus, we choose the MRC-based NER method [5] as the base model and improve its performance through self-training.

2.2 Entity Triggers

Entity Triggers [6] are defined as a set of words that help explain the entity recognition process in a sentence. When we recognize some entity in a sentence, we usually take certain words or phrases in the sentence as the basis for our judgment, even if it is a word we are not familiar with. In short, entity triggers can effectively help us understand the training process of the model and enable the model to summarize the information of entity categories better. This method was proposed by Lin et al. [6], and it achieved good results in few-shot settings by using labeled data with entity triggers. Fig. 3 presents such an example, where t_i denotes an entity and its corresponding trigger.

When it lacks enough annotation data, entity triggers may provide us supplementary information different from the original label information. It can be regarded as supplementary annotations in the case of insufficient annotation data, so as to help the model learn and summarize better from the limited annotation data. Therefore, we select relevant information of entity triggers as an auxiliary to compute the confidence of weak labeled data during self-training process, and it can effectively filter out noisy data and improve the performance of our model.

Trigger Extractor. Although annotating entity triggers manually may have high quality, it needs domain knowledge and high labor costs, which is not practical for NER tasks in few-shot settings. Therefore, we design a model for automatic extraction of triggers based on the AutoTrigger model proposed by Lee et al. [7]. We use SOC (Sampling and Occlusion) [8] algorithm to compute the context-independent importance of phrases, which can be used to extract triggers. SOC is a technique for model interpretation. The expression of the importance score of the phrase p in input sequence x is:

$$\phi(p, x) = \frac{1}{|S|} \sum_{\widehat{x}_\delta \in S} \left[s\left(x_{-\delta}; \widehat{x}_\delta\right) - s\left(x_{-\{\delta, p\}}; \widehat{x}_\delta; 0_p\right) \right] \tag{1}$$

where $s(x)$ denotes the predict score of the model, $x_{-\delta}$ denotes the sequence after masking a context of length N surrounding the phrase p from input sequence x, $\widehat{x}_\delta$ denotes the sequence of length N obtained according to sampling probability distribution $p(\widehat{x}_\delta | x_{-\delta})$ based on the pre-trained language model, 0_p denotes paddings for phrase p, and S denotes a collection of samples $\widehat{x}_\delta$ from a pre-trained language model. Therefore, the importance score of phrase p can be interpreted as the expectation of difference between predict scores after masking phrase p in all possible context $\widehat{x}_\delta$ of p, which can also eliminate the relationship between the importance score and the context of the phrase.

The process of automatic trigger extraction can be simply described as follows:

1) It first trains a classifier M_t based on annotation data D_L. For the input $x = \left(x^{(1)}, x^{(2)}, \ldots, x^{(n)}\right)$, the classifier M_t uses conditional probability $P(y|x)$ to denote its output, y is the corresponding label sequence. The predict score of target entity e can be expressed as the following formula:

$$s(x, e) = \frac{1}{|e|} \sum_{x^{(j)} \in e} P\left(y^{(j)} | x^{(j)}\right) \tag{2}$$

2) Then generate the candidate trigger set P according to the set of phrase nodes from the constituency parse tree, and calculate the importance score of its target entity for each phrase $p_i \in P$:

$$\phi(p_i, x, e) = \frac{1}{|S|} \sum_{\widehat{x}_\delta \in S} \left[s\left(x_{-\delta}, e; \widehat{x}_\delta\right) - s\left(x_{-\{\delta, p_i\}}, e; \widehat{x}_\delta; 0_{p_i}\right) \right] \tag{3}$$

3) For all candidate triggers $p_i \in P$, select top-K triggers with the highest score after computing the importance score.

2.3   Self-training Framework

Trigger Representation Learning. After extracting entity triggers, we train the model to learn the representation of triggers.

First, for the annotation data with triggers, we obtain the embedding of input sentence S and trigger t according to the method proposed by Lin et al. [9], denoted as g_s and g_t respectively. g_s is the weighted sum of token embeddings in the sentence, and g_t is the weighted sum of embeddings of triggers in the sentence. Then we learn the weight matrix by training in two tasks and obtain the trigger embedding. Fig. 2 shows the framework. For the first task, we learn trigger vectors by using entity types as supervision. The second task aims at making the trigger vector and sentence matched. The final loss is the weighted sum of the loss of these two tasks.

Confidence. In the iterative process of self-training, how to find and remove noisy data is critical. By selecting reliable weak annotation data, we can improve the quality of expanded labeled data, and then improve and model performance.

Based on trigger vectors learned in last subsection, we compute the distance $d = \|g_x - g_t\|_2$ between trigger t and weak annotation sentence x, and set the threshold λ. For the set of triggers $T_x = \left\{ t_x^{(1)}, t_x^{(2)}, \ldots \right\}$ satisfying $d < \lambda$, the corresponding entity type and quantity set is $E_x = \{(e_1, n_1), (e_2, n_2), \ldots, (e_k, n_k)\}$, where e_i denotes the corresponding entity type and n_i denotes the number of triggers belong to this entity type.

For weak annotation data (x, y), the annotation entity type is e_i and its entity type and quantity set is E_x, if the following conditions are satisfied, we will regard this weak annotation data as reliable one:

$$\frac{n_i}{\sum_{j=1}^{k} n_j} \geq \theta_1 \quad or \quad n_i \geq \theta_2 \tag{4}$$

where θ_1 and θ_2 are thresholds. For the reliable weak annotation data obtained after each iteration and the previous labeled data, we define the loss function in the next iteration as follows:

$$L_{ST} = \frac{1}{|D^L|} \sum_{(x,y) \in D^L} L\left(f\left(x\right), y\right) + \frac{\lambda}{|D^U|} \sum_{(x,y) \in D^U} L\left(f\left(x\right), y\right) \tag{5}$$

where $f\left(\cdot\right)$ denotes new trained model based on D^L and D^U, and λ_U denotes weight. Self-training is carried out iteratively according to the corresponding steps until reaches the maximum number of iterations or meets stop conditions.

3 Experiments

3.1 Datasets

We use two datasets CoNLL2003 [10] and BC5CDR [11] for experiments. CoNLL2003 is an English general domain dataset, including four named entities: Location, Organization, Person, and Miscellaneous. BC5CDR is an English dataset in the biomedical field, including two named entities: Disease and Chemical. Tagging-style annotation data in two datasets are transformed into corresponding MRC-style annotation data. The queries corresponding to the entity category are obtained from the annotation guide notes.

3.2 Baselines

We select the following models as baselines:

- BiLSTM-CRF [3]: A classical sequence labeling model.

Table 1. Results on CoNLL2003, where P and R denote Precision and Recall, respectively

Per.	BiLSTM-CRF			TMN			TMN+self-training			BERT-tagger			STM		
	P	R	F1	P	R	F1	P	R	F1	P	R	F1	P	R	F1
1%	41.88	17.62	24.81	71.43	54.27	61.68	74.13	53.53	**62.17**	31.53	30.47	30.99	52.92	62.18	57.18
3%	55.19	46.97	50.75	76.06	74.13	75.08	80.06	75.23	**77.57**	56.01	48.99	52.27	71.18	72.02	71.6
5%	70.26	53.92	61.02	80.38	79.13	79.75	80.45	81.14	**80.79**	59.46	57.74	58.59	76.49	79.52	77.98
7%	71.46	61.26	65.97	82.78	81.31	82.04	82.75	81.99	82.37	65.09	65.23	65.16	84.28	83.75	**84.01**
10%	75.41	70.43	72.83	84.55	82.43	83.48	84.55	82.59	83.56	69.18	71.88	70.5	85.28	85.16	**85.22**
13%	78.03	74.49	76.22	84.79	83.2	83.99	84.51	84.03	84.27	72.01	70.97	71.49	85.02	85.97	**85.49**
15%	79.37	76.15	77.73	85.12	83.47	84.29	86.02	83.29	84.63	73.48	73.19	73.33	84.96	86.33	**85.64**
17%	80.27	77.65	78.94	85.33	84.01	84.66	86.31	83.94	85.11	73.88	75.24	74.55	86.34	86.39	**86.36**
20%	83.11	77.2	80.05	85.5	85.64	85.57	86.32	85.47	85.89	75.26	77.14	76.19	87.23	86.51	**86.87**

- Trigger Matching Network (TMN) [6]: NER model based on manually labeled triggers.
- TMN with Self-training [6]: Self-training is adopted to TMN, and the confidence is computed based on MNLP proposed by Shen et al. [12].
- Bert-Tagger [4]: Sequence labeling model based on BERT.

3.3   Results and Analysis

Table 1 and Table 2 show the results in CoNLL2003 and BC5CDR respectively. It can be observed that, when training data is 1% of the dataset, the F1 value of BilSTM-CRF model is only 24.81%. Few training data leads to poor generalization ability of the model. Although with training samples become more, the model performance has been significantly improved a lot, there is still a big gap between BilSTM-CRF and our model (STM). The performance of Bert-Tagger model is similar to that of BilSTM-CRF. STM performs much better than BilSTM-CRF and Bert-Tagger in the case of few training samples. When compared with two TMN (+self-training) models based on trigger matching, the performance of STM is slightly poor when the training samples are less than 5%. The reason may be that when the sample size is small, the quality of extracted trigger is not high enough, and the query information imported to MRC model cannot be learned well. But when training samples reach 7% or above, the performance will be improved, and it has certain advantages when compared with TMN (+self-training). On the whole, when training samples are less than 20%, STM has a relatively good performance by importing external knowledge and mining information from limited training data. The disadvantage is that when the size of training data is too small (less than 5%), the model can not fully filter out noisy data because of the poor quality of extracted triggers, resulting in poor performance. Therefore, the model can be improved by improving the quality of extracted triggers for few-shot settings, such as transferring existing entity triggers to low-resource field.

The entity definition in the biomedical field is complex, and it's difficult to identify. Therefore, the overall model performance is much lower than that

Table 2. Results on BC5CDR

Percentage	TMN			TMN+self-training			STM		
	Precision	Recall	F1	Precision	Recall	F1	Precision	Recall	F1
1%	59.01	48.78	53.41	59	49.33	**53.73**	48.2	52.03	50.04
3%	66.35	57.24	61.46	65.42	59.23	62.17	60.21	73.44	**66.17**
5%	69.37	63.29	66.19	68.14	66.89	67.51	66.9	71.36	**69.06**
7%	70.29	67.89	69.07	71.46	67.7	69.53	73.17	72.91	**73.04**
10%	72.01	69.35	70.66	69.61	72.84	71.19	75.03	75.06	**75.04**
13%	73.16	70.61	71.86	75.14	69.56	72.24	77.4	76.56	**76.98**
15%	75.04	69.11	71.95	71.38	73.41	72.38	79.37	76.27	**77.79**
17%	74.72	71.01	72.81	74.13	73.64	73.88	77.63	78.22	**77.93**
20%	74.35	72.64	73.48	75.13	73.71	74.41	79.63	79	**79.31**

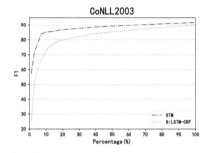

Fig. 4. Effect of varying percentage of training samples on CoNLL2003

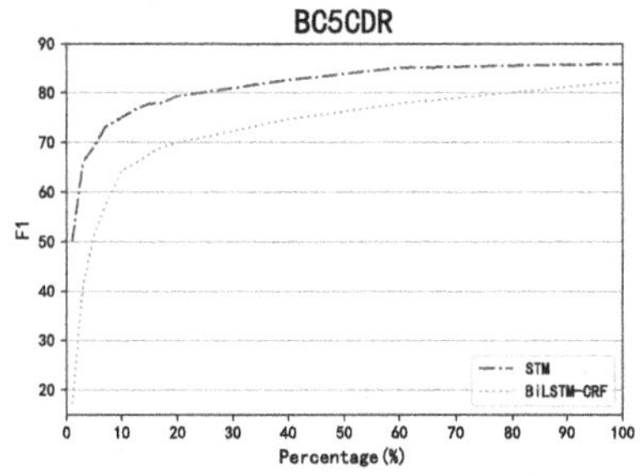

Fig. 5. Effect of varying percentage of training samples on BC5CDR

in CoNLL2003. Compared with the results in CoNLL2003, STM has a more significant advantage in BC5CDR (F1 value is about 4%–5% higher on average). The possible reason is that STM can make full use of the corresponding external knowledge for entities in the biomedical field by setting appropriate queries. In this way, the significant features of the entity category can be extracted, and the noisy data that is easily confused can be filtered out based on triggers, so the advantages are more obvious than in CoNLL2003.

Corresponding line charts are drawn for the performance of STM and BILSTM-CRF in different percentages of training data in CoNLL2003 and BC5CDR respectively, as shown in Fig. 4 and Fig. 5. It can be seen that, less training data, the greater advantage of STM compared with BiLSTM-CRF, the reason is that when the size of training data is small, it is hard for BiLSTM-CRF to learn the important features of corresponding entity category, which leads to poor generalization ability. However, the external knowledge introduced by STM and the information mined from different perspectives of limited training data lead to good generalization ability even if the size of training data is small.

The results of ablation experiments are shown in Table 3. It shows the results of STM, BERT-MRC model without self-training, and self-training based

Table 3. Ablation results of CoNLL2003

Model	Precision	Recall	F1
STM	52.92	62.18	57.18
MRC	48.53	61.76	54.35
STM without Triggers	61.94	44	51.45

MRC model without filtering out noisy data in only 1% training samples of CoNLL2003. After the introduction of entity triggers to filter out noisy data and expand training data with high-quality weak annotation data, the performance of STM (F1 value is 57.18%) improves a lot when compared with that of BERT-MRC model (F1 value is 54.35%). Without the process of filtering out noisy data, STM without Triggers only use weak annotation data to expand training data, although the size of training data has been increased, the quality falls and prediction error of the model will be accumulated, so the model performance falls when compared with F1 value of BERT-MRC model, it is reduced by about 3%. Therefore, it can be concluded that the process of filtering out noisy data by mining trigger information in training data is very important.

4   Conclusion

In this paper, we propose a self-training based NER method to improve the generalization ability of the model in the settings of few-shot. Our model uses MRC-based model as the base model and trains the model under the framework of self-training. The experimental results show that the proposed method outperforms the existing methods.

Acknowledgement. This work was supported by Science and Technology Committee Shanghai Municipality (Grant No. 19ZR1404900) and National Natural Science Foundation of China (Grant No. 61902074).

References

1. Krupka, G.R., Hausman, K.: IsoQuest Inc. 3900 Jermantown Ave., Suite 400 Fairfax, VA 22030 gkrupka@isoquest.com
2. Lafferty, J.D., McCallum, A., Pereira, F.C.N.: Conditional random fields: probabilistic models for segmenting and labeling sequence data. In: ICML (2001)
3. Ma, X., Hovy, E.H.: End-to-end sequence labeling via bi-directional LSTM-CNNs-CRF. In: ACL (1) (2016)
4. Devlin, J., Chang, M.W., Lee, K., et al.: BERT: pre-training of deep bidirectional transformers for language understanding. In: NAACL-HLT (1) (2019)
5. Li, X., Feng, J., Meng, Y., et al.: A unified MRC framework for named entity recognition. In: Proceedings of the 58th Annual Meeting of the Association for Computational Linguistics, pp. 5849–5859 (2020)

6. Lin, B.Y., Lee, D.H., Shen, M., et al.: TriggerNER: learning with entity triggers as explanations for named entity recognition. In: Proceedings of the 58th Annual Meeting of the Association for Computational Linguistics, pp. 8503–8511 (2020)
7. Lee, D.H., Kiran Selvam, R., Sarwar, S.M., et al.: AutoTriggER: named entity recognition with auxiliary trigger extraction. arXiv e-prints arXiv: 2109.04726 (2021)
8. Jin, X., Wei, Z., Du, J., et al.: Towards hierarchical importance attribution: explaining compositional semantics for neural sequence models. In: International Conference on Learning Representations (2019)
9. Lin, Z., Feng, M., dos Santos, C.N., et al.: A structured self-attentive sentence embedding. arXiv preprint arXiv:1703.03130 (2017)
10. Sang, E.F.T.K., De Meulder, F.: Introduction to the CoNLL-2003 shared task: language-independent named entity recognition. Development **922**, 1341 (1837)
11. Li, J., Sun, Y., Johnson, R.J., et al.: BioCreative V CDR task corpus: a resource for chemical disease relation extraction. Database (2016)
12. Shen, Y., Yun, H., Lipton, Z.C., et al.: Deep active learning for named entity recognition. In: International Conference on Learning Representations (2018)

Knowledge Graph

Reasoning Path Generation for Answering Multi-hop Questions Over Knowledge Graph

Yuxuan Xiang[1], Jiajun Wu[1], Tiexin Wang[1], Meng Wang[2], Tianlun Dai[1], Gaoxu Wang[3], Shidong Xu[1(✉)], and Jing Li[1]

[1] Nanjing University of Aeronautics and Astronautics, Nanjing 211100, China
shidongxu@nuaa.edu.cn
[2] Southeast University, Nanjing 211100, China
[3] Nanjing Hydraulic Research Institute, Nanjing 210000, China

Abstract. Multi-hop Knowledge Graph Question Answering (KGQA) aims to find the answer entity via a reasoning path consisting of multiple fact triples in the knowledge graph (KG). Most of end-to-end KGQA approaches only pay attention to answering one-hop simple questions and lack scalability and interpretability. Meanwhile, since the high cost for data annotations, the lack of intermediate supervision signals becomes a major challenge. To address these challenges, we propose a policy-based reinforcement learning model called RPGQA which converts the task of KGQA to a reasoning path generation task in the KG. Firstly, in order to improve the interpretability of the model, the agent in our model learns an effective policy to reason a path to the answer entity as the evidence for the question. Secondly, we design an algorithm for entity disambiguation during entity linking. After that, the topic entity in the question can be linked as the beginning of the reasoning path. Furthermore, we propose a reward shaping policy consisting of three parts to enhance intermediate supervision signals, which alleviates the problem of reward delay and sparsity of reward. Extensive experiments on multiple benchmark datasets have demonstrated the effectiveness of our model. RPGQA outperforms most of the state-of-art baselines on the multi-hop KGQA task.

Keywords: Knowledge Graph Question Answering · Reinforcement learning · Reward shaping · Reasoning path

1 Introduction

Question answering (QA) has always been an essential issue in the field of natural language processing and this task aims to provide people with the information they need automatically. Recently, people have higher requirements for the accuracy and interpretability of QA systems with the widespread application of artificial intelligence. Therefore, how to introduce knowledge to build QA systems is widely concerned [7]. Thanks to the creation of large-scale knowledge

B. Li et al. (Eds.): APWeb-WAIM 2022, LNCS 13422, pp. 195–209, 2023.
https://doi.org/10.1007/978-3-031-25198-6_16

graphs (KG), QA system can be armed with well-structured knowledge on specific and open domains. The goal of KGQA [21] is to find the answers based on the natural language from the KG which is a multi-relational graph with a number of triples, and the answers often correspond to entities in the KG.

For KGQA task, previous methods [1,26] only consider single-hop questions which need only one triple to answer, such as "Who is Barack Obama's wife?". In comparison, how to effectively answer multi-hop questions which need two or more triples is still a big challenge. For example, the question "Where does Barack Obama's wife live in?" relies on multiple fact triples (Barack Obama, Spouse, Michelle Obama) and (Michelle Obama, Places Lived, Chicago). The absence or error of any triple will result in the wrong answer. Therefore, these methods lack interpretability and cannot be used in medical, financial and other fields that require a complete reasoning evidence. How to design a method with interpretability is a challenge for multi-hop reasoning.

Another key challenge is how to generate a reasoning path leading to the answer entity without intermediate supervision signals. The training data for QA usually only provides question-answer pairs. Such weak supervision signals make it difficult to figure out the unknown reasoning steps pointing to the answer. As shown in Fig. 1, even if a reasoning path leading to the correct answer is obtained, this path may be spurious. Spurious reasoning paths [9] will mislead the learning process.

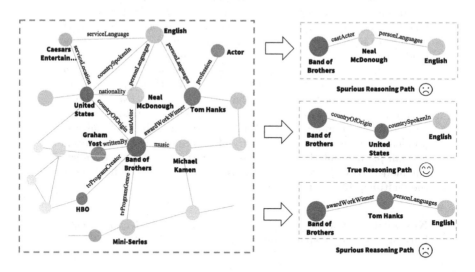

Fig. 1. An example of spurious paths in multi-hop KGQA. The question is "What is the official language of the country to which the movie Band of Brothers belongs?". Three reasoning paths can be obtained, their beginning and ending points are the same, two of which are spurious and one is true.

To address the challenges mentioned above, we propose a method of multi-hop KGQA by reasoning path generation (RPGQA). The method is based on reinforcement learning (RL) and is used to enhance intermediate supervision signals. We convert the multi-hop KGQA task into a path search task in the KG. Compared to deep learning-based approaches, our model is no longer a black box and all reasoning bases are available in the reasoning path,e.g.(Barack Obama, Spouse, Michelle Obama, Places Lived, Chicago). With the generation of reasoning paths, we can analyze the reasoning process and thus improve the interpretability and scalability of the model. Recent researches [8,12] with RL algorithm maximize the expected reward in KGQA in order to deal with the problem of insufficient training data. Feedback can only be received after reasoning for a multi-hop question, which causes sparse reward and reward delay. To alleviate the problem, we propose a reward shaping policy including two soft rewards: semantic reward and efficiency reward. With the reward shaping policy, the agent will get different rewards from the true path and the spurious path.

RPGQA takes the subject entity in the user's question as the source entity, and search along the hop-by-hop path of multiple triples < *head entity, relation, tail entity*> of the knowledge graph to find the answer entity or relation. We first use the semantic information of the question and labels of relations in the KG for entity linking [5] which is very important for accurately locating the source entity as the beginning point of the reasoning path. We use an entity re-ranking [28] algorithm for entity disambiguation in the process of entity linking. Bidirectional LSTMs (BiLSTM) and convolutional neural networks are fused to extract semantic features of questions and relations.

The main contributions of this paper are as follows:

– We propose a novel model based on reinforcement learning for multi-hop KGQA, which makes up for the interpretability in previous work. Reasoning paths can explain the process of getting answers and analyze correctness.
– With a reward shaping policy, we enhance intermediate supervision signals to alleviate the delayed and sparse reward problem during training.
– We build a semantic parsing based entity disambiguation framework during entity linking to avoid error propagation in reasoning process, which helps the agent identify the correct beginning point for reasoning.

2   Related Work

The approaches to solving multi-hop KGQA tasks can be generally classified into embedding-based and path-based approaches.

Embedding-based multi-hop KGQA approaches convert questions and answer candidates into semantic vector representations in low dimension vector space for operation. Previous work [22] embedding the question and KG triples to express the semantics of features. MC-CNN [4] further uses neural network models with stronger learning capabilities to learn information of answers but it does not adequately consider representations of answer candidates. Embed-KGQA [15] uses link prediction based on the KG embedding to alleviate the problem of incomplete data in multi-hop KGQA. GraftNet [19] and PullNet [18] combine external texts and KG based on the graph neural network (GNN). In recent years, in order to solve the problem of large-scale and incomplete KG, researches on the combination of logical rules and knowledge embedding [6,27] have attracted a lot of attention. Query2Box [13] and BetaE [14] express the query as a directed acyclic calculation graph, which indicates the steps to perform multi-hop KGQA.

Path based multi-hop KGQA approaches take the topic entity in the question as the source entity and then search the answer entity along with triples of the KG. DeepPath [23] applies reinforcement learning to path learning and encoded the state in the continuous space through the translation embedding method [29]. MINERVA [3] proposes an algorithm that treats the path to the correct answer entity as a sequential decision problem. To alleviate the problem of reward delay and sparsity of reward, reward shaping policy [9] that can enhance intermediate supervision signals is proposed. M-Malk [17] uses Monte Carlo tree search to overcome the challenge of sparse positive reward. SRN [11] proposes a potential-based reward shaping policy but ignores the importance of entity linking during reinforcement learning. Our work focuses both on entity linking and reward shaping policy and uses the feedback after each action to generate an interpretable reasoning path.

3   Task Definition

In this section, we introduce the notations used throughout the paper and formally define the task of KGQA. We use $\mathcal{E}$ to denote the set of entities and use $\mathcal{R}$ to denote the set of relations. We can construct a knowledge graph $\mathcal{G}$ which is a collection of atomic facts stored as triples $\langle e_1, r, e_2 \rangle$, where $e_1, e_2 \in \mathcal{E}$ are represented as the nodes and $r \in \mathcal{R}$ as the labeled edge between them. Formally, $\mathcal{F}$ is the set of all possible triples. We add the inverse relation of each relation edge, for example, $\langle e_2, r^{-1}, e_1 \rangle \in \mathcal{F}$ is the inverse relation triple of $\langle e_1, r, e_2 \rangle \in \mathcal{F}$, which to allow the agent to have a retreat path when exploring potential wrong decision. A self-loop edge $\langle e, loop, e \rangle \in \mathcal{F}$ is added to each entity node to give the agent the choice to stop at the current node e.

A question is expressed as a sequence of words $\mathcal{Q} = (q_1, q_2, ..., q_n)$ as input. The relation candidate r also has a textual representation as word sequence $r = (t_1, t_2, ..., t_n)$, e.g.$(place, of, birth)$. a is the answer node to $\mathcal{Q}$. $e \in \mathcal{E}$ represents the entity node in the knowledge graph $\mathcal{G}$. Firstly we design an algorithm to find the correct source entity e_s which is the subject entity of $\mathcal{Q}$. Then, we define the query condition $\sigma = (e_s, \mathcal{Q})$ and train an agent to return entities from $\mathcal{E}$ as answers to a given question. Historical track $\mathcal{H}_\sigma = (e_q, r'_1, e_1, ..., e_{n-1}, r'_n, a)$ can be used as an evidence chain to enhance interpretability.

4   Methodology

In this section, we introduce our multi-hop KGQA model in Fig. 2. First, we propose a deep neural model to learn question and relation representations. Next, we introduce the entity disambiguation module in the process of entity linking. Then, we introduce the reinforcement learning external environment and reward shaping policy. Finally, we describe our policy network and training.

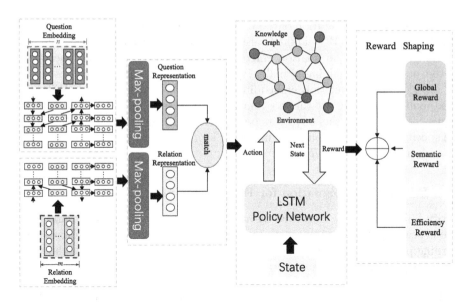

Fig. 2. Overview of our model. Different colored rectangular containers indicate diverse embeddings as dense vectors. Circles in each container represent the features in vector.

4.1 Question and Relation Representation Learning

In our work, we split each question and relation into word sequences. Then, each token is transformed to its word embedding and two BiLSTMs are used to capture contexts for more precise semantics. At last, we get question representation $\mathcal{Q}^r$ and relation representation $\mathcal{R}^r$.

Take question embedding input as an example, we first replace the topic entity's mention with a token $\langle e \rangle$ in question $\mathcal{Q}$ after entity linking by the entity linker. Then, we denote $e(q_i)$ as the word embedding of word q_i. Left context $l_c(q_i)$ and right context $r_c(q_i)$ are dense vectors used to capture semantics:

$$l_c(q_i) = f\left(W^{(l)}l_c(q_{i-1}) + W^{(sl)}e(q_{i-1})\right), \tag{1}$$

$$r_c(q_i) = f\left(W^{(r)}r_c(q_{i+1}) + W^{(sr)}e(q_{i+1})\right), \tag{2}$$

$W^{(l)}$ and $W^{(r)}$ are matrices that convert the context to the next hidden layer. $W^{(sl)}$ and $W^{(sr)}$ are matrices that are used to respectively fuse the left and right context of the word and the semantics of the current word. f is a non-linear activation function. For word q_i, we define a new form of representation x_i^q that consists of three parts:

$$x_i^q = [l_c(q_i); e(q_i); r_c(q_i)]. \tag{3}$$

$$y_i^q = \tanh\left(W^{(t)}x_i^q + b^{(t)}\right), \tag{4}$$

where x_i^q is the representation of the word q_i and $W^{(t)}$ is the matrix of the linear transformation. Both $W^{(t)}$ and $b^{(t)}$ are parameters obtained after learning.

When inputting the relation embedding, we perform the same operations as the question embedding. Finally, the question embedding becomes $Y^q = \{y_1^q, ..., y_n^q\}$ and the relation embedding becomes $Y^r = \{y_1^r, ..., y_m^r\}$. With the max-pooling layer, we get the most important semantic information in the two sequences. These two sets of vectors become the final representation $\mathcal{Q}^r$ and $\mathcal{R}^r$. We specify a rule to compute the matching score of each relation r for question $\mathcal{Q}$ as $S_r(\mathbf{r}; \mathcal{Q}) = \cos(\mathcal{Q}^r, \mathcal{R}^r)$.

4.2   Entity Disambiguation for Entity Linking

Entity linking is easily disturbed by ambiguous entities. So entity disambiguation is necessary to get an accurate beginning state for subsequent reinforcement learning part. Our work uses the existing entity linker [24] to generate the top-k entities for question $\mathcal{Q}$, then operates an entity re-ranking algorithm to disambiguate entities.

In this part, we first select all relations or relation chains in the KG within the two-hop range around the entity candidate generated by the entity linker as relation candidates $\mathcal{R}_e$ and compute their scores given question $\mathcal{Q}$ as input. After that, we get the top-t relations $\mathcal{R}_{\mathcal{Q}}^t$ in descending order of score to re-rank the original entity candidates, then take the intersection of the two sets as the relation $r \in R_{\mathcal{Q}}^t \cap R_e$ with the high confidence score for each entity candidate e. Next, we re-rank the entities with the re-rank score $S(e; \mathcal{Q})$ as follows:

$$S(e; \mathcal{Q}) = \alpha \cdot S_l(e; \mathcal{Q}) + (1 - \alpha) \cdot \max_{r \in R_{\mathcal{Q}}^t \cap R_e} S_r(r; \mathcal{Q}), \tag{5}$$

where $S_l(e; \mathcal{Q})$ is the original entity linker score of entity e and α is a constant in the range of 0 to 1. At last, the entity with the highest score is selected as the source entity of the question.

4.3   Reinforcement Learning Formulation

The reinforcement learning formulation can be viewed as a Markov decision process (MDP) [20]. The environment of the knowledge graph $\mathcal{G}$ is defined as a five-tuple $(\mathcal{S}, \mathcal{O}, \mathcal{A}, \mathcal{P}, \mathcal{R})$, where $\mathcal{S}$ is the state space, $\mathcal{O}$ is the observation space, $\mathcal{A}$ is the action space, $\mathcal{P}$ and $\mathcal{R}$ respectively represent the state transition function and reward function. The main components of MDP will be explained in detail below.

States. The state space $\mathcal{S}$ contains all valid combinations of knowledge graph node information. Each state $S_t = (e_t, \mathcal{Q}, e_s, a) \in \mathcal{S}$ at step t not only includes the location of the entity node e_t where the agent is currently located but also includes global information $(\mathcal{Q}, e_s)$ where $\mathcal{Q}$ is the question and e_s is the source entity. a is the answer at the current step. Initial state $S_I \in \mathcal{S}$ and termination state $S_T \in \mathcal{S}$ are represented as $S_I = (e_s, \mathcal{Q}, e_s, a)$ and $S_T = (a, \mathcal{Q}, e_s, a)$.

Observation. The agent cannot know the answer to the question during the reasoning process so the complete state is invisible to the agent. The reasoning agent knows its location e_t at step t and the input query condition $(\mathcal{Q}, e_s)$. Therefore, we can set the observation function as $O(S_t) = (e_t, \mathcal{Q}, e_s)$.

Action. The action space $\mathcal{A}$ is the combination of effective edges in the knowledge graph. At step t, the set of candidate actions $\mathcal{A}(S_t)$ is composed of all the outgoing edges of the current location e_t where the agent is located in. Concretely, $A_t = \{(r, e) | (e_t, r, e) \in \mathcal{G}\}$.

Transition. The state S_t and the reward R_t at step t only depend on the predecessor states and actions. In other words, when the values of the predecessor state and action are given, the specific values of these random variables, the probability of appearing at step t is the state transition probability. The transition function $\mathcal{P}$ is defined as $\mathcal{P}(S_t, A_t) = (e', \mathcal{Q}, e_s, a)$, where $A_t = (r', e')$ and $S_t = (e_t, \mathcal{Q}, e_s, a)$.

Rewards. At step t, the agent performs the action A_t, it will accumulate the reward as $R_t \in \mathcal{R}$. In the traditional knowledge graph path reasoning with reinforcement learning, a binary reward function which has limitations is generally used. We propose a reward shaping policy to improve it, which will be detailed elaborated in Sect. 4.4.

4.4   Reward Shaping

In knowledge graph $\mathcal{G}$, the reward for spurious reasoning paths of the binary reward method may be the same as the one for the true reasoning path as Fig. 1. So we propose the reward shaping policy which consists of three parts.

First, we define a semantic reward based on the relation label of each path. We use the module that is mentioned in Sect. 4.1 and Sect. 4.2, and then evaluate the similarity between the relation representation $\mathcal{R}^r$ and the question representation $\mathcal{Q}^r$ through the cosine function. On the one hand, the cosine similarity is highly interpretable and easy to calculate. On the other hand, $\cos(x) \in [-1, 1]$, the result can be either the reward for the correct path or the penalty for the wrong exploration path. We define the semantic reward as:

$$R_{\text{SEMANTIC}} = \begin{cases} \cos(Q^r, R^r), & t > 1, \\ 0, & t = 1, \end{cases} \tag{6}$$

where t is the time step.

Meanwhile, we tend to take an efficient path to reach the answer entity. The efficiency reward which decreases as the path grows is defined as follows:

$$R_{\text{EFFICIENCY}} = \frac{\beta}{\text{length}(\mathcal{H}_\sigma)}, \tag{7}$$

where β is a parameter that we can adjust during training and $\mathcal{H}_\sigma$ is the historical track defined in Sect. 3. The length of $\mathcal{H}_\sigma$ is the number of relation edges in it. We only calculate the efficiency reward at the end of each query.

The last part of the reward is the binary reward called global reward:

$$R_{\text{GLOBAL}} = \begin{cases} +1, & e_t = a \\ 0, & e_t \neq a. \end{cases} \tag{8}$$

Final reward R of the question Q will consist of the above three parts, and the calculation formula is as follows:

$$R = R_{\text{GLOBAL}} + R_{\text{EFFICIENCY}} + \sum_{t=1}^{N} (R_t)_{\text{SEMANTIC}}, \tag{9}$$

where N is a constant that represents the length of the path. In a query, the reward for each subsequent step is -1 when the global reward is obtained. It makes the agent tend to stay in the answer entity node by doing self-loop after getting the global reward, instead of continuing to walk for the higher reward.

4.5   Policy Network

The policy network takes the current state information S_t and the action in the last step A_{t-1} as input. The purpose is to learn a parameterized policy $\pi = (\pi_1, \pi_2, ..., \pi_n)$ and output the probability distribution of candidate actions as $\pi_t : S_t \rightarrow P\{A(S_t)\}$, where S_t is the current state.

Formally, for the question $Q = (q_1, q_2, ..., q_n)$ with n words, we convert it into the question representation of d-dimension $Q^r \in R^d$ as mentioned in Sect. 4.1. The probability of reward changes with time and the agent encodes the historical trajectory $H_t = \{H_{t-1}, A_{t-1}, 0_t\}$ which is about the sequence of previous observations and actions at step t as a continuous representation with the BiLSTM network as follows:

$$H_t^r = \text{BiLSTM}\left(H_{t-1}^r, R_{t-1}^r\right), \tag{10}$$

where H_{t-1}^r is the output of the recurrent neural network in the previous step $t-1$ and R_{t-1}^r is the relation representation in A_{t-1}. In addition, H_0^r and R_0^r are both zero vectors.

Then, we use a similar approach as the one mentioned in Sect. 4.1. Like Eq. 3, we use a BiLSTM to get the sequence of question embedding $Y^q = \{y_1^q, ..., y_n^q\}$. But in the sequence Y^q, we use the original topic entity instead of token $\langle e \rangle$. We calculate the similarity T_m between relation embedding R^r and question word y_m^q:

$$T_m = W' \cdot (R^r \odot y_m^q) + b', \tag{11}$$

where $W' \in R^d$ and $b' \in R$ are parameters and $\odot$ denotes matrix multiplication. Then we pass the result through a SoftMax layer to get an attention distribution on Y^q as follows:

$$\alpha_i = \frac{\exp(T_i)}{\sum_{m=1}^{n} \exp(T_m)}, \tag{12}$$

and the weighted sum of these vectors is obtained by the following formula:

$$\rho_{q.r} = \sum_{i=1}^{n} \alpha_i \cdot y_i^q. \tag{13}$$

Based on the historical trajectory embedded H_t^r, the policy network decides to select an action from all available actions $A(S_t)$. The semantic score for each action A_t is calculated through a perceptron. We input the historical trajectory embedding H_t^r, the current observation O_t and the semantic score of the query relation $\rho_{q.r}$ into two nonlinear feedforward neural network layers, and then use a SoftMax layer to calculate the probability of each action being selected at step t as the policy network π_t:

$$\pi_t = \text{softmax} \left(\text{M}_A \left(\text{W}_2 \text{ReLU} \left(\text{W}_1 \left[\text{H}_t; \text{O}_t; \rho_{q.r} \right] + \text{b}_1 \right) + \text{b}_2 \right) \right), \tag{14}$$

where M_A is a matrix that consists of all action representations and π_t is the probability distribution of all action candidates. W_1 and W_2 are feedforward network weights. b_1 and b_2 are biases.

5   Experiments and Results

5.1   Datasets

1. **MetaQA** [30] is a large-scale KGQA dataset which contains more than 400k movie questions. It consists of up to 3-hop complex multi-hop questions. For our experiments, we used the "Vanilla" version and the "Vanilla-EU" (EU stands for topic entity unlabeled) of questions and the knowledge graph provided with WikiMovies, which includes about 43k entities and 135k triples.
2. **WebQuestionsSP** [25] is a QA dataset composed of 4737 natural language questions and their answers with Freebase [2]. Due to the scale of Freebase being too large, we use the subgraph of Freebase knowledge graph generated by GRAFT-NET [19] which includes about 528k entities (Table 1).

Table 1. Statistics of MetaQA and WebQuestionsSP and knowledge graphs used in experiments. #Entity and #Relation represent the number of entities and relations of KGs for our KGQA tasks.

Dataset	Train	Dev	Test	#Entity	#Relation
MetaQA 1-hop	96,106	9,992	9,947	43,233	9
MetaQA 2-hop	118,960	14,872	14,872	43,233	9
MetaQA 3-hop	114,196	14,274	14,274	43,233	9
WebQuestionsSP	2,848	250	1,639	528,617	513

5.2   Baselines

In order to verify the effectiveness of our method, we selected some baseline methods to compare with RPGQA.

- **KV-MemNN** [10] uses texts as the external knowledge which extracts the information in texts in the form of key-value pairs as KG triples.
- **MINERVA** [3] proposes a reinforcement learning approach to model the state space based on the question. It takes the path to correct answer entities as a sequential optimization problem based on the structure of the KG.
- **VRN** [30] uses a variational learning algorithm for question answering over the knowledge graph. This method is also under weak supervision.
- **GraftNet** [19] proposes a question answering approach based on open domain, which integrates text and KG.
- **SRN** [11] proposes a reward shaping policy based on potential, which can accelerate the convergence speed of the training algorithm.
- **R-GCN** [16] proposes relational graph convolutional networks for the task of knowledge graph completion. It is mainly designed for multi-relation data.

5.3   Training and Implementation Details

In the process of training, our goal of training is to maximize the expected rewards when we perform question answering tasks over the knowledge graph:

$$J(\theta) = E_{(\mathcal{Q},a) \sim D} \left[E_{(A_1, A_2, \cdots, A_T) \sim \pi_\theta} \left[\sum_{t=1}^{T} \eta^{t-1} R\left(S_t, A_t\right) \mid (\mathcal{Q}, a) \right] \right], \quad (15)$$

$J(\theta)$ is the gradient of a certain performance index. D is the dataset for training and $(\mathcal{Q}, a)$ is a set of question answering data in it. We set η as a discount factor in the range 0 to 1 to make the proportion of reward gradually decrease with the increase of the number of steps. In order to encourage the agent to sample diverse paths during training, we add the value of ϵ in ϵ-greedy algorithm, the learning rate and so on as hyper-parameters to better maintain the balance between exploration and exploitation.

In our work, we use the pre-trained model to initialize each word embedding. Then we tune the hyper-parameters on development sets. The hidden dimension of the BiLSTM in the policy network as history encoder is set to 300 and the dropout rate is 0.3. We use ADAM as the optimizer, meanwhile, we tune the initial learning rate λ within {0.01, 0.05, 0.1, 0.5, 1.0}. For the reinforcement learning algorithm, we set the discount factor $\eta = 0.9$ and tune the parameter β in the efficiency reward within (0.5, 1.0). In greedy search, we set the value of ϵ gradually increasing during training and finally to 1. We tune parameter β and learning rate λ in Fig. 3 and choose the optimal solution in training.

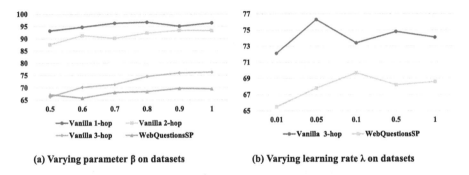

(a) Varying parameter β on datasets   (b) Varying learning rate λ on datasets

Fig. 3. Performance tuning of our model.

5.4   Results and Analysis

We use Hits@1 as the index for evaluating the effect of the multi-hop KGQA model. Regard question answering as a task of answer ranking, Hits@1 refers to the proportion that the correct answer is the first one in the final ranking. The detailed results of MetaQA dataset are listed in Table 2.

Table 2. Test results on the MetaQA dataset(Hits@1 in percent)

	Vanilla 1-hop	Vanilla 2-hop	Vanilla 3-hop	Vanilla-EU 1-hop	Vanilla-EU 2-hop	Vanilla-EU 3-hop
KV-MemNN	93.5	84.3	53.8	85.2	80.8	35.2
MINERVA	96.3	92.9	55.2	87.7	89.1	36.1
VRN	**97.5**	89.9	62.5	82.0	75.6	38.3
SRN	97.0	**95.1**	75.2	88.4	91.2	49.2
RPGQA	96.7	93.4	**76.3**	**93.6**	**91.8**	**52.6**

For the "Vanilla" version of MetaQA, since the topic entities in questions have been labeled, this version is mainly used to evaluate the logical reasoning ability of the method. Although our method (RPGQA) does not achieve the best results of "Vanilla 1-hop" and "Vanilla 2-hop", the gap with the optimal baseline is only about 1%. RPGQA achieves the best result of "Vanilla 3-hop".

For the "Vanilla-EU" version of MetaQA dataset, the performance of all baselines becomes worse for the lack of topic entity labels. However, RPGQA is less affected compared with baselines, for we specially design the entity disambiguation module to take topic entity linking as the foundation. Therefore, entity linking and reward shaping policy both play important roles in multi-hop question answering under weak supervision signals.

In order to further evaluate the performance of RPGQA on datasets in different domains and scales, we use Hits@1 and F1-score to measure the performance

Table 3. Test results on WebQuestionsSP dataset (Hits@1 and F1 in percent)

	Hits@1	F1
R-GCN	37.2	30.5
KV-MemNN	46.7	38.6
GraftNet	67.8	62.8
RPGQA	**69.7**	**64.3**

of our method and baselines on WebQuestionsSP dataset. The detailed results are listed in Table 3. As shown in Table 3, our method (RPGQA) based on reinforcement learning performs better than all baselines in the test set. Facts have proved that for complex questions with multiple hops, the system of generating reasoning path with reinforcement learning has more advantages in answer quality.

5.5   Ablation Studies

We conduct further ablation studies to prove the effectiveness of each part. Since the topic entities that have been labeled on "Vanilla" version of MetaQA, we only study on "Vanilla-EU" version of MetaQA and WebQuestionsSP.

Table 4. Ablation study of RPGQA (Hits@1 in percent)

	Vanilla-EU 1-hop	Vanilla-EU 2-hop	Vanilla-EU 3-hop	WebQuestionsSP
GR	87.6	84.2	35.5	58.6
GR+ER	87.9	85.7	42.1	61.3
GR+SR	88.8	86.9	48.3	64.0
GR+EDM	91.3	89.7	39.2	66.5
RSP	89.2	87.6	51.4	66.2
RPGQA	93.6	91.8	52.6	69.7

We use the entity disambiguation module (EDM) and the two additional soft rewards including semantic reward (SR) and efficiency reward (ER) as variables to compare the effectiveness of the experiment. The compared question answering methods include: (1) GR: the model using only global reward (GR); (2) GR + SR: the model using global reward and semantic reward; (3) GR + ER: the model using global reward and efficiency reward; (4) RSP: the model using complete reward shaping policy without entity disambiguation module; (5) GR + EDM: the model using global reward and entity disambiguation module.

As the results that are shown in Table 4, efficiency reward hardly works when answering 1-hop questions, but it can improve the effect of GR by 18.6%

when answering 3-hop questions. Semantic reward improves the baseline effect by 36.1% on 3-hop questions. It proves that the performance of semantic reward is greater than the one of efficiency reward on multi-hop question answering. RSP improves the performance of GR by 55.7%, which proves that it is necessary to use reward shaping policy under weak supervision signals. On the other hand, the entity disambiguation module in our model plays an important role in mutil-hop KGQA.

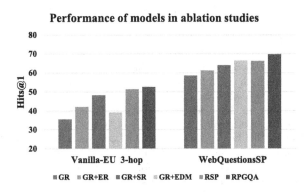

Fig. 4. Performance comparison on Vanilla-EU 3-hop and WebQuestionsSP.

To compare the performance of the above models more intuitively, we selected two typical datasets in Fig. 4. The performance fluctuation on the dataset Vanilla-EU 3-hop is significantly greater than that on the dataset WebQues-tionsSP. We consider the main reason for this situation is the difference in the difficulty of questions in the two datasets. The performance difference of these models is mainly reflected in the processing of complex questions.

6   Conclusion

In this paper, we present a novel reinforcement learning model RPGQA which generates the reasoning path to enhance intermediate supervision signals and then answers multi-hop questions. We employ the entity re-ranking algorithm for entity linking to find the correct source entity as the beginning point of the path. We design a reward shaping policy from both semantic and efficiency to alleviate the delayed and sparse reward problem. Experimental results on two QA datasets validate the effectiveness of the model on multi-hop KGQA.

References

1. Bast, H., Haussmann, E.: More accurate question answering on freebase. In: Proceedings of the 24th ACM International on Conference on Information and Knowledge Management, pp. 1431–1440 (2015)

2. Bollacker, K., Evans, C., Paritosh, P., Sturge, T., Taylor, J.: Freebase: a collaboratively created graph database for structuring human knowledge. In: Proceedings of the 2008 ACM SIGMOD International Conference on Management of Data, pp. 1247–1250 (2008)

3. Das, R., et al.: Go for a walk and arrive at the answer: reasoning over paths in knowledge bases using reinforcement learning. In: 6th International Conference on Learning Representations (2017)

4. Dong, L., Wei, F., Zhou, M., Xu, K.: Question answering over freebase with multi-column convolutional neural networks. In: Proceedings of the 53rd Annual Meeting of the Association for Computational Linguistics and the 7th International Joint Conference on Natural Language Processing, pp. 260–269 (2015)

5. Gao, H., et al.: CSIP: enhanced link prediction with context of social influence propagation. Big Data Res. **24**, 100217 (2021)

6. Guo, S., et al.: Knowledge graph embedding preserving soft logical regularity. In: Proceedings of the 29th ACM International Conference on Information and Knowledge Management, pp. 425–434 (2020)

7. Li, X., Hu, S., Zou, L.: Knowledge based natural answer generation via masked-graph transformer. World Wide Web **25**(3), 1403–1423 (2022)

8. Liang, C., Berant, J., Le, Q., Forbus, K.D., Lao, N.: Neural symbolic machines: learning semantic parsers on freebase with weak supervision. In: Proceedings of the 55th Annual Meeting of the Association for Computational Linguistics, pp. 23–33. Association for Computational Linguistics (2017)

9. Lin, X.V., Socher, R., Xiong, C.: Multi-hop knowledge graph reasoning with reward shaping. In: Proceedings of the 2018 Conference on Empirical Methods in Natural Language Processing, Brussels, Belgium, 31 October– 4 November, 2018, pp. 3243–3253 (2018)

10. Miller, A., Fisch, A., Dodge, J., Karimi, A.H., Bordes, A., Weston, J.: Key-value memory networks for directly reading documents. In: Proceedings of the 2016 Conference on Empirical Methods in Natural Language Processing, pp. 1400–1409 (2016)

11. Qiu, Y., Wang, Y., Jin, X., Zhang, K.: Stepwise reasoning for multi-relation question answering over knowledge graph with weak supervision. In: Proceedings of the 13th International Conference on Web Search and Data Mining, pp. 474–482 (2020)

12. Qiu, Y., et al.: Hierarchical query graph generation for complex question answering over knowledge graph. In: Proceedings of the 29th ACM International Conference on Information and Knowledge Management, pp. 1285–1294 (2020)

13. Ren, H., Hu, W., Leskovec, J.: Query2box: reasoning over knowledge graphs in vector space using box embeddings. In: 8th International Conference on Learning Representations. OpenReview.net (2020)

14. Ren, H., Leskovec, J.: Beta embeddings for multi-hop logical reasoning in knowledge graphs. In: Advances in Neural Information Processing Systems 33: Annual Conference on Neural Information Processing Systems 2020 (2020)

15. Saxena, A., Tripathi, A., Talukdar, P.: Improving multi-hop question answering over knowledge graphs using knowledge base embeddings. In: Proceedings of the 58th Annual Meeting of the Association for Computational Linguistics, pp. 4498–4507 (2020)

16. Schlichtkrull, M., Kipf, T.N., Bloem, P., van den Berg, R., Titov, I., Welling, M.: Modeling relational data with graph convolutional networks. In: Gangemi, A., et al. (eds.) ESWC 2018. LNCS, vol. 10843, pp. 593–607. Springer, Cham (2018). https://doi.org/10.1007/978-3-319-93417-4_38

17. Shen, Y., Chen, J., Huang, P.S., Guo, Y., Gao, J.: M-walk: learning to walk over graphs using Monte Carlo tree search. In: Advances in Neural Information Processing Systems 31: Annual Conference on Neural Information Processing Systems 2018, pp. 6787–6798 (2018)
18. Sun, H., Bedrax-Weiss, T., Cohen, W.W.: PullNet: open domain question answering with iterative retrieval on knowledge bases and text. In: Proceedings of the 2019 Conference on Empirical Methods in Natural Language Processing and the 9th International Joint Conference on Natural Language Processing, pp. 2380–2390 (2019)
19. Sun, H., Dhingra, B., Zaheer, M., Mazaitis, K., Salakhutdinov, R., Cohen, W.W.: Open domain question answering using early fusion of knowledge bases and text. In: Proceedings of the 2018 Conference on Empirical Methods in Natural Language Processing, pp. 4231–4242 (2018)
20. Sutton, R.S., Barto, A.G.: Reinforcement Learning: An Introduction. MIT Press, Cambridge (2018)
21. Wang, M., Wang, H., Li, B., Zhao, X., Wang, X.: Survey of key technologies of new generation knowledge graph. J. Comput. Res. Dev., 1–18 (2022). (Chinese)
22. Wu, J., Li, B., Ji, Y., Tian, J., Xiang, Y.: Text-enhanced knowledge graph representation model in hyperbolic space. In: Li, B., et al. (eds.) ADMA 2022. LNCS (LNAI), vol. 13088, pp. 137–149. Springer, Cham (2022). https://doi.org/10.1007/978-3-030-95408-6_11
23. Xiong, W., Hoang, T., Wang, W.Y.: DeepPath: a reinforcement learning method for knowledge graph reasoning. In: Proceedings of the 2017 Conference on Empirical Methods in Natural Language Processing, pp. 564–573 (2017)
24. Xu, K., Reddy, S., Feng, Y., Huang, S., Zhao, D.: Question answering on freebase via relation extraction and textual evidence. In: Proceedings of the 54th Annual Meeting of the Association for Computational Linguistics (2016)
25. Yih, W.t., Chang, M.W., He, X., Gao, J.: Semantic parsing via staged query graph generation: question answering with knowledge base. In: Proceedings of the 53rd Annual Meeting of the Association for Computational Linguistics and the 7th International Joint Conference on Natural Language Processing of the Asian Federation of Natural Language Processing, pp. 1321–1331 (2015)
26. Yin, W., Yu, M., Xiang, B., Zhou, B., Schütze, H.: Simple question answering by attentive convolutional neural network. In: COLING 2016, 26th International Conference on Computational Linguistics, Proceedings of the Conference: Technical Papers, pp. 1746–1756 (2016)
27. Yu, H., Li, H., Mao, D., Cai, Q.: A relationship extraction method for domain knowledge graph construction. World Wide Web **23**(2), 735–753 (2020). https://doi.org/10.1007/s11280-019-00765-y
28. Yu, M., Yin, W., Hasan, K.S., Santos, C.d., Xiang, B., Zhou, B.: Improved neural relation detection for knowledge base question answering. In: Proceedings of the 55th Annual Meeting of the Association for Computational Linguistics, pp. 571–581 (2017)
29. Zhang, Y., et al.: Fine-grained evaluation of knowledge graph embedding model in knowledge enhancement downstream tasks. Big Data Res. **25**, 100218 (2021)
30. Zhang, Y., Dai, H., Kozareva, Z., Smola, A.J., Song, L.: Variational reasoning for question answering with knowledge graph. In: Proceedings of the Thirty-Second AAAI Conference on Artificial Intelligence, (AAAI-18), The 30th Innovative Applications of Artificial Intelligence (IAAI-18), and the 8th AAAI Symposium on Educational Advances in Artificial Intelligence (EAAI 2018), pp. 6069–6076 (2018)

E^3L: Experience Enhanced Entity Linking for Question Answering Over Knowledge Graphs

Zhirong Hou[1], Meiling Wang[2], Min Li[2], and Ying Li[1(✉)]

[1] Peking University, Beijing, China
{hou.zhirong,li.ying}@pku.edu.cn
[2] ICBC Technology Co. Ltd., Beijing, China
{wangml,lim02}@tech.icbc.com.cn

Abstract. Entity linking is a pivotal factor for building robust Question Answering systems over Knowledge Graphs (KGQA), and representation of entities occupies an important position for tackling entity linking for questions. To alleviate the deficiency of entity descriptions that contextual knowledge is insufficient for entity representations, we introduce entity experiences as a new text-style contextual knowledge source to enrich entity representations, and propose an Experience Enhanced Entity Linking framework called as E^3L. For the modeling of entity experiences, we derive embeddings using the entity mentions occurring in history questions, and design an attention-based retriever to capture key information relative to user questions. Then the distilled entity experiences are integrated with entity descriptions to enhance entity representations, and question representations are refined with a multi-level attention mechanism. Finally, entity linking is improved with the entity representations and the refined representations of questions. Experimental results on end-to-end benchmark datasets demonstrate that our approach achieves state-of-the-art F1-score, and provides an effective way to improve test performance for universal models using entity experiences without fine-tuning.

Keywords: Question Answering over Knowledge Graphs · Entity linking · Mention detection · Entity disambiguation · Entity experience

1 Introduction

Question Answering systems over Knowledge Graphs (KGQA), which answers users' natural language questions with a Knowledge Graph (KG), has emerged as a promising technique to provide unified and user-friendly access to KGs [6]. Entity Linking for KGQA (ELQA), which links user questions with the entities available in KGs, is crucial for building robust KGQA, since answers must be connected to the entity mentions in the questions via some path over KGs [9]. ELQA typically involves two basic tasks: mention detection and entity

B. Li et al. (Eds.): APWeb-WAIM 2022, LNCS 13422, pp. 210–224, 2023.
https://doi.org/10.1007/978-3-031-25198-6_17

Table 1. Results for linking an example question to WikiData entities using entity introductory texts. Entity introductory texts in parentheses are from Wikipedia articles.

Question	What nba teams has shaq played for?
GOLD result	Shaq→ "Shaquille O'Neal" (Shaquille O'Neal, known commonly as "Shaq", is an American former professional basketball player.)
PRED result	Shaq→ "Tupac Shakur" (Tupac Shakur, also known by his stage names 2Pac and Makaveli, was an American rapper and actor.)

disambiguation [8, 11]. Mention detection extracts mention spans in the questions and entity disambiguation links these mentions to their corresponding entities in a KG. Taking the question "what nba teams has shaq played for?" in Table 1 as an example, the mention detection task tries to identify "shaq" as a mention and the entity disambiguation task makes attempt to link it to the appropriate WikiData entity.

The main challenge for ELQA is that real-world user questions are typically short texts in which contextual information is scarce, leading to mention misidentification and linkage error on ambiguous questions [9, 10]. For example, when processing the question "what nba teams has shaq played for?" without any background knowledge, the word "played" could be explained as "giving a show" or "sporting". This ambiguity causes difficulty in deciding whether "shaq" is an actor or an athlete. To tackle the challenge, numerous studies have adopted the approaches that enrich entity representations with descriptive texts or knowledge graph context. In detail, descriptive texts, including entity titles and introductory texts, are used to enrich conceptual and linguistics knowledge [1, 8, 11, 13, 14], while knowledge graph context such as entities, relations and attributes are employed to enhance contextual knowledge [1, 11]. Between the two approaches, encodings of descriptive texts are more accessible but lack of contextual knowledge. As shown in Table 1, mention "shaq" could be linked to entity "Tupac Shakur" rather than "Shaquille O'Neal" when only entity introductory texts are utilized. The problem can be alleviated by applying history questions to enrich contextual knowledge of entities. The intuition is that humans generally understand questions by obtaining experiences from similar or related history questions, which extends contextual knowledge of certain entities. For example, from a previous question "when did shaq come to the nba?" with mention "shaq" linked to entity "Shaquille O'Neal", human can get hints that "shaq" in the question of Table 1 may also be linked to "Shaquille O'Neal", because the contextual knowledge of "Shaquille O'Neal" is enriched with "come to the nba" from the previous question. As a text-style contextual knowledge source, history questions are more accessible compared with knowledge graph context, and the idea could become effective when history questions being accumulated.

Based on the above idea, we regard the occurrences of each entity in history questions as its experiences and propose an Experience Enhanced Entity Linking framework (E^3L) for KGQA. In the framework, we focus on the modeling of entity experiences and its utilization to enhance entity representations and question representations. In detail, for the modeling of entity experiences, we first derive embeddings using the entity mentions occurring in history questions, and then capture key information relative to user questions with an attention-based retriever. The entity representations are produced by integrating entity experiences and entity descriptions, and question representations are refined with a multi-level attention mechanism. In the follow-up stage, the resulting representations are used for mention detection and entity disambiguation by training with joint optimization. We conduct experimental evaluations on the end-to-end ELQA benchmark datasets WebQSP$_{EL}$ and GraphQ$_{EL}$ [8]. Results of model comparison demonstrate that our approach achieves state-of-the-art ELQA F1-score, and provides an effective way to improve test performance for universal models using entity experiences without fine-tuning. In addition, results of ablation study and case study demonstrate the effectiveness of our design on entity experience modeling and utilization.

The contributions of this paper are threefold.

(1) We propose to take entity experience as a new accessible knowledge source to improve ELQA. In our knowledge, this is the first work that introduces entity experience into ELQA modeling.
(2) We propose a framework for entity experience modeling including encoding and retrieval, and the framework guarantees the applicability of entity experience to ELQA in practice.
(3) We propose mechanisms for enhancement of entity representations and question representations by utilizing entity experiences, which benefit ELQA effectively.

2   Related Work

With respect to the representation of entities for ELQA task, researchers have studied the solutions of utilizing descriptive texts and knowledge graph context.

Firstly, majority of studies employ entity labels and titles as basic features for entity representations, and other entity descriptions such as introductory texts are also used to enrich entity encodings. Sorokin and Gurevych [11] extract character-level features for entity labels. Banerjee et al. [1] employ fastText embeddings of Wikidata descriptions. Driven by recent advances in pre-training technologies, Wu et al. [13] and Li et al. [8] apply BERT model [4] as the encoder of Wikipedia descriptions and integrate general language knowledge into the representations of entities.

Besides, researchers have augmented entity representations with knowledge graph context such as entities, relations and attributes in knowledge graphs. Sorokin and Gurevych [11] use the embeddings of entities and relations that are connected to entity candidates, and knowledge graph embeddings are trained with TransE algorithm [2]. Banerjee et al. [1] employ pre-computed and pre-indexed embeddings learned by TransE algorithm over Wikidata.

Between the two kinds of approaches, representations of descriptive texts, which enrich conceptual and linguistics knowledge for entities, are more accessible but lack of contextual knowledge. In this paper, we explore to enrich contextual knowledge for entity representations by applying history questions, which are more accessible compared with knowledge graph context.

3   Framework

Definition 1 (Entity Experience). *Given a set of entity candidates $\mathcal{E} = \{e_i | 1 \leq i \leq N_{\mathcal{E}}\}$, experiences of e_i is defined as its occurrences in the history questions:*

$$Q(e_i) = \{(q_{ij}, start_{ij}, end_{ij}) | 1 \leq j \leq N_i^Q \wedge 1 \leq start_{ij} \leq end_{ij} \leq N_{q_{ij}}\}$$

where $N_i^Q = |Q(e_i)|$, q_{ij} is a history question containing $N_{q_{ij}}$ tokens, and $start_{ij}$ and end_{ij} are the start location and end location for the mention of e_i in q_{ij}.

For example, entity "Shaquille O'Neal" occurs in question "when did shaq come to the nba?" as mention "shaq", and both the start location and end location are 3.

Definition 2 (ELQA Task). *Given a user question $q = (x_1, \cdots, x_{N_q})$ containing N_q tokens and the set of entity candidates $\mathcal{E} = \{e_i\}$, where e_i has a set of experiences $Q(e_i)$ and a descriptive text $d(e_i)$, the goal of ELQA is to output a list of tuples, $(e, [start, end])$, whereby $e \in \mathcal{E}$ is the entity corresponding to the mention span from start-th to end-th token in q.*

Taking entity "Shaquille O'Neal" as an example, its descriptive text is an introductory text from Wikipedia articles, as shown in Table 1.

Aiming to solve the ELQA task, our proposed framework E^3L contains five modules as shown in Fig. 1. *Question Basic Encoder* module derives initial token-level encodings for user questions. For each entity candidate, *Entity Encoder* module encodes its entity experiences and entity description and integrates them to produce entity representation. We employ a tri-encoder architecture for *Question Basic Encoder* and *Entity Encoder*, where the encoder of entity experiences shares parameters with *Question Basic Encoder*. *Entity Encoder* computes encodings of entity experiences based on mentions occurring in history question texts and utilizes an attention-based retriever to capture important information relative to user questions. With the representations of all the entity candidates, *Question Representation Refinement* module refines question representations based on a multi-level attention mechanism. Finally, *Mention Detection* module computes the likelihood scores of mention spans, and *Entity Disambiguation* module computes the likelihood distribution over all the entity candidates conditioned on mentions, and they are trained with joint optimization. In the following subsections, we detail the design of E^3L.

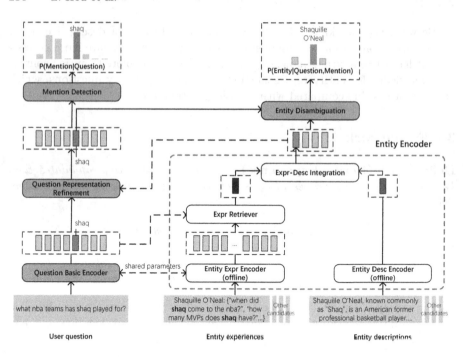

Fig. 1. Architecture of E^3L.

3.1 Question Basic Encoder

For the user question $q = (x_1, \cdots, x_{N_q})$, we apply an encoder T^q to encode the context of q into token-level encodings $\mathbf{q} = (\mathbf{x}_1, \cdots, \mathbf{x}_{N_q})$,

$$(\mathbf{x}_1, \cdots, \mathbf{x}_{N_q}) = T^q (x_1, \cdots, x_{N_q}) \tag{1}$$

where $\mathbf{x}_j \in \mathbb{R}^h, 1 \leq j \leq N_q$. Various encoders can be considered for T^q. In this paper we apply BERT pretrained model for T^q, and the input token sequence to T^q is constructed as $[CLS]x_1 \cdots x_{N_q}[SEP]$.

3.2 Entity Encoder

For each entity $e_i \in \mathcal{E}$, which has a set of experiences $Q(e_i)$ and a descriptive text $d(e_i)$, firstly we encode $Q(e_i)$ and $d(e_i)$ into vector $\mathbf{e}_i^Q \in \mathbb{R}^h$ and $\mathbf{e}_i^d \in \mathbb{R}^h$ respectively, and then compute entity representation $\mathbf{e}_i \in \mathbb{R}^h$ with $\mathbf{e}_i^Q$ and $\mathbf{e}_i^d$:

$$\mathbf{e}_i = f\left(\mathbf{e}_i^Q, \mathbf{e}_i^d\right) \tag{2}$$

where $f(\cdot)$ is a function that combines two vectors into one. In this paper, we define $f(\cdot)$ based on vector addition:

$$f\left(\mathbf{e}_i^Q, \mathbf{e}_i^d\right) = \beta \cdot \mathbf{e}_i^Q + \alpha \cdot \mathbf{e}_i^d \tag{3}$$

where α and β are hyperparameters and $0 \le \alpha \le 1, 0 \le \beta \le 1$. We describe details on how to compute $\mathbf{e}_i^d$ and $\mathbf{e}_i^Q$ in the remainder of this section.

Encoding of Entity Experiences. For each entity $e_i \in \mathcal{E}$ and its experiences

$$Q(e_i) = \{(q_{ij}, start_{ij}, end_{ij}) | 1 \le j \le N_i^Q \wedge 1 \le start_{ij} \le end_{ij} \le N_{q_{ij}}\},$$

we first apply the question encoder T^q to encode the context $q_{ij} = \left(x_{i1}, \cdots, x_{iN_{q_{ij}}}\right)$ into token-level encodings $\left(\mathbf{x}_{i1}, \cdots, \mathbf{x}_{iN_{q_{ij}}}\right)$,

$$\left(\mathbf{x}_{i1}, \cdots, \mathbf{x}_{iN_{q_{ij}}}\right) = T^q \left(x_{i1}, \cdots, x_{iN_{q_{ij}}}\right) \tag{4}$$

where $\mathbf{x}_{it} \in \mathbb{R}^h, 1 \le t \le N_{q_{ij}}$. Then we obtain mention representation $\mathbf{e}_{ij}^Q \in \mathbb{R}^h$ for the occurrence $(q_{ij}, start_{ij}, end_{ij})$ of e_i by averaging $\mathbf{x}_{start_{ij}}, \cdots, \mathbf{x}_{end_{ij}}$:

$$\mathbf{e}_{ij}^Q = \frac{1}{(end_{ij} - start_{ij} + 1)} \sum_{t=start_{ij}}^{end_{ij}} \mathbf{x}_t \tag{5}$$

where $\mathbf{e}_{ij}^Q$ represents the mention $[start_{ij}, end_{ij}]$ and the context surrounding it. Thus we get a vector sequence $\mathbf{e}_{i1}^Q, \cdots, \mathbf{e}_{iN_i^Q}^Q$ for the entity experiences of e_i, and next we capture key information from the vector sequence by experience retrieval.

Retrieval of Entity Experiences. Intuitively, $\mathbf{e}_{i1}^Q, \cdots, \mathbf{e}_{iN_i^Q}^Q$ represent different aspects of e_i's usage context in history questions and are related to q in different degrees. For example, entity "Shaquille O'Neal" occurs in history question "when did shaq come to the nba?" and "how many MVPs does shaq have?" as mention "shaq". Compared with the latter question, the occurrence of "Shaquille O'Neal" in the former question is more important for understanding "what nba teams has shaq played for?", because "come to the nba" is more contextual. Formally, we first obtain a sentence-level representation $\mathbf{q}^c \in \mathbb{R}^h$ for q, and then distill key information from $\mathbf{e}_{i1}^Q, \cdots, \mathbf{e}_{iN_i^Q}^Q$ with attention mechanism.

Firstly, we obtain $\mathbf{q}^c$ with a reduction function $red^q(\cdot)$:

$$\mathbf{q}^c = red^q(\mathbf{x}_1, \cdots, \mathbf{x}_{N_q}) \tag{6}$$

where we choose $red^q(\cdot)$ as the average over all the vectors to catch more comprehensive information. Then we compute the related weights of $\mathbf{e}_{i1}^Q, \cdots, \mathbf{e}_{iN_i^Q}^Q$ to $\mathbf{q}^c$ and obtain the representation $\mathbf{e}_i^Q$ of e_i's experiences:

$$\mathbf{e}_i^Q = \sum_{t=1}^{N_i^Q} w_{it} \mathbf{e}_{it}^Q \tag{7}$$

where

$$\left(w_{i1}, \dots, w_{iN_i^Q}\right) = \text{softmax}\left(\mathbf{q}^c \cdot \mathbf{e}_{i1}^Q, \dots, \mathbf{q}^c \cdot \mathbf{e}_{iN_i^Q}^Q\right) \tag{8}$$

Encoding of Entity Descriptions. For each entity $e_i \in \mathcal{E}$ and its descriptive text $d(e_i) = (x_{i1}, \cdots, x_{iN_i^d})$ containing N_i^d tokens, we apply an encoder T^d to encode $d(e_i)$ into sequence of vectors $(\mathbf{x}_{i1}, \cdots, \mathbf{x}_{iN_i^d})$,

$$\left(\mathbf{x}_{i1}, \cdots, \mathbf{x}_{iN_i^d}\right) = T^d \left(x_{i1}, \cdots, x_{iN_i^d}\right) \tag{9}$$

where $\mathbf{x}_{it} \in \mathbb{R}^h, 1 \le t \le N_i^d$. Then the sequence of vectors $\left(\mathbf{x}_{i1}, \cdots, \mathbf{x}_{iN_i^d}\right)$ is transformed into vector $\mathbf{e}_i^d \in \mathbb{R}^h$ by a reduction function $red^d(\cdot)$:

$$\mathbf{e}_i^d = red^d \left(\mathbf{x}_{i1}, \cdots, \mathbf{x}_{iN_i^d}\right) \tag{10}$$

In this paper, we apply BERT pre-trained model for encoder T^d, its input token sequence is constructed in the same manner as T^q, and then we choose $red^d(\cdot)$ to be the last layer of the output of the $[CLS]$ token.

3.3   Question Representation Refinement

We refine the representation of user question q with the representations of entity candidates $\mathcal{E}$. Intuitively, when considering the knowledge of $\mathcal{E}$, each token of q has different importance for the ELQA task, and meanwhile each entity in $\mathcal{E}$ has different importance for understanding each token of q. We draw on the concept of attention over attention [3] and propose a multi-level attention mechanism to model the ideas.

Given the initial token-level representations $\mathbf{q} = (\mathbf{x}_1, \cdots, \mathbf{x}_j, \cdots, \mathbf{x}_{N_q})$ for q and entity representations $\mathbf{e}_1, \cdots, \mathbf{e}_i, \cdots, \mathbf{e}_{N_{\mathcal{E}}}$ for $\mathcal{E}$, where $\mathbf{x}_j \in \mathbb{R}^h$, $\mathbf{e}_i \in \mathbb{R}^h$, $1 \le j \le N_q$ and $1 \le i \le N_{\mathcal{E}}$, we first calculate a pair-wise matching matrix $\mathbf{m} \in \mathbb{R}^{N_{\mathcal{E}}*N_q}$ between entity representations and the initial token-level representations, and obtain individual entity-level attentions (as Eq. 12) and individual question-level attentions (as Eq. 13) respectively:

$$\mathbf{m}_{i,j} = \mathbf{e}_i^T \cdot \mathbf{x}_j \tag{11}$$

$$\mathbf{w}_t^e = \text{softmax}(\mathbf{m}_{1,t}, \ldots, \mathbf{m}_{N_{\mathcal{E}},t}), \forall x_t \in q \tag{12}$$

$$\mathbf{w}_t^q = \text{softmax}(\mathbf{m}_{t,1}, \ldots, \mathbf{m}_{t,N_q}), \forall e_t \in \mathcal{E} \tag{13}$$

Secondly, we calculate importance distributions over entities when looking from the whole question. In detail, we average all the $\mathbf{w}_t^e$ to get an entity-level attention $\mathbf{w}^e \in \mathbb{R}^{N_{\mathcal{E}}}$:

$$\mathbf{w}^e = \frac{1}{N_q} \sum_{t=1}^{N_q} \mathbf{w}_t^e \tag{14}$$

Thirdly, we calculate importance distributions over question tokens when considering all the entities with weights. In detail, we obtain an attention matrix $\mathbf{w}^q \in \mathbb{R}^{N_{\mathcal{E}}*N_q}$ with the individual question-level attentions:

$$\mathbf{w}^q = [\mathbf{w}_1^q, \ldots, \mathbf{w}_t^q, \ldots, \mathbf{w}_{N_{\mathcal{E}}}^q] \tag{15}$$

Regarding $\mathbf{w}^e$ as the importance weights of individual question-level attentions in $\mathbf{w}^q$, we calculate the importance distributions over question tokens $\mathbf{w} \in \mathbb{R}^{N_q}$:

$$\mathbf{w} = \mathbf{w}^{q^T}\mathbf{w}^e \tag{16}$$

Finally, we compute the refined token-level representations $\tilde{\mathbf{q}} = (\tilde{\mathbf{x}}_1, \cdots, \tilde{\mathbf{x}}_{N_q})$ for q:

$$\tilde{\mathbf{x}}_j = (1 + \mu \mathbf{w}_j)\mathbf{x}_j, 1 \le j \le N_q \tag{17}$$

where μ is a scaling factor and $0 \le \mu \le 1$.

3.4 Mention Detection and Entity Disambiguation

Base on the entity representations for $\mathcal{E}$ and the refined question representation $\tilde{\mathbf{q}}$ for q, we draw on the ideas of ELQ model [8] to implement mention detection and entity disambiguation.

For mention detection, we first compute scores for each token x_j in q being the start(s_l), the end(s_r) and part(s_p) of a mention:

$$s_l(j) = \mathbf{W}_l^T \tilde{\mathbf{x}}_j, \quad s_r(j) = \mathbf{W}_r^T \tilde{\mathbf{x}}_j, \quad s_p(j) = \mathbf{W}_p^T \tilde{\mathbf{x}}_j \tag{18}$$

where $\mathbf{W}_l, \mathbf{W}_r, \mathbf{W}_p \in \mathbb{R}^h$ are learnable vectors. Then we compute the likelihood score of each span $[start, end]$ being an entity mention in q up to length L:

$$p([start, end]) = \sigma \left(s_l(start) + s_r(end) + \sum_{t=start}^{end} s_p(t) \right) \tag{19}$$

where $1 \le start \le end \le \min(start + L - 1, N_q)$.

For entity disambiguation, we first compute the representation $\tilde{\mathbf{x}}_{[start,end]} \in \mathbb{R}^h$ for mention span $[start, end]$ by averaging $\tilde{\mathbf{x}}_{start}, \cdots, \tilde{\mathbf{x}}_{end}$:

$$\tilde{\mathbf{x}}_{[start,end]} = \frac{1}{(end - start + 1)} \sum_{t=start}^{end} \tilde{\mathbf{x}}_t \tag{20}$$

Then we compute the matching score between the mention and each entity $e \in \mathcal{E}$:

$$s(e, [start, end]) = \mathbf{e}^T \tilde{\mathbf{x}}_{[start,end]} \tag{21}$$

Finally we compute the likelihood distribution over all the entities in $\mathcal{E}$ conditioned on the mention:

$$p(e \mid [start, end]) = \frac{\exp(s(e, [start, end]))}{\sum_{e' \in \mathcal{E}} \exp(s(e', [start, end]))} \tag{22}$$

3.5   Training and Inference

We jointly train mention detection and entity disambiguation by optimizing the sum of their losses. The loss of mention detection is defined as

$$\mathcal{L}_{\text{MD}} = -\frac{1}{N_{mc}} \sum_{\substack{1 \leq start \leq end \leq \\ \min(start+L-1, N_q)}} (y_{[start,end]} \log p([start, end])$$
$$+ (1 - y_{[start,end]}) \log(1 - p([start, end]))) \tag{23}$$

whereby $y_{[start,end]}$ is the label for mention span, and $y_{[start,end]} = 1$ if $[start, end]$ is a gold mention and 0 otherwise. N_{mc} is the total number of mention candidates and its value is selected in the same manner as ELQ model.

The loss of entity disambiguation is defined as

$$\mathcal{L}_{\text{ED}} = -\log p(e_g \mid [start, end]) \tag{24}$$

whereby e_g is the gold entity corresponding to mention span $[start, end]$.

In details for training and inference:

(1) Encodings of entity experiences are updated with the gold mentions of training questions at each training epoch, and the entity experiences derived from a training question are excluded from its own inference.
(2) Off-line generated entity descriptions and encodings of entity experiences are cached and reused for the inference of each input question to speed up prediction.
(3) For question representation refinement, we refine the representation for each question with its gold entities at the training stage, and top κ entity candidates at the inference stage, where κ is a hyper-parameter.
(4) For entity disambiguation of each mention, we first recall K closest entity candidates using encodings of entity descriptions, and then re-rank them using the whole encodings of entities.

4   Experiments

We conduct experiments on the end-to-end benchmark datasets WebQSP$_{\text{EL}}$ and GraphQ$_{\text{EL}}$ [8] to evaluate our approach. WebQSP$_{\text{EL}}$ and GraphQ$_{\text{EL}}$ are derived from publicly available QA datasets WebQSP [15] and GraphQuestions [12]. WebQSP contains questions that were collected from web search logs, and GraphQuestions was created by collecting manual paraphrases for automatically generated questions. Based on WebQSP and GraphQuestions, Sorokin and Gurevych [11] compile two new datasets for entity disambiguation on questions by extracting gold entities and mapping them to Wikidata. By annotating all mention boundary labels, Li et al. [8] further extend them and create WebQSP$_{\text{EL}}$ and GraphQ$_{\text{EL}}$ for end-to-end ELQA. Table 2 shows the statistics of WebQSP$_{\text{EL}}$ and GraphQ$_{\text{EL}}$.

Table 2. Dataset statistics of WebQSP$_{EL}$ and GraphQ$_{EL}$. #**Q** and #**M** indicate the number of questions and entity mentions, respectively. #**E** indicates the proportion of mentions whose groundtruth entities have experience questions in WebQSP$_{EL}$ train data.

Data	Train			Test		
	#Q	#M	#E	#Q	#M	#E
WebQSP$_{EL}$	2974	3242	59%	1603	1806	57%
GraphQ$_{EL}$	2089	2253	–	2075	2229	13%

We evaluate the performance of ELQA approaches using precision, recall and F1-score of entity linking. Following the definitions of Li et al. [8], a ELQA prediction is correct only if the groundtruth entity is identified and the predicted mention span overlaps with the groundtruth span.

In the following subsections, the first part is to compare our approach with existing important models, then we conduct two ablation studies, and at last we carry out case study.

4.1 Model Comparison

Baselines. The compared baselines for our approach include the followings.

(1) TAGME [5] is a lightweight and on-the-fly entity linking system popular for many downstream QA tasks.
(2) VCG [11] is a jointly optimized neural architecture for ELQA. It derives entity embeddings using entity labels and knowledge graph context.
(3) BLINK [13] is an entity disambiguation model with pre-specified mention boundaries. It derives entity embeddings using descriptive texts.
(4) ELQ [8] is the current state-of-the-art model on WebQSP$_{EL}$ and GraphQ$_{EL}$. It employs a BERT-based bi-encoder and also derives entity embeddings using descriptive texts.

Experimental Settings. We employ BERT$_{Large}$[1] as the basic model for tri-encoder, and reuse encodings of entity descriptions released by [8]. We evaluate our framework on WebQSP$_{EL}$ and GraphQ$_{EL}$ test data under two settings.

(1) WebQSP$_{EL}$ training setting. We implement our framework and train it on WebQSP$_{EL}$ using batch size 128 and question context window size of 20 tokens, and then predict on test data. We refer to the evaluation results as E^3L.
(2) Wikipedia training setting. We implement our framework based on the inference process of Wikipedia-trained ELQ model[2] and then directly predict on

---

[1] Model available at https://huggingface.co/bert-large-uncased/tree/main.
[2] Code, data and model available at https://github.com/facebookresearch/BLINK/tree/master/elq.

test data using entity experiences, which are collected by encoding gold mentions of WebQSP$_{EL}$ train data with Wikipedia-trained ELQ model. We refer to the evaluation results as E^3L^*.

For both training and prediction, mention detection considers all the candidate spans up to length $L = 10$, and 10 closest entity candidates per mention span are retrieved by FAISS index [7] for re-ranking, i.e., $K = 10$, in which 3 entity candidates are used for question representation refinement, i.e., $\kappa = 3$. Weight α and β are set as 0 and 0.1 for deriving entity encodings used in question representation refinement, and 1.0 and 0.01 in entity disambiguation. Scaling factor μ is set as 0.01 for refining question representations with entity encodings.

Experimental Results. Table 3 displays the evaluation results of E^3L, E^3L^* and baseline models on WebQSP$_{EL}$ and GraphQ$_{EL}$ test data. We find that:

(1) Among all the models trained on WebQSP$_{EL}$, E^3L achieves the best precision, recall and F1-score on WebQSP$_{EL}$ test data, and achieves the best precision and F1-score on GraphQ$_{EL}$ test data. Especially, E^3L obtains significant improvement on precision, showing that entity experience modeling benefits precision effectively when training on question-specific data.

(2) Compared with the baseline models trained on Wikipedia, E^3L^* achieves the best precision, recall and F1-score on both WebQSP$_{EL}$ and GraphQ$_{EL}$, demonstrating that the effects of universal models can be improved effectively using entity experiences even without training.

The reason for above improvement is that contextual knowledge introduced by entity experiences helps to reduce mention misidentification and linkage error. We also observe that recall decreases for E^3L on GraphQ$_{EL}$ test data. The possible reason is that some noise introduced from WebQSP$_{EL}$ train data leads to mention missing or linkage error on GraphQ$_{EL}$ test data.

In sum, the experimental results demonstrate that our approach achieves state-of-the-art ELQA F1-score, and provides an effective way to improve universal models without fine-tuning, which is of great application value.

4.2   Ablation Study

We conduct ablation studies under the WebQSP$_{EL}$ training setting as Sect. 4.1.

Ablation Study for Entity Experience. We evaluate (1) whether and how the use of entity experience in question representation refinement and entity disambiguation contributes to our full model and (2) whether the use of entity description on question representation refinement can improve our model. The experiments ablate three components in detail.

Table 3. Results of model comparison on test data. Highest scores per setting are underlined. For baselines, we follow the results reported in [8].

Training setting	Model	WebQSP$_{EL}$			GraphQ$_{EL}$(zero-shot)		
		Prec	Recall	F1	Prec	Recall	F1
WebQSP$_{EL}$	VCG	82.4	68.3	74.7	54.1	30.6	39.0
	ELQ	90.0	85.0	87.4	60.1	<u>57.2</u>	58.6
	E^3L	<u>91.8</u>	<u>85.7</u>	<u>88.6</u>	<u>71.7</u>	54.3	<u>61.8</u>
Wikipedia	TAGME	53.1	27.3	36.1	49.6	36.5	42.0
	BLINK	82.2	79.4	80.8	65.3	61.2	63.2
	ELQ	86.1	81.8	83.9	69.8	69.8	69.8
	E^3L^*	<u>86.8</u>	<u>82.7</u>	<u>84.7</u>	<u>70.1</u>	<u>70.1</u>	<u>70.1</u>

Table 4. Results of ablation study for entity experience on test data.

Training setting	Model	WebQSP$_{EL}$			GraphQ$_{EL}$(zero-shot)		
		Prec	Recall	F1	Prec	Recall	F1
WebQSP$_{EL}$	E^3L	91.8	85.7	88.6	71.7	54.3	61.8
	w/o Expr$_{qrr}$	91.5	84.5	87.9	71.6	52.1	60.3
	w/o Expr$_{ed}$	91.7	84.0	87.7	70.7	52.6	60.3
	w Desc$_{qrr}$	91.3	84.7	87.9	70.5	52.2	60.0

(1) w/o Expr$_{qrr}$, where entity experiences are not used for question representation refinement, and output encodings of *Question Basic Encoder* module are used for mention detection and entity disambiguation directly.

(2) w/o Expr$_{ed}$, where entity experiences are not used for entity disambiguation, and only encodings of entity descriptions are used for entity disambiguation.

(3) w Desc$_{qrr}$, where weight α of entity descriptions is set as 0.1 for deriving entity representations used in question representation refinement.

Table 4 displays the evaluation results on WebQSP$_{EL}$ and GraphQ$_{EL}$ test data. We find that ELQA performance drops on all the test data if we remove entity experiences from question representation refinement and entity disambiguation, demonstrating that our design around the utilization of entity experience to improve entity representation and question representation for end-to-end ELQA is effective. We find that ELQA performance drops if entity descriptions are used in question representation refinement, which is probably because descriptive texts of entities bring more noise.

Ablation Study for Attention Mechanism. We evaluate whether and how the attention mechanisms for experience retrieval and question representation refinement contribute to our full model. The experiments ablate two components in detail.

Table 5. Results of ablation study for attention mechanisms on test data.

Training setting	Model	WebQSP$_{EL}$			GraphQ$_{EL}$ (zero-shot)		
		Prec	Recall	F1	Prec	Recall	F1
WebQSP$_{EL}$	E^3L	91.8	85.7	88.6	71.7	54.3	61.8
	w/o Att$_{expr}$	91.6	83.9	87.6	70.6	51.5	59.5
	w/o Att$_{qrr}$	92.0	84.1	87.9	71.0	52.6	60.4

(1) w/o Att$_{expr}$, where attention is not performed in the retrieval of entity experiences, and all the entity experiences have equal importance weights to the current user question.

(2) w/o Att$_{qrr}$, where attention is not performed to compute importance distributions over entities in the question representation refinement, and all the entities have equal importance weights for refining question representation.

Table 5 displays the evaluation results on WebQSP$_{EL}$ and GraphQ$_{EL}$ test data. We find that ELQA performance drops if we remove attention mechanisms, and the removal on experience retriever causes much more decline in performance. The results demonstrate the necessity to distinguish the importance of entities and their experiences for our approach.

4.3 Case Study

Table 6 displays three typical test cases for comparing E^3L and ELQ model. Case 1 shows that E^3L links mention "new york" to entity "New York City" correctly with the similar usage context of "New York City" in a history question, but ELQ model cannot. Case 2 shows that E^3L detects mention "ohio" and links it to entity "Ohio" correctly with the similar usage context of the other entity "Governor" in a history question, but ELQ model misses the mention. The two examples demonstrate that E^3L effectively corrects errors by mining the usage context of entities and the implicit relations between entities from history questions.

Case 3 is a bad case, where E^3L does not detect mention "hudson". The possible reason is that the context from all the history questions is less related or even noisy, which reduces the recall of ELQA. We will further tackle this bad case and improve our approach in the future work.

Table 6. Examples for case study. Red represents incorrect entity linking.

Case 1	Question	Which competition events occurred in new york?
	ELQ	New york-> "New York (state)"
	E^3L	New york-> "New York City"
	History questions	What tv shows are taped in new york city? new york city-> "New York City"
Case 2	Question	Who is governor of ohio 2011?
	ELQ	Governor of ohio-> "Governor"
	E^3L	Governor-> "Governor", ohio-> "Ohio"
	History questions	Who is the governor of california 2010? Governor-> "Governor", california-> "California"
Case 3	Question	On the hudson, there is what kind of bridge?
	ELQ	Hudson-> "Hudson River"
	E^3L	[]
	History questions	1. Where does hudson river start? hudson river-> "Hudson River" 2. Where's the hudson river? hudson river-> "Hudson River"

5   Conclusion

In the context of KGQA, we introduce entity experience to enrich entity representations and propose an experience enhanced framework E^3L for entity linking in this paper. In the framework, we focus on the modeling of entity experience and its utilization to the enhancement of entity representations and question representations for ELQA. With experimental evaluations, we demonstrate that E^3L outperforms previous state-of-the-art models on F1-score and has great application value. In addition, we prove that our design on entity experience modeling is effective to the enhancement of ELQA.

Despite being originally designed for ELQA, E^3L has the potential for implementation on broader context of short texts, such as tweets and search queries. This contribution helps to improve the performance of short text understanding. To improve the recall of E^3L, a further study on entity representation strategy is needed.

References

1. Banerjee, D., Chaudhuri, D., Dubey, M., Lehmann, J.: PNEL: Pointer network based end-to-end entity linking over knowledge graphs. In: Pan, J.Z., et al. (eds.) ISWC 2020. LNCS, vol. 12506, pp. 21–38. Springer, Cham (2020). https://doi.org/10.1007/978-3-030-62419-4_2
2. Bordes, A., Usunier, N., Garcia-Durán, A., Weston, J., Yakhnenko, O.: Translating embeddings for modeling multi-relational data. In: 26th International Conference on Neural Information Processing Systems, pp. 2787–2795 (2013)
3. Cui, Y., Chen, Z., Wei, S., Wang, S., Liu, T., Hu, G.: Attention-over-attention neural networks for reading comprehension. In: 55th Annual Meeting of the Association for Computational Linguistics, pp. 593–602 (2017)
4. Devlin, J., Chang, M.W., Lee, K., Toutanova, K.: Bert: Pre-training of deep bidirectional transformers for language understanding. In: 2019 Conference of the North American Chapter of the Association for Computational Linguistics: Human Language Technologies, pp. 4171–4186 (2019)
5. Ferragina, P., Scaiella, U.: Fast and accurate annotation of short texts with Wikipedia pages. IEEE Softw. **29**(1), 70–75 (2011)
6. Gu, Y., et al.: Beyond IID: Three levels of generalization for question answering on knowledge bases. In: Web Conference 2021, pp. 3477–3488 (2021)
7. Johnson, J., Douze, M., Jégou, H.: Billion-scale similarity search with gpus. IEEE Trans. Big Data (2019)
8. Li, B.Z., Min, S., Iyer, S., Mehdad, Y., Yih, W.T.: Efficient one-pass end-to-end entity linking for questions. In: 2020 Conference on Empirical Methods in Natural Language Processing, pp. 6433–6441 (2020)
9. Singh, K., Lytra, I., Radhakrishna, A.S., Shekarpour, S., Vidal, M.E., Lehmann, J.: No one is perfect: Analysing the performance of question answering components over the dbpedia knowledge graph. J. Web Semant. **65**, 100594 (2020)
10. Singh, K., et al.: Why reinvent the wheel: Let's build question answering systems together. In: 2018 World Wide Web Conference, pp. 1247–1256 (2018)
11. Sorokin, D., Gurevych, I.: Mixing context granularities for improved entity linking on question answering data across entity categories. In: 7th Joint Conference on Lexical and Computational Semantics, pp. 65–75 (2018)
12. Su, Y., et al.: On generating characteristic-rich question sets for QA evaluation. In: 2016 Conference on Empirical Methods in Natural Language Processing, pp. 562–572 (2016)
13. Wu, L., Petroni, F., Josifoski, M., Riedel, S., Zettlemoyer, L.: Scalable zero-shot entity linking with dense entity retrieval. In: 2020 Conference on Empirical Methods in Natural Language Processing, pp. 6397–6407 (2020)
14. Yang, Y., Chang, M.W.: S-mart: Novel tree-based structured learning algorithms applied to tweet entity linking. In: 53rd Annual Meeting of the Association for Computational Linguistics and 7th International Joint Conference on Natural Language Processing, pp. 504–513 (2015)
15. Yih, W.T., Richardson, M., Meek, C., Chang, M.W., Suh, J.: The value of semantic parse labeling for knowledge base question answering. In: 54th Annual Meeting of the Association for Computational Linguistics, pp. 201–206 (2016)

M2R: From Mathematical Models to Resource Description Framework

Chenxin Zou[1], Xiaodong Li[1(✉)], Pangjing Wu[1], and Haoran Xie[2]

[1] College of Computer and Information, Hohai University, Nanjing 210098, China
zoucx@hhu.edu.cn, xiaodong.c.li@outlook.com
[2] Department of Computing and Decision Sciences, Lingnan University,
Hong Kong 999077, China
hrxie@ln.edu.hk

Abstract. Domain-specific knowledge graphs usually have requirements for deeper and more accurate knowledge. Existing knowledge graphs in academics mainly focus on authors, abstracts, keywords, and citations, which help explore themes of papers and analyze relationships between different papers. However, these contents are summarizations and only reveal shallow meanings, not involving cores of scientific papers. Mathematical models, ignored by existing knowledge graphs, are what authors really want to express through papers. Knowledge from mathematical models makes it possible to use knowledge graphs for mathematical derivation, not just literal reasoning. To model this knowledge, we propose a knowledge graph construction framework, named M2R, from Mathematical Models to Resource Description Framework. Mathematical models are usually described in formulae. We first identify formula positions according to pre-defined rules and find out contexts explaining variables in the formulae. Next, we split the formulae and related contexts from PDF papers in the form of images, and employ optical character recognition to identify image contents. Then, regular expressions designed based on sentence patterns are used to extract variable symbols and variable explanations. Finally, the formulae are regarded as relations between the variables to form triples whose subjects and objects are the variables, and predicates are the formulae. Similar triples are fused to generate a final knowledge graph. Experimental results demonstrate that precision of the formula extraction is up to 76.97%. Besides, a convincing case study shows that we can effectively extract formulae and related variables, and construct a knowledge graph about mathematical models of scientific papers.

Keywords: Knowledge graph construction · Scientific papers · Mathematical models · Formulae · Variables

1 Introduction

Knowledge graph, whose early idea comes from the semantic web [4], is regarded as a technical method to describe relations between everything in the world by graph models [3]. It plays an important role in many intelligent applications such as intelligent Q&A [19], big data analysis [22], and interpretability of machine learning [20]. In domain-specific knowledge graphs, there are usually strong requirements for depth and accuracy of knowledge. Existing knowledge graphs in academics pay close attention to authors, abstracts, keywords, and citations, which only convey some basic information about papers. For example, authors and citations help identify relationships between papers by judging whether there are the same authors; abstracts and keywords give brief summarizations to papers. These knowledge graphs do not cover the cores of papers and ignore what papers would like to express, i.e., mathematical models. These models report principles and details for solving academic problems, which are the keys to papers. Besides, relationships discovered from the models imply mathematical meanings that are helpful for formal reasoning between papers. To incorporate this knowledge from mathematical models, we propose a knowledge graph construction framework, named M2R, from mathematical models to Resource Description Framework (RDF). In the M2R, we first identify positions of formulae and related contexts. Next, we split the formulae and contexts from PDF papers in the form of images, and recognize image contents by optical character recognition. Then, variable symbols and variable explanations, such as shown in Fig. 1, are extracted from contexts according to sentence patterns. Finally, we regard the formulae as relations between variables to form triples and fuse similar triples from multiple papers to construct a final knowledge graph. Experimental results demonstrate that we can effectively extract formulae and related variables to form a knowledge graph. The knowledge graph based on mathematical models can help researchers quickly find accurate mathematical relations between variables, making it possible to do some related calculations automatically according to formulae, and also be helpful in model reasoning and discovery of new research areas.

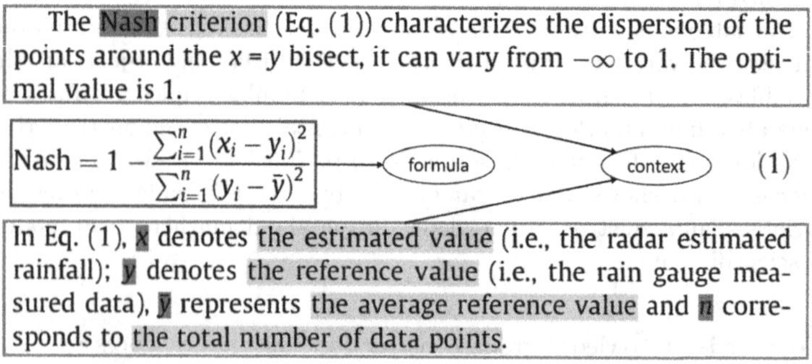

The Nash criterion (Eq. (1)) characterizes the dispersion of the points around the $x = y$ bisect, it can vary from $-\infty$ to 1. The optimal value is 1.

$$\text{Nash} = 1 - \frac{\sum_{i=1}^{n}(x_i - y_i)^2}{\sum_{i=1}^{n}(y_i - \bar{y})^2} \quad \text{formula} \quad \text{context} \quad (1)$$

In Eq. (1), x denotes the estimated value (i.e., the radar estimated rainfall); y denotes the reference value (i.e., the rain gauge measured data), $\bar{y}$ represents the average reference value and n corresponds to the total number of data points.

Fig. 1. An example of what we extract from PDF papers. Formula is in the blue box. Contexts are in the green boxes. Variable symbols are highlighted in red and variable explanations in yellow. (Color figure online)

Our main contributions are summarized as follows:

(1) We take the mathematical model knowledge ignored by domain-specific knowledge graphs in the academic field into consideration and model it by knowledge graphs.
(2) We propose a novel unified knowledge graph construction method, i.e., the M2R, which is applicable to construct a different domain-specific knowledge graph with mathematical model knowledge.
(3) The M2R realizes a complete and low-coupling prototype construction process from PDF papers to a knowledge graph, and each part of it can be easily replaced with a probably better method.
(4) Experimental results demonstrate that the M2R can effectively extract mathematical model knowledge and construct a knowledge graph about it.

The remaining sections of this paper are arranged as follows. Section 2 introduces some related works about mathematical knowledge management and knowledge graph construction in detail. Section 3 presents the M2R framework. Section 4 reports experiments on a collected dataset and gives a case study. Section 5 concludes the paper and indicates the future work.

2   Related Work

Mathematical models that the M2R focuses on belong to mathematical knowledge. As early as 2004, mathematical knowledge management had attracted extensive attention. It was regarded as an interdisciplinary field, and its aim was to better manage mathematical knowledge [11]. However, few works leverage knowledge graphs to model this knowledge, which may result in it not being fully used. To fill this gap, the M2R is proposed to construct such a knowledge graph about mathematical models. The related works of mathematical knowledge management and knowledge graph construction are discussed in detail as follows.

2.1   Mathematical Knowledge Management

For a fine-grained analysis of mathematical knowledge management, Carette et al. [7] summarized 6 perspectives of it and 25 topics that were usually discussed in previous works. One of these perspectives was digital, which focused on handling mathematical knowledge by computers. Elizarov et al. [10] proposed a mathematical knowledge analytics and management digital ecosystem OntoMath. Mathematical objects (e.g., formulae) were extracted directly from mathematical papers in LaTeX and used as basic classes in ontology. However, papers in LaTeX are not available in every field, which brings some challenges to knowledge extraction. To address this issue, Zanibbi et al. [27] gave a survey on the recognition and retrieval of mathematical expressions. For a mathematical formula recognition system, there were usually 3 forms of input: vector graphics, strokes, and images. Expression detection, symbol extraction, layout analysis, and mathematical content interpretation were the four keys of the system.

Kacem et al. [14] proposed an automatic formula extraction method based on fuzzy logic and propagation of context. Fuzzy logic was used to remove ambiguities, and propagation of context was done to group symbols properly into units. In addition, Phong et al. [13] proposed two methods to detect mathematical variables in PDF documents. One was a rule-based method in which they designed five rules and detected variables according to font, glyph, and bounding box information extracted from documents. The other was a method combing convolutional neural networks (CNNs) and machine learning algorithms. They first used the pre-trained CNNs to extract features of word images obtained by text image segmentation. Then support vector machine and k-nearest neighbors were used for classification. Similarly, Yu et al. [26] identified symbols and characters in terms of font and bounding box information. Then the syntax tree was built based on a structural analysis of a formula. There were also many deep learning extraction methods [12], but they were costly in data preparation and model optimization.

2.2   Knowledge Graph Construction

Knowledge graph construction is to organize knowledge from different sources by graph models. The whole construction process can be divided into four important parts: knowledge extraction, knowledge fusion, knowledge completion, and knowledge graph generation. There are usually different emphases in different construction methods. Elhammadi et al. [9] designed a pipeline construction method, which combined multiple extraction technologies with the financial dictionary they built to extract information from financial news for knowledge graphs. They paid more attention to knowledge extraction. Among methods for knowledge extraction, deep learning models [15,25] were popular, and joint models [1,24] were used to eliminate error propagation. Bosselut et al. [5] proposed commonsense transformers. They transferred implicit knowledge in deep pre-trained language models to explicit knowledge. It helped complete and extend commonsense knowledge graphs. Al-Khatib et al. [2] proposed an end-to-end method to construct argumentation knowledge graphs, which emphasized the simplicity of methods. Martinez-Rodrigue et al. [17] introduced open information extraction methods to simplify knowledge querying and representation. They focused more on the availability of generated knowledge graphs.

The M2R aims to construct knowledge graphs about mathematical model knowledge from scientific papers, which is similar to educational and scientific knowledge graph construction. In the educational field, Wang et al. [23] proposed a rule and semantic-based method. It first recognized prior/reference terms. Then prior terms were filtered through rules, and reference terms were classified according to sentence semantics for coreference resolution. Chen et al. [8] designed a construction system that extracted concepts of subjects or courses by the neural sequence labeling algorithm and identified relations between these concepts based on probabilistic association rules. The generated knowledge graph only described whether there were mathematical relations but did not specify what these relations were. In the scientific field, Buscaldi et al. [6] combined some state-of-the-art methods to extract entities and relations and adopted Leven-shtein string similarity and hierarchical clustering algorithms for merging. Luan et al. [16] proposed a unified model with multi-task setups for cross-sentence relation extraction. Ren et al. [18] divided relations into multiple types for fine-grained extraction. Tosi et al. [21] mined internal relations between concepts through semantic analysis of texts. However, these knowledge graphs only focus on authors, abstracts, keywords, and citations. Besides, the extracted relations contain only basic information. Compared with the above methods, the M2R provides a complete construction method for knowledge graphs about mathe-matical model knowledge. The relations contain deeper mathematical meanings. The M2R fills the gap that existing methods ignore and makes it possible for mathematical reasoning.

3   M2R Framework

Different from previous works, the M2R framework focuses on mathematical model knowledge. Its workflow is shown in Fig. 2. In the M2R, we first prepro-cess papers and get an image for each page. Next, a rule-based method is used to segment formula images and related context images. The image contents are obtained by optical character recognition. Then, regular expressions designed based on sentence patterns are used to extract variable symbols and correspond-ing explanations from contexts. Finally, variable symbols and explanations are combined as variable units, and formulae are used as relations between vari-ables to form triples. We also use the cosine similarity to fuse similar triples and construct the final knowledge graph.

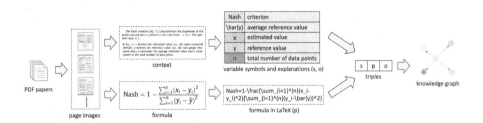

Fig. 2. Workflow of the M2R framework.

3.1   Task Definition

The M2R aims to construct a knowledge graph about mathematical models. Mathematical models are usually in scientific papers. The input of the framework is a set of papers. To facilitate subsequent knowledge extraction, each paper is transformed into images. The input papers are further represented as a set of images, where an image corresponds to a page. Mathematical model knowledge (i.e., formulae and variables) is extracted from each image and stored in the form of LaTeX. The obtained formulae and variables are constructed into triples according to the inclusion between them. Similar triples are fused to form a final knowledge graph, which is also the output of the framework. Specifically, the M2R is divided into four modules:

(1) The preprocessing module converts each paper into multiple images;
(2) The image segmentation and identification module extracts formula images and context images according to pre-defined rules. Then it further identifies image contents through optical character recognition;
(3) The variable extraction module regards a variable symbol and related explanation as a variable unit. Both of them are extracted based on designed regular expressions;
(4) The triple generation and fusion module takes mathematical formulae as relations between two variable units to generate triples. We adopt the cosine similarity for knowledge fusion to generate the final knowledge graph.

3.2   Preprocessing

Most papers are stored in PDF format. It is difficult for machines to understand and utilize this kind of knowledge. So we take LaTeX format as an intermediate expression between papers and knowledge graphs. Existing methods can easily convert LaTeX to PDF, but it is difficult to reverse directly. Images can serve as a bridge from PDF to LaTeX. We further process images to obtain what we want. In summary, PDF papers are transformed into images, and then mathematical model knowledge in LaTeX is extracted from these images. A PDF paper can be represented as a tuple $\langle M, F, C, V, S, E \rangle$, where,

- $M = \{m_i \mid$ the image of ith page$\}$;
- $F = \{f_{ij} \mid$ the jth formula in image $m_i\}$;
- $C = \{c_{ij} \mid$ the context of formula $f_{ij}\}$;
- $V = \{v_{ij}^k \mid$ the kth variable unit in context $c_{ij}\}$;
- $S = \{s_{ij}^k \mid$ the variable symbol of variable unit $v_{ij}^k\}$;
- $E = \{e_{ij}^k \mid$ the variable explanation of variable unit $v_{ij}^k\}$.

3.3   Image Segmentation and Identification

A mathematical model is expressed by a set of formulae in papers. After converting PDF papers to images, we further segment the images to obtain formula and context images by a rule-based method according to formula features. These features are as follows,

- There are usually numerical indices on the right side of formulae.
- Formulae are usually centered.
- Formulae are presented only once in the same paper.
- The distance between symbols in formulae is bigger than that in body.

According to the above features, we design different rules to screen formula images in two steps. The first step is to roughly distinguish formula images and non-formula images, using rules as follows,

– Consecutive white pixels on both sides of images.

$$lW > \alpha_1, \quad rW > \alpha_1, \quad ||lW - rW| - \alpha_2| < \alpha_3, \tag{1}$$

where lW and rW are the numbers of consecutive white pixel columns on the left and right sides, respectively. α_* is a parameter.
– Ratio of black and white pixels.

$$\alpha_4 < ratio < \alpha_5. \tag{2}$$

– Height and width.
$$w > h, \quad \alpha_6 < h < \alpha_7, \tag{3}$$

where w and h are the width and height of an image.
– Content. Formulae contain characters that are non-letter and non-number. We collect 10 standard number images and 52 standard letter images in advance. These standard images are shown in Fig. 3.

0 1 2 3 4 5 6 7 8 9

a b c d e f g h i j k l m

n o p q r s t u v w x y z

A B C D E F G H I J K L M

N O P Q R S T U V W X Y Z

Fig. 3. Standard number images and standard letter images collected by us.

The formula images are further segmented into character images according to the connectivity of characters. We resize the character images to be consistent with the collected images and calculate similarity cs_1 between them based on the distributions of black and white pixels,

$$cs_1 = \frac{\sum_{i=1}^{ch} \sum_{j=1}^{cw} f(i,j)}{ch \times cw}, \tag{4}$$

$$f(i,j) = \begin{cases} 1, & \text{if pixels in } i\text{th row and } j\text{th column are the same,} \\ 0, & \text{otherwise,} \end{cases} \tag{5}$$

where cw and ch are the width and height of a segmented character image. If the similarity between a character image and all collected images is less than a threshold γ, this character will be non-letter and non-number.

If an image meets the above rules, it is considered as a formula image.

The second step is to remove a small number of impurities such as statistical images and citation images from the first step. The rules used are as follows,

- Same image. If the similarity between two images is greater than α_8, then they are not formula images.
- Number of long consecutive white pixel column segments. If the number is greater than 3, the image is not a formula image.
- Length of consecutive black pixel columns and rows. If the length meets the following conditions, the image is not a formula image,

$$XB_{\max} > \alpha_9, \quad YB_{\max} < \alpha_{10}, \tag{6}$$

where $XB_{\max}$ and $YB_{\max}$ are the maximum numbers of consecutive black pixel columns and rows.
- Average height of characters. If the average height is greater than α_{11}, the image is not a formula image.

After formula images are extracted, the related context images are obtained according to the positions of formula images. Optical character recognition is used to identify image contents. Formulae are directly used as predicates of triples. As for contexts, we need to further extract variable symbols and explanations from them.

3.4   Variable Extraction

Sentences describing the variables in a formula usually have fixed patterns. Therefore, we design regular expressions according to these patterns and use them to extract variable symbols and explanations. The common fixed sentence patterns describing variables are,

- "variable symbol" is "variable explanation".
- "variable explanation" ("variable symbol").

Regular expressions are designed based on the above patterns. With the help of regular expressions, we can extract variable symbols and corresponding explanations from each sentence effectively. There may be parameters in formulae, but we also regard them as special variables. The purpose is to make all symbols and corresponding explanations included in the knowledge graph, which is convenient for numerical calculation and reasoning by the knowledge graph.

3.5   Triple Generation and Fusion

Variable symbols and explanations are regarded as variable units. Formulae are used as relations between two units. We take variable units as subjects and objects, and formulae as predicates to form triples. In addition, we use cosine similarity to measure the similarity between two variables for triple fusion. Firstly, a temporary dictionary is constructed according to two variable explanations $E_1 = \{w_1^1, w_2^1, \ldots, w_{n_{E_1}}^1\}$ and $E_2 = \{w_1^2, w_2^2, \ldots, w_{n_{E_2}}^2\}$,

$$D = E_1 \cup E_2, \tag{7}$$

where w_* is a word of two variable explanations in its original form and stop words are removed in advance. Then, we generate vectors $\overrightarrow{Vec_1} = \{x_1^1, x_2^1, \ldots, x_{|D|}^1\}$ and $\overrightarrow{Vec_2} = \{x_1^2, x_2^2, \ldots, x_{|D|}^2\}$ for variable explanations based on the dictionary,

$$x_i = \begin{cases} 1, \text{ if a variable explanation contains } i\text{th word in the dictionary,} \\ 0, \text{ otherwise,} \end{cases} \tag{8}$$

where $|D|$ is the number of words in the dictionary. Finally, we calculate the cosine similarity cs_2 between two vectors,

$$cs_2 = \frac{\overrightarrow{Vec_1} \cdot \overrightarrow{Vec_2}}{\|\overrightarrow{Vec_1}\|\|\overrightarrow{Vec_2}\|}. \tag{9}$$

If the similarity is greater than a threshold β, the two variables are the same. According to the similarity of variables, we remove similar triples and complete the construction of a knowledge graph based on mathematical model knowledge.

In summary, the M2R is mainly based on some classical methods, such as the rule-based method for formula extraction and the regular expressions for variable extraction. These methods may not be better than learning-based methods in performance, but they are able to solve problems quickly and are capable of relatively good performance. Besides, the M2R is a prototype framework that realizes a whole construction process from mathematical models in PDF papers to a knowledge graph. It is low-coupling, so the methods for tasks in the M2R can be easily replaced if there are better ones.

4   Experiments

In this section, we use the proposed M2R framework to conduct experiments on a collected dataset of water conservancy papers. Besides, we give a case study to show a knowledge graph constructed based on a random paper. We also build a search application based on it and compare the application with others to illustrate its specific advantages.

4.1   Dataset

The dataset used in this paper is constructed by ourselves. The aim of the M2R is to construct a domain-specific knowledge graph. We collect papers about flood forecasting in the water conservancy field to form a water conservancy paper dataset. There are 400 papers in the dataset. To ensure the quality of samples, these papers are collected from three top journals: Journal of hydrology, Water Resources Research, and Advances in Water Resources. We only use the keyword "flood forecast" to retrieve papers and further select related papers through abstracts. This dataset is used as the input to the framework.

4.2   Experimental Setup

First, we use the PDFBox[1] tool to transform PDF papers into images. Next, the rule-based method is used to segment formula images and corresponding context images. Each image is black & white. The parameters of the rules are set as $\alpha_1 = 50$, $\alpha_2 = 120$, $\alpha_3 = 20$, $\alpha_4 = 0.015$, $\alpha_5 = 0.1$, $\alpha_6 = 30$, $\alpha_7 = 350$, $\alpha_8 = 0.5$, $\alpha_9 = 500$, $\alpha_{10} = 200$, $\alpha_{11} = 16$, and $\gamma = 0.8$. Further, we use the optical character recognition to obtain image contents in LaTeX. Then, two regular expressions are designed to extract variable symbols and explanations,

$$(.*?)(is|denotes|represents| =)(.*?)\$, \tag{10}$$

$$\backslash(([A - Z]+)\backslash). \tag{11}$$

Finally, triples are constructed, and the similarity between variables is calculated for knowledge fusion, where the similarity threshold is set as $\beta = 0.8$.

The M2R involves some external tools, but it is not dependent on these tools. It is low-coupling, and these tools can be replaced with other better solutions. We set many parameters in the M2R so that it can be extended more easily. Values of the parameters are determined according to performance on a small number of random papers. Besides, we do not give baselines to compare with the M2R because there are few similar methods for this special knowledge about mathematical models. It is meaningless to forcibly transfer some slightly related methods to solve the problem we focus on. Therefore, an additional case study is given to support the M2R.

4.3   Experimental Results

The final knowledge graph we have constructed is shown in Fig. 4. There are 1607 entities and 1154 relations, and a total of 8063 triples are formed in the knowledge graph. Experimental results demonstrate that the knowledge graph is capable of a certain scale and availability. Especially, formula extraction is an important part of the M2R framework, so we randomly select some papers to evaluate this part individually. More than 500 formulae in these papers are evaluated in the experiment. The precision, recall, and F1 scores of formula extraction are 76.97%, 74.75%, and 75.84%, demonstrating that the M2R can extract mathematical model knowledge effectively.

---

[1] https://pdfbox.apache.org/.

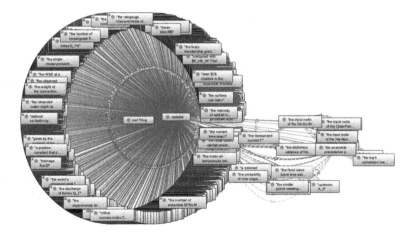

Fig. 4. The final knowledge graph constructed by our M2R framework.

4.4   Case Study

We randomly select a paper and use the proposed M2R framework to build a micro knowledge graph (mKG). The formula and related contexts are shown in Fig. 1. To display the mKG, we store the generated knowledge graph as an OWL ontology file and show it with protégé tool[2] which is shown in Fig. 5. Variable units: "the estimated value:x", "the reference value:y", and "the average reference value:\bar{y}" are successfully extracted and used as variable nodes in the knowledge graph. The formula, whose URI is "http://hydrology/#Nash=1-\ f rac{\sum_{i=1}^{n}(x_i-y_i)^2}{\sum_{i=1}^{n}(y_i-\bar{y})^2} ", is used as a mathematical relation between every two units.

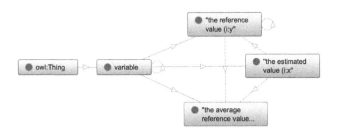

Fig. 5. Generated knowledge graph shown in protégé.

To illustrate the effectiveness of the knowledge graph, we use Elasticsearch[3] to build a simple search engine (S-mKG). The knowledge graph is served as

[2] https://protege.stanford.edu/.

[3] https://www.elastic.co/cn/elasticsearch/.

the supporting data of this search engine. Elasticsearch helps automatically construct indexes for entities and relations. We input "estimated value" as searching keywords and obtain different returned results shown in Fig. 6 from S-mKG, Wikipedia[4], Microsoft Bing[5], and Baidu[6]. It is worth noting that these search engines are all based on knowledge graphs and what they return depends on the knowledge in knowledge graphs. The S-mKG can give an accurate mathematical relation. If values of the variables are known, it will be easy to calculate the *Nash*. However, the results from other well-known search engines cannot even show a complete variable, because they do not consider mathematical model knowledge in their supporting knowledge graphs. The knowledge graph constructed by the M2R may not be better than these knowledge graphs, but it fills the gap of mathematical knowledge in knowledge graphs, making it possible to use knowledge graphs for mathematical reasoning.

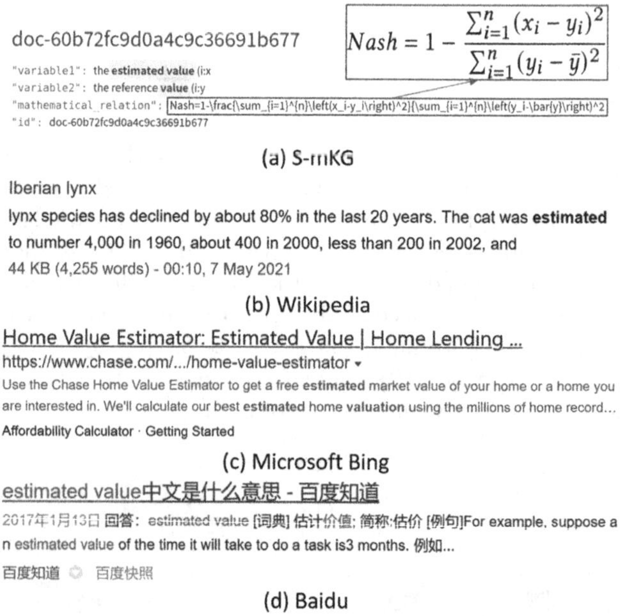

doc-60b72fc9d0a4c9c36691b677

$$Nash = 1 - \frac{\sum_{i=1}^{n}(x_i - y_i)^2}{\sum_{i=1}^{n}(y_i - \bar{y})^2}$$

"variable1": the **estimated value** (i:x
"variable2": the reference value (i:y
"mathematical relation": Nash=1-\frac{\sum_{i=1}^{n}\left(x_i-y_i\right)^2}{\sum_{i=1}^{n}\left(y_i-\bar{y}\right)^2
"id": doc-60b72fc9d0a4c9c36691b677

(a) S-mKG

Iberian lynx

lynx species has declined by about 80% in the last 20 years. The cat was **estimated** to number 4,000 in 1960, about 400 in 2000, less than 200 in 2002, and

44 KB (4,255 words) - 00:10, 7 May 2021

(b) Wikipedia

Home Value Estimator: Estimated Value | Home Lending ...
https://www.chase.com/.../home-value-estimator ▾
Use the Chase Home Value Estimator to get a free **estimated** market value of your home or a home you are interested in. We'll calculate our best **estimated** home **valuation** using the millions of home record...
Affordability Calculator · Getting Started

(c) Microsoft Bing

estimated value中文是什么意思 - 百度知道
2017年1月13日 回答: estimated value [词典] 估计价值; 简称:估价 [例句]For example, suppose a n estimated value of the time it will take to do a task is3 months. 例如...

百度知道 ◌ 百度快照

(d) Baidu

Fig. 6. The query results from (a) S-mKG, (b) Wikipedia, (c) Microsoft Bing, and (d) Baidu.

5   Conclusions

This paper proposes a knowledge graph construction framework based on mathematical model knowledge. This knowledge is the core of scientific papers but is

---

[4] https://en.wikipedia.beta.wmflabs.org/wiki/Main_Page.

[5] https://cn.bing.com/.

[6] https://www.baidu.com/.

ignored by previous knowledge graphs. The advantages of our framework lie in its flexibility and automation. It is applicable to the construction of knowledge graphs with mathematical models in different fields, and the whole construction process does not require human participation. The experiments and the case study demonstrate that our framework can extract formulae and variables from papers effectively, and construct a feasible knowledge graph.

In future work, we will concentrate on the extraction of variable symbols and explanations, and the transformation from PDF to LaTeX, which will make the overall framework more integrated and automated. Besides, we will explore how to reason and solve mathematical problems in terms of our generated knowledge graph.

Acknowledgments.. This work was supported in part by the National Natural Science Foundation of China under Grant No. 61602149, and in part by the Fundamental Research Funds for the Central Universities, China under Grant No. B210202078.

References

1. Adel, H., Schütze, H.: Global normalization of convolutional neural networks for joint entity and relation classification. In: Proceedings of the 2017 Conference on Empirical Methods in Natural Language Processing, pp. 1723–1729 (2017)
2. Al-Khatib, K., Hou, Y., Wachsmuth, H., Jochim, C., Bonin, F., Stein, B.: End-to-end argumentation knowledge graph construction. In: Proceedings of the AAAI Conference on Artificial Intelligence, vol. 34, pp. 7367–7374 (2020)
3. Amit, S.: Introducing the knowledge graph: Things, not strings. Official Google Blog (2012)
4. Berners-Lee, T., Handler, J., Lassila, O.: The semantic web. Sci. Am. **284**(5), 34–43 (2003)
5. Bosselut, A., Rashkin, H., Sap, M., Malaviya, C., Celikyilmaz, A., Choi, Y.: Comet: commonsense transformers for automatic knowledge graph construction. In: Proceedings of the 57th Annual Meeting of the Association for Computational Linguistics, pp. 4762–4779 (2019)
6. Buscaldi, D., Dessì, D., Motta, E., Osborne, F., Reforgiato Recupero, D.: Mining scholarly publications for scientific knowledge graph construction. In: European Semantic Web Conference, pp. 8–12 (2019)
7. Carette, J., Farmer, W.M.: A review of mathematical knowledge management. In: International Conference on Intelligent Computer Mathematics, pp. 233–246 (2009)
8. Chen, P., Lu, Y., Zheng, V.W., Chen, X., Yang, B.: Knowedu: A system to construct knowledge graph for education. IEEE Access **6**, 31553–31563 (2018)
9. Elhammadi, S., et al.: A high precision pipeline for financial knowledge graph construction. In: Proceedings of the 28th International Conference on Computational Linguistics, pp. 967–977 (2020)
10. Elizarov, A., Kirillovich, A., Lipachev, E., Nevzorova, O.: Digital ecosystem ontomath: Mathematical knowledge analytics and management. In: International Conference on Data Analytics and Management in Data Intensive Domains, pp. 33–46 (2016)

11. Farmer, W.M.: MKM: A new interdisciplinary field of research. ACM SIGSAM Bullet. **38**(2), 47–52 (2004)
12. Gao, L., Yi, X., Liao, Y., Jiang, Z., Yan, Z., Tang, Z.: A deep learning-based formula detection method for pdf documents. In: 2017 14th IAPR International Conference on Document Analysis and Recognition (ICDAR), vol. 01, pp. 553–558 (2017). https://doi.org/10.1109/ICDAR.2017.96
13. Hai Phong, B., Manh Hoang, T., Le, T.L., Aizawa, A.: Mathematical variable detection in pdf scientific documents. In: Intelligent Information and Database Systems, pp. 694–706 (2019)
14. Kacem, A., Belaïd, A., Ben Ahmed, M.: Automatic extraction of printed mathematical formulas using fuzzy logic and propagation of context. Int. J. Docum. Anal. Recogn. **4**(2), 97–108 (2001)
15. Li, X., Feng, J., Meng, Y., Han, Q., Wu, F., Li, J.: A unified MRC framework for named entity recognition. In: Proceedings of the 58th Annual Meeting of the Association for Computational Linguistics, pp. 5849–5859 (2020)
16. Luan, Y., He, L., Ostendorf, M., Hajishirzi, H.: Multi-task identification of entities, relations, and coreference for scientific knowledge graph construction. In: Proceedings of the 2018 Conference on Empirical Methods in Natural Language Processing, pp. 3219–3232 (2018)
17. Martinez-Rodriguez, J.L., López-Arévalo, I., Rios-Alvarado, A.B.: Openie-based approach for knowledge graph construction from text. Exp. Syst. Appl. **113**, 339–355 (2018)
18. Ren, F., et al.: Techkg: A large-scale Chinese technology-oriented knowledge graph. arXiv preprint arXiv:1812.06722 (2018)
19. Saha, A., Pahuja, V., Khapra, M., Sankaranarayanan, K., Chandar, S.: Complex sequential question answering: Towards learning to converse over linked question answer pairs with a knowledge graph. In: Proceedings of the AAAI Conference on Artificial Intelligence, vol. 32 (2018)
20. Song, W., Duan, Z., Yang, Z., Zhu, H., Zhang, M., Tang, J.: Explainable knowledge graph-based recommendation via deep reinforcement learning. arXiv preprint arXiv:1906.09506 (2019)
21. Tosi, M.D.L., dos Reis, J.C.: Scikgraph: A knowledge graph approach to structure a scientific field. J. Inf. **15**(1), 101109 (2021)
22. Wang, H., et al.: Ripplenet: Propagating user preferences on the knowledge graph for recommender systems. In: Proceedings of the 27th ACM International Conference on Information and Knowledge Management, pp. 417–426 (2018)
23. Wang, T., Li, H.: Coreference resolution improves educational knowledge graph construction. In: 2020 IEEE International Conference on Knowledge Graph (ICKG), pp. 629–634 (2020)
24. Wang, Y., Yu, B., Zhang, Y., Liu, T., Zhu, H., Sun, L.: Tplinker: Single-stage joint extraction of entities and relations through token pair linking. In: Proceedings of the 28th International Conference on Computational Linguistics, pp. 1572–1582 (2020)
25. Wei, Z., Su, J., Wang, Y., Tian, Y., Chang, Y.: A novel cascade binary tagging framework for relational triple extraction. In: Proceedings of the 58th Annual Meeting of the Association for Computational Linguistics, pp. 1476–1488 (2020)
26. Yu, B., Tian, X., Luo, W.: Extracting mathematical components directly from pdf documents for mathematical expression recognition and retrieval. In: Advances in Swarm Intelligence, pp. 170–179 (2014)
27. Zanibbi, R., Blostein, D.: Recognition and retrieval of mathematical expressions. Int. J. Docum. Anal. Recogn. (IJDAR) **15**(4), 331–357 (2012)

KEP-Rec: A Knowledge Enhanced User-Item Relation Prediction Model for Personalized Recommendation

Lisha Wu, Daling Wang[(⊠)], Shi Feng, Yifei Zhang, and Ge Yu

School of Computer Science and Engineering, Northeastern University, Shenyang, China
{wangdaling,fengshi,zhangyifei,yuge}@cse.neu.edu.cn

Abstract. For more accurate, diversified and interpretable personalized recommendation, the joint consideration of user-item interaction information and side information in knowledge graph has become a research hotspot. Traditional models based on collaborative filtering usually have cold start and sparse problems. The existing recommendation model based on knowledge graph can enrich the representation of users and items by using graph structure information from the knowledge graph, and make it more interpretable. Although the efforts have achieved a certain performance improvement, they consider all entities in knowledge graph globally for all users, and the aggregation strategy is single. In this paper, we propose KEP-Rec, a **K**nowledge **E**nhanced User-Item Relation **P**rediction Model for Personalized **Rec**ommendation. For a given target user and candidate item, KEP-Rec represents the user and item with enhanced information by knowledge graph for predicting the interacted probability between them and further personalized recommendation. In detail, KEP-Rec takes into account the changes in preferences of specific users and the differences in user perception of relations. Based on the idea of collaborative filtering, KEP-Rec selects an extended entity set of the items relevant with target user and candidate item as the initial set to propagate in knowledge graph. Moreover, KEP-Rec sets an item-aware attention mechanism to consider the interaction of candidate items with different weights given by target user's historical preferences to realize the diverse representation of the user preferences. In the propagation process alone knowledge graph, the relation embedding is considered for target user to achieve personalization. Empirical results on three real datasets of music, books, and movies show that KEP-Rec significantly outperforms state-of-the-art methods.

Keywords: User-Item Relation Prediction · Personalized recommendation · User and item embedding · Knowledge graph · Entity propagation

1 Introduction

At present, the recommender system (RS) becomes an effective approach to solve how users can efficiently obtain items they are interested in under the situation of "information overload". An effective traditional recommendation method is collaborative filtering (CF), which represents users and items as vectors, and models the historical interactions

© The Author(s), under exclusive license to Springer Nature Switzerland AG 2023
B. Li et al. (Eds.): APWeb-WAIM 2022, LNCS 13422, pp. 239–254, 2023.
https://doi.org/10.1007/978-3-031-25198-6_19

between users and items as a matrix through operations such as inner product or neural network. However, CF-based models generally suffer from cold start and sparsity issues. In order to solve these problems, using knowledge graph as auxiliary information into the recommender system has attracted researchers. Knowledge graph (KG) is a heterogeneous graph, in which nodes are entities and edges represent relations between entities. The items and their attributes in the recommender system can be mapped to KG's entities which makes it easy to learn more about the relationship between items. Integrating user information into KG can capture more diverse user preferences. The existing KG-based recommender system models can be divided into embedding-based, path-based, and comprehensive methods [5]. The embedding-based approach enriches the item or user's presentation by directly using KG, but it ignores the connectivity of entities in KG. The path-based algorithm is used to explore multiple meta-paths between user and item in KG to infer the user preferences, but it will cost highly for artificially setting paths, and the rich structural information stored in KG is ignored. Both of them cannot mine and utilize well enough the comprehensive correlation between user and item. So comprehensive methods based on GNN have been proposed. For example, ripple2vec [25] proposes to implement node embedding by constructing a context graph via a new defined ripple distance over ripple vectors. GNN can directly model the high-order connectivity between entities. This high-order connection contains rich semantics that can refine the entity representation by leveraging the entity's multi-hop neighbor information.

Although these methods have improved interpretability and performance, they still have three limitations:

(1) Incorporating knowledge graph can contain more information, but there are fewer items that can align entities in the knowledge graph due to the user-item interaction sparse data. so that is not enough to mine more user-preferred entities.

(2) Directly using the items interacted by a user and mapping them to the knowledge graph can obtain more knowledge information, but this information is based on the user's historical behavior. In fact, a user's preference may be changeable. How to dynamically measure a user's diversified preference based on his history for the current candidate item is the key to recommender systems.

(3) Using GNN can refine the entity representation, which shows that using neighborhood information can greatly promote the completion of recommendation tasks. However, this method ignores that different users have different weight for the same relation in the graph. This weight can also be understood as different users' different cognition of entities and relations in the knowledge graph.

In order to solve the above problems, we propose KEP-Rec, a **K**nowledge **E**nhanced User-Item Relation **P**rediction Model for Personalized **Rec**ommendation that considers preference changes and collaborative interaction for specific users. For a given target user and candidate item, KEP-Rec represents the user and item by knowledge graph for predicting the interacted probability between them and further personalized recommendation. Similar with CKAN [18], KEP-Rec analyzes the items related with target user and the users related with candidate item and more relevant items. However, KEP-Rec considers the influence of candidate items on user preferences and pays different attention to the relation of the target users in KG by two attention mechanisms.

In general, our contributions are the following:

(1) We propose KEP-Rec model, an end-to-end collaborative knowledge propagation user-item representation framework for user-item relation predicting and further personalized recommendation.
(2) We construct a diversified representation of user preferences, that is, the impact of candidate items on users' historical preferences has changed the user's preferences. Moreover, the user-specific relation perception can obtain the preference of the target user to the knowledge relations.
(3) We conducted empirical experiments in three real-world recommendation scenarios datasets, and the results showed that our KEP-Rec is better than the existing state-of-the-art baselines.

2 Related Work

Now, recommender systems based on KG have been implemented in three ways: embedding-based methods, path-based methods and comprehensive methods [5].

2.1 Embedding-Based Recommendation

This method directly encodes entities into low-rank embeddings and uses the semantic information of the user/item in the KG for recommendation. The algorithm models can be divided commonly into two categories: translation distance models, such as TransE [1], TransH [19], TransR [9], and semantic matching models, such as DistMult [20]. The classic CKE [22] model integrates various auxiliary information in the framework of collaborative filtering which uses the TransR algorithm to encode item's structured knowledge and integrates the content knowledge to represent the item. Another model called CFKG [23] constructs a user-item knowledge graph. In that KG, user, item and their related attributes are regarded as entities, and the users' historical behaviors are regarded as a special type of relationship between entities. Based on a specific distance function, the model can learn user/item embeddings which is the implementation of semantic matching models.

2.2 Path-Based Recommendation

The common path-based method calculates the semantic similarity between entities on different paths to make recommendation. In HeteRec [21], L different types of meta-paths connecting users and items are defined. The similarity of items in each path is measured by PathSim [12] and formed L item extended matrices. Because the interaction matrices between users and items, multiplying with the item extended matrices can get L extended interaction matrices. Finally, the non-negative matrix factorization technique [3] is applied to obtain the latent vectors of users and items in different meta-paths which enriches users' and items' representations.

2.3 Comprehensive Methods

Both embedding-based and path-based methods only use partial information in knowledge graph. In order to make better use of knowledge graph, a comprehensive method is proposed to extend the embedding of the entity in the graph, directly model the high-order connectivity between entities, and use the entity's multi-hop neighbor information to refine the entity representation. RippleNet [15] is the first to propose the concept of preference propagation, which enriches user representations by aggregating the multi-hop neighbors of the user's historical interactive items. KGCN [16] is to aggregate the candidate item with its multi-hop neighbors to update the item's representation. In the aggregation process, the weight of the neighbor is jointly determined by the user and the candidate item, so that the user's preferences are implicitly existing in the entity representation. Another model, such as KGAT [17], combines user-item interaction matrix with a knowledge graph containing attribute information and uniformly represents users and items in the same graph. Then users and items are aggregated with their respective multi-hop neighbors in the graph to enrich their representations. Recently, CKAN [18] uses a heterogeneous propagation strategy to explicitly encode collaborative signals and knowledge associations and applies a knowledge-aware attention mechanism to distinguish the contributions of different neighbors.

Although the above methods achieve a certain performance improvement, they consider all entities in the knowledge graph at a global level and the aggregation strategy is single. Inspired by GARG [26], which takes full advantage of the collaborative, sequential and content-aware information, we first use collaborative information to expand the initial entity set of users and items. In particular, we think that for a specific target user, the relation between users, items, and entities in KG is not fully equal, so the KEP-Rec model considers the influence of candidate items on user preferences and pays different attention to the relation of the target users in KG.

3 Problem Definition

There are a set of M users and a set of N items are expressed as $U = \{u_1, u_2, ..., u_M\}$ and $V = \{v_1, v_2, ..., v_N\}$ respectively in a typical recommender system. We define the user-item interaction as a binary matrix $Y = \{y_{uv}|u \in U, v \in V\}$, where $y_{uv} = 1$ means that the user u ever interacted with the item v, such as clicking, collecting or purchasing; otherwise $y_{uv} = 0$. Note, the value of 1 for y_{uv} indicates that there is an explicit interaction between the user u and the item v, but doesn't necessarily mean u's preference over v.

In addition, we have a knowledge graph $\mathcal{G}=\{(h, r, t)|h, t\in\mathcal{E}, r\in\mathcal{R}\}$which consists of massive knowledge triples. Each triple (h, r, t) demonstrates that a relation r exists between head entity h and tail entity t. And the sets of entities and relations in $\mathcal{G}$ are denoted as $\mathcal{E}$ and $\mathcal{R}$. It is also mentioned in [27] that representing machine-interpretable statements in the form of subject-predicate-object triples is a mature practice for capturing the semantics of structured data. For example, the triple (*A Song of Ice and Fire, book.book.author, George Martin*) states the fact that George Martin writes the book "A Song of Ice and Fire". In this triple, *A Song of Ice and Fire* is head entity, *book.book.author* is relation, and *George Martin* is tail entity.

The problem we want to solve is defined as follows: given a knowledge graph $\mathcal{G}$ and the historical interaction matrix Y between user set U and item set V, for a target user $u_T \in U$ and a candidate item $v_C \in V$, predict the probability that u_T would interacts with v_C which has not interacted with before, i.e. $y_{u_T v_C} = 0$ in Y.

For solving above problem, we apply knowledge graph $\mathcal{G}$,, assume that every item $v_j \in V$ $(j = 1, 2, ..., N)$ in interaction matrix Y can be linked to a corresponding entity $e \in \mathcal{G}$ by the entity linking technique [2]. Moreover, based on U, V, Y, and $\mathcal{G}$,, we learn a prediction function $\hat{y}_{uv} = f(u_T, v_C | \Theta, Y, \mathcal{G})$,, where $\hat{y}_{uv}$ represents the probability of our prediction, and Θ represents the parameters of the function.

4   Methodology

Our proposed KEP-Rec framework is shown as Fig. 1, the model contains three main layers, we introduce them in detail in Sect. 4.1, 4.2, and 4.3 respectively.

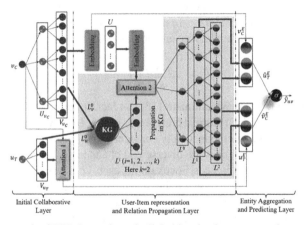

Fig. 1. The framework of KEP-Rec, where the light blue dot denotes user, the orange dot denotes item, the red dot denotes the entity in KG, and the layered color dots are the different embedding representations respectively. In the pink background, it is emphasized that the two initial entity sets are based on the relation-aware propagation process in KG, where L^i $(i = 1, 2, ..., k)$ represents the entity set after each layer of propagation.

4.1   Initial Collaborative Layer

The layer is shown as the left side in Fig. 1 and consists of three parts. The first part is to construct the initial entity set of target user u_T, the second part is to construct the initial entity set of candidate item v_C, and the third part is to consider the influence of v_C on the preference of u_T. Next, we will explain these three parts in detail.

The Target User's Initial Entity Set. Intuitively, the user's preferences can be reflected through the items which have interacted with, so we can consider these items as the user's initial entity set to express user preferences. From the user's point of view, starting from the entities in this initial entity set and propagating along the connections in the knowledge graph can reasonably expand the user's preference range and enrich the user's representation. For target user u_T, his interacted history with items is V_{u_T}, where $V_{u_T} \subset V, \forall v \in V_{u_T}$ satisfies $y_{u_r v} = 1$ in Y and the set size of V_{u_T} denoted as $m = |V_{u_T}|$. We define V_{u_T} as the initial entity set of target user u_T.

The Candidate-Item Initial Entity Set. User-based collaborative filtering [14] assumes that users who like the same items are similar. According to this assumption, we can consider that users who ever interacted with candidate items v_C have similar preferences. So, it is a great extension to use these items that the users have interacted with the initial entity set of v_C. As shown in the upper left of Fig. 1, we denote the user set interacted with v_C as U_{V_C} where $U_{V_C} \subset U, \forall u \in U_{v_c}$ satisfies $y_{uv_C} = 1$ in Y. Then, for every $u \in U_{V_C}$, we denote the item set interacted with u as V_u, where $V_u \subset V, \forall v \in V_u$ satisfies $y_{uv} = 1$ in Y. And then $\forall u \in U_{V_C}$, we take the union of all entity sets V_u constructed by u denoted as V_{v_C} and the set size of V_{v_C} denoted as $n = |V_{v_c}|$. Equally, for every $v = V_{v_C}$, the v interacted with at least a $u \in V_{v_c}$. Finally, we define V_{v_C} as initial entity set of candidate item v_C. In this case, v^E is the origin embedding of $v \in V$, and the candidate item v_C initial entity embedding is directly represented as:

$$v_C^E = \frac{\sum_{j=1}^{n} v_j^E}{n}, v_j \in V_{v_c} \tag{1}$$

The Item-Aware Attention Mechanism [13]. In fact, for all $v_i \in V_{v_C}$ $(i = 1, 2, ..., n)$ the items can only reflect the historical preferences of $u \in U_{V_C}$. As mentioned in [27], the existing recommendation methods mainly consider the representation of users as static feature sets. But user preferences are diverse and changeable, and the newly candidate items v_C may affect user preferences. For example, user A has watched *Avengers* and *Iron Man*, both of which belong to action movies. Then the user watching the romance movie will affect his preference for action movies. To characterize user's diverse interests, we use an item-aware attention network, i.e. Attention 1 in Fig. 1, to model the different impacts of the user's historical interaction items on the candidate item. Specifically, we apply a multi-layer perceptron to calculate the score between candidate item and historical items and SoftMax [10] function to calculate the normalized impact weight:

$$\pi(v_C \| v_i) = W_2(ReLu(W_1(v_C \| v_i) + b_1)) + b_2 \tag{2}$$

$$\pi(v_C \| v_i) = softmax \left(\frac{\pi(v_C \| v_i)}{\sum_{i=1}^{m} \pi(v_C \| v_i)} \right) \tag{3}$$

where $v_i \in V_{u_T}$ and $m = |V_{u_T}|$. Finally, we get the weighted initials user embedding:

$$u_T^E = \sum_{i=1}^{m} \pi(v_C \| v_i) * v_i \tag{4}$$

Through this item-aware attention mechanism to aggregate user historical interactive items, it is possible to realize the diverse user preferences, which better represents real-life scenarios in the selection process of items by users.

4.2 User-Item Representation and Relation Propagation Layer

The layer is shown as the middle side in Fig. 1. In this layer, we mainly accomplish two tasks: one is knowledge graph propagation, that is, start from the initial entity set V_{u_T} and V_{v_C} we defined above, and use relation links to obtain extended entities and triples; the other is constructing a user-specific relation attention mechanism.

For introducing knowledge graph $\mathcal{G}$, we define $L_u^0 = V_{u_T}$ and $L_v^0 = V_{v_C}$ as 0th layer entity set in $\mathcal{G}$. Based on the graph structure of $\mathcal{G}$, the entities L_u^0 and L_v^0 can be iteratively propagated along relations to reach more connected entities and we define the ith ($i = 1, 2, ..., k$, where k is the layer number propagated finally) set of entities as follows:

$$L^i = \{t^i | (h^{i-1}, r^i, t^i) \in \mathcal{G}, h^{i-1} \in L^{i-1}, with\ i = 1, 2, \ldots, k \tag{5}$$

After k hops propagation, we have $k + 1$ sets of entity sets L^i ($i = 1, 2, ..., k$).

The traditional graph convolutional network only propagates the information embedded by the entity itself and ignores the semantic information encoded in the relationship. In KG, an entity can be connected to multiple neighbors through various relationships, indicating different semantics. In our KEP-Rec model, the user-oriented personalized recommendation is embodied in that for different users, the KG entity should have different representations to characterize its embedding. For example, for the movie entity, some users may watch the movie *Iron Man* because the actor Robert is famous. Others watch the movie *The Fantasy Drifting of the Youth Pie* because the director is famous by Ang Lee, so the rich semantics encoded in the relationship is crucial to understanding different user intentions.

In G, we need to consider the entity feature representation when facing different users, that is, setting different weights for tail entities to reveal the different semantic information of different head entities and relations. Specifically, for a given entity, neighbors under each relation surrounding the entity are scored in a user-specific function that measures the influence of each entity's neighbors. After k^{th} layers propagations, we can get k^{th} set of tail entities. And we set a tail entity's embedding t_i^l with user-specific attentive weight as such:

$$t_i^l = A(h^l, r^l, u_{origin}^E)t_i^l \tag{6}$$

where h^l is the head entity's embedding, r^l is the relation's embedding, t^l is the tail entity's embedding of the lth triple, and u_{origin}^E is the user's origin embedding. $A(h^l, r^l, u_{origin}^E)$ is the attentive weight impacted by the head entity, the relation between head h^l and tail r^l and specific user u_{origin}^E which implemented by the two-layer neural network (see Attention 2 in Fig. 1):

$$a_0 = ReLu\left(W_0\left(h^l \| r^l \| u_{origin}^E\right) + b_0\right) \tag{7}$$

$$A(h^l, r^l, u_{origin}^E) = softmax(\sigma(ReLu(W_1 a_0 + b_1))) \tag{8}$$

Here we chose ReLU [6] and Sigmoid [7] as the activation functions. W and b are trainable weight matrices and characteristic parameters, and their respective subscripts indicate different layers. This user-specific relation attention mechanism uses the information from a given user, item, and relation to determine which neighbor has more information about the item while increasing personalized choices.

Based on the above, we use the initial entity set V_{u_T} and V_{V_C} for target user u_T and candidate item v_C respectively as the first head entity to be propagated in $\mathcal{G}$, then, the kth layer entities set embedding can be represented as: $t^k = \sum t_i^k, t_i^k \in Lk$. After integrating all entities k layers propagation, we can generate the final entity embedding of $t = \{t^1, t^2, \ldots, t^k\}$.

Finally, we combine initial entity set representations of target user u_T and candidate item v_C initial entity set representations and the entity sets representations that have been propagated in $\mathcal{G}$ to obtain user and item final representations as follow:

$$\hat{u}_T^E = \left\{ u_T^E, t_u^1, t_u^2, \ldots, t_u^L \right\} \tag{9}$$

$$\hat{v}_C^E = \{ v_C^E, t_v^1, t_v^2, \ldots, t_v^L \} \tag{10}$$

4.3   Entity Aggregation and Prediction Layer

The final step in this model is the prediction shown in the right side in Fig. 1, which aggregates entity itself representation and its multi-hop neighbors after k layers knowledge propagation. There are three common aggregators that we can aggregate multi-embeddings and we have implemented them in our model.

Sum Aggregator. The *Sum* aggregator sums multiple representations, followed by a nonlinear transformation:

$$agg_{sum} = \sigma(W \cdot (e_1 + e_2 + \cdots + e_n) + b) \tag{11}$$

Concat Aggregator. The *Concat* aggregator concatenates multiple representations, and then applies a nonlinear transformation:

$$agg_{concat} = \sigma(W \cdot (e_1 \| e_2 \| \ldots \| e_n) + b) \tag{12}$$

Pooling Aggregator. The *Pooling* aggregator takes the maximum value of multiple vectors as the same dimension, followed by a non-linear transformation:

$$agg_{pool} = \sigma(W \cdot maxpool(e_1, e_2, \ldots, e_n) + b) \tag{13}$$

According to our experimental results, we apply *Concat* Aggregator in KEP-Rec model. Based on the aggregators, we can further predict the interaction probability for target user u_T and candidate item v_C.

Model Prediction. Based on the target user' representation $\hat{u}_T^E$ and candidate item's representation $\hat{v}_C^E$, the predicted probability is calculated by $\hat{y}_{uv} = \sigma(\hat{u}_T^{E^T} \hat{v}_C^E)$, where $\sigma()$ is the sigmoid function.

Loss Function. For each user, we randomly select the same number of negative samples with positive samples to make sure the effectivity of model training. We define the loss function of the model KEP-Rec as follows:

$$\mathcal{L} = \sum_{u \in \cup} \left(\sum_{v \in \{v | (u,v) \in P^+\}} \mathcal{J}(y_{uv}, \hat{y}_{uv}) - \sum_{v \in \{v | (u,v) \in P^-\}} \mathcal{J}(y_{uv}, \hat{y}_{uv}) \right) + \lambda ||\Theta||_2^2$$

(14)

where $\mathcal{J}$ is the cross-entropy loss, P^+ is the positive sample while P^- means the negative sample. Θ is the model parameter set, and $||\Theta||_2^2$ is the L_2-regularizer that is parameterized by λ.

5 Experiments

In this section, we evaluate the KEP-Rec model in three real-world scenario datasets. Inspired by mostly related work and the paper [24] which discussed the value of experimentation and measurement, the experiments will answer the following research questions:

Q1: How does KEP-Rec perform compared with the state-of-the-art KG-based recommendation methods?

Q2: How do different parameters affect KEP-Rec?

5.1 Datasets

In order to verify the effectiveness of KEP-Rec in different application scenarios, we apply three general used datasets from different fields (movies, books, and music) in our experiments.

- **MovieLens-20M[1] (ML for short)**: This dataset is collected by GroupLens Research which obtained nearly 20 million rating (from 1 to 5) from 27,000 movies by 138 thousand users on the MovieLens website.
- **Book-crossing[2] (BC for short)**: This dataset is collected by Cai-Nicolas Ziegler from the Book-Crossing community (August to September 2004). It contains 278,858 users 1,149,780 ratings (from 1 to 10) for approximately 271,379 books.
- **Last.FM[3] (FM for short)**: This dataset contains the social networks and music artist information of two thousand users who listened to the online music system of Last.FM.

Since KEP-Rec aims to predict the interacted probability between target user and candidate item and make recommendation based on implicit feedback, we set a scoring threshold to convert the explicit feedback into implicit feedback. For MovieLens-20M, we set the positive score threshold to 4, and scores greater than 4 are positive samples

---

[1] Https://grouplens.org/datasets/movielens/.

[2] http://www2.informatik.uni-freiburg.de/.cziegler/BX/.

[3] https://grouplens.org/datasets/hetrec-2011/.

($y_{uv} = 1$ in Y), and vice versa ($y_{uv} = 0$ in Y). However, the Book-Crossing and Last.FM data are too sparse, and the above threshold settings are not suitable. So, for those two datasets, we set the items that the user interacts with as the positive samples. For negative samples, we randomly select items of the same size as the positive samples from items that the user has not interacted with.

In addition to the aforementioned user and item interaction datasets, we also need to choose sub-KGs of each dataset. For MovieLens-20M, Book-Crossing, and Last.FM, we choose sub-KGs from KG called Satori[4] from Microsoft. Each sub-KG is a subset of the entire KG, the confidence level is greater than 0.9. For the sake of simplicity, we exclude the items matched with multiple entities and those unmatched to any of entities. Table 1 summarizes the statistics of these experimental datasets.

Table 1. Statistics of Movie-Lens20M, Last.FM, and Book-Crossing, where avgI means the average interactions per user, and avgL means the average link per entity

	#users	#items	#interaction	#avgI	#entities	#relations	#triples	#avgL
ML	138,159	16,954	13,501,622	23	102,569	32	499,474	29
FM	1,872	3,846	42,346	98	9,366	60	15,518	4
BC	17,860	14,967	139,746	8	77,903	25	13,150	10

5.2 Baselines

In the experiments, we will compare our KEP-Rec model with the following baselines.

- **BPRMF** [11] uses matrix factorization based on Bayesian personalized ranking, which is based on the user's paired preferences as a single collaborative filtering method.
- **CKE** [22] combines a CF module with knowledge embedding, text embedding, and image embedding of items in a unified framework and jointly learn to make recommendations.
- **RippleNet** [15] is a state-of-the-art propagation-based model that uses a large number of entities related to the user's historical clicks to enrich the user's representation, so that the user's potential preferences can be propagated in KG. Then the click-through rate of the user-item pair is predicted.
- **KGAT** [17] uses embedded propagation to directly model high-connectivity between users and items. It applies TransR model to obtain the initial representation of the entity. Then it runs entity propagation from the entity itself along the relationship link in the knowledge graph. In the process of outward propagation, the information from the entity will iteratively interact with the multi-hop neighbor.
- **CKAN** [18] uses a heterogeneous propagation strategy to explicitly encode collaborative signals and knowledge associations, and applies a knowledge-aware attention mechanism to distinguish the contributions of different knowledge-based neighbors.

---

[4] https://searchengineland.com/library/bing/bing-satori.

5.3 Experimental Setup

In our experiment, for each dataset, 60% are randomly selected for training, 20% are for evaluation, and the rest 20% are for prediction. We evaluate our method in CTR prediction and top-K recommendation. CTR prediction generally refers to the click-through rate estimation task which is with an item, predicting the probability that the user clicks on this item. We adopt AUC to evaluate this performance. For top-K recommendation, we adopt F1@K to evaluate the performance. For optimization, we use ADAM [8] to optimize all models in training. We set the batch size to 1024 during training and use the default Xavier initialize [4] to initialize the parameters of the model.

We implement our model in PyTorch. The best hyper-parameters are obtained by grid search. We set the learning rate to be searched in $\{10^{-3}, 5 * 10^{-3}, 10^{-2}, 5 * 10^{-2}\}$. The embedding size is tuned among $\{16, 32, 64, 128, 256\}$. The coefficient of L_2 normalization is searched in $\{10^{-6}, 10^{-5}, 10^{-4}, 10^{-3}, 10^{-2}\}$. We search the set of size in $\{4, 8, 16, 32, 64\}$ for user and item embedding.

5.4 Experimental Results and Analysis

The experimental results and analysis are shown as follows.

Performance Comparison with Baselines (Q1). In this section, we present the results of performance comparisons among KEP-Rec and baselines. The results of all methods in CTR prediction and top-K recommendation are presented in Table 2 respectively.

Table 2. The result of AUC and F1 in CTR prediction comparing of different models

Dataset	ML		FM		BC	
Model	AUC	F1	AUC	F1	AUC	F1
BPRFM	0.958 (−2.6%)	0.914 (−2.7%)	0.756 (−9.7%)	0.701 (−8.6%)	0.658 (−9.5%)	0.611 (−6.6%)
CKE	0.927 (−5.7%)	0.874 (−6.7%)	0.747 (−10.6%)	0.674 (−11.3%)	0.676 (−7.7%)	0.623 (−5.4%)
RippleNet	0.976 (−0.8%)	0.927 (−1.4%)	0.776 (−7.7%)	0.702 (−8.5%)	0.721 (−3.2%)	0.647 (−3.0%)
KGAT	0.976 (−0.8%)	0.928 (−1.3%)	0.829 (−2.4%)	0.742 (−4.5%)	0.731 (−2.2%)	0.654 (−2.3%)
CKAN	0.976 (−0.8%)	0.929 (−1.2%)	0.842 (−1.1%)	0.769 (−1.8%)	0.753 (−0%)	0.673 (−0.5%)
KEP-Rec	**0.984**	**0.941**	**0.853**	**0.787**	**0.753**	**0.677**

From Table 2, we can be observed that:

(1) KEP-Rec consistently outperforms all baselines across mostly datasets in terms of all measures. More specifically, it achieves significant improvements over the

strongest baseline CKAN w.r.t. F1 by 1.2%, 1.8%, and 0.5% in MovieLens-20M, Last.FM and Book-Crossing respectively. That may because the two attention mechanisms played a big role in KEP-Rec. It is worth mentioning that on the MovieLens-20M dataset, the results of all baselines are at a higher value. It shows that more user interaction is conducive to make a better recommendation.

(2) The two path-based baselines, RippleNet and KGAT, are better than the CF-based method BRPFM and the embedding-based CKE, indicating that KG's graph structure information is helpful for recommendation. In addition, although RippleNet and KGAT achieved excellent performance, they still did not exceed KEP-Rec. This is because RippleNet neither incorporates user click history items into the user representation, nor does it introduce high-level connections, and KGAT does not mix GCN layer information, nor does it consider user preferences when collecting KG information.

(3) From the experimental results, the method based on KG propagation is higher than the pure CF-based BPRMF model on the three data sets of all evaluation indicators. This experimental result shows that the use of KG is very helpful for recommendation. However, it is worth noting that BPRMF outperforms CKE in some indicators, which means that modeling direct relationship in KG may not be able to make full use of the rich information encoded in KG and proved the effectiveness of high-level connectivity information. The second reason may be that CKE is aimed at multi-modal information, and only one of the graph structure information is used here, which leads to its poor performance.

(4) In addition, by comparing CKAN, KEP-Rec and KGAT, although both utilize high-level connectivity, CKAN and KEP-Rec outperforms KGAT. The possible reason for the analysis is that while modeling the item presentation, CKAN and KEP-Rec both further considers the collaborative signals in the interaction between the user and the product, thereby realizing the enrichment of the item. They explore the connections between users, item and entities through collaborative interaction and knowledge graphs.

Performance Comparison with Different Parameters (Q2). To get deep insights on different parameters of KEP-Rec, we investigate their impact in three datasets respectively. We first study the influence of layer numbers, and then examine the influence of knowledge graph embedding dimension. Finally, we analyze the influence of item attention and user-specific relationship propagation layer.

Impact of Number of Layers: The impact of different number of layers is shown in Table 3. We conducted experiments with the same other parameter settings. The results show that when layer number is 2, ML, and BC perform best; when layer number reaches 4, FM achieves better performance.

According to the analysis of different datasets, one possible reason for this phenomenon is that when there are more links in the knowledge graph of the dataset, long-distance propagation provides more supplementary knowledge information, but also brings more noise. However, in the case of a small amount of data in the knowledge graph of the dataset, deeper propagation can make use of knowledge information to a greater extent.

Table 3. F1 result of KEP-Rec with the different number of layer number.

Layer number	2	3	4
ML	**0.937**	0.934	0.934
FM	0.783	0.781	**0.787**
BC	**0.677**	0.673	0.671

Impact of Dimension of Embedding. We use the same dimensional parameters to embed the entities and relations in the KG, and compare the performance of KEP-Rec on all three data sets for different dimensions. The result is shown as Table 4.

Table 4. F1 result of KEP-Rec with dimension of embedding.

Embedding dimension	32	64	128	256
ML	0.941	0.941	0.941	0.941
FM	0.775	0.784	0.785	0.783
BC	0.668	0.669	0.671	0.672

From Table 4, it can be seen that the dimensional changes of KEP-Rec on the three datasets did not cause excessive fluctuations in the final evaluation index. This means that it has a strong tolerance for size selection, which reduces the dependence of the experiment on parameters and makes it easier to reproduce the experimental results.

Impact of Different Personalized Components. In order to verify the influence of item attention and user-specific relation propagation layer, we conducted ablation experiments of two sub-models. One is KEP-Rec that only removes item attention and we mark it as KEP-Rec$_{(-A)}$, and the other is to remove two parts at the same time, and we mark it as KEP-Rec $_{(-U-A)}$. The results are shown in Table 5.

Table 5. F1 result of KEP-Rec without different components. KEP-Rec $_{(-U-A)}$ means KEP-Rec without item-aware attention mechanism and user-specific relation layer. KEP-Rec $_{(-A)}$ means without user-specific relation propagation layer.

	KEP-Rec $_{(-U-A)}$	KEP-Rec$_{(-A)}$	KEP-Rec
ML	0.929	0.934	**0.941**
FM	0.773	0.778	**0.787**
BC	0.674	0.675	**0.677**

The result supports that item attention and user-specific relation propagation layer are both powerful determinants, and the combination of the two can be more completely encoded into the potential user/item vector representation.

Impact of Aggregators. In order to verify the influence of the aggregator on the results of our model, we chose **Sum**, **Pool** and **Concat** three aggregators, while keeping other conditions consistent. The experimental results are shown in Table 6.

Table 6. F1 result of KEP-Rec with different aggregators.

	Sum	Pool	Concat
ML	0.934	0.924	**0.941**
FM	0.736	0.733	**0.787**
BC	0.666	0.654	**0.677**

By analyzing the experimental results, we have the following observations that *Concat* is always better than *Sum* and *Pool*. This may be because, compared with the other two aggregators, the *Concat* aggregator can retain the information content of the embedded representation as much as possible without filtering and mixing. Based on the result, *Concat* aggregator is applied in our KEP-Rec model.

6   Conclusion

In this work, we propose KEP-Rec, a knowledge enhanced user-item relation prediction for users' diverse preferences representation, which is an end-to-end, user-oriented, and collaborative knowledge propagation prediction model of user-item relation for personalized recommendation. We construct a diverse representation of user preferences and set a user-specific relation attention mechanism to describe those relations between different users and the same item. Meanwhile, the high-order connectivity of the knowledge graph is used to finally obtain an enhanced representation of users and items. A large number of experiments have proved the superiority of KEP-Rec.

Since the proposed personalized component is aimed at entities existing in knowledge graph, the method can also be applied to fields related to graph structure, such as social networks. We believe that KEP-Rec can be widely used in related applications. In addition, the user's preference is actually also changed by time, and a real-going idea is to use CNN to join the time series in the model which is a viable direction.

Acknowledgement. The work was supported by National Natural Science Foundation of China (62172086, 61872074, 62106039).

References

1. Bordes, A., Usunier, N., García-Durán, A., Weston, J., Yakhnenko, O.: Translating Embeddings for Modeling Multi-relational Data. NIPS 2013, pp. 2787–2795 (2013)
2. David, N., Ian, H.: Witten: learning to link with wikipedia. CIKM 2008, pp. 509–518 (2008)
3. Ding, C., Tao Li, T., Jordan, M.: Convex and semi-nonnegative matrix factorizations. IEEE Trans. Pattern Anal. Mach. Intell. **32**(1), 45–55 (2010)
4. Glorot, X., Bengio, Y.: Understanding the difficulty of training deep feedforward neural networks. AISTATS 2010, pp. 249–256 (2010)
5. Guo, Q., et al.: A Survey on Knowledge Graph-Based Recommender Systems. CoRR abs/2003.00911 (2020)
6. Hahnloser, R., Sarpeshkar, R., Mahowald, M.A., et al.: Digital selection and analogue amplification coexist in a cortex-inspired silicon circuit. Nature **405**(6789), 947–951 (2000)
7. Han, J., Moraga, C.: The influence of the sigmoid function parameters on the speed of backpropagation learning. In: Mira, J., Sandoval, F. (eds.) IWANN 1995. LNCS, vol. 930, pp. 195–201. Springer, Heidelberg (1995). https://doi.org/10.1007/3-540-59497-3_175
8. Kingma, D., Ba, J.: Adam: A Method for Stochastic Optimization. ICLR (Poster) (2015)
9. Lin, Y., Liu, Z., Sun, M., Liu, Y., Zhu, X.: Learning Entity and Relation Embeddings for Knowledge Graph Completion. AAAI 2015, pp. 2181–2187 (2015)
10. Memisevic, R., Zach, C., Hinton, G., Pollefeys, M.: Gated Softmax Classification. NIPS 2010: 1603–1611 (2010)
11. Rendle, S., Freudenthaler, C., Gantner, Z., Schmidt-Thieme, L.: BPR: Bayesian Personalized Ranking from Implicit Feedback. UAI 2009, pp. 452–461 (2009)
12. Sun, Y., Han, J., Yan, X., Yu, P., Wu, T.: PathSim: meta path-based Top-K similarity search in heterogeneous information networks. Proc. VLDB Endow. **4**(11), 992–1003 (2011)
13. Vaswani, A., et al.: Attention is All you Need. NIPS 2017, pp. 5998–6008 (2017)
14. Wang, C., Blei, D.: Collaborative topic modeling for recommending scientific articles. KDD 2011, pp. 448–456 (2011)
15. Wang, H., et al.: RippleNet: Propagating User Preferences on the Knowledge Graph for Recommender Systems. CIKM 2018, pp. 417–426 (2018)
16. Wang, H., Zhao, M., Xie, X., Li, W., Guo, M.: Knowledge Graph Convolutional Networks for Recommender Systems. WWW 2019, pp. 3307–3313 (2019)
17. Wang, X., He, X., Cao, Y., Liu, M., Chua, T.: KGAT: Knowledge Graph Attention Network for Recommendation. KDD 2019, pp. 950–958 (2019)
18. Wang, Z., Lin, G., Tan, H., Chen, Q., Liu, X.: CKAN: Collaborative Knowledge-aware Attentive Network for Recommender Systems. SIGIR 2020, pp. 219–228 (2020)
19. Wang, Z., Zhang, J., Feng, J., Chen, Z.: Knowledge Graph Embedding by Translating on Hyperplanes. AAAI 2014, pp. 1112–1119 (2014)
20. Yang, B., Yih, W., He, X., Gao, J., Deng, L.: Embedding Entities and Relations for Learning and Inference in Knowledge Bases. ICLR (Poster) (2015)
21. Yu, X., et al.: Recommendation in heterogeneous information networks with implicit user feedback. RecSys 2013, pp. 347–350 (2013)
22. Zhang, F., Yuan, N., Lian, D., Xie, X., Ma, W.: Collaborative Knowledge Base Embedding for Recommender Systems. KDD 2016, pp. 353–362 (2016)
23. Zhang, Y., Ai, Q., Chen, X., Wang, P.: Learning over Knowledge-Base Embeddings for Recommendation. CoRR abs/1803.06540 (2018)
24. Liu, C.H.B., Chamberlain, B.P., McCoy, E.J.: What is the value of experimentation and measurement? Data Sci. Eng. **5**(2), 152–167 (2020). https://doi.org/10.1007/s41019-020-001 21-5

25. Luo, J., Xiao, S., Jiang, S., Gao, H., Xiao, Y.: ripple2vec: node embedding with ripple distance of structures. Data Sci. Eng. **7**(2), 156–174 (2022)
26. Wu, S., Zhang, Y., Gao, C., Bian, K., Cui, B.: GARG: anonymous recommendation of point-of-interest in mobile networks by graph convolution network. Data Sci. Eng. **5**(4), 433–447 (2020). https://doi.org/10.1007/s41019-020-00135-z
27. Liu, Y., Li, B., Zang, Y. et al.: A Knowledge-Aware Recommender with Attention-Enhanced Dynamic Convolutional Network. CIKM 2021, pp. 1079–1088 (2021)
28. Sikos, L.F., Philp, D.: Provenance-aware knowledge representation: a survey of data models and contextualized knowledge graphs. Data Sci. Eng. **5**(3), 293–316 (2020). https://doi.org/10.1007/s41019-020-00118-0

Probing the Impacts of Visual Context in Multimodal Entity Alignment

Yinghui Shi[1], Meng Wang[2(✉)], Ziheng Zhang[3], Zhenxi Lin[3], and Yefeng Zheng[3]

[1] School of Cyber Science and Engineering, Southeast University, Nanjing, China
`shiyinghui@seu.edu.cn`
[2] School of Computer Science and Engineering, Southeast University, Nanjing, China
`meng.wang@seu.edu.cn`
[3] Tencent Jarvis Lab, Shenzhen, China
`{zihengzhang,chalerislin,yefengzheng}@tencent.com`

Abstract. Multimodal entity alignment (MMEA) aims to identify equivalent entities across different multimodal knowledge graphs (KGs), and this topic has drawn increasing attention in recent years. Although the benefits of multimodal information have been observed, its negative impacts are non-negligible as injecting images without constraints brings much noise. It also remains unknown to what extent or under what circumstances visual context is truly helpful to the task. In this work, we employ graph structures and visual context to align entities in different multimodal KGs and propose to selectively combine feature similarities between cross-KG entities of these two aspects when making alignment decision. Specifically, we exploit image classification techniques and entity types to remove potentially un-useful images (visual noises) via generating entity mask vectors in the learning and inference processes. The extensive experiments have validated that the incorporation of selected visual context can substantially improve the MMEA. We also provide a thorough analysis about the impacts of the visual modality and discuss a few cases where injecting entity images induces misalignment.

Keywords: Multimodal entity alignment · Visual context · Knowledge graph

1 Introduction

Entity alignment (EA) is a task aiming to find entities from different knowledge graphs (KGs) that refer to the same real-world object. It plays an important role in KG construction and knowledge fusion as KGs are often independently created and suffer from incompleteness. Most existing models for EA leverage graph structures and/or side information of entities such as name and attributes along with KG embedding techniques to achieve alignment [17,26]. Several recent methods enrich entity representations by incorporating images, a natural component of entity profiles in many KGs such as DBpedia [10] and Wikidata [19], to address EA in a multimodal view [3,7,11].

B. Li et al. (Eds.): APWeb-WAIM 2022, LNCS 13422, pp. 255–270, 2023.
https://doi.org/10.1007/978-3-031-25198-6_20

Fig. 1. Thumbnail examples of the DBpedia entities where the images on top depict the equivalent entities, *Oakland,_California*, and the images below depict the equivalent entities, *Little_Mix*.

While experimental results have demonstrated that incorporating visual context benefits the EA task [3,11], it is worth noting that the use of entity images may introduce noises. An error analysis in EVA [11] points out that hundreds of source entities are correctly matched to their counterparts before injecting images but are mismatched with images present. Different visual representations of equivalent entities could be potential noises that induce mismatches, and there are various reasons for the visual inconsistency between two equivalent entities. One major reason is that entities naturally have multiple visual representations. As shown in Fig. 1, images (visual context) at left are dissimilar from their counterparts at right, yet they refer to same real-world entities. In addition, the incompleteness of visual data is also a challenging issue for multimodal EA, as reported in [11] that ca. 15–50% entities in the most commonly used benchmark DBP15K [14] are not provided with images.

The aforementioned observations raise a doubt: to what extent or under what circumstances is visual context truly helpful to the EA task? Is there a way to filter potential noises and better use entity images? To investigate the above issues, in this work, we propose Masked-MMEA, a novel framework capable of identifying and filtering potential visual noises for multimodal EA. Specifically, we utilize classification techniques and entity types (classes), common properties defined by the ontology of many KGs, to locate potential visual noises and meanwhile generate binary mask vectors which indicate whether an entity image should be filtered in the alignment learning or inference phase. Masked-MMEA learns structural and visual representations of entities separately, and then computes the corresponding similarities for each candidate pair where the mask vector is applied to determine the final similarity between two multimodal representations as a weighted sum.

In summary, our main contributions are three-fold: *(i)* To the best of our knowledge, we are the first to investigate the positive and negative aspects of incorporating visual context for EA. We provide insights on actual visual noises that tend to induce misalignment in the multimodal EA. *(ii)* We propose a novel framework, Masked-MMEA, to identify and filter potential visual noises for the multimodal EA, by utilizing classification techniques and ontologies of KGs. Extensive experiments have validated that the selective use of visual context benefits multimodal EA. *(iii)* We build a new dataset based on DBP15K which additionally includes ontological information and a full set of entity images. With the proposed dataset, we hope to facilitate the community in the development of multimodal learning approaches for KGs. The source code and datasets are publicly accessible at https://github.com/Shiyinghui/Masked-MMEA.

2   Related Work

Embedding-Based Entity Alignment. Embedding-based approaches for entity alignment (EA) can be generally divided into two categories: that only utilized graph structures and that used additional side information of entities [25,26]. Among the first category, MTransE [4] adopted TransE [1] to encode language-specific KGs in separate embedding spaces and learned a transformation to align counterpart entities across embeddings. IPTransE [28] and BootEA [15] embedded two KGs in a unified space and bootstrapped the labeled alignments iteratively. Among the second category, GCN-Align [21], JAPE [14] and AttrE [18] used attribute triples in the KGs to refine structural embeddings. MultiKE [24] explored more types of features. It learned entity embeddings from three different views including entity names, relations and attributes. HMAN [23] further exploited literal descriptions of entities to boost performance. UEA [27] utilized useful features from side information in an unsupervised framework to perform EA in the open world. Although some of the above approaches can achieve high accuracy on EA, the visual context has not been explored yet.

Multimodal KG Embeddings. In recent years, a few attempts have been made to incorporate entity images into KGs and build multimodal embeddings for EA. MMEA [3] applied TransE to learn structural embeddings for entities, and utilized image features to learn visual representations. It integrated multiple representations of entities via common space learning. HMEA [7] adopted the hyperbolic graph convolutional networks (HGCNs) to learn structural and visual embeddings of entities separately, then merged them in the hyperbolic space by a weighted Mobius addition. EVA [11] employed GCNs [20] to learn structural representations for entities, and used feed-forward networks to learn embeddings from image, relation and attribute features, respectively. Then it fused embeddings of different modalities by a trainable weighted concatenation. Although existing multimodal entity alignment approaches have shown promising performance, all of them ignored the negative impact of noisy data in entity images.

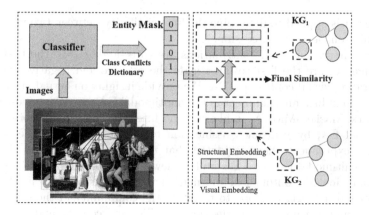

Fig. 2. The framework of Masked-MMEA, where the block on the left represents the 1st part, detailed in Sect. 3.1 and the block on the right represents the 2nd part, detailed in Sects. 3.2 and 3.3.

3   Method

We start with the task definition and notations. A KG is denoted as $G = (E, R, T, I)$, where E, R, T, I are the sets of entities, relations, triples and images, respectively. Given a source KG $G_1 = (E_1, R_1, T_1, I_1)$ and a target KG $G_2 = (E_2, R_2, T_2, I_2)$, multimodal entity alignment (MMEA) aims to find every pair (e_1, e_2) where $e_1 \in E_1$, $e_2 \in E_2$ and e_1 and e_2 refer to the same real-world object. Our approach, Masked-MMEA, can be divided into two parts, as illustrated in Fig. 2. The 1st part identifies entity images that are potential noises for MMEA and outputs a binary mask vector $\mathbf{M}$ to filter noises. In the 2nd part, $\mathbf{M}$ is applied in learning entity representations and calculating the similarity metrics between entities. These two parts will be detailed in Sect. 3.1 and Sects. 3.2 and 3.3, respectively.

3.1   Visual Noises Identification

We observe that in most cases visual representations of entities vary largely from a type to another, while they are less different within a type. Based on the findings, we take entity types as classes of images to train a classifier, and use it to identify images whose predicted class is semantically distant from their actual class, i.e., visual noises. To this end we obtain entity types and inter-class conflicts from the ontology of KGs, and design mask vectors to store identification results.

Entity Types. The ontology of KGs usually contains properties and hierarchical classes, and defines subsumption relationships between classes and class disjointness optionally [8]. Types (classes) are often organized in a hierarchical tree structure in the ontology of a KG, and an entity is often associated to a set

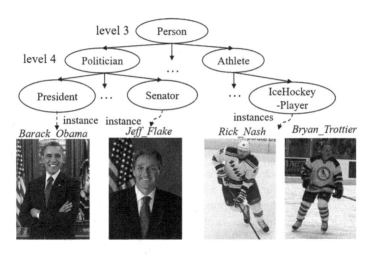

Fig. 3. An example of hierarchical classes.

of types. For example, as shown in Fig. 3, the entity *Barack Obama* in French DBpedia has four types declared: *Agent*, *Person*, *Politician* and *President*, with *Agent* as the most generic type and *President* as the most specific and a leaf node. As we observe that entities of fine-grained types like *President* and *Senator* are more semantically different than visually different, and therefore, we take the type of each entity at most at the fourth level (*Politician* in this example) as the label of its image. We also empirically find the choice of the fourth level, rather than the third or the fifth, yields better classification performance.

Inter-class Conflicts. To measure the semantic discrepancy between the predicted and real classes of an entity image, and inspired by OntoEA [22], we use a class conflict dictionary (CCD) to store the inter-class conflicts. Given two classes a and b, we set (a, b) as the key and $C[a, b]$ as the value, which represents the conflict degree between class a and class b. For better illustration, we let V denote the hierarchical class tree in which each node refers to a unique class and o the root (typically *owl:Thing*), and define S_x^c as the set of children (subclasses) of node x and S_x^d as the set of all the descendants of x in V, respectively. We assume that all subclasses of the root in V are mutually disjoint, which is in accordance with the design intent for the class hierarchy, and we regard any two descendants of two disjoint classes as disjoint. Let D denote the set of all disjoint class pairs, thus $D = \{(a, b) | a, b \in S_o^c, a \neq b\} \cup \{(a, b) | \forall c_1, c_2 \in S_o^c, a \in S_{c_1}^d, b \in S_{c_2}^d, c_1 \neq c_2\}$. Given two classes a and b, we firstly determine if $a \equiv b$ or $a \in S_b^d$ or $b \in S_a^d$, and set $C[a, b] = 0$ if they satisfy the condition, which ensures that a class does not conflict with itself or its descendant class, otherwise we look up D and set $C[a, b] = 1$ if $(a, b) \in D$, i.e., two disjoint classes are treated as conflicted. If neither of the above two conditions is met, we follow OntoEA and calculate $C[a, b]$ as:

$$C[a, b] = 1 - \frac{|S(a) \cap S(b)|}{|S(a) \cup S(b)|}, \tag{1}$$

where $S(a)$ and $S(b)$ denote the sets of classes passed by routing from a and b to the root class, respectively, and $|\cdot|$ denotes the set cardinality.

Entity Mask. We use $\mathbf{M} \in \mathbb{R}^{|E|}$ as an entity mask and denote $\mathbf{M}_{e_i}$ as the mask value of the i-th entity e_i in E. If the image of e_i is determined as potential noise, we set $\mathbf{M}_{e_i} = 0$, which means e_i is masked and its image should be filtered in the training or test phase; otherwise we set $\mathbf{M}_{e_i} = 1$. We initialize $\mathbf{M}$ with all zeros and update it iteratively. Specifically, given a conflict degree threshold λ, for $e \in E$, we feed its corresponding image to a classifier to obtain top k predictions (denoted as $p_1, \ldots, p_k$), and if the minimum conflict degree between the predictions and the actual class (denoted as g) of e is no greater than λ, i.e., $\min_{1 \leq i \leq k} \{C[p_i, g]\} \leq \lambda$, we reset the mask value of e to be 1.

3.2 Entity Embedding

To better analyze the impacts of visual context on MMEA, we only model two modalities in the entity embeddings, i.e., graph structures and visual context.

Structural Embedding. Graph convolutional networks (GCNs) have proven to be effective in capturing information from graph structures and have been used for embedding-based EA recently [17]. Formally, given as input the adjacency matrix $\mathbf{A}$ of a KG and randomly initialized feature matrix $\mathbf{H}^{(0)}$ of its entities, a multi-layer GCN iteratively updates entity representations from the i-th layer to the $(i + 1)$-th layer with the following propagation rule:

$$\mathbf{H}^{(i+1)} = \phi \left(\hat{\mathbf{D}}^{-\frac{1}{2}} \hat{\mathbf{A}} \hat{\mathbf{D}}^{-\frac{1}{2}} \mathbf{H}^{(i)} \mathbf{W}^{(i+1)} \right), \tag{2}$$

where $\hat{\mathbf{A}} = \mathbf{A} + \mathbf{I}$ and $\mathbf{I}$ is an identity matrix, $\hat{\mathbf{D}}$ is the diagonal degree matrix of $\hat{\mathbf{A}}$, $\mathbf{W}^{(i+1)}$ denotes learnable parameters in the $(i + 1)$-th layer and ϕ is the activation function ReLU. Following previous works [21,23], we adopt GCNs to encode the neighborhood information of entities and take the output of the last GCN layer as the structural embeddings.

Visual Embedding. We choose ResNet-152 [9] pre-trained on the ImageNet [6] recognition task as the initial image classifier and fine-tune it with our datasets for EA. The fine-tuning details are given in Sect. 4.1. The fine-tuned model is used to extract image features. We feed each image $i \in I$, through a forward pass and take the output of last layer before logits as its feature vector. Then we project the feature into a low-dimensional space by a linear transformation to obtain visual embedding $\mathbf{e}_v$:

$$\mathbf{e}_v = \mathbf{W}_v \cdot \text{ResNet}(i) + \mathbf{b}_v, \tag{3}$$

where $\mathbf{W}_v$ is the projection matrix and $\mathbf{b}_v$ is the bias vector.

3.3 Alignment Learning and Inference

This section presents details about alignment learning and inference with an entity mask. We integrate G_1 and G_2 as one KG and learn both structural embeddings and visual embeddings of entities in E_1 and E_2 in a unified space. For notations, we let E_s and E_t denote the sets of source entities and the corresponding target entities, respectively, where $E_s \subseteq E_1$, $E_t \subseteq E_2$ and $|E_s| = |E_t|$. We rearrange the elements in both sets in order that the i-th entity in E_s corresponds to the i-th in E_t. We denote P as the set of all aligned pairs, i.e., $P = \{(e_1, e_2) \mid e_1 \equiv e_2, e_1 \in E_s, e_2 \in E_t\}$, and $\mathbf{M} \in \mathbb{R}^{|E_s|+|E_t|}$ as the entity mask used to filter potential noisy images. The training and test sets are obtained by splitting P with a ratio r.

Alignment Learning. Let $\hat{E}_s$ and $\hat{E}_t$ denote the source entities and target entities in the training set, respectively. For the structural modality, we compute a similarity matrix $\mathbf{Sim}^{(r)} = \langle \hat{\mathbf{E}}_s^{(r)}, \hat{\mathbf{E}}_t^{(r)} \rangle \in \mathbb{R}^{|\hat{E}_s| \times |\hat{E}_t|}$, where $\hat{\mathbf{E}}_s^{(r)}$ ($\hat{\mathbf{E}}_t^{(r)}$) represents the structural embeddings of entities in $\hat{E}_s$ ($\hat{E}_t$), and each entry $\mathbf{Sim}_{ij}^{(r)}$ corresponds to the cosine similarity between the i-th entity in $\hat{E}_s$ and j-th in $\hat{E}_t$. To better punish hard negatives and mitigate the hubness problem [5], we choose HAL loss [12] as the objective function and apply it to obtain the loss of structural modality $\mathcal{L}^{(r)}$ and train the structural embeddings:

$$\mathcal{L}^{(r)} = \frac{1}{N} \sum_{i=1}^{N} (\frac{1}{\alpha} \log(1 + \sum_{m \neq i} e^{\alpha \mathbf{Sim}_{mi}^{(r)}})$$

$$+ \frac{1}{\alpha} \log(1 + \sum_{n \neq i} e^{\alpha \mathbf{Sim}_{in}^{(r)}}) - \log(1 + \beta \mathbf{Sim}_{ii}^{(r)})), \tag{4}$$

where α, β are temperature scales and N is the batch size. As for the visual modality, we keep aligned pairs in which both the mask values of the source entity and the target entity are ones, i.e., neither of the two entity images is considered as potential noise. Thus we firstly obtain a new set of alignment pairs $P' = \{(e_1, e_2) \mid e_1 \equiv e_2, e_1 \in \hat{E}_s, e_2 \in \hat{E}_t, \mathbf{M}_{e_1} = 1, \mathbf{M}_{e_2} = 1\}$ with P and $\mathbf{M}$, then we determine from P' the new sets of source entities and target entities, denoted by $\tilde{E}_s$ and $\tilde{E}_t$, respectively. Likewise, we compute a cosine similarity matrix $\mathbf{Sim}^{(v)} = \langle \tilde{\mathbf{E}}_s^{(v)}, \tilde{\mathbf{E}}_t^{(v)} \rangle \in \mathbb{R}^{|\tilde{E}_s| \times |\tilde{E}_t|}$ for the visual modality and obtain the loss $\mathcal{L}^{(v)}$ by replacing $\mathbf{Sim}^{(r)}$ with $\mathbf{Sim}^{(v)}$ in Eq. (4).

Inference. Given source entity set $\bar{E}_s$ and target entity set $\bar{E}_t$ used for inference, we compute $\mathbf{Sim}^{(r)} = \langle \bar{\mathbf{E}}_s^{(r)}, \bar{\mathbf{E}}_t^{(r)} \rangle$ and $\mathbf{Sim}^{(v)} = \langle \bar{\mathbf{E}}_s^{(v)}, \bar{\mathbf{E}}_t^{(v)} \rangle$, where $\mathbf{Sim}^{(r)}$, $\mathbf{Sim}^{(v)} \in \mathbb{R}^{|\bar{E}_s| \times |\bar{E}_t|}$ are cosine similarity matrices for the structural and visual modalities, respectively. Then we combine them by a weighted addition and a position mask $\mathbf{pos} \in \mathbb{R}^{|\bar{E}_s| \times |\bar{E}_t|}$ to obtain the fused similarity matrix $\mathbf{Sim}$. Specifically, the similarity score between the i-th entity e_i in $\bar{E}_s$ and the j-th e_j in $\bar{E}_t$, i.e., the (i, j) entry of $\mathbf{Sim}$, is computed as:

$$\mathbf{Sim}_{ij} = \begin{cases} w \cdot \mathbf{Sim}_{ij}^{(r)} + (1 - w) \cdot \mathbf{Sim}_{ij}^{(v)} & \text{if } \mathbf{pos}_{ij} = 1 \\ \mathbf{Sim}_{ij}^{(r)} & \text{otherwise} \end{cases}, \tag{5}$$

where $w \in (0, 1)$ is a hyper-parameter to balance the two modalities, and $\mathbf{pos}_{ij}$ is used to determine if their visual similarity should be considered which is defined as:

$$\mathbf{pos}_{ij} = \begin{cases} 1 & \text{if } \mathbf{M}_{e_i} = 1 \text{ and } \mathbf{M}_{e_j} = 1 \\ 0 & \text{otherwise} \end{cases} . \tag{6}$$

Equation (5) and (6) illustrate the principal idea of fusing the two modalities: for source entity e_i and candidate target entity e_j, the in-between similarity is predicted from both aspects of knowledge only when their images are regarded as potentially useful; otherwise it is solely based on the structural similarity. After obtaining **Sim**, we further use cross-domain similarity local scaling (CSLS) [5] to post-process it. Then for $e_i \in \bar{E}_s$, we retrieve the similarity scores of the i-th row in **Sim**, rank them in a descending order, and take the top ranked entity as the match.

4 Experiments

4.1 Experimental Settings

Dataset. We build our dataset based on DBP15K [14], which contains three bilingual subsets: Chinese-English (ZH-EN), Japanese-English (JA-EN), and French-English (FR-EN). Each subset has 15K aligned entity pairs. DBpedia has provided links of thumbnails for many entities, however, it does not cover all of them. Statistics show that ca. 50–85% entities in DBP15K have images [11]. To solve the problem of data incompleteness, for (almost) every entity without an image in DBP15K, we crawl web images from Bing Images search by using its name as the query and take the most relevant as its final image. The statistics of image coverage are presented in Table 1. To retrieve entity types, we query the classes of each entity with *rdf:type* via a public SPARQL endpoint.[1] We also obtain the subsumption relationships between classes which are explicitly defined by the *rdfs:subClassOf* property in the DBpedia ontology. The original DBP15K, images of entities, and the ontology altogether constitute the dataset used in this paper.

Table 1. Statistics of image coverage.

	FR-EN		JA-EN		ZH-EN	
	FR	EN	JA	EN	ZH	EN
Image covered (by DBpedia)	13,858	14,174	12,739	13,741	15,910	14,125
Image covered (by web source)	5,794	5,816	7,011	6,035	3,421	5,441
All entities	19,661	19,993	19,814	19,780	19,388	19,572

---

[1] http://dbpedia.org/sparql.

Classification. We collect unique entities from all three subsets of DBP15K, filter those either without a type or an image, and use the remaining entities E' as indices to retrieve their images and labels. For each split of DBP15K, we fine-tune a classifier based on the pre-trained ResNet152 [9], and build the test and training data from $E_s \cup E_t$ and $E'\backslash(E_s \cup E_t)$, respectively. We adopt stochastic gradient descent (SGD) to update parameters of classifiers with a learning rate of 0.001 and a momentum of 0.9. We set the batch size to 32 and the number of epochs for training to 25. During test, we obtain top 5 predictions for each entity image, and set the conflict degrees $\lambda = 0$ and $\lambda = 1$ to calculate its mask value, respectively. Note that when $\lambda = 1$, no image will be filtered. Numbers of images in training and test sets for each split and the classification accuracies are reported in Table 2.

Table 2. Entity image classification results on the DBP15K dataset.

	Training images	Test images	Classes	Hits@1	Hits@5
FR-EN	54,117	29,479	76	0.513	0.828
JA-EN	54,799	29,498	82	0.509	0.821
ZH-EN	55,146	28,979	82	0.480	0.805

Alignment Settings. We employ a three-layer GCN (including the input layer) and set the dimensions of the input, hidden and output layers to 400, 400 and 200, respectively. The dimension of the visual embeddings is set to 200. We train our model for 1,000 epochs and adopt AdamW to update parameters. The learning rate is set to 5×10^{-4}, and the weight decay is 10^{-2}. When calculating losses, we set $\alpha = 5$, $\beta = 10$ for $\mathcal{L}^{(r)}$, and $\alpha = 15$, $\beta = 10$ for $\mathcal{L}^{(v)}$. We set $w = 0.5$ during inference. Following conventions, we use 30% of the aligned pairs for training and the remaining for evaluation, and choose H@1 (Hits@1), H@10 (Hits@10) and mean reciprocal rank (MRR) as the evaluation metrics. For the most relevant baseline, EVA, and our methods, SimpleEA and Masked-MMEAs, we conduct five experiments with different random seeds and present the averaged results along with their standard deviations $\underset{\pm Stds.}{Means}$.

4.2 Classification Performance and Analysis

We collect classification results of all three datasets in DBP15K and merge them for general analysis. For better understanding, we take nodes at the second level of the hierarchical class tree as base classes, and then use them to group fine-grained types, i.e., image labels used in the classification experiments. Note that we additionally treat *Person* and *Organization*, which are subclasses of *Agent*, as two base classes, as they are drastically different in both semantics and visual representations. A total of 17 base classes are identified and including their descendants, the total number of classes is 76 for FR-EN and 82 for JA-EN

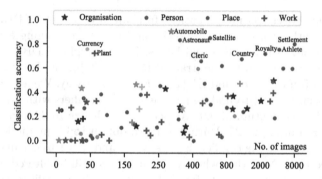

Fig. 4. The distribution of classification accuracy and the number of test images w.r.t. all classes. Each base class is denoted with a unique marker, and fine-grained types that share a common base class, such as *Royalty* and *Athlete*, are labeled with the same marker. The top-10 classes ranked by classification accuracy are explicitly annotated.

and ZH-EN (cf. Table 2). Among them, four base classes (together with their descendants) *Person, Organization, Work* and *Place* cover 92% of all test entities over three datasets. Figure 4 illustrates the distribution of accuracy and number of test (entity) images with respect to all classes.

We summarize the classification errors into two kinds: *(i)* the predicted class of an (entity) image and its true class are in the same group, i.e., one is the super class of the other or they are siblings or cousins, and *(ii)* the predicted class and the true class are disjoint. We find that without the first kind of errors, the accuracies of four base classes *Person, Place, Organization,* and *Work* rise from 0.53, 0.65, 0.36 and 0.31 to 0.91, 0.83, 0.51 and 0.52, respectively, which indicates that entities of *Person* or *Place* are more visually distinguishable, while entities of *Organization* and *Work* have less stable visual characteristics. By investigating the mispredictions, we identify several reasons that may explain the poor classification performance on many classes, which also provides insights into the quality of visual data used for MMEA. First, an image provided for an entity can be irrelevant to the entity itself. Second, the visual representations of entities of some classes are unstable. For example, entities of type *Single* or *Album* often have covers as their thumbnails, and these covers often vary widely from one to another depending on the design styles which are also easily misclassified into other classes like *Artist* and *Settlement*. Third, it is difficult to find accurate visual representations for conceptual entities, namely the entities referring to cognitive objects instead of physical objects. A typical type is *MusicGenre*, and its accuracy is as low as 0.03.

4.3   Alignment Results and Impacts of Entity Masks

To investigate the effectiveness of visual context, we develop a variant of our model denoted as SimpleEA by removing the visual components, and compare

Table 3. Entity alignment results on DBP15K. For fair comparison, the results of HMAN are from its variant that only uses training data in DBP15K as alignment signals, and the results of EVA are reproduced by only utilizing structural and visual context, as the setting of Masked-MMEA. λ_0 and λ_1 denote the corresponding results being obtained under $\lambda = 0$ and $\lambda = 1$, respectively. For EVA, SimpleEA, Masked-MMEA (λ_0) and Masked-MMEA (λ_1), Means$_{\pm \text{Stds.}}$ are shown.

Methods	FR-EN			JA-EN			ZH-EN		
	H@1	H@10	MRR	H@1	H@10	MRR	H@1	H@10	MRR
MTransE [4]	0.224	0.556	0.335	0.279	0.575	0.349	0.308	0.614	0.364
IPTransE [28]	0.333	0.685	0.451	0.367	0.693	0.474	0.406	0.735	0.516
JAPE [14]	0.324	0.667	0.430	0.363	0.685	0.476	0.412	0.745	0.490
GCN-Align [21]	0.373	0.745	0.532	0.399	0.745	0.546	0.413	0.744	0.549
SEA [13]	0.400	0.797	0.533	0.385	0.783	0.518	0.424	0.796	0.548
MuGNN [2]	0.495	0.870	0.621	0.501	0.857	0.621	0.494	**0.844**	0.611
HMAN [23]	0.543	0.867	–	0.565	**0.866**	–	0.537	0.834	–
AliNet [16]	0.552	0.852	0.657	0.549	0.831	0.645	0.539	0.826	0.628
MultiKE [24]	0.639	0.712	0.665	0.393	0.489	0.426	0.509	0.576	0.532
EVA [11]	<u>0.700</u> ±.005	<u>0.891</u> ±.005	<u>0.768</u> ±.004	<u>0.622</u> ±.004	<u>0.846</u> ±.008	<u>0.701</u> ±.005	<u>0.596</u> ±.007	<u>0.816</u> ±.008	<u>0.674</u> ±.007
SimpleEA	<u>0.504</u> ±.005	<u>0.826</u> ±.004	<u>0.616</u> ±.005	<u>0.505</u> ±.005	<u>0.797</u> ±.006	<u>0.608</u> ±.005	<u>0.479</u> ±.005	<u>0.772</u> ±.007	<u>0.582</u> ±.006
Masked-MMEA (λ_0)	<u>0.661</u> ±.007	<u>0.889</u> ±.004	<u>0.742</u> ±.006	<u>0.602</u> ±.004	<u>0.852</u> ±.006	<u>0.692</u> ±.004	<u>0.582</u> ±.006	<u>0.827</u> ±.008	<u>0.670</u> ±.007
Masked-MMEA (λ_1)	**0.712** ±.005	**0.901** ±.003	**0.779** ±.004	**0.627** ±.005	<u>0.858</u> ±.005	**0.711** ±.004	**0.612** ±.006	<u>0.837</u> ±.006	**0.693** ±.005

our full model Masked-MMEA with the variant and other baseline methods in Table 3.

From the results we can see that SimpleEA is comparable to other structure-based approaches, including MTransE, IPTransE, MuGNN, SEA and AliNet, and even surpasses two models using additional side information, JAPE and GCN-Align. Masked-MMEA (λ_1) slightly outperforms EVA, validating the effectiveness of our proposed approach for the EA part. More importantly, the results show that even under the strictest setting, our model Masked-MMEA (λ_0) gains 9.7–15.7% absolute improvement in Hits@1 over SimpleEA, and Masked-MMEA also outperforms MultiKE and HMAN, both of which leverage three kinds of side information. This demonstrates that the incorporation of the visual context can substantially improve the EA system.

To examine the effects of different entity masks, we choose different class conflict ratios: $\lambda \in \{0, 0.4, 0.67, 1\}$, in which $\lambda = 0$ corresponds to the strictest setting and $\lambda = 1$ is the no-masking setting where no entity images are filtered. Bigger λ indicates that more image pairs are involved and the visual context has more influence on alignment prediction during inference. Additionally, we design

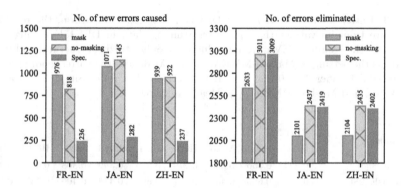

Fig. 5. Number of new errors caused (left) and number of errors eliminated (right) with the use of images on DBP15K. Different colors indicate the results from different settings.

Table 4. Alignment results under different settings. "No." denotes the number of entities with images.

Settings	FR-EN		JA-EN		ZH-EN	
	No.	Hits@1	No.	Hits@1	No.	Hits@1
Struct.	–	0.504	–	0.505	–	0.479
$\lambda = 0$	26,125	0.661	25,550	0.602	25,111	0.582
$\lambda = 0.4$	27,320	0.681	26,820	0.606	26,569	0.600
$\lambda = 0.67$	28,221	0.693	27,722	0.614	27,800	0.612
$\lambda = 1$	29,479	0.712	29,498	0.627	28,979	0.612
Spec.	28,681	0.771	28,387	0.701	28,079	0.683

a special mask based on the alignment result obtained when $\lambda = 1$. Specifically, we reset the mask value of an entity to be 0 if it is correctly matched by only structural similarity but is missed by a joint decision of the two modalities.

We conduct experiments under the above different settings and present the results in Table 4. As shown in Table 4, Hits@1 increases as λ is set larger and the no-masking setting outperforms the strictest setting by 2.5–5.1%. We consider that it is mainly attributed to the relatively low quality of visual data. As previously discussed, for some classes, the images of aligned entities which are visually similar, tend to be classified into irrelevant classes, thus being filtered as noises. However, this does not mean that filtering visual noises is useless, as we observe an average performance gain of 6.8% in Hits@1 with the special mask over the no-masking setting. We further analyze the change of errors after visual context is injected under three settings, i.e., the strictest (mask), the no-masking and the special (Spec.). As shown in Fig. 5, on all three datasets the use of special masks greatly reduces errors while retaining as much benefits as no-masking settings bring. The observations suggest such complexity of the problem that the model will not necessarily output better results with visual context considered. They also prove that the visual noise filtering is beneficial to the multimodal entity alignment. The key challenge lies in locating real visual noises. In the next part, we analyze and identity a few cases where entity images tend to induce misalignment based on experimental results.

4.4 How Visual Context Impacts Multimodal Entity Alignment

The incorporation of entity images can reduce thousands of errors; but on the other hand, it also brings in much noise leading to many new mismatches, as illustrated in Fig. 5. Overall, it improves the alignment performance. For the positive impact, we find that visual context is particularly helpful when structural information is insufficient to make correct alignment predictions. This finding is supported by the observation that among 3,011 newly aligned entity pairs under the no-masking setting on FR-EN, 78% of them have a summed degree below the mean value of the summed degrees of all aligned entity pairs (i.e., long-tailed entities), and a lower degree of an entity indicates less structural information available to learn reliable structural embeddings.

To gain some insights into the negative impact of injecting visual context, we take results of FR-EN as an example and collect new errors occurred under the no-masking setting. These new errors shed light on true visual noises that should be filtered. Among the 818 errors on FR-EN, 139 source entities have mask values of 0s, meaning that the top 1 predicted class of their image by the classifier is disjoint with their actual (entity) type, and that 139 errors could be reduced if these images are filtered. The remaining 679 errors are mostly about source entities with mask values of 1s, which we divide into three categories for detailed analysis: *(i)* The first category contains 436 source entities where both the mask values of their aligned counterparts and their predicted matches are 1s, and 80% of the mismatches are between entities of same or very close types, such as siblings, with *Person* and *Place* as two largest base classes. These mismatches are quite difficult to address because these entity types show relatively stable visual characteristics and the corresponding entity images are less visually distinguishable from those of the same types. *(ii)* The second category includes 154 source entities where one of the mask values of their aligned counterparts and their predicted matches is 0, indicating that inappropriate or inconsistent images induced mismatches and these errors could be avoided when the noises are excluded. *(iii)* Errors of the last category, making up about 9% of the total errors, are about source entities mismatched to entities without images, which means these images are not as useful as structural information in multimodal entity alignment.

5   Conclusions

This paper investigated impacts of incorporating visual context (entity images) for multimodal entity alignment. We proposed to selectively use entity images both in alignment learning and inference phase by filtering potential noises, which were identified using image classification techniques and the ontology of KGs, and conducted extensive experiments to examine our approach. We found that visual context overall is beneficial and that while challenging, filtering noises can further boost performance. At last we identified a few cases where images were noises to EA. We experimentally prove that selectively masking potential visual noises brings the most benefits to EA even though the results largely depend on the quality of visual data. It is also worth noting that by probing the impacts of visual context in MMEA, our work examines the quality of entity images in some multimodal KGs, which has not been inspected by existing studies. For future work, we will consider adapting the proposed masking technique to apply to other multimodal scenarios (e.g., multimodal entity linking). Besides, we will seek to devise methods for the visual noise detection based on automatic learning.

Acknowledgements. This work was supported by the National Key Research and Development Program of China with Grant No. 2021ZD0113202; the National Natural Science Foundation of China with Grant No. 61906037; the Fundamental Research Funds for the Central Universities.

References

1. Bordes, A., Usunier, N., García-Durán, A., Weston, J., Yakhnenko, O.: Translating embeddings for modeling multi-relational data. In: Proceedings of the 26th International Conference on Neural Information Processing Systems, pp. 2787–2795 (2013)
2. Cao, Y., Liu, Z., Li, C., Liu, Z., Li, J.Z., Chua, T.S.: Multi-channel graph neural network for entity alignment. In: Proceedings of the 57th Annual Meeting of the Association for Computational Linguistics, pp. 1452–1461 (2019)
3. Chen, L., Li, Z., Wang, Y., Xu, T., Wang, Z., Chen, E.: MMEA: Entity alignment for multi-modal knowledge graph. In: International Conference on Knowledge Science, Engineering and Management, pp. 134–147 (2020)
4. Chen, M., Tian, Y., Yang, M., Zaniolo, C.: Multilingual knowledge graph embeddings for cross-lingual knowledge alignment. In: Proceedings of the 26th International Joint Conference on Artificial Intelligence, pp. 1511–1517 (2017)
5. Conneau, A., Lample, G., Ranzato, M., Denoyer, L., J'egou, H.: Word translation without parallel data. In: International Conference on Learning Representation (2018)
6. Deng, J., et al.: ImageNet: a large-scale hierarchical image database. In: Proceedings of the IEEE Conference on Computer Vision and Pattern Recognition, pp. 248–255 (2009)
7. Guo, H., Tang, J., Zeng, W., Zhao, X., Liu, L.: Multi-modal entity alignment in hyperbolic space. Neurocomputing **461**, 598–607 (2021)
8. Hao, J., Chen, M., Yu, W., Sun, Y., Wang, W.: Universal representation learning of knowledge bases by jointly embedding instances and ontological concepts. In: Proceedings of the 25th ACM SIGKDD International Conference on Knowledge Discovery & Data Mining, pp. 1709–1719 (2019)

9. He, K., Zhang, X., Ren, S., Sun, J.: Deep residual learning for image recognition. In: Proceedings of the IEEE Conference on Computer Vision and Pattern Recognition, pp. 770–778 (2016)

10. Lehmann, J., et al.: DBpedia-a large-scale, multilingual knowledge base extracted from wikipedia. Semant. Web **6**(2), 167–195 (2015)

11. Liu, F., Chen, M., Roth, D., Collier, N.: Visual pivoting for (unsupervised) entity alignment. In: Proceedings of the AAAI Conference on Artificial Intelligence, pp. 4257–4266 (2021)

12. Liu, F., Ye, R., Wang, X., Li, S.: HAL: improved text-image matching by mitigating visual semantic hubs. In: Proceedings of the AAAI Conference on Artificial Intelligence, pp. 11563–11571 (2020)

13. Pei, S., Yu, L., Hoehndorf, R., Zhang, X.: Semi-supervised entity alignment via knowledge graph embedding with awareness of degree difference. In: The World Wide Web Conference, pp. 3130–3136 (2019)

14. Sun, Z., Hu, W., Li, C.: Cross-lingual entity alignment via joint attribute-preserving embedding. In: International Semantic Web Conference, pp. 628–644 (2017)

15. Sun, Z., Hu, W., Zhang, Q., Qu, Y.: Bootstrapping entity alignment with knowledge graph embedding. In: Proceedings of the 27th International Joint Conference on Artificial Intelligence, pp. 4396–4402 (2018)

16. Sun, Z., et al.: Knowledge graph alignment network with gated multi-hop neighborhood aggregation. In: Proceedings of the AAAI Conference on Artificial Intelligence, pp. 222–229 (2020)

17. Sun, Z., et al.: A benchmarking study of embedding-based entity alignment for knowledge graphs. Proc. VLDB Endow. **13**, 2326–2340 (2020)

18. Trisedya, B.D., Qi, J., Zhang, R.: Entity alignment between knowledge graphs using attribute embeddings. In: Proceedings of the AAAI Conference on Artificial Intelligence, pp. 297–304 (2019)

19. Vrandečić, D., Krötzsch, M.: Wikidata: a free collaborative knowledgebase. Commun. ACM **57**(10), 78–85 (2014)

20. Wang, W., Feng, D., Li, B., Tian, J.: ATextCNN model: a new multi-classification method for police situation (2020)

21. Wang, Z., Lv, Q., Lan, X., Zhang, Y.: Cross-lingual knowledge graph alignment via graph convolutional networks. In: Proceedings of the 2018 Conference on Empirical Methods in Natural Language Processing, pp. 349–357 (2018)

22. Xiang, Y., Zhang, Z., Chen, J., Chen, X., Lin, Z., Zheng, Y.: OntoEA: ontology-guided entity alignment via joint knowledge graph embedding. In: Findings of the Association for Computational Linguistics, ACL-IJCNLP 2021, pp. 1117–1128 (2021)

23. Yang, H.W., Zou, Y., Shi, P., Lu, W., Lin, J.J., Sun, X.: Aligning cross-lingual entities with multi-aspect information. In: Proceedings of the 2019 Conference on Empirical Methods in Natural Language Processing, pp. 4430–4440 (2019)

24. Zhang, Q., Sun, Z., Hu, W., Chen, M., Guo, L., Qu, Y.: Multi-view knowledge graph embedding for entity alignment. In: Proceedings of the 28th International Joint Conference on Artificial Intelligence, pp. 5429–5435 (2019)

25. Zhang, Y., Li, B., Gao, H.: Fine-grained evaluation of knowledge graph embedding model in multiple types of downstream tasks. Big Data Res. **25**, 100218 (2021)

26. Zhang, Z., et al.: An industry evaluation of embedding-based entity alignment. In: Proceedings of the 28th International Conference on Computational Linguistics, Barcelona, Spain, pp. 179–189 (2020)

270 Y. Shi et al.

27. Zhao, X., Zeng, W., Tang, J., Li, X., Luo, M., Zheng, Q.: Toward entity alignment
in the open world: an unsupervised approach with confidence modeling. Data Sci.
Eng. **7**, 1–14 (2022)
28. Zhu, H., Xie, R., Liu, Z., Sun, M.: Iterative entity alignment via joint knowledge
embeddings. In: Proceedings of the 26th International Joint Conference on Artificial Intelligence, pp. 4258–4264 (2017)

Incorporating Prior Type Information for Few-Shot Knowledge Graph Completion

Siyu Yao[1,2(✉)], Tianzhe Zhao[1,2], Fangzhi Xu[1,2], and Jun Liu[2,3]

[1] School of Computer Science and Technology, Xi'an Jiaotong University,
Xi'an, China
{cheryl,Leo981106}@stu.xjtu.edu.cn
[2] Shaanxi Provincial Key Laboratory of Big Data Knowledge Engineering,
Xi'an, China
liukeen@xjtu.edu.cn
[3] National Engineering Lab for Big Data Analytics, Xi'an, China

Abstract. Few-shot knowledge graph completion aims to infer unknown triple facts with only a small number of reference triples. Existing methods have shown a strong capability on this problem by combining knowledge representation learning and meta learning. They ignore prior knowledge in the few-shot scenario, while prior knowledge can boost useful information to handle the challenges brought by limited referenced instances. To address the above issue, we propose a few-shot knowledge graph completion model PiTI-Fs, with entity type information as prior knowledge in a two-module learning framework. In the prior knowledge learning module, we propose to extract a metagraph for capturing prior type information by entity clustering where entities in the same cluster are considered to have the same attribute. We pre-train the metagraph to learn the prior knowledge features and fuse them into the embeddings of entities. In the meta learning module, we introduce a transformer-based relation learner to model the interactions within reference entity pairs and implement an optimization-based meta learning paradigm to train our model. Our method outperforms most of baseline models for the few-shot knowledge graph completion task. The experimental results demonstrate the effectiveness of the proposed modules.

Keywords: Few-shot · Knowledge graph completion · Meta learning

1 Introduction

A knowledge graph (KG) structured as a directed multi-relational graph is composed of a large number of factual triples in the form (h, r, t). Real-world KGs, such as Freebase [1], Wikidata [24] and NELL [3] contain huge amounts of triples widely implemented for question answering [10,28], recommendation system [31], information retrieval [7]. However, most KGs still suffer the incompleteness issue. It arouses the interest of research in automatically predicting missing triples with the reference of existing ones, which is usually formulated as knowledge graph completion (KGC) [9,14,17].

ⓒ The Author(s), under exclusive license to Springer Nature Switzerland AG 2023
B. Li et al. (Eds.): APWeb-WAIM 2022, LNCS 13422, pp. 271–285, 2023.
https://doi.org/10.1007/978-3-031-25198-6_21

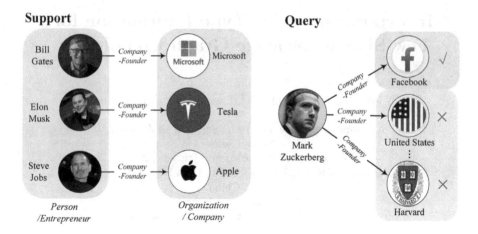

Fig. 1. An example of 3-shot knowledge graph completion.

Extensive research efforts have proven that knowledge graph embedding (KGE) is promising. KGE aims to represent entities and relations in a latent and low-dimensional embedding space. For example, TransE [2] introduces Euclidean distance to measure similarities between tail entities and head entities with a relation translation operation. Nevertheless, existing KGE models require sufficient training examples, while few-shot problems are quite common in current KGs. For example, about 10% of relations in Wikidata have no more than 10 triples [4]. Besides, in the real world, KGs are dynamically evolving. When new relations are added to a KG for the first time, few-shot scenarios usually occur that the number of triples associated with the added relation is quite small. Under such a circumstance, KGE methods are not effective due to the lack of training examples.

To cope with this problem, recent researches attempt to fit the KGE methods into the few-shot scenario. These methods apply KGE in the meta learning training paradigm. GMatching [27], FSRL [30] and FAAN [20] conduct few-shot KGC by incorporating the structural neighbor information into the metric matching. MetaR [4] follows MAML [8] to transfer relation by a gradient descent update. Despite the great success, they only attend to entity embeddings derived from the background graph and fail to explore the abundant prior knowledge (e.g., entity type attributes).

According to the observation from KGs, we capture a phenomenon that head and tail entities with the same relation tend to be grouped into two type attributes respectively.

As shown in Fig. 1, three supported factual triples with the relation *Company-Founder* form a 3-shot KGC example. Head entities, i.e. *Bill Gate*, *Elon Musk* and *Steven Jobs*, share the same attribute of *Person/Entrepreneur*. Similarly, tail entities in the example can be categorized as *Organization/Company*. Intuitively, type attributes play an important role in predicting the true tail entity *Facebook* for the missing fact (*Mark Zuckberg, Company-Founder*, ?).

Inspired by the above phenomenon, we propose to incorporate **Prior Type Information** for **Few-shot** KGC (PiTI-Fs), which is a two-module learning framework. In the prior knowledge learning module, motivated by Chung and Whang [5], we first cluster entities in the background graph (BG) based on the same affinity metric to form a metagraph (MG). Entities in the same cluster share the same type attributes. PiTI-Fs pre-trains the given BG and its corresponding MG to learn the entity embeddings and type embedding respectively, and then applies an aggregation function to represent entities. In the meta learning module, we introduce an optimization-based meta learning paradigm inspired by MetaR [4] to query missing facts in the few-shot scenario. Unlike MetaR which treats different observed entity pairs equally, we introduce a transformer-based relation encoder to model interactions between referenced triples for better few-shot relation representations.

The main contributions of the paper can be listed as follows:

(1) A unified method PiTI-Fs is proposed to capture prior type attributes to enrich entity representations. It introduces a transformer-based relation learner to adapt reference representations to different queries.
(2) The experimental results validate that our proposed model outperforms the existing baseline models in most cases. Furthermore, comparison experiments and variants analysis prove the effectiveness of each module in our method.

2   Related Work

2.1   Traditional Knowledge Graph Completion

The main methods of traditional KGC task are knowledge graph embedding-based models (KGE-based models), which represent entities and relations in low-dimension vector space. These methods can be divided into two categories:

Translation-Based Models. In translation-based models, relations are usually considered as transition or mapping operations from head entities to tail entities. TransE [2] is a typical work that interprets relations as translation operations between entity pairs in the embedding space. TransH [26] and TransR [16] are two extensions of TransE. TransH projects an entity embedding to relation-specific hyper-planes for dealing with 1-to-N, N-to-1, and N-to-N situations. Our of the same motivation, TransR projects entity embeddings by relation-specific matrices. RotatE [21] defines relations as rotations in complex vector spaces.

Semantic-Based Models. Semantic-based models derive the plausibility of triples via matching functions based on the latent semantics of the entire triple. DistMult [29], ComplEx [22], ConvE [6], and ConvKB [18] are excellent works of them. DistMult implements a bi-linear transformation to each components in the triple and applies the latent semantic similarity to score plausibility. ComplEx extends DistMult and uses vectors with complex values to represent entities and relations trying to exploit semantics in different ways. ConvE and ConvKB attempt to capture more expressive semantics with convolution operations.

These KGE-models heavily rely on the sufficient training triples, thus their performances are limited in the few-shot setting. Some recent advances of the knowledge graph embedding make full use of the strong representation ability of pre-trained language models [25,32]. As one of the representatives, StAR [25] has mentioned its applicability to the few-shot scenario, but its performance highly depends on the pre-trained language model whose training is very expensive.

In short, existing methods either lack the generalization capability to the few-shot settings, or cost intensive resources. Thus, we propose to tackle both issues in a unified framework for few-shot KGC.

2.2   Few-Shot Knowledge Graph Completion

Compared to the traditional KGC task, the few-shot KGC task faces the challenge that only quite a few triples can be referred to. Existing few-shot KGC methods can be grouped into two categories:

Metric-Based Models. GMatching [27] represents entities by leveraging neighbor node embeddings, and it introduces a matching processor with a memory mechanism to evaluate the similarity between query triples and reference triples. FSRL [30] and FAAN [20] share a similar idea of GMatching using the memory-based metric-match processor, but they respectively introduce a relation-aware attention mechanism and an adaptive attention mechanism to obtain better representations of entities in BG. Unlike the above three models relying on the BG, MetaP [11] directly implements convolution operations to capture the pattern of each triple and uses learned pattern representations as a matching metric.

Optimization-Based Models. Inspired by MAML [8], which is a typical few-shot learning framework, MetaR [4] transfers the relation meta information from few reference triples to incomplete ones through a fast gradient descent update procedure. Sharing a similar idea of MetaR and TransH, Niu [19] proposed MTransH to deal with the complex relations in the few-shot KGC task.

However, the above methods fail to exploit the huge potential of the type information existing in the entities. In this work, we stress much importance on the prior type information to benefit the few-shot KGC model.

3   Preliminary

In this section, we give formal definitions of the knowledge graph, the metagraph, the few-shot knowledge graph completion task, and the corresponding few-shot learning setting.

Definition 1. Knowledge Graph. A knowledge graph $\mathcal{G}$ is a multi-relational graph represented as a set of factual triples $\mathcal{G} = \{\mathcal{E}, \mathcal{R}, \mathcal{F}\}$, where $\mathcal{E}$, $\mathcal{R}$, and $\mathcal{F}$

denote the entity set, relation set and fact set of $\mathcal{G}$, respectively. $\mathcal{F} = \{(h, r, t) \in \mathcal{E} \times \mathcal{R} \times \mathcal{E}\}$ is a set of factual triples. In each triple, h, t denote the head entity and tail entity, and r denotes the relation.

Definition 2. Metagraph. Given a knowledge graph $\mathcal{G} = \{\mathcal{E}, \mathcal{R}, \mathcal{F}\}$, the extracted corresponding metagraph $\mathcal{MG}$ is also a knowledge graph as $\mathcal{MG} = \{\mathcal{C}, \tilde{\mathcal{R}}, \tilde{\mathcal{F}}\}$, where $\mathcal{C} \subseteq 2^{\mathcal{E}}$, $\tilde{\mathcal{R}} \subseteq \mathcal{R}$, $\tilde{\mathcal{F}} = \{(c_i, \tilde{r}, c_j) \in \mathcal{C} \times \tilde{\mathcal{R}} \times \mathcal{C}\}$. $\mathcal{C}$ is derived from $\mathcal{E}$ by a clustering algorithm. Entities grouped to the same cluster share the same type attributes.

Definition 3. Few-shot Knowledge Graph Completion. Few-shot knowledge graph completion is a specialized knowledge graph completion task in the few-shot scenario. Given a relation $r \in \mathcal{R}$ and a handle of corresponding factual triples $\mathcal{S}_r = \{(h_i, r, t_i) | h_i, t_i \in \mathcal{E}\}$, a few-shot knowledge graph completion $\mathcal{T}$ is to predict missing tail entities of incomplete triples $\mathcal{Q}_r = \{(h_j, r, ?)\}$. We denote the support set and query set as $\mathcal{S}_r, \mathcal{Q}_r$, respectively. And $|\mathcal{S}_r| = K$ suggests a K-shot knowledge graph completion task.

We follow the same few-shot settings proposed by Xiong et al. [27] and Chen et al. [4]. The training phase is based on a set of sampled tasks $\mathcal{T}_{train} = \{\mathcal{T}_i\}_{i=1}^M$, where each task formulated as $\mathcal{T}_i = \{\mathcal{S}_i, \mathcal{Q}_i\}$ is associated to an individual few-shot KGC task with its own support and query set. The testing phase is correspondingly composed of new tasks $\mathcal{T}_{test} = \{\mathcal{T}_j\}_{j=1}^N$. Relations in testing tasks are not seen during training. For relations with few triples to form training or testing tasks, we denote them as few-shot relations. We also assume that the background graph is a set of triples with high-frequency relations.

4 Method

In this section, we introduce our proposed model PiTI-Fs as shown in Fig. 2. The framework consists of two modules: (1) Prior knowledge learning module (Sect. 4.1). In this module, PiTI-Fs clusters entities in the given BG based on the same affinity metric to form a MG. PiTI-Fs pre-trains BG and its corresponding MG to learn the entity embeddings and type embeddings respectively, then applies an aggregation function to represent entities, as entities in the same cluster share the same type attributes. (2) Meta learning module (Sect. 4.2). In this module, PiTI-Fs introduces a transformer-based relation encoder to model interactions between referenced triples for better few-shot relation representations, and further queries the incomplete triples for few-shot KGC tasks.

4.1 Prior Knowledge Learning Module

In this module, we firstly propose to extract MG to capture prior type information via clustering the entities, and those in the same cluster are considered to have the same attribute.

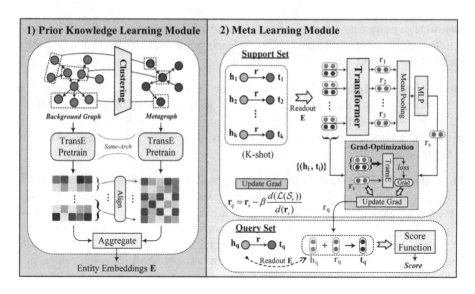

Fig. 2. The framework of PiTI-Fs.

Following the key idea from Chung and Whang [5], we extract a MG from the given BG. Hyperedges are introduced to connect entities which share the same head entity with the same relation or share the same tail entity with the same relation. Thus, BG can be converted into a hypergraph so that structurally similar entities are connected via hyperedges. The affinity of the two entities, i.e., e_i and e_j, is expected to be high when the set $\mathcal{N}$ of the hyperedges which contain them both is larger or the entity numbers d_n within these hyperedges are small. The affinity of e_i and e_j is defined as follows:

$$a_{ij} = \sum_{n \in \mathcal{N}} \frac{1}{d_n^2}, \tag{1}$$

and with the entity-level hypergraph normalized cut, the affinity is further refined as:

$$\hat{a}_{ij} = \frac{a_{ij}}{\sum_{k=0}^{|\mathcal{E}|} a_{ik}} + \frac{a_{ij}}{\sum_{k=0}^{|\mathcal{E}|} a_{kj}}. \tag{2}$$

Based on the refined affinity, an agglomeration hierarchical clustering with the average linkage strategy is applied to group similar entities into $\lfloor mp \rfloor$ clusters, where m is the number of entities, and $0 < p < 1$. Each cluster stands for a latent type attribute, and entities in the same cluster share the same type representations.

Each cluster is treated as an individual entity in MG. A triple $(c_i, \tilde{r}, c_j)$ will be added into MG based on the statistic connections between entities in c_i and c_j.

With MG extracted from the given BG, we pre-train BG and MG respectively using a well-known distance-based model TransE [2]. We follow the same setting based on the score function of TransE, i.e., $E(h, r, t) = ||\mathbf{h} + \mathbf{r} - \mathbf{t}||$. Therefore, each entity embedding consists of two parts: $\mathbf{e}_o$ and $\mathbf{e}_{type}$, where $\mathbf{e}_o$ denotes the entity embeddings from BG and $\mathbf{e}_{type}$ from MG.

Furthermore, we add two pre-trained vectors to be the initial representation of entities:

$$\mathbf{e} = \mathbf{e}_o + \mathbf{e}_{type}. \tag{3}$$

Specifically, for a triple (h, r, t), representations of h and t are formulated as:

$$\mathbf{h} = \mathbf{h}_o + \mathbf{h}_{type}, \tag{4}$$

$$\mathbf{t} = \mathbf{t}_o + \mathbf{t}_{type}. \tag{5}$$

We also try several other aggregation strategies to fuse the two embeddings, and we will discuss the performance of them in the next section.

4.2 Meta Learning Module

In this module, we aim to derive few-shot relation representations and infer the missing triples in an optimization-based meta learning paradigm motivated by MetaR [4]. Initially, unlike MetaR that equally treats all the referenced entity pairs in the support set for embedding relations, we introduce a transformer-based encoder to generalize the representation for few-shot relations considering the different importance of different entity pairs.

In specific, we first concatenate embeddings of the head entity and tail entity to form representations for each entity pair:

$$\mathbf{p}_i = [\mathbf{h}_i : \mathbf{t}_i], \tag{6}$$

where $h_i \in \mathbb{R}^d, t_i \in \mathbb{R}^d$ are obtained from the prior knowledge learning module denoting the embeddings of head entity and tail entity respectively. $[x : y]$ denotes the concatenation of vector x and y.

Then we use a transformer-block [23] to integrate the interactions between entity pairs and implement a multi-layer perception to get the general relation representation r_s after an average-pooling operation followed by the transformer encoder:

$$\mathbf{p}_i^l = \mathtt{Transformer}(\mathbf{p}_i^{l-1}), l = 1, 2, \cdots, L, \tag{7}$$

$$\mathbf{r}_s = \mathtt{MLP}(\mathtt{Pool}[\mathbf{p}_1{}^L, \mathbf{p}_2{}^L \cdots, \mathbf{p}_K{}^L]), \tag{8}$$

where $\mathbf{p}_i^l$ denotes the hidden state of after the l-th layer transformer, $\mathtt{Pool}$ is to average all entity pair representations encoded by the whole transformer. $\mathtt{MLP}$ is composed with several fully connected neural networks.

At the next step, similar to MetaR, we implement a fast update on the relation representation r_S and transfer it to the corresponding query set. We measure the scores of the referenced triples in the support set by the key idea of TransE:

$$E(h_i, r, t_i) = ||\mathbf{h}_i + \mathbf{r}_s - \mathbf{t}_i||, \tag{9}$$

Table 1. Statistics of the benchmark dataset. #Entity denotes the number of unique entities and #Relation denotes the number of all relations. #Task-Train/#Task-Valid/#Task-Test respectively denote the numbers of training, validation, and test relations.

Dataset	#Entity	#Relation	#Triple	#Task-Train	#Task-Valid	#Task-Test
NELL-One	68,545	358	181,109	51	5	11

where $|| \cdot ||$ represents the L2 norm of a vector, and we follow the similar margin loss in the support set which is defined as:

$$\mathcal{L}(\mathcal{S}_r) = \sum_{(h_i,r,t_i)\in\mathcal{S}_r} [\gamma + E(h_i,r,t_i) - E(h_i,r,t_i')]_+, \tag{10}$$

where $[x]_+ = max(0,x)$, and γ is a margin hyperparameter. $E(h_i,r,t_i')$ is the score of the negative triples by negative sampling the tail entities of $(h_i,r,t_i) \in \mathcal{S}_r$.

The transferred relation representation is next updated by the gradient descent adaptation:

$$\mathbf{r}_q = \mathbf{r}_s - \beta\frac{d(\mathcal{L}(\mathcal{S}_r))}{d(\mathbf{r}_s)}, \tag{11}$$

where β indicates the step size for gradient descent.

With the acquisition of the updated relation representation $\mathbf{r}_q$, we score the triples in the query set in the same way:

$$E(h_j,r,t_j) = ||\mathbf{h}_j + \mathbf{r}_q - \mathbf{t}_j||, \tag{12}$$

$$\mathcal{L}(\mathcal{Q}_r) = \sum_{(h_j,r,t_j)\in\mathcal{Q}_r} [\gamma + E(h_j,r,t_j) - E(h_j,r,t_j')]_+. \tag{13}$$

And our training objective is to minimize the loss summed of query loss for all samples testing tasks:

$$\mathcal{L} = \sum_{(\mathcal{S}_r,\mathcal{Q}_r)\in\mathcal{T}_{train}} \mathcal{L}(\mathcal{Q}_r). \tag{14}$$

5   Experiments

In this section, extensive experiments were conducted to compare our model with baseline methods in the few-shot KGC task. Comparison study and analysis are followed to demonstrate the effectiveness of the proposed method.

5.1   Dataset and Evaluation Metrics

NELL-One [27] is a benchmark dataset derived from NELL [3]. Its statistics are shown in Table 1.

Two traditional metrics are used to evaluate different models on the benchmark, i.e., MRR and Hits@N, where MRR is the mean reciprocal rank, and Hits@N is the proportion of correct entities ranked within top-N in KGC.

Table 2. Results of the few-shot KGC tasks on NELL-One. † Resulting numbers are reported from our re-implementation, and others are taken from the original papers. The **Bold** numbers are state-of-the-art performances while the <u>underline</u> numbers are the second best results of all.

NELL-One	MRR		Hits@10		Hits@5		Hits@1	
	1-shot	5-shot	1-shot	5-shot	1-shot	5-shot	1-shot	5-shot
GMatching [27]	.185	.176	.313	.294	.260	.233	.119	.113
FSRL [30]	.211	.153	.317	.319	.247	.212	.156	.073
MetaP [11]	<u>.232</u>	–	.330	–	.281	–	<u>.174</u>	–
FAAN [20] †	.174	**.279**	.332	<u>.419</u>	.249	**.362**	.099	**.200**
MetaR [4] †	.213	.231	<u>.335</u>	.358	<u>.283</u>	.313	.149	.164
PiTI-Fs	**.245**	<u>.262</u>	**.388**	**.427**	**.322**	<u>.351</u>	**.179**	<u>.179</u>

5.2 Baseline Models

We compare our model with several recent few-shot KGC models. GMatching [27] integrates neighbor information to represent entities and queries incomplete triples by a matching network. FSRL [30] introduces a relation-aware attention to encode neighbors and implements a LSTM-based aggregation network to model references. FAAN [20], which is the state-of-the-art method, applies a dynamic attention module to strengthen the representations of entities. MetaP [11] explores the pattern of triples as query metric which are encoded by convolution operations. MetaR [4] transfers the shared relation information and adapts a gradient-based fast training strategy.

5.3 Implementation Details

In the prior knowledge learning module, we set p to 0.7 for the MG construction. Sizes of pre-trained embeddings of both the BG and MG are set to 100. In the meta learning module, we set the number of transformer layers to one and the number of transformer heads to eight. We set $\gamma = 1$, $\beta = 5$. The number of positive and negative triples in a query set is three. During training, we apply mini-batch gradient descent with the batch size of 512. We use Adam [12] with the initial learning rate as 0.0005.

5.4 Main Results in Few-Shot Knowledge Graph Completion

Table 2 shows the results of two few-shot KGC tasks, i.e. 1-shot and 5-shot, on NELL-One.

It is observed that our method is able to achieve state-of-the-art performances on some tasks while competitive performances on all the rest. Specially, the results of our method outperform MetaR which proves the effectiveness of incorporating prior type information. The improvement is especially significant in terms of Hits@10. And in 1-shot situation, our method outperforms all other

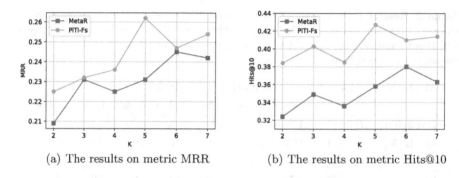

(a) The results on metric MRR   (b) The results on metric Hits@10

Fig. 3. Impact of different few numbers.

methods. Furthermore, compared with FAAN, our model is still competitive for 5-shot with much less training time and even better for 1-shot.

5.5 Impact of Different Few-Shot Size

We conduct experiments to analyze the impact of the few-shot size K. Figure 3 reports the performance of our model and the main baseline model MetaR on NELL in different settings of K. According to the figure:

(1) Our model consistently outperforms MetaR under different few-shot size K, demonstrating the effectiveness of the proposed model for few-shot KGC.
(2) We observe that referring more existing triples will not always achieve improvements in the few-shot scenario. It shows the performance under the few-shot scenario depends heavily on the quality of sampled instances. However, PiTI-Fs shows the great potential to bridge such gaps and brings more robustness compared with MetaR.

5.6 Comparison over Different Relations

We conduct experiments to evaluate the performance of our model on different relations in NELL-one under the 5-shot scenario. Table 3 shows the comparison results between our model and the advanced optimization-based baseline MetaR. We observe the overall trend that it is more difficult to make precise predictions when the number of candidate entities increases. Our model has better performances in most cases especially for those relations with lots of candidates such as *geopliticalLocationOfPerson*, while our model achieves 57.6% and 60.0% improvements over MetaR by the metric MRR and Hit@10 respectively.

5.7 Analysis for Impact of Components in PiTI-Fs

Impact of the Initial Embeddings for Meta Learning Module. To further analyze the impact of different representations learned from the prior knowledge learning module, we design three comparison situations, including a) a

Table 3. Results of MetaR and PiTI-Fs for each relation in NELL-One testing data. #Candidates denotes the size of candidate entity set. The **Bold** numbers are the better results regarding each relation.

Relations	# Candidates	MRR		Hit@10	
		MetaR	PiTI-Fs	MetaR	PiTI-Fs
sportsGameSport	123	.971	**.972**	.971	**.971**
athleteInjuredBodypart	299	.275	**.324**	.312	**.391**
animalSuchAsInvertebrate	786	**.268**	.264	.546	**.554**
automobilemakerDealersInCountry	1,084	.533	**.580**	.813	**.923**
SportSchoolIncountry	2,100	**.526**	.517	**.663**	.622
politicianEndorsesPolitican	2,160	.212	**.233**	.262	**.357**
agriculturalProductFromCountry	2,222	.159	**.170**	**.385**	.363
producedBy	3,174	**.303**	.292	.543	**.577**
automobilemakerDealersInCity	5,716	.053	**.075**	**.156**	.133
teamCoach	10,569	.075	**.121**	.176	**.223**
geopoliticalLocationOfPerson	11,618	.118	**.186**	.145	**.232**

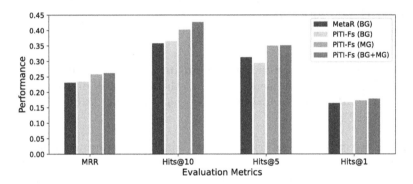

Fig. 4. Impact of the initial embeddings for meta learning module. (Color figure online)

strong optimization-based baseline MetaR, b) applying the pre-trained embeddings from BG, and c) applying the pre-trained embeddings from MG. The results are shown in Fig. 4:

(1) Our method incorporating type attributes (marked as the red bar) performs significantly better than initializing with only one pre-trained representations (marked as the orange bar and the green bar). This demonstrates that type attribute information is of great benefit in the few-shot scenario.
(2) Compared to the situation with only the pre-trained embeddings from BG (marked as the orange bar), straightly applying the MG pre-trained embeddings (marked as the green bar) has better performance by all metrics. This matches the conclusion derived from Chung and Whang [5].

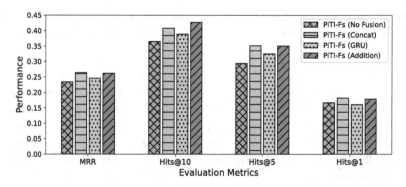

Fig. 5. Impact of aggregation strategy for fusing type attribute representations. (Color figure online)

(3) Compared to MetaR, our PiTI-Fs has better performance under the fair situation that they all pre-trained on the BG only (marked as the blue bar and the orange bar respectively). It shows that our proposed transformer-based relation learner has a stable capacity to generalize the few-shot relation representations from the given instances.

Impact of Aggregation Strategy for Fusing Type Attribute Representations. We employ experiments to explore the influence of different strategies fusing type attribute representations in the prior knowledge learning module. As shown in Fig. 5:

(1) Three selected aggregation ways can bring improvements by all evaluation metrics compared to our method without fusing type representations (marked as orange crossed bar). This further illustrates that type attributes can well alleviate the few-shot issue.

(2) The concatenation operation of two pre-trained representations (marked as green lined bar) demonstrates strong competitiveness by the metric MRR, Hits@5 and Hits@1 compared to the addition strategy (marked as red slashed bar). However, these two strategies both have great potential to restore the type semantics. Considering the implementation of concatenation operation will double the embedding dimension and enlarge the computational space, we determine to adapt the addition strategy (introduced in Sect. 4.1) to fuse the type embeddings.

Discussion of Transformer Settings. We conduct experiment to explore the effects of transformer layers and heads numbers. We visualize this effects by marking with blue boxes shown in Fig. 6. Obviously, we can observe that one layer with eight heads contributes most to both MRR and Hits@10. We also notice that there is no need to apply much more transformer layers. It illustrates

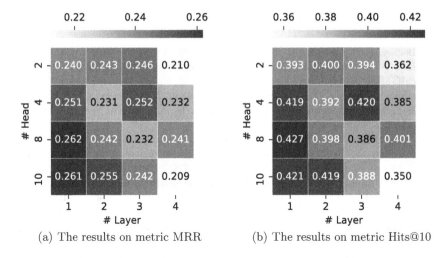

(a) The results on metric MRR   (b) The results on metric Hits@10

Fig. 6. Comparison of different numbers of transformer layers and heads.

that our PiTI-Fs does not rely heavily on the stacking of the transformer layers, which helps reduce parameters.

6   Conclusion

In this paper, we present a method PiTI-Fs for few-shot KGC with the assistance of incorporating prior type information. PiTI-Fs proposes to enhance entity representations with derived type attributes and to represent few-shot relations by a transformer-based encoder. Comprehensive experiments show the competitiveness of our method in the few-shot scenario and the effectiveness of the components in PiTI-Fs.

In the future, we will explore more representative prior knowledge to tackle the few-shot issues. In addition, we also consider to extend logic rules and contrastive learning for few-shot KGC, which have been widely used in many other knowledge graph researches [13,15].

Acknowledgments. This work was supported by National Key Research and Development Program of China (2020AAA0108800), National Natural Science Foundation of China (62137002, 61937001, 62192781, 62176209, 62176207, 62106190, and 62250009), Innovative Research Group of the National Natural Science Foundation of China (61721002), Innovation Research Team of Ministry of Education (IRT_17R86), Consulting research project of Chinese academy of engineering "The Online and Offline Mixed Educational Service System for 'The Belt and Road' Training in MOOC China", "LENOVO-XJTU" Intelligent Industry Joint Laboratory Project, CCF-Lenovo Blue Ocean Research Fund, Project of China Knowledge Centre for Engineering Science and Technology, Foundation of Key National Defense Science and Technology Laboratory (6142101210201), the Fundamental Research Funds for the Central Universities (xhj032021013-02, xzy022021048, xpt012022033).

References

1. Bollacker, K., Bollacker, K., Evans, C., Paritosh, P., Sturge, T., Taylor, J.: Freebase: a collaboratively created graph database for structuring human knowledge. In: SIGMOD Conference (2008)
2. Bordes, A., Usunier, N., Garcia-Duran, A., Weston, J., Yakhnenko, O.: Translating embeddings for modeling multi-relational data. In: Advances in Neural Information Processing Systems (NIPS) (2013)
3. Carlson, A., Betteridge, J., Kisiel, B., Settles, B., Hruschka, E.R., Mitchell, T.M.: Toward an architecture for never-ending language learning. In: Twenty-Fourth AAAI Conference on Artificial Intelligence (AAAI) (2010)
4. Chen, M., Zhang, W., Zhang, W., Chen, Q., Chen, H.: Meta relational learning for few-shot link prediction in knowledge graphs. In: Proceedings of the 2019 Conference on Empirical Methods in Natural Language Processing and the 9th International Joint Conference on Natural Language Processing (EMNLP-IJCNLP), pp. 4216–4225 (2019)
5. Chung, C., Whang, J.J.: Knowledge graph embedding via metagraph learning. In: Proceedings of the 44th International ACM SIGIR Conference on Research and Development in Information Retrieval (SIGIR), pp. 2212–2216 (2021)
6. Dettmers, T., Minervini, P., Stenetorp, P., Riedel, S.: Convolutional 2D knowledge graph embeddings. In: Proceedings of the AAAI Conference on Artificial Intelligence (AAAI) (2018)
7. Dietz, L., Kotov, A., Meij, E.: Utilizing knowledge graphs for text-centric information retrieval. In: The 41st International ACM SIGIR Conference on Research and Development in Information Retrieval (SIGIR), pp. 1387–1390 (2018)
8. Finn, C., Abbeel, P., Levine, S.: Model-agnostic meta-learning for fast adaptation of deep networks. In: Proceedings of the 34th International Conference on Machine Learning, pp. 1126–1135. International Conference on Learning Representation (ICLR) (2017)
9. Guo, X., Gao, H., An, Y., Zou, Z.: Diversified top-k querying in knowledge graphs. In: Web and Big Data - 4th International Joint Conference, APWeb-WAIM, pp. 319–336 (2020)
10. Ji, Y., Li, B., Liu, Y., Zhang, Y., Cai, K.: Multi-space knowledge enhanced question answering over knowledge graph. In: Web and Big Data - 5th International Joint Conference, APWeb-WAIM, pp. 135–140 (2021)
11. Jiang, Z., Gao, J., Lv, X.: MetaP: meta pattern learning for one-shot knowledge graph completion. In: Proceedings of the 44th International ACM SIGIR Conference on Research and Development in Information Retrieval (SIGIR), pp. 2232–2236 (2021)
12. Kingma, D.P., Ba, J.L.: Adam: a method for stochastic optimization. In: 3rd International Conference on Learning Representations (ICLR) (2015)
13. Liang, S., Shao, J., Zhang, D., Zhang, J., Cui, B.: DRGI: deep relational graph infomax for knowledge graph completion. IEEE Trans. Knowl. Data Eng. (TKDE) (2021)
14. Lin, Q., et al.: Incorporating context graph with logical reasoning for inductive relation prediction. In: The 45th International ACM SIGIR Conference on Research and Development in Information Retrieval (SIGIR), pp. 893–903 (2022)
15. Lin, Q., et al.: Contrastive graph representations for logical formulas embedding. IEEE Trans. Knowl. Data Eng. (TKDE) (2021)

16. Lin, Y., Liu, Z., Sun, M., Liu, Y.: Learning entity and relation embeddings for knowledge graph completion. In: AAAI (2015)
17. Liu, B., Wang, X., Liu, P., Li, S., Wang, X.: PAIRPQ: an efficient path index for regular path queries on knowledge graphs. In: Web and Big Data - 5th International Joint Conference, APWeb-WAIM, pp. 106–120 (2021)
18. Nguyen, D.Q., Nguyen, T.D., Nguyen, D.Q., Phung, D.: A novel embedding model for knowledge base completion based on convolutional neural network. In: NAACL-HLT, pp. 327–333. Association for Computational Linguistics (2018)
19. Niu, G., et al.: Relational learning with gated and attentive neighbor aggregator for few-shot knowledge graph completion. In: Proceedings of the 44th International ACM SIGIR Conference on Research and Development in Information Retrieval (SIGIR), pp. 213–222 (2021)
20. Sheng, J., et al.: Adaptive attentional network for few-shot knowledge graph completion. In: Proceedings of the 2020 Conference on Empirical Methods in Natural Language Processing (EMNLP), pp. 1681–1691 (2020)
21. Sun, Z., Deng, Z.H., Nie, J.Y., Tang, J.: RotatE: Knowledge graph embedding by relational rotation in complex space. In: International Conference on Learning Representations (ICLR) (2019)
22. Trouillon, T., Welbl, J., Riedel, S., Gaussier, E., Bouchard, G.: Complex embeddings for simple link prediction. In: Proceedings of The 33rd International Conference on Machine Learning (ICML), pp. 2071–2080 (2016)
23. Vaswani, A., et al.: Attention is all you need. In: Advances in Neural Information Processing Systems (NIPS) (2017)
24. Vrandecic, D., Krötzsch, M.: Wikidata: a free collaborative knowledgebase. Commun. ACM 57, 78–85 (2014)
25. Wang, B., Shen, T., Long, G., Zhou, T., Chang, Y.: Structure-augmented text representation learning for efficient knowledge graph completion. In: WWW (2021)
26. Wang, Z., Zhang, J., Feng, J., Chen, Z.: Knowledge graph embedding by translating on hyperplanes. In: Proceedings of the AAAI Conference on Artificial Intelligence (AAAI) (2014)
27. Xiong, W., Yu, M., Chang, S., Guo, X., Wang, W.Y.: One-shot relational learning for knowledge graphs. In: Proceedings of the 2018 Conference on Empirical Methods in Natural Language Processing (EMNLP), pp. 1980–1990 (2018)
28. Xu, F., Liu, J., Lin, Q., Pan, Y., Zhang, L.: Logiformer: a two-branch graph transformer network for interpretable logical reasoning. In: Proceedings of the 45th International ACM SIGIR Conference on Research and Development in Information Retrieval (SIGIR), pp. 1055–1065 (2022)
29. Yang, B., Yih, W.t., He, X., Gao, J., Deng, L.: Embedding entities and relations for learning and inference in knowledge bases. In: International Conference on Learning Representations (ICLR) (2015)
30. Zhang, C., Yao, H., Huang, C., Jiang, M., Li, Z., Chawla, N.V.: Few-shot knowledge graph completion. In: Proceedings of the AAAI Conference on Artificial Intelligence (AAAI), pp. 3041–3048 (2020)
31. Zhang, F., Yuan, N.J., Lian, D., Xie, X., Ma, W.Y.: Collaborative knowledge base embedding for recommender systems. In: KDD (2016)
32. Zhang, Y., Li, B., Gao, H., Ji, Y., Chen, W.: Fine-grained evaluation of knowledge graph embedding model in multiple types of downstream tasks. Big Data Res. 25, 100218 (2021)

Answering Why-Not Questions on GeoSPARQL Queries

Yin Li[1,2(✉)] and Bixin Li[1]

[1] School of Computer Science and Engineering, Southeast University, Nanjing, China
leein121999@126.com

[2] Jiangsu JARI Information Technology, Lianyungang 222006, China

Abstract. Nowadays geo-spatial knowledge graph is expanding gradually in Location Bases Services (LBS) to improve the search relevancy as well as to present background information about points of interests. They allow answering complex GeoSPARQL queries efficiently by returning a subset of records that match the query. Now consider if a query does not return a record that you believe should be returned, a natural question is to ask for an explanation "why not?". In this study, we firstly formalize the why-not question on GeoSPARQL queries, then propose a novel framework called AWQG (Answering Why-Not Questions on GeoSPARQL), which is capable of answering why-not questions based on a penalty function. AWQG generates logical explanations to help users refine their initial queries at the levels of topological functions and spatial constraints. The experimental results show that the model provides high-quality explanations of why-not questions for GeoSPARQL queries efficiently.

Keywords: GeoSPARQL · Missing answers · Why-not · Spatial query

1 Introduction

With the rapid development of Semantic Web, Location Based Services (LBS) organize the rich geographic information into knowledge graphs to enhance the quality of the geo-spatial search [9] and points of interests (POIs) recommendation [3,19]. A geo-spatial knowledge graph is a graph which consists of over millions of geo-entities and their relationships. Moreover, this kind of graphs arrange data pieces as a set of triples, each of which follows the form of "subject - predicate - object" (e.g., Cinema - locatedIn - Wanda Plaza) where subjects and objects denote the nodes in the graph, and predicates (relations) denote the edges. Knowledge graph triples are traditionally accessed by using structured query languages, such as SPARQL [8]. The recent GeoSPARQL [15], defined by the Open Geospatial Consortium (OGC), extends SPARQL to represent geographic information and support spatial queries over knowledge graphs.

However, using GeoSPARQL queries requires users or dataset experts to be exactly aware of the query functions as well as to precisely perceive the

B. Li et al. (Eds.): APWeb-WAIM 2022, LNCS 13422, pp. 286–300, 2023.
https://doi.org/10.1007/978-3-031-25198-6_22

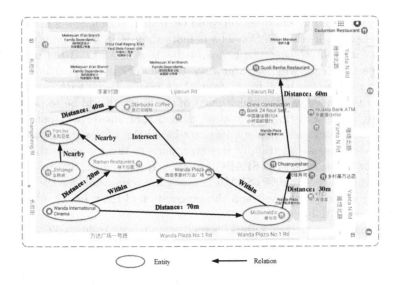

Fig. 1. Example geo-spatial knowledge graph

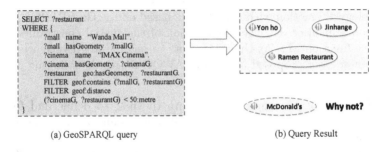

(a) GeoSPARQL query   (b) Query Result

Fig. 2. Example GeoSPARQL query and the query result

geo-spatial RDF data schema. Therefore, it is a common scenario where an inappropriate query returns the results that are at odds with the prior expectations of users. When a user finds that the query result does not contain the expected objects, a natural **Why-not** question would be asked subsequently. Figure 1 depicts a geo-spatial knowledge graph about a *Wanda Plaza* and the POIs around it. Consider the following motivating example.

EXAMPLE 1. After watching a movie in an *IMAX Cinema*, a user plans to find a restaurant nearby the cinema to take a dinner. Consequently, he may have a request, i.e., Give me all restaurants that are in the *Wanda Plaza* and within 50 m of the *IMAX Cinema*. However, after posing a GeoSPARQL query over the geo-spatial knowledge graph, as shown in Fig. 2, the user finds out that the *McDonald's* which he just passed by is failed to be shown in the result. Hence, the user may raise a why-not question, i.e., why is the *McDonald's* absent from the query result?

The answer to this why-not question could be: i) this *McDonald's* is missed in the knowledge graph, ii) this *McDonald's* is more than 50 m away from the *IMAX Cinema* or iii) this *McDonald's* is not in the *Wanda Plaza*. If the user asks a similar question, e.g., why the *Starbucks Coffee* does not appear in the query result, the reason may be different from the former[1]. Faced with such why-not questions, users have no idea which parts of the query should be responsible for the missing items. Existing explanation models that answer why-not questions fall into three categories, namely, instance-based [12,13], operator-based [1,4], and query-refinement-based [5,6]. However, none of them is applicable to solve the why-not questions on GeoSPARQL queries. In this paper, we aim to explain which parts of the original GeoSPARQL query are responsible for the contradictions between the query result and the expectation of the user. This helps users to refine their initial queries.

In this paper, we present a unified explanation model, called AWQG, for why-not questions on GeoSPARQL queries. AWQG answers why-not questions in the following procedures: (i) AWQG adopts query refinement algorithm to generate refined queries with approximate minimal changes based on a penalty function which balances the weights of topological functions and spatial constraints. (ii) For the modification of the topological functions, we define the penalty of replacing a topological function with another one based on the GeoSPARQL topological relation semantics. For the modification of the spatial constraints, we use the penalty function to relax the constraints through a progressive process.

In summary, our proposed framework AWQG makes the following contributions:

(i) To the best of our knowledge, we are the first to define and formalize the why-not questions on GeoSPARQL queries and analyze the underlying causes of them.
(ii) A unified explanation model is proposed to answer why-not questions on GeoSPARQL queries based on a penalty function. The explanation model generates explanations by modifying topological functions and spatial constraints.
(iii) An efficient query refinement algorithm is proposed to quickly select the optimal explanation from a set of candidate explanations using a bound-and-prune strategy.
(iv) Extensive experiments on real-world datasets are conducted to evaluate the performance of the proposed framework AWQG. The results illustrate that AWQG is capable of performing well in both effectiveness and efficiency.

2 Related Work

We briefly review the research studies related to our work in the following three aspects: explanation models for why-not questions, Geo-spatial POI Recommendations, and provenance for SPARQL queries.

---

[1] The *McDonald's* is more than 50 m away from the *Wanda Plaza*. The *Starbucks Coffee* is not in the *Wanda Plaza*.

Three types of models can be used to answer why-not questions: instance-based [12,13], operator-based [1,4], and query-refinement-based [5,6]. Answering why-not questions have also received attention recently in other research fields. Cate et al. [16] introduce an ontology-based model for explaining why-not questions on conjunctive queries. Calvanese et al. [2] leverages abductive reasoning to answer why-not questions on data represented by a DL-Lite ontology. As for graph databases, Saiful et al. [14] proposes a query refinement model to address the problem of why-not questions in similar graph matching.

Geo-spatial POI recommender systems take into account local information and users' context for recommending nearby available POIs such as restaurants, cinemas, hospital, and pharmacies. Yin et al. [18] implemented location content aware recommender systems to provide user with particular set of POIs. These systems were evaluated using large scale data set and results showed that the models had minimum error margin in successfully recommending POIs to users. However, these recommender systems concern social network data and recommend POIs based on both personal interests and local preferences. In this paper, we focus on topological relations and spatial constraints between different POIs which are organized into geo-spatial knowledge graphs.

Data provenance has been studied to understand why a particular item exists in query results [10]. For SPARQL queries, existing works [11,17] focus on explaining the provenance of data to evaluate the data quality and trustworthiness. Data provenance only can be used to answer why questions rather than why-not questions, on the contrary, answering why-not question concerns the items that do not appear in the query result.

3   Problem Analysis

3.1   Problem Formulation

Before defining the why-not question on the GeoSPARQL query, we follow the official GeoSPARQL standard [7] and briefly introduce the important notations employed in the remainder of this paper.

Definition 1 (POI). The GeoSPARQL standard defines the main class *geo:Feature*. A point of interests (POI) p is defined as an instance of the class *geo:Feature*, and it represents a specific site that can have spatial locations (e.g., a restaurant or a cinema).

Definition 2 (Geometry). The class *geo:Geometry* is a single root geometry class defined by the GeoSPARQL standard. Its instance geometry g is a representation of the spatial locations of a POI p. In addition, the geometry g is linked with the POI p by the standard property *geo:hasGeometry*.

Definition 3 (WktLiteral). A wktLiteral w is an instance of the standard class *geo:wktLiteral*. A wktLiteral w represents the detail geometrical information of a geometry g, which are linked by the standard property *geo:asWKT*.

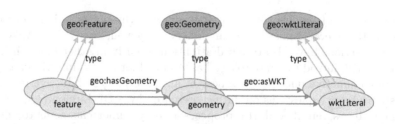

Fig. 3. The relationship of some essential terms in the GeoSPARQL standard

To be more specific, Fig. 3 illustrates the relationships among POI, geometry, and wktLiteral.

Definition 4 (Topological Function). A topological function can be denoted as $tf(g_1, g_2)$ where g_1 and g_2 are two geometries. The function $tf(g_1, g_2)$ returns TRUE if g_1 and g_2 satisfy the topological relationship of the function, otherwise returns FALSE.

Definition 5 (Spatial Constraint Function). A spatial constraint function can be denoted as $sp(g_1, g_2, u)$ where g_1 and g_2 are two geometries, and u is the distance. And the function $sp(g_1, g_2, u)$ can perform spatial constraint operations on the geo-spatial data.

Table 1 illustrates the definitions of the GeoSPARQL query functions exploited in this paper.

Definition 6 (Geo-spatial RDF Term). Let P be a set of POIs, G be a set of geometries, W be a set of wktLiterals, and R be a set of properties. A geo-spatial RDF term is a member of the set $T = P \cup G \cup W \cup R$.

Definition 7 (Geo-spatial knowledge graph). A geo-spatial RDF triple (*subject, predicate, object*) is a member of the set $(P \cup G) \times R \times (P \cup G \cup W)$. A geo-spatial knowledge graph DS is a finite set of geo-spatial RDF triples.

Definition 8 (Triple Pattern). A triple pattern t is a member of the set $(P \cup G \cup V) \times (R \cup V) \times (P \cup G \cup W \cup V)$. V is a set of query variables, and it is disjoint from T. A triple pattern is similar to a geo-spatial RDF triple but allows the usage of variables for the subject, the predicate, and the object.

Definition 9 (Basic Graph Pattern). A BGP (Basic Graph Pattern) B is a finite set of triple patterns, and $B = \{t_1, ...t_n\}$.

Definition 10 (GeoSPARQL Query). A GeoSPARQL query Q is a tuple (E, DS, QF), where E is a GeoSPARQL algebra expression, DS is a geo-spatial knowledge graph and QF is a query form. A GeoSPARQL algebra expression E consists of a BGP B and a query function set GQF.

Table 1. Definitions of the standard GeoSPARQL query functions

tf/sp	Query functions	Definitions
tf_1	$equals(g_1, g_2)$	To return TRUE if g_1, g_2 have identical coordinate values
tf_2	$disjoint(g_1, g_2)$	To return TRUE if the intersection of g_1, g_2 is an empty set
tf_3	$intersects(g_1, g_2)$	To return TRUE if the intersection of g_1, g_2 is not an empty set
tf_4	$touches(g_1, g_2)$	To return TRUE if the points common to g_1, g_2 do not intersect the interiors of them
tf_5	$crosses(g_1, g_2)$	To return TRUE if the intersection of g_1, g_2 is an interior geometry whose dimension is one less than the maximum dimension of g_1, g_2
tf_6	$within(g_1, g_2)$	To return TRUE if g_1 is within g_2
tf_7	$contains(g_1, g_2)$	To return TRUE if g_1 contains g_2
tf_8	$overlaps(g_1, g_2)$	To return TRUE if the intersection of g_1, g_2 is a geometry different from them but of the same dimension
sp	$distance(g_1, g_2, u)$	To return the shortest distance u between any two points in g_1, g_2

The BGP B is evaluated to match geo-spatial knowledge graphs in DS. The query form QF (SELECT, CONSTRUCT, ASK, DESCRIBE) exploits the matched graphs to provide the query result. GQF is the core of the GeoSPARQL query. In GQF, topological functions establish the topological relationship of the geo-spatial RDF terms in B, while spatial constraint functions claim the spatial constraints.

Definition 11 (Why-Not Question). Given a GeoSPARQL query Q on the geo-spatial knowledge graph DS, $Q(DS)$ presents the query result. Let v be a variable that appears in the GeoSPARQL query form QF of the query Q, and a POI p be a solution of v. We define a why-not question as a mapping $v \rightarrow p$ that indicates why the geo-spatial POI p does not appear in $Q(DS)$. Considering EXAMPLE 1, the query result does not contain the *McDonald's*. Consequently, the why-not question raised by the user can be denoted as *?restaurant $\rightarrow$ McDonald's*.

Definition 12 (Explanation). An explanation GQF' presents the reason for a why-not question $v \rightarrow p$ on the GeoSPARQL query Q. The explanation GQF' is a modified query function set. The query result of the refined GeoSPARQL query Q' whose query function set is GQF' should contain the expected answer p. And the set of explanations of the why-not question $v \rightarrow p$ can be denoted as $SGQF'$.

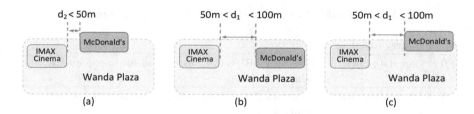

Fig. 4. Three why-not scenarios of EXAMPLE 1

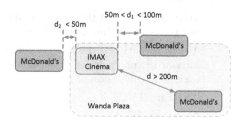

Fig. 5. A particular scenario of EXAMPLE 1

3.2   Analysis of Why-Not Questions

The result of our investigation shows that there are multidimensional causes that can lead to the occurrence of why-not questions on GeoSPARQL queries, such as inappropriate geo-spatial features, the incompleteness of the geo-spatial knowledge graph, and spelling errors of users. In this study, we focus on inappropriate geo-spatial features in the GeoSPARQL queries.

By analyzing the evaluation of GeoSPARQL queries over geo-spatial knowledge graphs, we find out that **questionable restrictive topological functions** and **restrictive spatial constraints** are two main reasons for why-not questions on GeoSPARQL queries. And the absence of expected answers in the query result may be caused by one of them or both of them.

Consider the why-not question $?restaurant \rightarrow McDonald's$ for the query in EXAMPLE 1. There is a topological function $contains(?mallWKT, ?restaurantWKT)$ in the GeoSPARQL algebra expression. Only if the geometrical relationship between the geometries denoted as $?mallWKT$ and $?restaurantWKT$ is $contains$, will the function return TRUE. However, the geometrical relationship between the *IMAX Cinema* and the *McDonald's* maybe not $contains$. In this case, the reason for the why-not question is the questionable restrictive topological function. A possible scenario of this case is illustrated in Fig. 4(a). In this scenario, the geometrical relationship between the *IMAX Cinema* and the *McDonald's* is *overlaps* instead of *contains*. The corresponding explanation of the why-not question shall be $overlaps(?mallWKT, ?restaurantWKT)$, $distance(?cinemaWKT, ?restaurantWKT) < 50\}$. In EXAMPLE 1, restrictive spatial constraints can also be to blame for the absence of the *McDonald's* in the query result. Fig. 4(b) illustrates a scenario of this case where the distance between the *IMAX Cinema* and the *McDonald's* is more than 50 m. Hence, the spatial constraint function $distance(?cinemaWKT, ?restaurantWKT)$ returns a number larger than 50, and it will remove the *McDonald's* from the query result. In this scenario, the explanation of the why-not question shall be $\{contains(?mallWKT, ?restaurantWKT),$ $distance(?cinemaWKT, ?restaurantWKT)$ $<$ $100\}$. Figure 4(c) illustrates another scenario where questionable restrictive topological functions and restrictive spatial constraints are both the reasons for the why-not question in EXAMPLE 1. In this scenario, the explanation of the why-not question shall be $\{overlaps(?mallWKT, ?restaurantWKT),$ $distance(?cinemaWKT, ?restaurantWKT) < 100\}$.

Figure 5 illustrates a particular scenario of the why-not question in EXAM-PLE 1. In this scenario, there are three *McDonald's* around the *IMAX Cinema*, and we can get three different explanations. How to efficiently and accurately select the optimal one from the candidate explanations becomes a key issue of this problem.

4 Model Explanation

4.1 Penalty Function

Considering the two reasons for why-not questions on GeoSPARQL queries, AWQG generates an explanation by modifying topological functions and spatial constraints. As we analyzed in Sect. 3.2, there may be a set of candidate explanations of a why-not question. Hence, AWQG needs to evaluate the quality of the generated explanations. In doing so, AWQG employs a penalty function as follows:

$$Penalty(GQF, GQF') = \lambda \sum_{j=1}^{m} \frac{\Delta u_j}{u_j} + (1 - \lambda) \sum_{i=1}^{n} \Delta t f_i \qquad (1)$$

The penalty function is based on a hypothesis, the less modification of the query functions, the better the explanation. Here, λ is a user preference for the modification of spatial constraints versus topological functions. The first term in the penalty function represents the penalty of modifying spatial constraints. u_j is the distance of the j-th spatial constraint function in the original query function set GQF. Δu_j is the modification of u_j in the generated explanation GQF'. The second term in the penalty function represents the penalty of modifying topological functions. $\Delta t f_i$ is the penalty of replacing the i-th topological function in GQF with another topological function in GQF' while the parameters of the topological function are reserved. And the penalty of replacing a topological function with another one is defined considering the topological semantic difference of their corresponding relationships. Table 2 illustrates the detail of the modification penalties of topological functions.

Table 2. Penalties of modifying topological functions

	tf_1	tf_2	tf_3	tf_4	tf_5	tf_6	tf_7	tf_8
tf_1	0	1	0.5	0.9	0.8	0.5	0.5	0.5
tf_2	1	0	1	0.5	1	1	1	1
tf_3	0.5	1	0	0.5	0.5	0.4	0.4	0.4
tf_4	0.9	0.5	0.5	0	0.8	0.8	0.8	0.6
tf_5	0.8	1	0.5	0.8	0	0.7	0.7	0.4
tf_6	0.5	1	0.4	0.8	0.7	0	1	0.4
tf_7	0.5	1	0.4	0.8	0.7	1	0	0.4
tf_8	0.5	1	0.4	0.6	0.4	0.4	0.4	0

4.2   Basic Algorithm

Given an RDF dataset DS, a GeoSPARQL query $Q = (E, DS, QF)$ and a why-not question $v \to p$. The GeoSPARQL algebra expression E consists of a BGP B and a query function set GQF. GQF consists of n topological functions denoted as $tf_i()$, $i = 1, ..., n$ and m spatial constraint functions denoted as $sp_j()$, $j = 1, ..., m$. AWQG modifies the query functions to generate a modified query function set GQF' and the corresponding modified GeoSPARQL query $Q' = (E', DS, QF)$. If $Q'(DS)$ contains the expected answer p, GQF' is a candidate explanation, and $GQF' \in SGQF'$. Among all the candidate explanations in the set $SGQF'$, the optimal explanation is the one with minimum penalty.

Algorithm 1 illustrates the basic algorithm for the generation of the optimal explanation. In the input, $SGQF$ contains all the possible query function sets whose topological functions are modified, and spatial constraints are reserved. In line 3 and line 11, Q_i and Q'_i are the GeoSPARQL queries which contain GQF_i and GQF'_i as query function sets, respectively.

4.3   Bound-and-Prune Algorithm

There are mainly two obstacles that make the basic algorithm impracticable. Firstly, for a GeoSPARQL query Q contains n topological functions, there are 9^n modified GeoSPARQL query function sets in $SGQF$. The time consumption of the basic algorithm is proportional to 9^n. If n is too large, the basic algorithm will be time-consuming even infeasible. Secondly, if the GeoSPARQL query Q_i that contains the modified query function set GQF_i cannot retrieve the expected answer p, the basic algorithm will modify the distance of the spatial constraint functions in GQF_i until there is no probability to get p by further modification of the distance. The issue is that estimating the possibility of retrieving p by further modification of spatial constraint distance is too hard to manipulate under existing conditions.

To address the above bottlenecks, we propose Algorithm 2 that employs the current minimum penalty mp_c as an upper bound. In the input, $SGQF$ is com-

---

Algorithm 1. Basic algorithm

Input: the original GeoSPARQL query Q, the why-not question $v \to p$, the original
query function set $GQF = \{tf_1(), ..., tf_n(), sp_1(), ..., sp_m()\}$, a set of modified query
function sets $SGQF = \{GQF_1, ..., GQF_{9^n}\}$, the preset step size s

Output: the optimal explanation $GQF' = \{tf_1'(), ..., tf_n'(), sp_1'(), ..., sp_m'()\}$

1: Initialization: $GQF' = \varnothing$, $SGQF' = \varnothing$, $s = 1000$;
2: **for all** $GQF_i \in SGQF$ **do**
3:     **if** $p \in Q_i(DS)$ **then**
4:         $SGQF'.add(GQF_i)$;
5:     **else**
6:         **if** it's possible to get p **then**
7:             modify parameters of $sp_j() \in GQF_i$ by s to get GQF_i';
8:         **else**
9:             continue;
10:        **end if**
11:        **if** $p \in Q_i'(DS)$ **then**
12:            $SGQF'.add(GQF_i')$;
13:        **else**
14:            go to 6;
15:        **end if**
16:    **end if**
17: **end for**
18: **for all** $GQF_i'' \in SGQF'$ **do**
19:     **if** $s > Penalty(GQF, GQF_i'')$ **then**
20:         $GQF' = GQF_i''$;
21:         $s = Penalty(GQF, GQF_i'')$;
22:     **end if**
23: **end for**
24: **return** GQF'

---

posed of all the possible query function sets whose topological functions are
modified, and spatial constraints are reserved. In line 4, Q_i is the GeoSPARQL
query consisting of query function set GQF_i in OP. In line 8, GQF_i' is generated by modifying parameters of spatial constraint functions in GQF_i. And in
line 10, Q_i' is the GeoSPARQL query containing query function set GQF_i'. In
Algorithm 2, mp_c is the upper bound. The algorithm employs mp_c to prune modification cases of topological functions at the beginning of the loop, which makes
the algorithm capable of generating explanations for cumbersome GeoSPARQL
queries. And for each possible modification case of topological functions, we try
to generate the explanation whose penalty is local minimum in the shortest
time by interrupting the modification process when the current penalty is bigger
than the current upper bound, which can significantly improve the efficiency of
AWQG. Compared to the basic algorithm, the bound-and-prune algorithm is
more effective and more efficient.

Algorithm 2. Bound-and-prune algorithm

Input: the original GeoSPARQL query Q, the why-not question $v \to s$, the original query function set $GQF = \{tf_1(), ..., tf_n(), sp_1(), ..., sp_m()\}$, a set of modified query function sets $SGQF = \{GQF_1, ..., GQF_{9^n}\}$, the preset step size p

Output: the optimal explanation $GQF' = \{tf'_1(), ..., tf'_n(), sp'_1(), ..., sp'_m()\}$

1: Initialization: $GQF' = \varnothing$, $p_c = 1000$;
2: **for all** $GQF_i \in SGQF$ **do**
3:      **if** $Penalty(GQF, GQF_i) < p_c$ **then**
4:          **if** $s \in Q_i(DS)$ **then**
5:              $p_c = Penalty(GQF, GQF_i)$;
6:              $GQF' = GQF_i$;
7:          **else**
8:              modify parameters of $sp_j() \in GQF_i$ by p to get GQF'_i;
9:              **if** $Penalty(GQF, GQF'_i) < p_c$ **then**
10:                 **if** $s \in Q'_i(DS)$ **then**
11:                     $p_c = Penalty(GQF, GQF'_i)$;
12:                     $GQF' = GQF'_i$;
13:                 **else**
14:                     go to 8;
15:                 **end if**
16:             **else**
17:                 continue;
18:             **end if**
19:         **end if**
20:     **else**
21:         continue;
22:     **end if**
23: **end for**
24: **return** GQF'

---

5   Experiments

In this section, we evaluate our proposed explanation model, AWQG, in terms of effectiveness and efficiency.

Datasets: Three real-world datasets are used in the experiments: GeoNames (geographic objects around the world), EURO (points of interests in Europe), and GN (geographic names in the United States). GeoNames contains over 10 million geographic objects, among which 48,139 objects are cities with more than 5,000 people; EURO is a set of 162,033 objects of locations in Europe, such as hotels and interests; and GN is a dataset of 1,868,821 geographic objects in the United States. Objects in the three datasets are all described with location coordinates and rich geographic descriptions.

Query Set: There is no benchmark available for the evaluation. Therefore, we construct sixty GeoSPARQL queries (divided into three groups, each group contains twenty queries) for volunteers to search information on the datasets. The first group of queries is related to cities around the world; the second group of

Table 3. Statistics of the why-not questions

Dataset	#why-not GeoSPARQL queries	#why-not questions	#why-not caused by *reason1*	#why-not caused by *reason2*	#why-not caused by both reasons
GeoNames	20	37	8	25	4
EURO	20	34	14	18	2
GN	20	33	9	19	5

queries is about interests in Europe; the third group of queries is about geographic names in the United States. In the analysis of Sect. 3.2, the absence of expected answers in the query result could blame *reason1* (questionable restrictive topological functions), *reason2* (restrictive spatial constraints) or both of them. The query set should include diverse queries to cover all kinds of why-not scenarios. Hence, the queries are designed to include topological functions and spatial constraint functions, as well as different combinations of them. The statistics of the why-not questions obtained from volunteers are shown in Table 3.

Evaluations: In the experiments, the effectiveness of AWQG was evaluated by (i) measuring the precision of the modified GeoSPARQL queries generated by AWQG; (ii) measuring the user satisfaction on the explanations of why-not questions. The efficiency was evaluated by (i) measuring the average time of generating an explanation; (ii) performing the same queries and why-not questions on different scale datasets. The experiments were conducted on a PC with an Intel Core i7 3.40 GHz CPU and 14 GB memory running Windows 10.

5.1   Effectiveness Evaluation

In this section, we measure the effectiveness of AWQG by using the precision metric. The primary requirement of an explanation model of why-not questions is that the query result of the modified GeoSPARQL query should contain the expected missing answers. Meanwhile, the modified GeoSPARQL query should be as precise as possible in terms of its query result. Any additional irrelevant results should be minimized. We propose the precision metric as follows to measure the precision of the modified GeoSPARQL queries generated by AWQG.

$$Pre(Q, Q') = 1 - \frac{|Q'(D) - S - Q'(D) \cap Q(D)|}{|Q'(D)|}. \tag{2}$$

In Eq. 2, Q is the original GeoSPARQL query, and Q' is the modified GenSPARQL query generated by AWQG. $Q(D)$ and $Q'(D)$ are the query results of Q and Q', respectively, and S is the missing answer set.

Table 4 reports the average precision values of modified GeoSPARQL queries in three groups, where ave_Pre_a is the average precision value of all the modified GeoSPARQL queries in a group. ave_Pre_t, ave_Pre_n and ave_Pre_b are average precision values of the modified GeoSPARQL queries generated by considering

Table 4. The average precision values of modified GeoSPARQL queries

Group	ave_Pre_a	ave_Pre_t	ave_Pre_n	ave_Pre_b
1	0.78	0.73	0.81	0.69
2	0.77	0.74	0.81	0.62
3	0.82	0.86	0.83	0.71

reason1, *reason2* and both reasons, respectively. For total 104 why-not questions in three groups, the average precision value of the modified GeoSPARQL queries generated by AWQG is 0.79, which means that AWQG has good quality in terms of precision metrics. Furthermore, the average precision value of modified GeoSPARQL queries generated by modifying spatial constraints is 0.82. And it is bigger than the average precision value of modified GeoSPARQL queries generated by modifying topological functions, which is 0.77. Hence, modifying topological functions may introduce more additional irrelevant items in the query result than modifying spatial constraints.

5.2   Efficiency Evaluation

In this section, we first measure the execution time required by AWQG to generate explanations on GeoNames, EURO, and GN. Then the performances were compared in terms of answering the same why-not questions in different scale datasets.

Efficiency Evaluation Performance of Modifying Topological Functions We measure the modification time for 31 why-not questions caused by *reason1* in each dataset, as shown in Fig. 6. The experimental result shows that the modified topological functions for explaining why-not questions can be generated within 2 s on each dataset. The average running time is 0.615 s (GeoNames), 0.84 s (EURO) and 1.16 s (GN), and the maximum time spent is 0.89 s (GeoNames), 1.10 s (EURO) and 1.68 s (GN). The time consumption is acceptable for users to obtain explanations.

Performance of Modifying Spatial Constraints. We measure the modification time required by AWQG for 62 why-not questions caused by *reason2* in each dataset, as shown in Fig. 7. The experimental result shows that the modified spatial constraints for explaining why-not questions can be generated within 10 s on each dataset. The average running time is 4.74 s (GeoNames), 5.66 s (EURO) and 7.37 s (GN), and the maximum is 9.31 s (GeoNames), 9.43 s (EURO) and 9.30 s (GN). The time consumption is quite tolerable considering that users are eventually provided an explanation.

Performance of Modifying Both Two Kinds of Query Functions. We measure the modification time required by AWQG for 11 why-not questions caused by both two reasons in each dataset, as shown in Fig. 8. The experimental

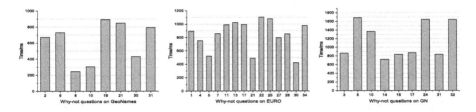

Fig. 6. Execution time on why-not questions caused by *topological functions*

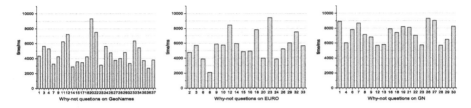

Fig. 7. Execution time on why-not questions caused by *spatial constraints*

result shows that the modified GeoSPARQL queries for a why-not question can be generated in less than 10 s on each dataset, the average running time is 5.51 s (GeoNames), 7.79 s (EURO) and 9.28 s (GN), and the maximum time spent is 6.25 s (GeoNames), 7.92 s (EURO), 9.75 s (GN). The time consumption is still acceptable.

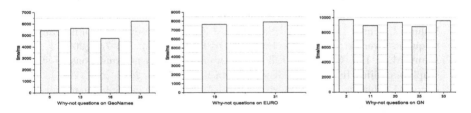

Fig. 8. Execution time on why-not questions caused by both *topological functions* and *spatial constraints*

6   Conclusion

In this paper, we formalized the problem of answering why-not questions on GeoSPARQL queries and proposed a model called AWQG to generate explanations for such why-not questions. AWQG generates explanations considering both topological functions and spatial constraints. An efficient bound-and-prune algorithm is designed to address the bottlenecks of the basic algorithm. The results on both effectiveness and efficiency prove that AWQG could generate high-quality explanations within a reasonable time.

References

1. Bidoit, N., Herschel, M., Tzompanaki, K.: Query-based why-not provenance with NedExplain. In: Extending Database Technology (EDBT) (2014)
2. Calvanese, D., Ortiz, M., Simkus, M., Stefanoni, G.: Reasoning about explanations for negative query answers in DL-Lite. J. Artifi. Intell. Res. **48**, 635–669 (2013)
3. Chang, B., Jang, G., Kim, S., Kang, J.: Learning graph-based geographical latent representation for point-of-interest recommendation. In: Proceedings of the 29th ACM International Conference on Information and Knowledge Management, pp. 135–144 (2020)
4. Chapman, A., Jagadish, H.: Why not? In: Proceedings of the 2009 ACM SIGMOD International Conference on Management of Data, pp. 523–534. ACM (2009)
5. Chen, L., Li, Y., Xu, J., Jensen, C.S.: Direction-aware why-not spatial keyword top-k queries. In: 2017 IEEE 33rd International Conference on Data Engineering, pp. 107–110. IEEE (2017)
6. Chen, L., Xu, J., Lin, X., Jensen, C.S., Hu, H.: Answering why-not spatial keyword top-k queries via keyword adaption. In: 2016 IEEE 32nd International Conference on Data Engineering, pp. 697–708. IEEE (2016)
7. Consortium, O.G., et al.: OGC GeoSPARQL-A geographic query language for RDF data. OGC Candidate Implementation Standard 2 (2012)
8. World Wide Web Consortium, et al.: SPARQL 1.1 overview (2013)
9. World Wide Web Consortium, et al.: RDF 1.1 concepts and abstract syntax (2014)
10. Cui, Y., Widom, J.: Lineage tracing for general data warehouse transformations. VLDB J. Int. J. Very Large Data Bases **12**(1), 41–58 (2003)
11. Damásio, C.V., Analyti, A., Antoniou, G.: Provenance for SPARQL queries. In: Cudré-Mauroux, P., et al. (eds.) ISWC 2012. LNCS, vol. 7649, pp. 625–640. Springer, Heidelberg (2012). https://doi.org/10.1007/978-3-642-35176-1_39
12. Herschel, M., Hernández, M.A.: Explaining missing answers to SPJUA queries. Proc. VLDB Endow. **3**(1–2), 185–196 (2010)
13. Huang, J., Chen, T., Doan, A., Naughton, J.F.: On the provenance of non-answers to queries over extracted data. Proc. VLDB Endow. **1**(1), 736–747 (2008)
14. Islam, M.S., Liu, C., Li, J.: Efficient answering of why-not questions in similar graph matching. IEEE Trans. Knowl. Data Eng. **27**(10), 2672–2686 (2015)
15. Perry, M., Herring, J.: OGC GeoSPARQL-A geographic query language for RDF data. OGC Implementation Standard, September 2012
16. Ten Cate, B., Civili, C., Sherkhonov, E., Tan, W.C.: High-level why-not explanations using ontologies. In: Proceedings of the 34th ACM SIGMOD-SIGACT-SIGAI Symposium on Principles of Database Systems, pp. 31–43. ACM (2015)
17. Theoharis, Y., Fundulaki, I., Karvounarakis, G., Christophides, V.: On provenance of queries on semantic web data. IEEE Internet Comput. **15**(1), 31–39 (2011)
18. Yin, H., Wang, W., Wang, H., Chen, L., Zhou, X.: Spatial-aware hierarchical collaborative deep learning for POI recommendation. IEEE Trans. Knowl. Data Eng. **29**(11), 2537–2551 (2017)
19. Zhou, F., Yin, R., Zhang, K., Trajcevski, G., Zhong, T., Wu, J.: Adversarial point-of-interest recommendation. In: The Web Conference, pp. 3462–34618 (2019)

Multi-Information-Enhanced Knowledge Embedding in Hyperbolic Space

Jiajun Wu[1], Qian Zhou[2(✉)], Yuxuan Xiang[1], Tianlun Dai[1], Hua Dai[2], Hao Wen[1], and Qun Yang[1]

[1] Nanjing University of Aeronautics and Astronautics, Nanjing 211100, China
[2] Nanjing University of Posts and Communications, Nanjing 210003, China
zhouqian@njupt.edu.cn

Abstract. Knowledge Graph Representation Learning(KGRL) aims to map entities and relationships into a low-dimensional dense vector space. Most of the existing models focus only on the information of the triple when doing representation learning, ignoring the rich external semantic information. At the same time, these models consider entities and relations as static and single representations, so the knowledge represent ability is poor. Accordingly, we propose a novel knowledge graph representation model which enhanced knowledge graph embedding with multi-information. Firstly, our model carries out text enhancement and hyperbolic space embedding of triples in the knowledge graph respectively; Secondly, we concatenate the enhanced vector. Then, the concatenated vector through two transformation layer to fuse the semantic information and spacial information. Finally, we use the fused information to learn the context information through the Transformer coding layer, which will dynamically produce the final representation of the entity based on its context. Experimental results show that our model has a great improvement over other models. In the link prediction task, the evaluation protocol Hits@10 and MRR in the public dataset FB15k improve by 28.4% and 29.5% compared with the translation model. Compared with state-of-the-art model, the improvement is 2.5%, 6.3%.

Keywords: Knowledge Graph · Representation learning · Transformer

1 Introduction

Knowledge Graph is an important technique for structuring knowledge. To store and utilize structured knowledge efficiently, Knowledge Graph such as: WordNet (Miller 1995) [13], DBpedia (Auer et al. 2007) [1], Freebase (Bollacker et al. 2008) [2] and other classical Knowledge Graph are constructed by combining expert annotation and computer automatic annotation. Knowledge Graph store huge numbers of structured data in the form of triples. These triples are usually represented as (h, r, t), where h represents the head entity, t represents the tail entity, and r represents the relation.

B. Li et al. (Eds.): APWeb-WAIM 2022, LNCS 13422, pp. 301–314, 2023.
https://doi.org/10.1007/978-3-031-25198-6_23

Large-scale knowledge graph has problems of low computational efficiency and sparse data. In addition, knowledge graph completion still far away from complete due to the constantly emerging new knowledge. At present, more and more scholars aim to project knowledge graph into low-dimensional and continuous vector space, so as to improve the computational efficiency of knowledge graph and alleviate the problem of data sparsity. These representation models could be divided into three types, Translation Models [3,9,11,21], Neural Network Models [4,10] and Tensor Decomposition Models [6,19,25]. Among them, translation model is the most classical method, which has received a large number of attention and application because it only needs fewer parameters in training and achieves better knowledge expression effect. Though these methods have achieved excellent performance, they ignore the rich external semantic information as well as the contextual information of the knowledge graph structure. In this way, some methods using these external semantic information become an active research for knowledge graph completion. Ruobing Xie [24] proposed a knowledge representation learning method TKRL based on entity hierarchy type embedding, and Wang [22] proposed TEKE for knowledge enhancement by using text description information of entities. In addition to considering the description information of entities, some scholars consider the information of the entity's neighbour nodes. The survey of Provenance-Aware Knowledge Representation [16] also mentioned the improvement of context to knowledge graph representation learning.

All the above models are modeled in Euclidean space. However, due to geometric constraints, these models often require high dimensions for knowledge representation. Some scholars begin to explore different Spaces for modeling. ManifoldE [23] maps entities and relations to manifold space. Kolyvakis et al. [7] proposed HyperKG, which embed entities and relations into hyperbolic space, and use the structural features of hyperbolic space to capture the hidden information of data to improve knowledge representation ability. In the latest research, Sun et al. [18]. Proposed the RotatE model, which represents entities into complex space and relations into rotation translation. However, the RotatE model does not take into account the context information of graphs when encoding, making them ineffective in dealing with complex relations.

According to the issue mentioned above, we propose a novel model to embed knowledge graph with multi-information. Inspired by multi-modal information fusion, we propose an effective method to fuse external semantic information and spatial structure information into the same space. At the same time, we take the context entity information into the account and consider the training process of the fact triplet as a Seq2Seq process, so that entities and relations are not a single static model.

We summarize the main contributions of our work as follows:

- We propose a novel method for information fusion of external semantic information and hyperbolic space information, which strengthens the constraint of context information learning.

- Instead of the previous methods that allow a single static representation for each entity or relations, we treat triple, entity and relation as a sequence and a context respectively, which is good for handling complex relationships.
- Extensive experiments have done to show our superiority in dealing with complex relationships. Compared with the state-of-art models, our model has significant advantages.

2   Related Work

Representation Learning of Knowledge Graph is to represent entities and relations in knowledge base as dense low-dimensional vectors. The classic model is the translation model, Bordes et al. [3] regard the process in which the head entity relates to the tail entity through relations as the translation process, and then measure the rationality of each triplet with a score function. After that, a series of translation models are proposed [11,21]. Another is the tensor factorization methods such as DisMult [25] and SimplE [6]. While these methods have few parameters and easy to model, they ignore the structural information of the knowledge graph itself and the abundant external corpus information which fail to improve the ability of dealing with complex relationships. In addition to some of the models mentioned above, many scholars have proposed models integrating external information. The first category is the research of embedding space. Many scholars believe that Euclidean space is not suitable for knowledge representation modeling due to geometric limitations. They try to transform the representation space into manifold space [23], hyperbolic space [7], complex space [18], and the latest research maps entities and relations in the knowledge graph to Quaternion space. The second category is the research on external semantic information, which integrates the information of entity type and entity description information into the representation of entity to improve the representation ability of entity. The third category is the study of knowledge graph structure. Quan Wang et al. proposed a contextualized knowledge graph embedding model (COKE) [20], arguing that the representation of entities and relations should change with context. They utilize the transformer to exploit the contextualized and dynamic representations for entities and relations. Besides Liu et al. [8,12] also used Knowledge graph to enhance the item in recommend system,arguing that they use attention-enhanced dynamic convolutional network to enhance the item. Based on Coke's and attention-enhanced idea, our paper further uses attention mechanism to integrate external entity description information to eliminate semantic deviation in multi-dimensional space, improve the ability to deal with complex relations.

3   Preliminary

In this section, we will introduce some important concepts covered in this article, including the definition of knowledge graphs and the concept of Hyperbolic Space. Finally we will give the definition of our work.

3.1 Knowledge Graph

Knowledge Graph is defined as a directed graph used to store structured information about real-world entities and facts. In this paper, given a Knowledge Graph $G = \{(h, r, t)\} \subset E \times \mathbb{R} \times E$, where E and $\mathbb{R}$ denote the entity set and relation set respectively. A fact triplet (h, r, t) describes a head entity $h \in E$ connected to a tail entity $t \in E$ through a relationship $r \in \mathbb{R}$. After knowledge graph representation learning, the head vector is represented by $\boldsymbol{h}$, the relation vector by $\boldsymbol{r}$, and the tail entity vector by $\boldsymbol{h}$.

3.2 Hyperbolic Geometry

Here we briefly give some concepts of Hyperbolic Space geometry, a more detailed explanation can be found in review article [14]. Hyperbolic space has the ability to reflect data hierarchy and has a large space capacity. In this paper, we choose the Poincaré-ball model, which has the feasibility of gradient optimization in this task. Here we define a Poincaré sphere by a Riemannian manifold (B^d, g_x), where $B^d = \{x \in \mathbb{R}^d, \|x\| < 1\}$ is a d-dimensional Poincaré ball and g_x is the Riemannian metric tensor. For two points $\boldsymbol{u}$ and $\boldsymbol{v}$ in hyperbolic space, their distance function can be expressed as:

$$d(\boldsymbol{u}, \boldsymbol{v}) = \text{arcosh}\left(1 + 2\frac{\|\boldsymbol{u} - \boldsymbol{v}\|^2}{(1 - \|\boldsymbol{u}\|^2)(1 - \|\boldsymbol{v}\|^2)}\right) \tag{1}$$

This formula is symmetric, and as nodes move from the origin towards the poincaré ball boundary, their distance will increase exponentially. This will provide more space for representation learning of entities.

Vector translation in poincaré ball model is also different from in Euclidean space. It is generally defined by Mobius addition, and its formula is as follows:

$$\boldsymbol{u} \oplus \boldsymbol{v} = \frac{(1 + 2\langle \boldsymbol{u}, \boldsymbol{v}\rangle + \|\boldsymbol{v}\|^2)\boldsymbol{u} + (1 - \|\boldsymbol{u}\|^2)\boldsymbol{v}}{1 + 2\langle \boldsymbol{u}, \boldsymbol{v}\rangle + \|\boldsymbol{v}\|^2\|\boldsymbol{u}\|^2} \tag{2}$$

3.3 Task Definition

The task of knowledge graph completion is to predict unknown nodes based on existing nodes and relations in the knowledge graph G. For example, predicting the tail entity with the given entities and relation$(h, r, ?)$. Specifically, the task is to design a scoring function that assign a higher score for positive sample triple than the negative sample triple.

4 Our Method

The goal of this work is to learn a model that can encode complex relations such as 1-N, N-1 and N-N. As depicted in Fig. 1, our model contains four key parts:

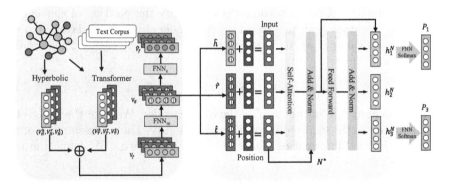

Fig. 1. An architecture overview of our model. The framework of the model contains four key modules: (1)Textual Embedding; (2)Hyperbolic Translation; (3)Information Fusion Network; (4)Contextualized Embedding

(1) Textual Embedding: Based on our approach in previous papers, we use external descriptive text for semantic enhancement of entities through encoders.

(2)Hyperbolic Translation: We translate the fact triplet of the knowledge graph into hyperbolic space and this will be the spatial information as the part of information fusion.

(3) Information Fusion Network: We extract features by maximizing the correlation between inputs from multiple information sources.

(4) Contextualized Embedding: For entities and relations after information fusion, we use Transformer to encode and obtain contextual representations of entities and relationships.

4.1   Textual Context Embedding

In order to associate entities in the knowledge graph with external text corpus, we employ an entity linking tool AIDA for entity annotation. Referring to the co-occurrence network used by TEKE [22], we obtain a unique description text for each entity. Given a knowledge graph G and an external text corpus $T = \{S_1, S_2, \cdots, S_n\}$, where S_i represents the text statements in the external text corpus. We represent the input sequence as $S = (w_0, w_2, \cdots, w_n)$, where w_0 represents the [CLS] tag and the other elements represent words in the text statement. For each word in the input, we pre-process the position information to the representation of each word as follows:

$$\mathbf{w}_i^0 = \mathbf{w}_i^{ele} + \mathbf{w}_i^{pos} \tag{3}$$

The main function of position coding is to identify the position of elements in a sequence. We stack multiple Transformer encoders into coding layers and use the output of each layer as the input of the next layer of encoders. The encoding formula is as follows:

$$\mathbf{w}_i^l = \text{Transformer} - \text{Encoder} \left(\mathbf{w}_i^{l-1} \right) \quad l = 1, 2, \cdots, L \tag{4}$$

where $\mathbf{w}_i^l$ denotes the output of L-th layer of encoder. We take the [CLS] tag representation as the final representation of the entity. The representation of the entity has learned information about the description text. We define the text enhanced entity as V_T.

$$V_T = \mathbf{w}_0^L \tag{5}$$

4.2 Hyperbolic Translation

Given a triple of a knowledge graph, the approach of translation models usually treats the relationship as a translation from head entity to tail entity. The biggest problem with these translation models in Euclidean space is that they require high dimensions to achieve adequate representation of entities. The spatial properties of hyperbolic space make it more accurate to express the data of hierarchy and topological structure. Hyperbolic space is curved, and the carrying capacity of the space increases exponentially with the distance from the center. Hence, vectors that require higher dimensions in Euclidean space can be simply represented in lower dimensions in hyperbolic space. The related geometric theory of hyperbolic space can be found in Sect. 3.2.

Based on previous work in this paper, we use hyperbolic relation-specific transformations on entity representations and this will lead to high complexity overhead. In our exploration of hyperbolic space, we refer to the translation method used in HyperKA [17] and treat entity relations as translations like TransE, avoiding complex overhead. Therefore, our distance function is defined for a triple $\nu = (h, r, t)$:

$$S(\nu) = d \left(V_H^h \oplus V_H^t, V_H^r \right) \tag{6}$$

where V_H^h, V_H^r, V_H^t denote the embedding for h, r, t respectively. We minimize the following constrastive learning loss:

$$\text{Loss} = \sum_{(\nu) \in G} \sum_{(\nu') \in G^-} [\gamma + S(\nu) - S(\nu')]_+ \tag{7}$$

where ν' denotes the randomly generated negative sample training data in knowledge graph. γ is the margin and we hope $S(\nu') ¿ \gamma$.

Finally, we define the vector after hyperbolic TransE as V_H.

4.3 Information Fusion

Our previous work uses relations-specific hyperbolic transformation on the text-enhanced representation of triples in the knowledge graph, which achieves better representation than text-enhanced and hyperbolic representation alone. However experimental results show that this leads to high complexity overhead. Inspired by Multi-modal Machine Learning, we obtain the enhanced triplet $\left(V_T^h, V_T^r, V_T^t\right)$ and the hyperbolic space mapped triplet $\left(V_H^h, V_H^r, V_H^t\right)$ through text enhancement and hyperbolic TransE respectively. We consider these two triples from different sources or forms of information as two modals. Due to different modals show different ways and view things from different perspectives, there will be some crossover in the fusion and process, which will lead to redundancy of information. But this also brings complementary features that are more expressive than a single feature. The text enhanced entity has the description information of the entity itself, and the semantic information is more complete. Entities after Hyperbolic TransE pay more attention to the hierarchical structure of the entity itself, making it more expressive. Inspired by our previous work and the traditional feature fusion algorithm [26–28], we propose a novel method to fuse the two different representations of fact triples in the knowledge graph. We consider enhanced triplet $\left(V_T^h, V_T^r, V_T^t\right)$ and hyperbolic TransE triplet $\left(V_H^h, V_H^r, V_H^t\right)$ as the input. Then we first concatenate the entity and relation separately to obtain a triple $\left(V_f^h, V_f^r, V_f^t\right)$. Then we reduce the dimension of head entity, relation entity and tail entity respectively through a FNN layer to $\left(V_E^h, V_E^r, V_E^t\right)$. Then, we use this to reconstruct the originally concatenated triple $\left(\hat{V}_f^h, \hat{V}_f^r, \hat{V}_f^t\right)$. We use two Feed-forward Neutral Network as the transformation layer to produce the fused vector. Referring to [15], we used the mean square error as the loss function of multi-modal feature fusion. Taking the head entity as an example, the MSE loss function is as follows:

$$Loss_f = \left\| \hat{V}_f^h - V_f^h \right\|^2 \tag{8}$$

Through the Information Fusion part, we consider the middle triple $\left(V_E^h, V_E^r, V_E^t\right)$ as the information fusion representation. In section4.4 we convert $\left(V_E^h, V_E^r, V_E^t\right)$ to $\left(\hat{h}, \hat{r}, \hat{t}\right)$ for symbolic simplicity.

4.4   Contextualized Embedding

Context information can help entities generate their representations dynamically, making them better able to handle complex relationships. Inspired by Quan Wang et al., we refer to the most basic form of graph context in their model to learn about contextual information embedding. We learn context information by masked multi-head self-attention. The idea of Transformer and Bert is well known to most people and the specific formulas are not shown here. The more detail of formula can check the paper [5].

Since we get the information fusion triple $\left(\hat{h}, \hat{r}, \hat{t}\right)$, we see this triple as the input element of the Transformer. For each input triple we add the position embedding and then obtain the final input:

$$h_i^0 = \hat{h} + \hat{h}^{pos} \tag{9}$$

After through the Transformer Decoder, we obtain:

$$\hat{h}_i^N = \text{Transformer}\left(\hat{h}_i^{N-1}\right), \quad N = 1, 2, \cdots, N, \tag{10}$$

During training, inspired by Coke [20], we create two training instances. For a triple (h, r, t), one is to replace t and the other is to replace h, then we put this triple into the Transformer encoder to generate a new sequence $\left(\hat{h}_1^n, \hat{h}_2^n, ...\hat{h}_n^N\right)$. After a feedforward layer and a standard softmax classification layer, we get the target predicted entity:

$$\mathbf{P}_1 = \text{softmax}\left(\mathbf{W}f\left(h_1^N\right)\right) \tag{11}$$

$$\mathbf{P}_n = \text{softmax}\left(\mathbf{W}f\left(h_n^N\right)\right) \tag{12}$$

where $\mathbf{W} \in \mathbb{R}^{V \times D}$ is the weight embedding matrix, D is the hidden size, V is the entity size. As to Coke, we use cross-entropy between the one-hot label y_1/y_n and the prediction P_1/P_n as training loss:

$$Loss(X) = -\sum_t y_t \log p_t \tag{13}$$

where y_t and p_t are the t-th components of y_1/y_2 and P_1/P_2. As same as the model Coke, we use a label smoothing strategy to lessen this restriction.

5   Experiment

We measure our model's effectiveness of representations in link prediction. The subsequent section describe our complete experimental setup, including baselines, datasets, evaluation metrics and experimental analysis.

5.1   Datasets

We choose four widely used knowledge graph datasets and an external text corpus to train and evaluate the model, which are described as follows:

Knowledge Graph Dataset. The complete knowledge graph is very large, so its subset is generally used to evaluate the performance of knowledge graph representation learning method. We adopt four KG datasets, namely, FB15k, WN18, FB15K-237 and WN18RR. FB15k and WN18 are subsets of Freebase and WordNet respectively. The detailed statistical data is shown in Table 1.

Text Corpus. The text corpus is introduced in (Zhigang Wang et al., 2016), sampling from Wikipedia. This corpus consists mainly of unstructured natural language documents composed of natural language statements. Wang et al. deleted the external text corpus and improved the quality of the corpus. Finally, the pre-processed text corpus and entities on KG are linked, and the statistical data results are shown in Table 2.

Table 1. KG dataset statistical information in experiment.

Dataset	FB15k	WN18	FB15k-237	WN18RR
Entities	14951	40,493	14541	40493
Relations	1245	18	237	11
Train	483,142	141,442	272,115	86,835
Valid	50,000	5,000	17,535	3,034
Test	59,071	5,000	20,466	3,134

Table 2. Text corpus and KG entity alignment result statistics.

KG	Entities	Annotated entities	Word stems
WN18	40,943	32,249	1,529,251
FB15K	14,951	14,405	744,983

5.2 Evaluation Protocol

Link prediction task is a commonly used knowledge graph representation learning evaluation method. Given a correct triplet (h, r, t), the task of link prediction is to predict the missing h or t in the case of $(?, r, t)$ or $(h, r, ?)$. In the evaluation part, h or t in each triplet (h, r, t) in the test data set is replaced with **Mask**, and then this sequence is put into our model to obtain the distribution of all entities, and then conduct the descending order according to the predicted value, and finally get the rank of the predicted entity. In this paper, Mean Reciprocal Rank(MRR) and Hist@n are used as evaluation indexes of the model. MRR is the average of the reciprocal ranking of the correct entities in all test samples in the test triples. Hits@n means the proportion of the valid test triples ranking in top n predictions. Higher MRR or higher Hits@n indicate better performance.

5.3 Baslines

To verify the express ability of our model, We compare our model with several state-of-art methods, including method that in Euclidean Space(TransE [3], TransR [11]), method that in Complex Space(RotatE [18]), method that in Hyperbolic Space(HyperKG [7]) and method that use context information and external information (TEKE [22], Coke [20]).

- **TransE** is the first vector translation based model by assuming that the head entity embedding should be close to the tail entity embedding after relational embedding translation.
- **TransR** is a variant of TransE. TransR improves the modeling capability of complex relationships by embedding entities and relationships into different semantic Spaces through mapping matrices.
- **TEKE** uses textual description information of entities and context of relations to improve the effect of knowledge embedding.

- **RotatE** extends TransE's ideas by representing entities into complex space. RotatE regards relations as rotations from head entity to tail entity.
- **CoKE** proposes that the representation of entity and relation should represent dynamically depending on the context in which they are expressed. Triples are converted into sequential inputs through Transformer model.

Table 3. Link prediction results on FB15k and WN18 dataset.

Evaluation Protocol	FB15K				WN18			
	MRR	Hits@1	Hits@3	Hits@10	MRR	Hits@1	Hits@3	Hits@10
Method that in Euclidean Space:								
TransE	0.523	0.476	0.528	0.572	0.541	0.513	0.549	0.615
TransR	0.556	0.484	0.544	0.605	0.548	0.52	0.569	0.642
TEKE	0.672	0.612	0.634	0.735	0.677	0.623	0.643	0.734
Coke	0.744	0.691	0.737	0.842	0.818	0.825	0.876	0.894
Method that in Complex Space:								
RotatE	0.712	0.708	0.714	0.732	0.757	0.73	0.759	0.787
Method that in Hyperbolic Space:								
HyperKG	0.687	0.656	0.693	0.634	0.739	0.724	0.735	0.766
Our method	0.807	0.732	0.785	0.867	0.824	0.838	0.887	0.906

5.4   Main Results

To verify the effectiveness of our model, we compare with several representative and widely used representation learning models. The experimental results of FB15K and WN18 are shown in Table 3. Part of the experimental results refer to the original paper, part is the result of self-tuning. Compared with the original paper of Baseline, there is a gap between our results and the Baseline, which may be due to data filtering. The optimal model in Baseline is CoKE. This illustrates the validity of considering that entities are dynamically generated depending on their context. On datasets FB15K and WN18, our model outperforms other models in terms of evaluation indicators. On FB15K, Hist@10 and MRR increased by 2.5% and 6.3%, respectively. This shows that it is useful to add more semantic and spatial considerations when doing dynamic contextual representations.

5.5   Complex Relation Study

Focus on the complex relations especially for the issue of 1-1, 1-N, N-1 and N-N, we conduct experiments on different relation categories. We choose the FB15k-237 as our datasets due to its abundant multi-relations and denser graph structure. We choose Rotate [18] as the baseline. From Fig. 2, we can find that our model is superior to the Baseline model in handling complex relationships. Compared with RotatE, our model has similar prediction performance in simple relationships 1-1 and 1-N. But in the case of complex relationships N-1and N-N, the predictive power of our model improved. Significantly. This indicates that considering contextual information is beneficial to improve the ability of our model to deal with complex relations.

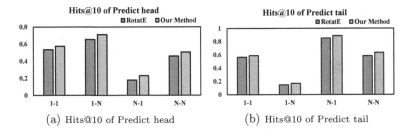

(a) Hits@10 of Predict head   (b) Hits@10 of Predict tail

Fig. 2. Complex relation test on FB15k-237 dataset. Following the [21], the relations are divided into four categories: one-to-one (1-1), one-to-many (1-N), many-to-one (N-1), and many-to-many (N-N).

5.6   Ablation Study

We have given the comparisons and analysis about different models and our model has achieved a significant improvement over all baselines. In this section, in order to further analyze the enhancement effect of hyperbolic spatial information and descriptive text information on dynamically generated entity representation, we conducted an ablation experiment on the proposed model. We do the experiment on FB15k-237 and WN18RR dataset. Our model adds external semantic information and hyperbolic space embedded information modules on the basis of CoKE. In this section, we mainly analyze the improvement effect of external information embedding on the model.

Textual Embedding. In this section, we analyze the importance of external text and how the external semantic information affect the performance of our model. As we can see from Fig. 3. Compared with TEACH(Contextualized only), the performance of the model with external text improved. We can observe that, with the help of external text information, our model can improve input constraint. A possible reason is that the fusion of external text information can enhance the semantic information of the input. It will not lose semantic information when doing context training later.

Hyperbolic Translation. Compared with no hyperbolic spatial information, the effect of hyperbolic spatial information is obviously improved. We think it is the properties of hyperbolic space that capture the hidden information in the triplet and improve the representation ability. It can be seen that the effect of adding external text description information is better than adding hyperbolic spatial information only, but the two kinds of information do not completely coincide. Adding description information and spatial information at the same time can add constraints to the representation of dynamically generated entities and relations in context and make it more accurate.

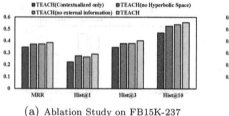
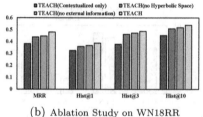

(a) Ablation Study on FB15K-237   (b) Ablation Study on WN18RR

Fig. 3. Performance comparisons of information fusion model. Contextualized only; information fusion without hyperbolic space; Information fusion without external information.

6   Conclusion

Most of the existing knowledge graph representation models often ignore the context information of entities in the knowledge graph. This will lead to Knowledge Graph Representation Learning(KGRL) difficult to resolve complex relations. The existing Internet has a large amount of description text of entities, which helps to improve the ability of entity representation. In this work, We propose a new model, which considers the influence of external semantic information and different contexts on entity representation. We also consider the ability of hyperbolic space to capture hidden information, which makes it more suitable for knowledge representation learning tasks of knowledge graph. To fuse the external description information and the information captured in hyperbolic space, we propose a novel information fusion method for information aggregation. Experimental results show that adding external information constraints can improve the dynamic context representation of entities and relations.

As future work, we will simultaneously consider the loss in the process of knowledge fusion training and the loss in the process of context information training, because limit the fused information may not be well suited to downstream mission objectives. Finally, we will focus more on the coding of knowledge graph relational schema, including improving the coding capabilities of symmetric/antisymmetric, inverse, composite and subrelation schema.

Acknowledgements. This work was supported partly by the National Key R&D Program of China(2020YFB1708100), National Natural Science Foundation of China(62172351), the 14th Five-Year Plan "Civil Aerospace Pre-research Project of China (D020101), Fundamental Research Funds for the Central Universities(NS2019001), the Fund of Prospective Layout of Scientific Research for NUAA(Nanjing University of Aeronautics and Astronautics.

References

1. Auer, S., Bizer, C., Kobilarov, G., Lehmann, J., Cyganiak, R., Ives, Z.: Dbpedia: A nucleus for a web of open data. In: The semantic web, pp. 722–735 (2007)
2. Bollacker, K., Evans, C., Paritosh, P., Sturge, T., Taylor, J.: Freebase: a collaboratively created graph database for structuring human knowledge. In: Proceedings of the 2008 ACM SIGMOD International Conference on Management of Data, pp. 1247–1250 (2008)
3. Bordes, A., Usunier, N., Garcia-Duran, A., Weston, J., Yakhnenko, O.: Translating embeddings for modeling multi-relational data. In: Advances in Neural Information Processing Systems, vol. 26 (2013)
4. Dettmers, T., Minervini, P., Stenetorp, P., Riedel, S.: Convolutional 2d knowledge graph embeddings. In: Thirty-second AAAI Conference on Artificial Intelligence (2018)
5. Devlin, J., Chang, M.W., Lee, K., Toutanova, K.: BERT: Pre-training of deep bidirectional transformers for language understanding. In: Proceedings of the 2019 Conference of the North American Chapter of the Association for Computational Linguistics: Human Language Technologies, Volume 1 (Long and Short Papers). pp. 4171–4186, Minneapolis, Minnesota (2019)
6. Kazemi, S.M., Poole, D.: Simple embedding for link prediction in knowledge graphs. In: Bengio, S., Wallach, H.M., Larochelle, H., Grauman, K., Cesa-Bianchi, N., Garnett, R. (eds.) Advances in Neural Information Processing Systems, vol. 31. In: Annual Conference on Neural Information Processing Systems 2018, NeurIPS 2018, December 3–8, 2018, Montréal, Canada, pp. 4289–4300 (2018)
7. Kolyvakis, P., Kalousis, A., Kiritsis, D.: Hyperkg: Hyperbolic knowledge graph embeddings for knowledge base completion. arXiv preprint arXiv:1908.04895 (2019)
8. Li, B.H., Liu, Y., Zhang, A.M., Wang, W.H., Wan, S.: A survey on blocking technology of entity resolution. J. Comput. Sci. Technol. **35**(4), 769–793 (2020)
9. Li, Z., Liu, X., Wang, X., Liu, P., Shen, Y.: Transo: a knowledge-driven representation learning method with ontology information constraints. World Wide Web, pp. 1–23 (2022)
10. Li, Z., Wang, X., Li, J., Zhang, Q.: Deep attributed network representation learning of complex coupling and interaction. Knowl.-Based Syst. **212**, 106618 (2021)
11. Lin, Y., Liu, Z., Sun, M., Liu, Y., Zhu, X.: Learning entity and relation embeddings for knowledge graph completion. In: Twenty-ninth AAAI conference on artificial intelligence (2015)
12. Liu, Y., Li, B., Zang, Y., Li, A., Yin, H.: A knowledge-aware recommender with attention-enhanced dynamic convolutional network. In: Demartini, G., Zuccon, G., Culpepper, J.S., Huang, Z., Tong, H. (eds.) CIKM '21: The 30th ACM International Conference on Information and Knowledge Management, Virtual Event, Queensland, Australia, November 1–5, 2021, pp. 1079–1088. ACM (2021)
13. Miller, G.A.: Wordnet: a lexical database for english. Commun. ACM **38**(11), 39–41 (1995)
14. Peng, W., Varanka, T., Mostafa, A., Shi, H., Zhao, G.: Hyperbolic deep neural networks: A survey. arXiv preprint arXiv:2101.04562 (2021)
15. Sahu, G., Vechtomova, O.: Adaptive fusion techniques for multimodal data. In: Merlo, P., Tiedemann, J., Tsarfaty, R. (eds.) Proceedings of the 16th Conference of the European Chapter of the Association for Computational Linguistics: Main Volume, EACL 2021, Online, April 19–23, 2021, pp. 3156–3166 (2021)

16. Sikos, L.F., Philp, D.: Provenance-aware knowledge representation: a survey of data models and contextualized knowledge graphs. Data Sci. Eng. 5(3), 293–316 (2020)

17. Sun, Z., Chen, M., Hu, W., Wang, C., Dai, J., Zhang, W.: Knowledge association with hyperbolic knowledge graph embeddings. In: Webber, B., Cohn, T., He, Y., Liu, Y. (eds.) Proceedings of the 2020 Conference on Empirical Methods in Natural Language Processing, EMNLP 2020, Online, November 16–20, 2020, pp. 5704–5716 (2020)

18. Sun, Z., Deng, Z., Nie, J., Tang, J.: Rotate: Knowledge graph embedding by relational rotation in complex space. In: 7th International Conference on Learning Representations, ICLR 2019, New Orleans, LA, USA, May 6–9, 2019 (2019)

19. Trouillon, T., Welbl, J., Riedel, S., Gaussier, É., Bouchard, G.: Complex embeddings for simple link prediction. In: International Conference on Machine Learning, pp. 2071–2080. PMLR (2016)

20. Wang, Q., et al.: Coke: Contextualized knowledge graph embedding. arXiv preprint arXiv:1911.02168 (2019)

21. Wang, Z., Zhang, J., Feng, J., Chen, Z.: Knowledge graph embedding by translating on hyperplanes. In: Proceedings of the AAAI Conference on Artificial Intelligence, vol. 28 (2014)

22. Wang, Z., Li, J., Liu, Z., Tang, J.: Text-enhanced representation learning for knowledge graph. In: Proceedings of International Joint Conference on Artificial Intelligent (IJCAI), pp. 4–17 (2016)

23. Xiao, H., Huang, M., Zhu, X.: From one point to a manifold: Knowledge graph embedding for precise link prediction. arXiv preprint arXiv:1512.04792 (2015)

24. Xie, R., Liu, Z., Sun, M., et al.: Representation learning of knowledge graphs with hierarchical types. In: IJCAI, pp. 2965–2971 (2016)

25. Yang, B., Yih, W., He, X., Gao, J., Deng, L.: Embedding entities and relations for learning and inference in knowledge bases. In: Bengio, Y., LeCun, Y. (eds.) 3rd International Conference on Learning Representations, ICLR 2015, San Diego, CA, USA, May 7–9, 2015, Conference Track Proceedings (2015)

26. Yin, H., Yang, S., Song, X., Liu, W., Li, J.: Deep fusion of multimodal features for social media retweet time prediction. World Wide Web 24(4), 1027–1044 (2021)

27. Zadeh, A., Chen, M., Poria, S., Cambria, E., Morency, L.: Tensor fusion network for multimodal sentiment analysis. In: Palmer, M., Hwa, R., Riedel, S. (eds.) Proceedings of the 2017 Conference on Empirical Methods in Natural Language Processing, EMNLP 2017, Copenhagen, Denmark, September 9–11, 2017, pp. 1103–1114 (2017)

28. Zhao, X., Jia, Y., Li, A., Jiang, R., Song, Y.: Multi-source knowledge fusion: a survey. World Wide Web 23(4), 2567–2592 (2020). https://doi.org/10.1007/s11280-020-00811-0

Knowledge Graph Entity Alignment Powered by Active Learning

Jiayi Pan and Weiguo Zheng[✉]

School of Data Science, Fudan University, Shanghai, China
{20210980084,zhengweiguo}@fudan.edu.cn

Abstract. Considering the diversity and heterogeneity of different knowledge graphs, it is necessary to logically establish a comprehensive, accurate and unified knowledge repository. We design a framework by importing active learning strategies to neural network models for entity alignment, aiming to create informative seeds for more efficient entity alignment models with lower annotation cost. The model measures the benefit of an entity being selected from the two aspects of its uncertainty and influence. Extensive experiments are conducted on two benchmark datasets, and the results show that our method achieves significant improvement over the existing models.

Keywords: Knowledge graph · Entity alignment · Active learning · Neural networks

1 Introduction

Knowledge graph fusion is an important link from knowledge graph construction to knowledge graph-based intelligence. Through knowledge graph fusion, the relevant knowledge in various graph systems can be organically complemented and integrated, and a comprehensive, accurate and systematic knowledge graph entity description can be established. The main purpose of entity alignment is to determine whether two or more objects from different sources refer to the same object in the real world.

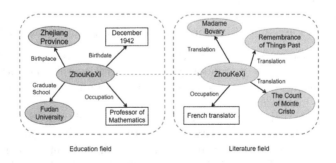

Fig. 1. An example of knowledge graph entity alignment in different domains

Current mainstream machine alignment models use embedding-based methods to extract semantic information about entities in knowledge graph, whether based on translation models that understand relationships as translations from their head entities to their tail entities, or graph neural networks that capture structural information about entity network neighbours to form entity representations, these methods are based on training the model with a sample set of seed-aligned entities, and the alignment effectiveness of the model is influenced by the quality of seed alignment data.

However, many entities can only get limited attention and have low embedding expression ability. So it can be seen the importance of adding human intervention to the machine alignment model. For example, Fig. 1 shows the knowledge graphs of the education field and the literature field. Due to the differences of knowledge graphs in specific fields, the machine cannot learn similarity information from entity neighbors. If the annotations are made by experts in the field of literature, it can be easily recognized that the two central entities actually refer to the same person.

This paper makes the following contributions:

- By applying active learning strategies to the entity alignment task, the most valuable and informative data are picked up for annotation;
- We design a new sample selection strategy by considering the uncertainty and influence of the entity;
- We summarize and analyze the characteristics of the aligned neural network suitable for active learning strategy.

2   Method

2.1   Model Framework

The task of this paper is to align entities across different knowledge graphs. A knowledge graph KG can be represented by $KG = (E, R, T)$, where E denotes a set of entities, R denotes a set of relations, and $T \subseteq E \times R \times E$ denotes a set of triples. The entity alignment task is to find such an entity pair $A = \{(e_1, e_2) \in E_1 \times E_2 | e_1 \sim e_2\}$ given two knowledge graphs KG_1 and KG_2 whose sets of entities are E_1 and E_2 respectively, with $\sim$ here indicating that both entities refer to the same object in the real world.

Figure 2 shows the overall framework of entity alignment based on active learning, where the annotation process is iterated with model training.

(1) The entity sample set is first loaded by the given dataset, and the initialization is generally done by randomly selecting entities from the entity sample set to form the training set, and the machine alignment model is trained to obtain the matching scores.

(2) Based on the calculated matching scores, a sample selection strategy is designed to obtain the entity pairs with the greatest gain, which are given to the experts for annotation, with aligned entity pairs labelled as $L^+ = (e_1 \in E_1, e_2 \in E_2)$ and unmatched entities labelled as $L^- = (e_1 \in E_1, e_2 \in null)$.

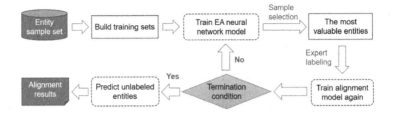

Fig. 2. A holistic framework for entity alignment based on active learning

(3) Add the resulting annotated data to the training set to train the alignment model again, and if a termination condition is achieved, the trained alignment model is used to predict the unannotated entities to obtain the entity alignment results.

(4) If the termination condition is not yet satisfied, the entity pair with the highest return in the sample set is selected for annotation and used to train the alignment model according to the sample selection strategy until the termination condition is achieved.

2.2 Sample Selection

Based on Uncertainty Reduction for Entities. For each entity in KG_1 the matching scores for all unmatched entities in KG_2 is obtained by the machine entity alignment model and is denoted as $EN(e_1, e_2)$. To obtain the entities that the current machine alignment model is least confident of matching, a marginal-based uncertainty measure is used.

$$U(e_1) = -(EN(e_1, e_2^*) - EN(e_1, e_2^{**})), \tag{1}$$

where e_2^* and e_2^{**} denote the entity e_1 with the largest and second largest matching score in KG_2 respectively. The smaller the largest matching score and the second largest matching score, the greater the uncertainty of the e_1 match.

Based on Entity's Influence on Its Neighbours. Since entities in the knowledge graph are interconnected, it is important to consider both the impact of the entity itself and the impact of the entity on its neighbours. Based on this, the final entity influence $Inf(e_1)$ can be defined as the amount of uncertainty that an entity can help its neighbours to eliminate.

$$Inf\left(e_1^i\right) = \alpha \sum_{e_1^i \to e_1^j, e_1^j \in \mathcal{N}_{out}^i} w_{ij} Inf\left(e_1^j\right) + (1-\alpha)\frac{U\left(e_1^i\right)}{\sum_{e_1 \in en_1} U(e_1)}, \tag{2}$$

where $\mathcal{N}_{out}^i$ denotes all entities pointed to by entity e_1^i, i.e., outgoing neighbours of e_1^i, and parameter α is used to balance the effect of entity e_1^i on context with the marginal uncertainty of e_1^i after normalisation. If e_1^i is connected to e_1^j, then $w_{ij} > 0$, otherwise $w_{ij} = 0$. For each entity e_1^j, assuming that its inbound neighbours can help remove all uncertainty, then there is $\sum_{e_1^i \to e_1^j, e_1^i \in \mathcal{N}_j^{in}} w_{ij} = 1$. Written in the form of matrix operations, it is

$$U = \alpha W Inf + (1 - \alpha)\frac{U}{|U|}, \tag{3}$$

where W represents the influence matrix between entities, U is the measurement uncertainty vector, and Inf is the entity influence vector. The annotation process for active learning is an iterative process.

2.3 Model Learning

Based on the cross-language embedding model [2], TransE is used to embed each language, and the entities and relationships of multiple languages are encoded into independent spatial structures.

The model based on aggregating distant neighbors [7] aggregates the information of direct neighbors and distant neighbors, adopts an attention mechanism and a gate mechanism to compose the output description of multiple aggregation functions.

The semi-supervised entity alignment model [4], which utilizes a small amount of label data and a large amount of unlabeled entity information for alignment, considers that the knowledge graph entity alignment is affected by the entity degree. Models based on relational path learning learn through relational paths [3], and without losing local relational information of entities, paths can provide more relational dependency information than triples.

3 Experimental Study

3.1 Experimental Setting

Dataset. There are two data sets in this experiment, each dataset contains two knowledge graphs and aligned entity pairs, both obtained from OpenEA [8]. EN-DE-15K V1 is a cross-lingual dataset of English and German. D-W-15K V1 is a monolingual dataset of English, both datasets have 15,000 entities.

Baseline Sampling. Several of the more classical sample selection methods were selected for comparison.

(1) Random sampling
    The random sampling strategy selects a random set of entities from those prepared for matching for annotation.
(2) PageRank
    PageRank is a link analysis algorithm that determines the centrality of an entity object by taking into account its degree and the significance of neighbouring entities.
(3) Uncertainty sampling
    Based on the marginal uncertainty measure to be used as the sampling criterion, entities with high uncertainty are selected for labelling.

Table 1. Alignment results of different sampling strategies on dataset D-W-15K V1

EA model	Alinet			SEA			RSN4EA		
Metrics	HIT@1	HIT@5	NDCG@10	HIT@1	HIT@5	NDCG@10	HIT@1	HIT@5	NDCG@10
Rand	62.280	81.453	0.736	52.053	74.320	0.661	58.733	76.560	0.697
PageRank	54.213	82.200	0.711	44.520	75.093	0.635	52.067	79.067	0.688
Uncertainty	**75.813**	88.800	0.836	53.680	76.147	0.679	66.347	80.493	0.754
ALEA	73.520	**90.280**	**0.841**	**54.227**	**77.04**	**0.689**	**70.067**	**87.747**	**0.807**

Table 2. Alignment results of different sampling strategies on dataset EN-DE-15K V1

EA model	MTransE			Alinet			SEA		
Metrics	HIT@1	HIT@5	NDCG@10	HIT@1	HIT@5	NDCG@10	HIT@1	HIT@5	NDCG@10
Rand	**49.160**	**70.733**	0.631	69.333	83.973	0.783	63.747	83.173	0.762
PageRank	40.493	65.000	0.566	57.853	79.040	0.714	52.600	77.427	0.686
Uncertainty	24.400	38.133	0.336	82.347	91.853	0.880	61.293	80.520	0.736
ALEA	48.907	68.440	**0.642**	**86.800**	**95.413**	**0.920**	**69.253**	**84.680**	**0.791**

Evaluation Indicators. The evaluation metrics used in this paper are HIT@K and NDCG@K, which are commonly used in entity alignment tasks.

In the formula for calculating the degree of entity influence, we set the parameter $\alpha = 0.1$, and the batch size is set to 100 when comparing the experimental effects of different benchmark models.

3.2 Results

Comparison with Baseline Models. The entity influence sampling strategy based on the knowledge graph structure proposed in this paper is applied to the models MTransE, SEA, Alinet and RSN4EA respectively and compared with the benchmark sampling strategy. The experimental results obtained on this dataset are shown in Table 1.

(1) Overall, Alinet achieves better experimental results than the other two models. It illustrates the need to utilize entity neighborhood structures as an important breakthrough to improve the overall effectiveness of entity alignment.

(2) ALEA has the most obvious improvement effect on the RSN4EA model. RSN4EA improves the ability to capture long-term relationship dependencies of entities, and the ALEA strategy effectively addresses the problem lack of seed alignment data.

We applied the model on the machine algorithms MTransE, Alinet and SEA, and the experimental results obtained are shown in Table 2.

(1) The uncertainty sampling effect is poor. It seems that relying only on the uncertainty based on the margin is not feasible.

(2) The better alignment results of Alinet and SEA models after applying the ALEA strategy indicates that the strategy can exert its effect on cross-language knowledge graphs.

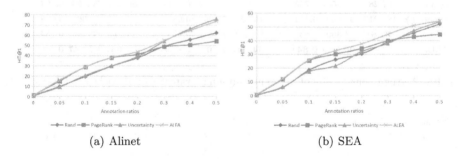

(a) Alinet   (b) SEA

Fig. 3. Experimental results of sampling strategies with different labelling ratios (D-W-15K V1)

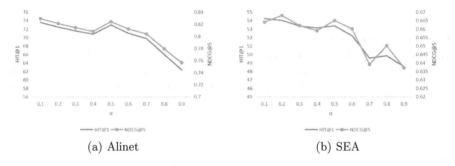

(a) Alinet   (b) SEA

Fig. 4. Experimental results of ALEA strategy with different α (D-W-15K V1)

Experimental Results at Different Labelling Ratios. We obtained the experimental results for different annotation ratios, as shown in Fig. 3. The ALEA strategy achieves good results in all annotation ratios and is similar to PageRank when the annotation ratio is less than 15%, but the advantages are more obvious as the amount of annotation increases.

Analysis of the parameter α. We apply our strategy to the dataset D-W-15K V1 using the Alinet model and the SEA model and obtained the results shown in Fig. 4. It can be observed that the strategy in this paper is not particularly sensitive to α, and the experimental results for the two models are somewhat different.

Analysis of the Parameter Batchsize. Considering that different batch sizes may have an impact on the training effect, we applied the ALEA strategy to the models Alinet and SEA respectively. The results are presented in Fig. 5.

The improvement effect of ALEA on RSN4EA is the most obvious, the ALEA strategy effectively solves the current challenges of the model and improves the overall performance of entity alignment.

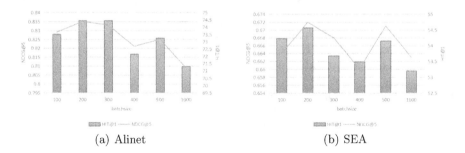

(a) Alinet   (b) SEA

Fig. 5. Experimental results of ALEA strategy with different batchsize(D-W-15K V1)

4   Related Work

Currently, the approaches to entity alignment include the traditional methods based on the calculation of similarity features, as well as embedding-based methods. In the translation model, Bordes et al. [1] propose the TransE model, which focuses on the relational triad and treats relations as translations from their head entity to their tail entity. To more adequately represent entity information, some translation models consider the inclusion of attribute information [6,10,11].

Graph neural networks use node features and graph-like structures to learn the representation vectors of nodes. The graph neural network uses a neighbourhood aggregation strategy to iteratively update the representation of a node. The heterogeneity of the knowledge graph structure is a major challenge for entity alignment. For complex relational information commonly found in multi-relational knowledge graphs, some graph neural network-based entity alignment methods consider the role of relationship types in feature aggregation. AVR-GCN [9] proposes a vector relational graph convolutional network to simultaneously learn embeddings of graph entities and relations in multi-relational networks. Yao et al. proposed a new relationship-aware neighbourhood matching model, RNM [12], designing an iterative framework to perform positive interactions between entity alignment and relationship alignment in a semi-supervised manner.

However, there are still some challenges in applying the current methods to large knowledge graphs. Active learning strategies have been widely used in many machine learning fields Qian et al. [5] applied active learning to traditional entity alignment algorithms. With the wide application of deep learning, applying the active alignment algorithm to the neural network-based entity alignment becomes a new entry point.

5 Conclusion

In this paper, we propose a framework for entity alignment based on active learning methods to design effective entity alignment strategies to create information-rich seed alignments to obtain more effective entity alignment models with lower annotation costs. Four representative machine alignment algorithms were selected for experimental analysis, in order to further explore the adaptability of the proposed strategy. Extensive experiments are conducted to confirm the effectiveness of our method.

Acknowledgement. This work was supported by National Natural Science Foundation of China (Grant No. 61902074) and Science and Technology Committee Shanghai Municipality (Grant No. 19ZR1404900).

References

1. Bordes, A., Usunier, N., Garcia-Duran, A., Weston, J., Yakhnenko, O.: Translating embeddings for modeling multi-relational data. In: Advances in Neural Information Processing Systems, vol. 26 (2013)
2. Chen, M., Tian, Y., Yang, M., Zaniolo, C.: Multilingual knowledge graph embeddings for cross-lingual knowledge alignment. arXiv preprint arXiv:1611.03954 (2016)
3. Guo, L., Sun, Z., Hu, W.: Learning to exploit long-term relational dependencies in knowledge graphs. In: International Conference on Machine Learning, pp. 2505–2514. PMLR (2019)
4. Pei, S., Yu, L., Hoehndorf, R., Zhang, X.: Semi-supervised entity alignment via knowledge graph embedding with awareness of degree difference. In: The World Wide Web Conference, pp. 3130–3136 (2019)
5. Qian, K., Popa, L., Sen, P.: Active learning for large-scale entity resolution. In: Proceedings of the 2017 ACM on Conference on Information and Knowledge Management, pp. 1379–1388 (2017)
6. Sun, Z., Hu, W., Li, C.: Cross-lingual entity alignment via joint attribute-preserving embedding. In: d'Amato, C., et al. (eds.) ISWC 2017. LNCS, vol. 10587, pp. 628–644. Springer, Cham (2017). https://doi.org/10.1007/978-3-319-68288-4_37
7. Sun, Z., et al.: Knowledge graph alignment network with gated multi-hop neighborhood aggregation. In: Proceedings of the AAAI Conference on Artificial Intelligence. vol. 34, pp. 222–229 (2020)
8. Sun, Z., et al.: A benchmarking study of embedding-based entity alignment for knowledge graphs. arXiv preprint arXiv:2003.07743 (2020)
9. Ye, R., Li, X., Fang, Y., Zang, H., Wang, M.: A vectorized relational graph convolutional network for multi-relational network alignment. In: IJCAI, pp. 4135–4141 (2019)
10. Zhang, Q., Sun, Z., Hu, W., Chen, M., Guo, L., Qu, Y.: Multi-view knowledge graph embedding for entity alignment. arXiv preprint arXiv:1906.02390 (2019)
11. Zhao, X., Zeng, W., Tang, J., Li, X., Luo, M., Zheng, Q.: Toward entity alignment in the open world: an unsupervised approach with confidence modeling. Data Sci. Eng. **7**(1), 16–29 (2022)
12. Zhu, Y., Liu, H., Wu, Z., Du, Y.: Relation-aware neighborhood matching model for entity alignment. arXiv preprint arXiv:2012.08128 (2020)

POSE: A Positional Embedding Model for Knowledge Hypergraph Link Prediction

Zirui Chen, Xin Wang$^{(\boxtimes)}$, Chenxu Wang, and Zhao Li

College of Intelligence and Computing, Tianjin University, Tianjin, China
{zrchen,wangx,cxwang1998,lizh}@tju.edu.cn

Abstract. Link prediction in knowledge hypergraphs has been widely recognized as crucial for various downstream tasks of knowledge-enabled applications, from question answering to recommender systems. However, most current approaches are directly extended from binary relation of the knowledge graph to n-ary relation, thus cannot capture entities' role and positional information in each n-ary tuple. To accommodate the transformation of relations from binary to n-ary in the knowledge hypergraph, in this work, we propose POSE, which exploits the semantic properties of tuples at both role and position levels. POSE explores an embedding space with basis vectors and represents the role and positional information of entities through a linear combination, which promotes similar representations for entities with related roles and the same positions. Then, a relation matrix is further employed to capture the compatibility of both information with all associated entities, and a scoring function is used to measure the plausibility of tuples composed of entities with specific roles and positions. Meanwhile, POSE achieves full theoretical expressiveness and predictive efficiency. Experimental results show that POSE achieves an average improvement of 4.1% on MRR compared to state-of-the-art knowledge hypergraph embedding methods. Our code is available at https://github.com/zirui-chen/POSE.

Keywords: Link prediction · Knowledge hypergraph · Postional embedding

1 Introduction

Knowledge graphs describe real-world knowledge in the form of triples, i.e., (h, r, t), where r is a binary relation while h and t are head and tail entities. As a more expressive generalization of the knowledge graph, the *knowledge hypergraph* attracts more attention due to its generality in modeling real-world scenarios. In the Freebase, more than one-third of entities participate in non-binary relations [11], and 61% of relations are non-binary [18], which raises the importance of investigating how knowledge hypergraphs can be leveraged to enhance various downstream tasks such as link prediction and node classification.

B. Li et al. (Eds.): APWeb-WAIM 2022, LNCS 13422, pp. 323–337, 2023.
https://doi.org/10.1007/978-3-031-25198-6_25

Unlike binary relations, *n-ary relations* describe relationships involving more than two entities and contain more complex semantics. As in the example in Fig. 1, an oval represents a *tuple*; a circle represents an entity. The entities in a tuple are ordered, each entity has a different *role* at a different *position*, and the semantics of the tuple is determined by all the entities involved. The significance of roles and positions for modeling knowledge hypergraphs is evident from this example. However, there is no existing work that uses role and positional information on knowledge hypergraph modeling.

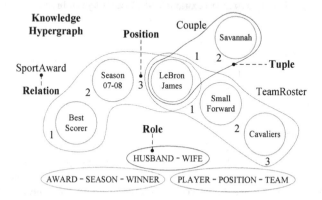

Fig. 1. An example of Lebron James in the form of the knowledge hypergraph. Each relation is composed of entities with corresponding semantic roles at different positions; the roles can be explicit (as in this figure) or implicit, while the positions are explicit.

Several works have aimed at the link prediction task on knowledge hypergraphs. However, existing methods [11–13,18,29] largely ignore the importance of roles and positions, and still follow the way of binary modeling in the knowledge graph. In particular, these methods embed *n*-ary relations and entities into a low-dimensional space without distinguishing the specific order of entities, and measured the plausibility of tuples based on these embeddings. For example, m-TransH [11] and RAE [12] both extend the knowledge graph model TransH [8] by projecting entities onto relation-specific hyperplanes for tuple plausibility scoring, but with weak expressivity [13,19]; while HypE [18] and GETD [13] extend the knowledge graph model SimplE [19] and TuckER [20], respectively. However, these models ignore the role and position semantics completely [17]. On the other hand, the role information has been adopted by NaLP [14], HINGE [16], and NeuInfer [17], all of which utilize neural networks to measure the tuple plausibility. However, these models only learn the role semantics of tuples and do not consider the effect of entity position on semantics. As far as we know, there has been no model that considers both the role semantics and positional information of each entity in the knowledge hypergraph.

Therefore, we have identified that, to fully represent knowledge hypergraph in an expressive manner, the following requirements should be met: firstly, the complex semantics of tuples in roles and positions aspects should be considered during modeling process, including the semantic relationship among roles, positions, and entity compatibility. Secondly, it is sufficiently expressive to represent all types of relations. To the best of our knowledge, none of the existing methods satisfy the above two requirements.

In this paper, we focus on the features of both role and position in each tuple of knowledge hypergraphs, and propose a <u>POS</u>itional <u>E</u>mbedding model with full expressive for the knowledge hypergraph. Different from the previous knowledge hypergraph embedding methods, POSE introduces a latent space for roles and positions, where entities with related roles and the same positions should have similar representations. Furthermore, POSE learns a relation matrix for each relation to capture its compatibility with all related entities. We further devise a scoring function for efficient prediction. The critical insight of POSE is to model the knowledge hypergraph in terms of roles and positions.

The contributions of this paper are as follows:

- A novel knowledge hypergraph embedding model, named POSE, aims to predict links in knowledge hypergraphs. POSE strengthens the importance of roles and positions in n-ary relations, and learns the latent space as well as the relation matrix of roles to capture semantic relatedness and compatibility, respectively.
- We prove that POSE is fully expressive for knowledge hypergraphs, which can model all patterns of relations without any limitation.
- Extensive experiments are conducted on six representative datasets, demonstrating that POSE achieves state-of-the-art performance on knowledge hypergraph datasets and comparable performance on knowledge graph datasets.

The rest of this paper is organized as follows. We introduce preliminaries in Sect. 3. Detailed description of POSE is provide in Sect. 4. The theoretical analysis of the full expression and complexity is presented in Sect. 5. Then, we report the experimental results in Sect. 6. Finally, we conclude this paper in Sect. 7.

2   Related Work

Our algorithm is conceptually related to previous models in knowledge graphs and recent models in knowledge hypergraphs, which can be classified into two categories.

2.1   Link Prediction in Knowledge Graphs

The most typical tensor decomposition-based method is RESCAL [1], which associates knowledge graphs with three-way tensors of head entities, relations, and tail entities. The learned entity and relation embeddings are used to reconstruct the tensors by minimizing the reconstruction error. Similarly, Complex [2] associates each relation with a matrix of head and tail entities, decomposed and learned as RESCAL. The main problem for the generalization of tensor methods is that a single model can only be trained and used for a certain arity of relations, while POSE can predict multiple arities of relations simultaneously.

Translation-based methods can be traced back to TransE [4,24]. It treats each valid triple as a translation from a head entity to a tail entity through their relation. Subsequently, several improved methods based on TransE were proposed over time [5–8,31]. Among them, TransH [8] introduces the relation-specific hyperplane. Entities are projected onto the relational hyperplane before translation, but such methods can only model symmetric relations, while our method can model any pattern of relations with full expressiveness.

Neural network-based methods model the effectiveness of triples. For example, ConvKB [9] treats each triple as a three-column matrix. This matrix is fed into convolutional layers, followed by fully connected layers to generate an effectiveness score. Nathani [10] further proposed a generalized graph attention model as an encoder to capture neighborhood features and applied ConvKB as a decoder. Such methods usually incur high time complexity, while the complexity of our model is linear with time and space.

2.2   Link Prediciton in Knowledge Hypergraphs

Since binary relations simplify the complexity of real-world facts, some recent studies have tried to represent and predict links in knowledge hypergraphs, primarily through embedding-based methods. These studies represent n-ary facts as tuples with predefined relations and generalize binary relation methods to n-ary cases.

m-TransH [11] and RAE [12] generalize TransH [8], which is a translation embedding model for binary relations, but these two models are not fully expressive and cannot model asymmetric relations. The influence of role and position on tuple semantics is not considered. NaLP [14] and HINGE [16] are different approaches that directly represent n-ary facts as attribute-value pairs and then model the associations between these attributes and values. However, in these methods, properties of n-ary facts are assumed to be equally important, which is not the case in real-world scenarios. Instead, we use the tuple form where different entities have different importance, which is closer to reality.

RAM [15] and NeuInfer [17] consider incorporating entity role information into embedding, which use tensor decomposition-based and neural network-based methods to measure the tuple plausibility, respectively. However, these models only learn the role semantics while do not particularly consider the impact of the positional information. Currently, there is no existing work that considers entity role and positional semantics in knowledge hypergraph modeling. Thus we utilize both role and positional information to enhance the performance of knowledge hypergraph link prediction.

Table 1. Notations and explanations.

Notation	Explanation
$\mathcal{H}$	Knowledge hypergraph
$e, \mathcal{E}$	Entity and entity set
$r, \mathcal{R}$	Relation and relation set
α	Arity of relation
$t, \mathcal{T}, \mathcal{T}_T, \mathcal{T}_O, \mathcal{T}_H$	Tuple and four kinds of tuple sets
L	Latent space size of role
d	Embedding dimension
ρ^r	Role of relation r
ϕ	Scoring function
σ	Element-wise softmax function
$\mathbf{e}, \mathbf{c}$	Embedding of entity and role
$\mathbf{b}$	Role latent vector
$\mathbf{w}$	Weight vector of role latent vector
$\mathbf{B}$	Basis matrix of relation
$\mathbf{R}$	Relation matrix
$\langle \cdot \rangle$	Multi-linear product

3 Preliminaries

This section presents the preliminaries of the knowledge hypergraph and the link prediction task. The notations used in our paper are summarized in Table 1.

Definition 1 (Knowledge Hypergraph). *A knowledge hypergraph is defined as $\mathcal{H} = (\mathcal{E}, \mathcal{R}, \mathcal{T}_O)$, where $\mathcal{E}$, $\mathcal{R}$, and $\mathcal{T}_O$ is a finite set of entities, relations, and observed tuples, respectively. $t_i = r(\rho_1^r : e_1, \rho_2^r : e_2, ..., \rho_\alpha^r : e_\alpha)$ denotes a tuple where $r \in \mathcal{R}$ is a relation, each $e_i \in \mathcal{E}$ is an entity, i is the position index, each ρ_i^r is the corresponding role of relation r, and α is the non-negative integral arity of the relation r.*

After clarifying the definition of the knowledge hypergraph, we give the definition for the task of link prediction in knowledge hypergraphs.

Definition 2 (Link Prediction in Knowledge Hypergraphs). *Let $\mathcal{T}$ denote all tuples set, $\mathcal{T}_O \subseteq \mathcal{T}_T \subseteq \mathcal{T}$ indicate the relationship among the set of observed, all ground truth, and all tuples, respectively. The hidden tuples set $\mathcal{T}_H$ is the differences between $\mathcal{T}$ and $\mathcal{T}_O$. Given the observed tuples $\mathcal{T}_O$, the aim of link prediction in knowledge hypergraphs is to predict the labels of the hidden tuples $\mathcal{T}_H$.*

4   The POSE Model

Our proposed method, named POSE, models knowledge hypergraphs from the role and position level, enabling semantic relatedness of roles and positions by exploiting a latent space. A relation matrix captures the compatibility among roles, positions, and all associated entities. Finally, a multi-linear product is adopted for plausibility measure, achieving full expressiveness. An overview of POSE is illustrated in Fig. 2.

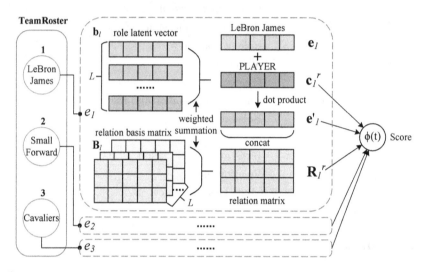

Fig. 2. Overview of POSE. Each entity generates an entity embedding, combines it with the role embedding to integrate the role semantics. Then implements the incorporation of the positional semantics through concatenation operation. The relation matrix is produced by combining the basis matrices of a relation. Finally, entity embedding, role embedding, and relation matrix of each entity are fed into the scoring function to calculate the confidence score of the tuple.

4.1   Latent Space for Roles and Positions

Since one entity may correspond to multiple positions and roles in a knowledge hypergraph dataset, such as the entity LeBron James in Fig. 1, the multi-embedding mechanism [23] is designed for entities and maps each entity $e_i \in \mathcal{E}$ to multiple embeddings. Let $\mathbf{e}_i \in \mathbb{R}^{m \times d}$ denote entity embedding, m be the layers of multi-embedding, d be the embedding dimension.

Inspired by sharing feature information of training examples in machine learning [30, 32, 33], in order to utilize semantic information about the positions and roles of the entities, a latent space is built for roles with L role latent vectors $\mathbf{b}_l \in \mathbb{R}^d, l = 1, 2, ..., L$. First of all, the role embedding $\mathbf{c}_i^r$ is computed by a combination of role latent vectors:

$$\mathbf{c}_i^r = \sum_{l \in L} \mathbf{b}_l \cdot \sigma(\mathbf{w}_i^r)[l] \tag{1}$$

where $\mathbf{w}_i^r \in \mathbb{R}^L$ is the weight vector of the role latent vector, known as the role weights. As a result, semantic relatedness is parameterized implicitly by role weights, while the weight vector needs to be normalized by the element-wise softmax function σ for all l and $l' \in L$:

$$\sigma(\mathbf{w}_i^r)[l] = \exp(\mathbf{w}_i^r[l]) / \sum_{l' \in L} \exp(\mathbf{w}_i^r[l']) \tag{2}$$

After obtaining the role embedding, it further assigning various semantics to entities that at different positions. To be specific, the entity embedding and the role embedding are multiplied to obtain an embedding $\mathbf{e}_i' = \mathbf{e}_i \cdot \mathbf{c}_i^r$ that incorporates the role semantics. Moreover, the i-th positional semantics is combined by the concatenation function cat:

$$\mathbf{e}_i' = (\mathbf{e}_i^1, \mathtt{cat}(\mathbf{e}_i^2, m \cdot d/\alpha), ..., \mathtt{cat}(\mathbf{e}_i^\alpha, m \cdot d \cdot (\alpha - 1)/\alpha)) \tag{3}$$

where $\mathtt{cat}(\mathbf{v}, x)$ shifts vector $\mathbf{v}$ to the left by x steps.

4.2 Relation Matrix

The relations in knowledge hypergraphs consist of entities at different positions and corresponding roles. To measure the degree of compatibility among the positions, roles, and all participated entities, the roles at each position in the relation are learned with a relation matrix. For a relation $r \in \mathcal{R}$, the relation matrix for the role at the i-th position is represented by $\mathbf{R}_i^r \in \mathbb{R}^{\alpha \times m}$, where the j-th row $\mathbf{R}_i^r[j, :]$ denotes the compatibility with multi-embedding of the j-th position entity. With a designed latent space of positions and roles, the relation matrix can be learned as follows for all $i \in \alpha$:

$$\mathbf{R}_i^r = \sum_{l \in L} \sigma(\mathbf{w}_i^r)[l] \cdot \sigma(\mathbf{B}_l) \tag{4}$$

where $\mathbf{B}_l \in \mathbb{R}^{\alpha \times m}$ is the basis matrix of relation linked with latent vector of role $\mathbf{b}_l$ in the latent space. The entire basis matrix is also normalized by σ. The basis relation matrix $\mathbf{B}_l$ is aligned with the latent role vector $\mathbf{b}_l$, which are used to compute for role embeddings and relation matrices.

4.3 Scoring Function

The scoring function employs a multi-linear product approach to calculate the confidence of the knowledge hypergraph tuple, which can effectively improve the performance and introduce fewer parameters, making the training more efficient. For each tuple $t_i = r(\rho_1^r : e_1, \rho_2^r : e_2, ..., \rho_\alpha^r : e_\alpha)$, the score of tuple is calculated by the following equation:

$$\phi(t) = \sum_{i \in \alpha} \langle \mathbf{c}_i^r, \mathbf{R}_i^r[1,:] \mathbf{e}_1, ..., \mathbf{R}_i^r[\alpha,:] \mathbf{e}_\alpha \rangle \tag{5}$$

where $\mathbf{R}_i^r[1,:] \mathbf{e}_1$ captures the compatibility between the role ρ_i^r and i-th entity e_i, i.e., the multi-embedding of e_i is weighted by the elements of $\mathbf{R}_i^r[1,:]$. Each summation term of the multi-linear product is the compatibility of the entity with the corresponding role at a different position.

4.4   Model Training

Generally, the knowledge hypergraph only provides positive examples, while negative examples need to be sampled by some way. Based on the scoring function designed above, the training loss and the learning target of the model are designed in the following way. For each positive tuple $t \in \mathcal{T}_O$, the negative samples are obtained by replacing the entity linked with ρ_i^r. The strategy generalizes from the ones in the binary case:

$$\mathcal{S}_t^{(i)} = \bigcup_{i \in \alpha} \{e_1, ..., \hat{e}_i, ..., e_\alpha \notin \mathcal{T}_O \mid \hat{e}_i \in \mathcal{E}, \hat{e}_i \neq e_i\} \tag{6}$$

Furthermore, an instantaneous multi-class log-loss is adopted and an optimizer of an empirical risk is formulated as follows:

$$\min_{\mathbf{e}_i \in E, \mathbf{b}_i \in B, \mathbf{w}_i^r \in W, \mathbf{B}_i \in \mathcal{B}} \sum_{t \in \mathcal{T}_O} \sum_{i \in \alpha} -\log \left[\exp(\phi(t)) / (\exp(\phi(t)) + \sum_{t' \in \mathcal{S}_t^{(i)}} \exp(\phi(t'))) \right] \tag{7}$$

where the set E, B, W, and $\mathcal{B}$ contains all elements of $\mathbf{e}_i$, $\mathbf{b}_i$, $\mathbf{w}_i^r$, and $\mathbf{B}_i$, respectively, the softmax loss guarantees that exactly one correct sample is learned among the candidates.

Algorithm 1 is the training process of POSE. For each tuple sampled from a knowledge hypergraph, its negative sample is obtained at first. Next, the embeddings and the relation matrix is computed. Then the confidence score of this sampled tuple is calculated. Finally, POSE is trained in mini-batch to minimize the above empirical risk formulation.

5   Theoretical Analysis

The POSE model is fully expressive, indicating that the model can correctly learn any valid n-ary relation in the knowledge hypergraph without being restricted to a specific pattern of relations. Given any ground truth tuples in the knowledge hypergraph, at least one embedding assignment of the model can correctly separate valid tuples from invalid ones. Furthermore, the POSE model can achieve linear time and space complexity. Its embedding dimension constraint is presented in Theorem 1, and the complexity analysis is conducted.

Algorithm 1: Training procedure for POSE

    Input : Observed tuples $\mathcal{T}_O$, iteration count n_{iter}, mini-batch size m_b,
            latent space size L
    Output: Role embedding, entity embedding, relation matrices

1 **for** $t = 1, ..., n_{\text{iter}}$ **do**
2   Sample a mini-batch $\mathcal{T}_{\text{batch}} \in \mathcal{T}_O$ of size m_b;
3   **for** *each tuple* $t \in \mathcal{T}_{\text{batch}}$ **do**
4   Construct negative samples for tuple t;
5            $\mathbf{c}_i^r \leftarrow$ compute role embeddings using (1);
6            $\mathbf{e}_i' \leftarrow$ compute entity embeddings using (3);
7            $\mathbf{R}_i^r \leftarrow$ compute relation matrices using (4);
8   Update learnable parameters w.r.t. gradients based on the whole
         objective in (7);

Theorem 1. *For any ground truth over entities $\mathcal{E}$ and relations $\mathcal{R}$ of the knowledge hypergraph containing $\eta \geq 1$ ground truth tuples, there exists a POSE model with the embedding dimension $d = \eta$, the multiplicity of entity embedding $m = \max_{r \in \mathcal{R}} \alpha$, and the latent space size $L = \eta$, which accurately represents the ground truth tuple.*

Proof. Let $\mathcal{T}_T$ be the set of all ground truth tuples in the knowledge hypergraph with $\rho = |\mathcal{T}_T|$. Then, the statement of Theorem 1 is equivalent to assigning parameters entity embeddings E, role basis vectors B, role weights W, and relation basis matrices $\mathcal{B}$ to POSE. Under the conditions of embedding dimension $d = \eta$, multiplicity of entity embeddings $m = \max_{r \in \mathcal{R}} \alpha$, and latent space size $L = \eta$, the scoring function can be expressed as follows:

$$\phi(t) = \begin{cases} > 0, & \text{if } t \in \mathcal{F}^{\text{all}} \\ = 0, & \text{if } t \notin \mathcal{F}^{\text{all}} \end{cases} \text{, for } t := \{\rho_1^r : e_1, ..., \rho_\alpha^r : e_\alpha\} \tag{8}$$

when $\eta \geq 1$, for each entity $e \in \mathcal{E}$ with multiple embeddings $\mathbf{e} \in \mathrm{R}^{m \times d}$, $\mathbf{e}[i, j]$ is set to 1 if the entity e involves with the i-th role of the j-th tuple in $\mathcal{T}_T$, and to 0 otherwise. As for the latent space, an identity matrix $\mathbf{I}_L$ is concatenated by the role latent vectors $[\mathbf{b}_1, ..., \mathbf{b}_l]$. The form of the relation basis matrix is $\mathbf{B}_i = [\mathbf{I}_\alpha, 0] \in \{0, 1\}^{\alpha \times m}$. Since a identity matrix is a group of latent vectors for R^η, the role weights $\{\mathbf{W}_i^r\}$ can be assigned to satisfy that $\mathbf{c}_i^r[j] = 1$ if the relation r involves with the j-th tuple in $\mathcal{T}_T$, and $\mathbf{c}_i^r = 0$ otherwise. Then the confidence score of the j-th ground truth tuple can be calculated by POSE through the following equation:

$$\phi(t) = \sum_{i \in \alpha} \langle \mathbf{c}_i^r, \mathbf{R}_i^r[1, :]\mathbf{e}_1, ..., \mathbf{R}_i^r[\alpha, :]\mathbf{e}_\alpha \rangle \tag{9}$$

and each summation term is equal to 1, the score for t is $\phi(t) = \alpha > 0$

    As for $\phi(t) = 0$, assume there exists a false tuple $t \notin \mathcal{T}_T$, $\phi(t) > 0$. Based on this assumption, there is at least one position j to ensure that $\mathbf{c}_i^r[j] = 1$ and the j-th elements of $\mathbf{R}_i^r[1,:]\mathbf{e}_1, ..., \mathbf{R}_i^r[\alpha,:]\mathbf{e}_\alpha$ are all equal to 1. However, this can only happen when entities $e_1, ..., e_\alpha$ and relation r appear in the j-th tuple of $\mathcal{T}_T$ simultaneously, then $t \in \mathcal{T}_T$, which contradicts the initial assumption. So that when $t \notin \mathcal{T}_T$, $\phi(t) = 0$. $\square$

For the time complexity, since our scoring function uses a multi-linear product, the linear time complexity is $\mathcal{O}(d)$. For space complexity, since the arity of relations in the knowledge hypergraph is rarely higher than 6 (as shown in Table 3), the assignment of parameter m will not exceed 3. If let m_α be the maximum arity of relation in the knowledge hypergraph, m_e be the number of entities, m_r be the number of relations, the parameters spent on the role latent vector, the basis matrix of relation, and the role weight vector is at most $\mathcal{O}(m_e d + Lm_r m_\alpha + Ld + Lmm_\alpha) = \mathcal{O}(m_e d + Lm_r m_\alpha)$. Thus the POSE model remains linear in both time and space.

6   Experiments

The performance of POSE was tested on two kinds of benchmarks. Section 6.1 summarizes the experimental setups, such as datasets and baselines. All experiments in Sect. 6.2 were conducted to predict hidden tuples or hidden triples.

Table 2. Dataset statistics. The size of train, valid, and test represent the number of triples or tuples, respectively.

Dataset	#entities	#relations	#train	#valid	#test
FB15k	14,951	1,345	483,142	50,000	59,071
WN18	40,943	18	141,442	5,000	5,000
FB15k-237	14,541	237	272,115	17,535	20,466
WN18RR	40,493	11	86,835	3,034	3,134
JF17K	29,177	327	77,733	–	24,915
FB-AUTO	3,388	8	6,778	2,255	2,180

6.1   Experiment Settings

Datasets. The experiments on link prediction were conducted on six datasets. The knowledge hypergraph dataset JF17K was proposed by Wen et al. [11], while FB-AUTO was proposed by Fatemi et al. [18]. As no validation set was proposed for JF17K, we randomly selected 20% of the train set as validation. Four standard knowledge graph benchmarks, i.e., WN18, FB15k, WN18RR, and FB15k-237 were used for link prediction in knowledge graphs. The detailed statistics of the datasets are summarized in Table 2, and the number of tuples with different arities are summarized in Table 3.

Baselines. For link prediction in knowledge hypergraphs, we compare POSE with state-of-the-art approaches, including RAE [12], NaLP [14], HINGE [16], NeuInfer [17], HypE [18], and RAM [15]. In addition, GETD [13] can only model single-arity knowledge hypergraphs and therefore is not included in the comparison. As for link prediction in knowledge graphs, we compared POSE with several baselines, including TransE [21], DistMult [24], ComplEx [2], SimplE [19], RotatE [25], TuckER [20], HAKE [27], and DualE [28].

Table 3. The number of tuples with different arities in the datasets.

Dataset	#arity=2	#arity=3	#arity=4	#arity=5	#arity=6
FB15k	592,213	0	0	0	0
WN18	151,422	0	0	0	0
FB15k-237	310,116	0	0	0	0
WN18RR	93,003	0	0	0	0
JF17K	56,332	34,550	9,509	2,230	37
FB-AUTO	3,786	0	215	7,212	0

Evaluation Metrics. Two evaluation metrics were employed to compare the performance of different link prediction methods: Mean Reciprocal Rank (MRR) and Hit@K, where H@K is in %, and all results in Sect. 6.2 are rounded. Two metrics above are measured by ranking a test tuple t within a set of replaced tuples. For each tuple in the test set and each position i in the tuple, $|\mathcal{E}| - 1$ replaced tuples are generated by replacing the entity e_i with each entity in $\mathcal{E} \backslash \{e_i\}$.

Table 4. Results of link prediction on knowledge hypergraph datasets.

Model	JF17K				FB-AUTO			
	MRR	Hit@1	Hit@3	Hit@10	MRR	Hit@1	Hit@3	Hit@10
RAE	0.396	0.312	0.433	0.561	0.703	0.614	0.764	0.854
NaLP	0.310	0.239	0.334	0.450	0.672	0.611	0.712	0.774
HINGE	0.473	0.397	0.490	0.618	0.678	0.765	0.706	0.765
NeuInfer	0.451	0.373	0.484	0.604	0.737	0.700	0.755	0.805
HypE	0.507	0.421	0.550	0.669	0.804	0.774	0.824	0.856
RAM	0.539	0.463	0.573	0.690	0.830	0.803	0.851	0.876
POSE* (Ours)	0.521	0.442	0.551	0.664	0.823	0.794	0.840	0.862
POSE (Ours)	**0.545**	**0.469**	**0.582**	**0.706**	**0.856**	**0.821**	**0.876**	**0.895**

6.2 Results

Link Prediction in Knowledge Hypergraphs. From Table 4, we observe that POSE improves the MRR on the FB-AUTO dataset by at most 2.6%.

All three models, RAE, NaLP, and HINGE, ignore both positional and role semantics when modeling the knowledge hypergraphs. RAE is based on the generalization of the translation model TransH applied to the knowledge hypergraphs, which is not fully expressive and can only model the symmetric relations, making the prediction performance a large gap compared with the current state-of-the-art models. NaLP and HINGE split a tuple into a primary tuple and some auxiliary key-value pair attributes. This manner of splitting tuples ignores the semantic information of positions and roles, making more information lost from the modeling process and leading to lower prediction performance.

HypE is a generalization of the SimplE model, which variations the position differences but not the role. The experimental results of RAM (only considering roles) outperform that of HypE, which further illustrates the importance of role semantics in link prediction. While NeuInfer and RAM model the knowledge hypergraph using neural networks and tensor decomposition-based methods, respectively, which consider the difference in role semantics but not the positional information. The lack of utilizing positional information causes the worse experimental results of POSE, which fully justifies the importance of positional information for knowledge hypergraph modeling.

To further demonstrate the effectiveness of positional information in link prediction tasks, we conducted an ablation study that does not consider the positional information, marked as POSE* in Table 4. We can see that the experimental results of POSE* are still better than HypE that only considers role semantics, but worse than that of POSE, verifying the importance of positional and role information in knowledge hypergraph modeling.

Table 5. Results of Link Prediction on different arities of knowledge hypergraph datasets.

Model	JF17K					FB-AUTO		
	2	3	4	5	6	2	4	5
GETD	**0.339**	**0.583**	**0.751**	0.746	0.350	0.524	0.237	0.786
RAM	0.337	0.578	0.736	0.805	0.697	0.557	0.456	0.904
POSE (Ours)	0.334	0.577	0.739	**0.813**	**0.708**	**0.572**	**0.477**	**0.912**

Link Prediction with Different Arities. In Table 5, we directly predict a tuple of different arities after training on the entire knowledge hypergraph dataset. The RAM model considers the role semantics but not the positions, and GETD is extended based on the TuckER model, which only considers the positions while missing the role information.

POSE achieves the best performance on all arities of relations in the FB-AUTO dataset, while on the JF17K, POSE achieves the best performance on high-arity relations, which is mainly due to the low-arity data noise on high-arity predictions during the training process. In general, POSE improves by an average of 4.1% on all arities compared with RAM, which can be seen that considering both positional and role information is of significance for the link prediction tasks in knowledge hypergraphs.

Table 6. Results of Link Prediction on Knowledge Graph Datasets.

Model	WN18			FB15k			WN18RR			FB15k-237		
	MRR	Hit@1	Hit@10	MRR	Hit@1	Hit@10	MRR	Hit@1	Hit@10	MRR	Hit@1	Hit@10
TransE	0.495	0.113	0.943	0.463	0.297	0.749	0.226	–	0.501	0.294	–	0.46
DistMult	0.822	0.728	0.936	0.654	0.546	0.824	0.430	0.390	0.490	0.241	0.155	0.419
ComplEx	0.941	0.939	0.947	0.727	0.660	0.838	0.440	0.410	0.510	0.247	0.158	0.428
SimplE	0.942	0.939	0.947	0.727	0.838	0.660	–	–	–	–	–	–
RotatE	0.949	0.944	**0.959**	0.797	0.746	0.884	0.476	0.428	0.571	0.338	0.241	0.533
TuckER	**0.953**	**0.949**	0.958	0.795	0.741	**0.892**	–	–	-	–	–	–
HAKE	–	–	–	–	–	–	**0.497**	**0.452**	**0.582**	0.346	0.250	0.542
DualE	0.951	0.945	0.961	0.790	0.734	0.881	–	–	–	–	–	–
POSE (Ours)	0.943	0.940	0.949	**0.801**	**0.751**	0.877	0.496	0.449	0.577	**0.349**	0.248	**0.544**

Link Prediction in Knowledge Graphs. POSE achieves state-of-the-art results on MRR and Hit@1, and achieves comparable results on FB15k. Such results validate that POSE can have comparable performance to the binary relation on the knowledge graph datasets and that the design that considers both position and (implicit) role is equally applicable to the knowledge graphs.

7   Conclusion

In this paper, we propose a link prediction model POSE for knowledge hypergraphs, which learns the embedding representation from both role and position levels. Leveraging the latent space for entity semantic relatedness of role and position, and relation matrix for entity compatibility achieves precise accuracy for link prediction, full expressiveness, and more generalized modeling of knowledge hypergraphs. The experimental results on both knowledge hypergraph datasets and four knowledge graph datasets demonstrate the superiority and robustness of POSE.

Acknowledgements. This work is supported by the National Key R&D Program of China (2020AAA0108504) and National Natural Science Foundation of China (61972275).

References

1. Nickel, M., Tresp, V., Kriegel, H.-P.: A three-way model for collective learning on multi-relational data. In: Proceedings of the 28th International Conference on Machine Learning, pp. 809–816 (2011)
2. Trouillon, T., Welbl, J., Riedel, S., Gaussier, E., Bouchard, G.: Complex embeddings for simple link prediction. In: Proceedings of the 33rd International Conference on Machine Learning, pp. 2071–2080 (2016)
3. Ding, B., Wang, Q., Wang, B., Guo, L.: Improving knowledge graph embedding using simple constraints. In: Proceedings of the 56th Annual Meeting of the Association for Computational Linguistics, pp. 110–121 (2018)
4. Bordes, A., Usunier, N., GarciaDuran, A., Weston, J., Yakhnenko, O.: Translating embeddings for modeling multirelational data. In: Proceedings of the 26th International Conference on Neural Information Processing Systems, pp. 2787–2795 (2013)
5. Lin, Y., Liu, Z., Sun, M., Liu, Y., Zhu, X.: Learning entity and relation embeddings for knowledge graph completion. In: Proceedings of the 29th AAAI Conference on Artificial Intelligence, pp. 2181–2187 (2015b)
6. Xiao, H., Huang, M., Zhu, X.: TransG: A generative model for knowledge graph embedding. In: Proceedings of the 54th Annual Meeting of the Association for Computational Linguistics, pp. 2316–2325 (2016)
7. Ebisu, T., Ichise, R.: TorusE: Knowledge graph embedding on a Lie group. In: Proceedings of the 32nd AAAI Conference on Artificial Intelligence, pp. 1819–1826 (2018)
8. Wang, Z., Zhang, J., Feng, J., Chen, Z.: Knowledge graph embedding by translating on hyperplanes. In: Proceedings of the 28th AAAI Conference on Artificial Intelligence, pp. 1112–1119 (2014)
9. Nguyen, D.Q., Nguyen, T.D., Nguyen, D.Q., Phung, D.: A novel embedding model for knowledge base completion based on convolutional neural network. In: Proceedings of the 16th Annual Conference of the North American Chapter of the Association for Computational Linguistics: Human Language Technologies, pp. 327–333 (2018)
10. Nathani, D., Chauhan, J., Sharma, C., Kaul, M.: Learning attention-based embeddings for relation prediction in knowledge graphs. In: Proceedings of the 57th Annual Meeting of the Association for Computational Linguistics, pp. 4710–4723, Florence, Italy (2019)
11. Wen, J., Li, J., Mao, Y., Chen, S., Zhang, R.: On the representation and embedding of knowledge bases beyond binary relations. In: Proceedings of the Twenty-Fifth International Joint Conference on Artificial Intelligence, pp. 1300–1307 (2016)
12. Zhang, R., Li, J., Mei, J., Mao, Y.: Scalable instance reconstruction in knowledge bases via relatedness affiliated embedding. In: Proceedings of the 2018 World Wide Web Conference, pp. 1185–1194 (2018)
13. Liu, Y., Yao, Q., Li, Y.: Generalizing tensor decomposition for n-ary relational knowledge bases. In: Proceedings of The Web Conference. pp. 1104–1114 (2020)
14. Guan, S., Jin, X., Wang, Y., Cheng, X.: Link prediction on n-ary relational data. In: Proceedings of the 2019 World Wide Web Conference, pp. 583–593 (2019)
15. Liu, Y., Yao, Q., Li, Y.: Roleaware modeling for n-ary relational knowledge bases. arXiv preprint arXiv:2104.09780 (2021)
16. Rosso, P., Yang, D., Cudr'eMauroux, P.: Beyond triplets: Hyper-relational knowledge graph embedding for link prediction. In: Proceedings of The Web Conference, pp. 1885–1896 (2020)

17. Guan, S., Jin, X., Guo, J., Wang, Y., Cheng, X.: NeuInfer: Knowledge inference on n-ary facts. In: Proceedings of the 58th Annual Meeting of the Association for Computational Linguistics, pp. 6141–6151 (2020)
18. Fatemi, B., Taslakian, P., Vazquez, D., Poole., D.: Knowledge Hypergraphs, Prediction Beyond Binary Relations. In: IJCAI (2020)
19. Kazemi, S.M., Poole, D.: Simple embedding for link prediction in knowledge graphs. In: NeurIPS (2018)
20. Balazevic, I., Allen, C., Hospedales, T: TuckER: Tensor Factorization for Knowledge Graph Completion. In: EMNLP, pp. 5188–5197 (2019)
21. Bordes, A., Usunier, N., Garcia-Duran, A., Weston, J., Yakhnenko, O.: Translating Embeddings for Modeling Multi-relational Data. In: NeurIPS (2013)
22. Lacroix, T., Usunier, N., Obozinski, G.: Canonical Tensor Decomposition for Knowledge Base Completion. In: ICML (2018)
23. Tran, H.N., Takasu, A.: Analyzing Knowledge Graph Embedding Methods from a Multi-embedding Interaction Perspective. arXiv preprint arXiv:1903.11406 (2019)
24. Wang, Q., Mao, Z., Wang, B., Guo, L.: Knowledge graph embedding: a survey of approaches and applications. TKDE **29**12, 2724–2743 (2017)
25. Sun, Z., Deng, Z.-H., Nie, J.-Y., Tang, J.: RotatE: Knowledge Graph Embedding By Relational Rotation in Complex Space. In: ICLR (2019)
26. Rossi, A., Barbosa, D., Firmani, D., Matinata, A., Merialdo, P.: Knowledge graph embedding for link prediction: a comparative analysis. ACM Trans. Knowl. Discov. Data **15**(2), 1–49 (2021). https://doi.org/10.1145/3424672
27. Zhang, Z., Cai, J., Zhang, Y., Wang, J.: Learning Hierarchy-Aware Knowledge Graph Embeddings for Link Prediction. In: Proceedings of the AAAI Conference on Artificial Intelligence, vol. 34, No. 03, pp. 3065–3072 (2020)
28. Cao, Z., Xu, Q., Yang, Z., Cao, X., Huang, Q.: Dual quaternion knowledge graph embeddings. Proc. AAAI Conf. Artif. Intell. **35**(8), 6894–6902 (2021)
29. Peng, Y., Choi, B., Xu, J.: Graph learning for combinatorial optimization: a survey of state-of-the-art. Data Sci. Eng. **6**(2), 119–141 (2021). https://doi.org/10.1007/s41019-021-00155-3
30. Wawrzinek, J., Pinto, J.M.G., Wiehr, O., Balke, W.-T.: Exploiting latent semantic subspaces to derive associations for specific pharmaceutical semantics. Data Sci. Eng. **5**(4), 333–345 (2020). https://doi.org/10.1007/s41019-020-00140-2
31. Zhang. F., Wang, X., Li, Z., Li, J.: Transrhs: A representation learning method for knowledge graphs with relation hierarchical structure. In: Proceedings of the Twenty-Ninth International Conference on International Joint Conferences on Artificial Intelligence, pp. 2987–2993 (2021)
32. Zhu, M., Shen, D., Xu, L., Wang, X.: Scalable multi-grained cross-modal similarity query with interpretability. Data Sci. Eng. **6**(3), 280–293 (2021). https://doi.org/10.1007/s41019-021-00162-4
33. Wu, S., Zhang, Y., Gao, C., Bian, K., Cui, B.: GARG: anonymous recommendation of point-of-interest in mobile networks by graph convolution network. Data Sci. Eng. **5**(4), 433–447 (2020). https://doi.org/10.1007/s41019-020-00135-z

Machine Learning

TraVL: Transferring Pre-trained Visual-Linguistic Models for Cross-Lingual Image Captioning

Zhebin Zhang[1], Peng Lu[2(✉)], Dawei Jiang[1], and Gang Chen[1]

[1] College of Computer Science and Technology, Zhejiang University, Hangzhou, China
{zhebinzhang,jiangdw,cg}@zju.edu.cn
[2] Institute of Computing Innovation, Zhejiang University, Hangzhou, China
lupeng@zjuici.com

Abstract. Visual-Linguistic (VL) pre-training is gaining increasing interest due to its ability to learn generic VL representations that can be used for downstream cross-modal tasks. However, the lack of large-scale and high-quality parallel corpora makes VL pre-training impractical for low-resource languages. Therefore, it is desirable to leverage existing well-trained English VL models for cross-modal tasks in other languages. But a basic approach suffers from its inability to capture the semantic correlation between different modalities and insufficient utilization of the hierarchical representations of VL models. In this work, we propose TraVL, a novel framework for transferring pre-trained VL models for cross-lingual image captioning. To enforce the semantic alignment during modality fusion, TraVL employs joint attention that constructs the key-value pair by concatenating the visual and linguistic representations. To fully exploit the hierarchical visual information, we develop an adjacent layer-fusion mechanism that allows each decoder layer to attend to the encoder's multilayer representations with similar semantics. Experiments on a Chinese image-text dataset show that TraVL outperforms state-of-the-art captioning models and other transfer learning methods.

Keywords: Artificial neural network · Visual-linguistic model · Image captioning

1 Introduction

Recent years have witnessed the rapid growth of pre-training techniques for visual-linguistic (VL) models [18,20,28,36]. By training on large-scale parallel image-text corpora, pre-trained VL models can learn generic multimodal representations of the input image-text pairs and be fine-tuned to adapt to cross-modal tasks such as image-text retrieval, image captioning, etc. This approach has benefited the vision-language community by advancing the state of the arts in various VL tasks.

B. Li et al. (Eds.): APWeb-WAIM 2022, LNCS 13422, pp. 341–355, 2023.
https://doi.org/10.1007/978-3-031-25198-6_26

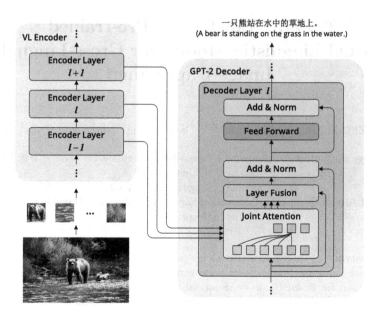

Fig. 1. The architecture of TraVL framework.

VL pre-training generally requires large-scale datasets containing significant amounts of image-text pairs. For example, the commonly used Conceptual Captions dataset [27] contains 3.3 million images annotated with English captions. Although these pre-training methods are not exclusive to English, they can be impractical for low-resource languages where image-text corpora are relatively scarce. Therefore, it is desirable to explore transferring the knowledge of these pre-trained VL models from English to other languages. This study can benefit researchers of different languages who lack either the data or the computational resources to perform VL pre-training.

To achieve this goal, we first conduct some preliminary studies into transferring a VL model, namely, VLP [36], for cross-lingual image captioning. We experiment on the COCO-CN dataset [19], which is an extension of MS-COCO [21] with annotations in Chinese. We explore three different approaches to the cross-lingual transfer of the VLP model, including a straightforward fine-tuning method, a knowledge distillation method, and a feature-based method. The third method, which employs VLP for feature extraction and relies on a pre-trained language model (i.e., GPT-2 [25]) for caption generation, achieves the best performance. But it still suffers from two limitations. First, the VLP model, like other state-of-the-art VL models, exhibits a single-stream architecture, where the concatenated visual and linguistic representations are jointly transformed via self attention during pre-training. Fusing the visual representations into GPT-2 using conventional cross attention fails to capture the semantic correlation between the visual features and the linguistic features. Second, given that VLP's multilayer representations capture a rich hierarchy of visual information, using only the highest-level representation results in the underutilization of this information.

To address the above challenges, we propose TraVL, a novel framework to transfer pre-trained VL models for cross-lingual image captioning. As shown in Fig. 1, TraVL employs the VL model as an encoder to extract a hierarchy of visual information from the image regions and uses the GPT-2 decoder to generate captions in the target language based on the extracted information. To address the aforementioned limitations, we incorporate two novel mechanisms into TraVL. First, joint attention is proposed in place of cross attention for modality fusion. By using the concatenation of the visual and linguistic representations as the key-value pair, joint attention enforces the semantic alignment between different modalities. Second, instead of only using the encoder's highest-level feature, an adjacent layer-fusion mechanism is employed to make better utilization of hierarchical information carried by the multilayer representations. Each decoder layer attends to a few aligned encoder layers with similar semantics and different attention outputs are dynamically aggregated based on the visual and linguistic context. We evaluate TraVL on subsets of COCO-CN with different sizes. Quantitative and qualitative analysis shows that TraVL produces better captions compared to state-of-the-art captioning models and other transfer learning methods. We also perform ablation studies to justify the architectural design of TraVL.

In summary, we make the following contributions in this paper:

- We propose TraVL, a framework for transferring the knowledge acquired by VL models from one language to another. TraVL allows researchers of different languages to leverage powerful English VL models to generate image captions in their own languages using limited in-domain data.
- We propose two novel mechanisms for TraVL, including a joint-attention mechanism that enforces the semantic alignment of different modalities and an adjacent layer-fusion mechanism that fully exploits the hierarchical visual representations.
- We compare TraVL with state-of-the-art captioning models and different transfer learning methods. Experiments on subsets of COCO-CN with different sizes show that TraVL achieves the best performance.

2   Related Work

Visual-Linguistic Pre-training. Inspired by the success of pre-trained language models such as BERT [5] and GPT-2 [25], much research attention has been attracted to VL pre-training which aims at learning generic multimodal representations of image-text pairs. Such representations are embedded with fused information from both modalities that can inform a variety of VL tasks including visual question answering, image captioning, etc. Based on a Transformer [30] backbone, VL models adopt either a multi-stream architecture [23] or a single-stream architecture [20,28,36]. Models belonging to the former encode the visual and linguistic features separately and employ co-attention modules to fuse them.

In the second architecture, the visual and linguistic features are concatenated and jointly transformed using self-attention modules.

To obtain contextualized multimodal representations, BERT-like objectives are typically used for the pre-training, such as masked language modeling, masked region modeling, and cross-modal matching. More recently, VL pre-training approaches based on contrastive learning [12,18] have been proposed and show enhanced performance.

Image Captioning. Early approaches to image captioning either rely on predefined templates [7,16] or formulate the problem as a retrieval task [10,14]. The former has the disadvantage that the captions are too simple and rigid, while the latter cannot generate novel captions. With the advancement of neural networks, much research attention has been paid to deep-learning-based image captioning methods. Inspired by the success of neural machine translation, these methods typically use a CNN encoder for feature extraction and an RNN decoder for caption generation. The encoded features can be a single vector [31], a grid of CNN features [34], or a set of image regions extracted using object detectors [1]. On the decoding side, in addition to the commonly used RNN and its variants, convolutional networks [2] and Transformer networks [4,11] have also been explored for caption generation.

Cross-Lingual image captioning aims to transfer the knowledge learned in a source language to generate captions in a target language. Although multilingual captioning models have been proposed [6,29], they require the use of images annotated in both languages. To address the lack of target-language training data, some studies [24,32] propose to train with corpora from both languages by using a shared image encoder and separate language decoders. Another study [17] proposes to construct target-language datasets using machine-translated sentences and employ a sentence fluency classifier to guide the training process. In contrast to existing work, we focus on the transfer learning of pre-trained VL models by using only a small amount of target-language training data.

3   Preliminary Studies

Previous studies [18,20,36] have shown that VL pre-training is highly effective in learning generic cross-modal representations that can improve the performance of image captioning models. However, pre-training VL models from scratch requires a large number of image-text pairs, which can be impractical for low-resource languages. It is interesting to investigate whether it is possible to reuse off-the-shelf VL models, which are pre-trained with English corpora, for developing captioning models in other languages. To this end, we first conduct some preliminary research.

We experiment with the VLP model which is based on a single-stream architecture and is pre-trained using both bidirectional and seq2seq objectives [36]. The following approaches are explored to transfer the VLP model for cross-lingual image captioning.

Table 1. Results of the preliminary studies on a subset of COCO-CN with 1000 training examples.

Model	BLEU-4	METEOR	ROUGE_L	CIDEr
Fine-Tuning	12.7	18.7	41.3	100.3
Fine-Tuning & K.D	12.6	18.0	40.6	97.6
Feature-Based	**14.3**	**19.8**	**41.6**	**110.1**

- **Fine-Tuning.** We modify the embedding layer and classification layer of VLP to accommodate to the Chinese vocabulary and initialize the rest part with weights pre-trained on English corpora. The model is fine-tuned on COCO-CN using the seq2seq objective.
- **Fine-Tuning & Knowledge Distillation.** Inspired by [33], we introduce Chinese BERT [5] and employ knowledge distillation to allow the VLP model to learn about Chinese from BERT during fine-tuning. Specifically, a few positions of each caption sentence are masked, indicated by P_m, and a distillation loss is calculated which measures the Kullback-Leibler distance between the predictive distributions of VLP and BERT, formally given as

$$\mathcal{L}_{KD} = \sum_{i \in P_m} \sum_{k \in \mathcal{V}} P(x_i = k|\boldsymbol{x}) \log \frac{P(x_i = k|\boldsymbol{x})}{P(y_i = k|\boldsymbol{y})}, \tag{1}$$

where $\mathcal{V}$ is the vocabulary, and $\boldsymbol{x}$ and $\boldsymbol{y}$ are the predictive distributions of VLP and BERT, respectively. The distillation loss adds to the total loss function which is minimized during training.
- **A Feature-Based Approach.** Inspired by [3], we introduce a generative pre-trained language model, i.e., Chinese GPT-2 [25], to guide the generation of Chinese captions. VLP is used as a feature extractor that encodes image regions into hidden representations, which are fused into GPT-2 via cross-attention modules to provide visual information during decoding. We keep the VLP encoder frozen and only optimize the weights of the GPT-2 decoder.

We experiment with the above approaches on COCO-CN which is a Chinese image-text parallel dataset. Detailed descriptions of the experimental setups are presented in Sect. 5. The results are shown in Table 1. Compared with directly fine-tuning, the use of knowledge distillation hardly improves model performance. We attribute this to the large discrepancy in distributions between the target corpus and BERT's predictions. Although BERT is assumed to provide more contextualized information, such a discrepancy can disrupt the optimization process. In contrast, the feature-based approach yields much better results, which shows that such a paradigm can make better use of the pre-trained knowledge. This motivates us to seek a more effective feature-based approach to the cross-lingual transfer of VL models.

4   TraVL

4.1   Overview

To fully exploit pre-trained VL models for cross-lingual image captioning, we propose a novel framework named TraVL consisting of a VLP encoder and a GPT-2 decoder, as shown in Fig. 1. TraVL takes as input an image-text pair denoted by $\{X, Y\}$, where $X = (x_1, \ldots, x_M)$ is a set of M image regions extracted from raw images using bottom-up and top-down attention [1], and $Y = (y_1, \ldots, y_N)$ is a set of N tokenized caption subwords.

The VLP encoder contains L identical attention layers that progressively transform X into contextualized visual representations. We denote by $R_l^{vis} = (r_{l,1}^{vis}, \ldots, r_{l,M}^{vis})$ the hidden representations extracted from the l-th VLP layer. The GPT-2 decoder, also composed of L layers, predicts the caption tokens Y in an auto-regressive manner. The hidden representations of the l-th GPT-2 layer are denoted as $R_l^{lan} = (r_{l,1}^{lan}, \ldots, r_{l,N}^{lan})$.

The feature-based approach, as described in Sect. 3, employs conventional cross-attention modules to fuse VLP's highest-level representations into the GPT-2 decoder and then performs fine-tuning on the target dataset. Designed on top of this basic approach, TraVL is equipped with two innovative mechanisms, namely, joint attention and adjacent layer fusion. Within each decoder layer, the original self-attention and cross-attention modules are replaced by a single joint-attention module to perform modality fusion. And the adjacent layer-fusion mechanism allows VLP's multilayer representations to be fused into each decoder layer where different fusion results are dynamically aggregated. The remaining parts of the decoder layer remain unaltered.

4.2   Joint Attention

Cross-modal learning relies on modality-fusion mechanisms to aggregate information from different modalities. Conventionally, cross-attention modules [30] are used to adapt the source modality to the target modality. In its simplest form, the attention operation tasks as input a query matrix (Q), a key matrix (K), and a value matrix (V). And the output is calculated as

$$\texttt{Attn}(Q, K, V) = \texttt{softmax}(\frac{QK^T}{\sqrt{d}})V, \tag{2}$$

In the case of image captioning, the visual representations R^{vis}, which serve as the source modality, are projected into K and V, while the linguistic representations R^{lan}, which serve as the target modality, are projected into Q, given as

$$Q = W^Q R^{lan}, \quad K = W^K R^{vis}, \quad V = W^V R^{vis}. \tag{3}$$

In TraVL, we propose to employ joint attention instead of cross attention for modality fusion. The rationale behind this design choice is based on the observation of how VLP constructs its visual representations. In the pre-training stage, given its single-stream architecture, VLP keeps the visual and linguistic representations at the same semantic level by transforming the concatenation of both modalities via self attention. The use of cross attention, however, fails to capture the semantic correlation and can lead to a mismatch in the architecture between the pre-training stage and the fine-tuning stage. Intuitively, joint attention mimics the way of jointly attending to both modalities as in self attention but only keeps the target modality as the query. TraVL employs joint attention to make up for the architectural discrepancy and to enforce the semantic alignment between different modalities.

Formally, in the joint-attention module, both the key and value matrices are constructed by concatenating the projections of R^{vis} and R^{lan}, while the query is projected from R^{lan}, given as

$$Q = W^Q R^{lan}, \quad K = W_1^K R^{vis} \| W_2^K R^{lan}, \quad V = W_1^V R^{vis} \| W_2^V R^{lan}, \quad (4)$$

where $\|$ is the concatenation operator. For simplicity, we use $\texttt{Attn}_{\texttt{joint}}(R^{lan}, R^{vis})$ to denote the result of joint-attention operation on the source modality R^{vis} and the target modality R^{lan}.

4.3   Adjacent Layer Fusion

Previous studies [13] indicate that different Transformer layers encode complementary features and that the features of adjacent layers are closely related. Hence, it's desirable to explore utilizing the hierarchical information carried by VLP's multiple layers instead of only focusing on its highest-level representations. Layer-wise coordination [9], which bridges encoder and decoder layers in the same semantic level, can make use of more fine-grained source information. On top of this idea, we propose a novel adjacent layer-fusion mechanism that allows each decoder layer to attend to VLP's multilayer representations with close related semantics and aggregate different attention outputs. The workflow of the fusion module within the decoder layer is depicted in Fig. 2, which proceeds following four steps.

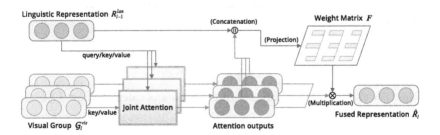

Fig. 2. Illustration of adjacent layer fusion within the l-th decoder layer. The module is parameterized by k which is the number of encoder layers that each decoder attends to. Unless otherwise stated, k is set to 3.

*Step*1 : We extract from the encoder the visual representations to be fused into each decoder layer. Specifically, for the l-th decoder layer, the visual representations are extracted from a fixed number of k adjacent encoder layers centered at the l-th encoder layer. We term the extracted representations a visual group, denoted as $\mathcal{G}_l^{vis} = \{R_{l-k'}^{vis}, \ldots, R_l^{vis}, \ldots, R_{l+k'}^{vis}\}$ where $k = 2k' + 1$. Note that the visual groups of the lowest and highest decoder layers contain fewer than k representations.

*Step*2 : Within each decoder layer, given the input linguistic representation as R_{l-1}^{lan}, we iterate over its visual group and perform joint attention on R_{l-1}^{lan} and each visual representation. For $R_{l'}^{vis} \in \mathcal{G}_l^{vis}$, the output of the joint-attention operation is denoted as

$$\bar{R}_{l'} = \texttt{Attn}_{\texttt{joint}}(R_{l-1}^{lan}, R_{l'}^{vis}) = (\bar{r}_{l',1}, \bar{r}_{l',2}, \ldots, \bar{r}_{l',N}). \tag{5}$$

After this step, we obtain k different attention outputs $\{\bar{R}_{l'}\}_{l'=1}^k$ for each decoder layer.

*Step*3 : To aggregate all the attention outputs, we compute a weight matrix F that is dynamically dependent on the visual and linguistic context. Specifically, the matrix F is derived by projecting the concatenation of the R_{l-1}^{lan} and the attention outputs,

$$
\begin{aligned}
F &= W^F \left[\bar{R}_{l-k'} \| \cdots \| \bar{R}_{l+k'} \| R_{l-1}^{lan} \right] \\
&= \begin{bmatrix}
f_{l-k',1} & f_{l-k',2} & \cdots & f_{l-k',N} \\
\vdots & \vdots & \ddots & \vdots \\
f_{l+k',1} & f_{l+k',2} & \cdots & f_{l+k',N}
\end{bmatrix} \in \mathbb{R}^{k \times N},
\end{aligned} \tag{6}
$$

where the concatenation is performed along the embedding dimension , $W^F \in \mathbb{R}^{k \times (k+1)d}$ is a trainable parameter matrix, and d is the dimension number. Each column of F contains k scalars corresponding to the weights assgined to k different attention outputs.

*Step*4 : The weight matrix F is used to combine the attention ouputs. At each position $n \in [1, N]$, we first normalize the weight matrix along the columns using softmax function, given by

$$\alpha_{l',n} = \frac{\exp(f_{l',n})}{\sum_{p=l-k'}^{l+k'} \exp(f_{p,n})}. \tag{7}$$

And then a weighted sum of the attention outputs is calculated as

$$\hat{r}_n = \sum_{l'=l-k'}^{l+k'} \alpha_{l',n} \bar{r}_{l',n}. \tag{8}$$

The final output of the layer fusion module is represented as $\hat{R}_l = (\hat{r}_1, \ldots, \hat{r}_N)$.

5 Experiments

5.1 Datasets

We experiment on the COCO-CN dataset [19] which contains 20342 images annotated with manually written Chinese sentences. The dataset is partitioned into training, validation, and test sets with 18342, 1000, and 1000 images, respectively. To evaluate the models on datasets with different sizes, we randomly draw out 1000, 2000, and 5000 image-text pairs from the original training set to construct training subsets, which we denote as COCO-CN-1000, COCO-CN-2000, and COCO-CN-5000, respectively. We use only the human-written captions for evaluation, excluding the translated sentences.

To avoid the performance difference induced by word segmentation strategies, we simply use character-level tokenization to preprocess the captions. Instead of building a new vocabulary based on the corpus, we adopt the same vocabulary that has been used to pre-train the Chinese GPT-2 model with a size of 21128 tokens.

5.2 Experimental Setup

We implement TraVL on top of the VLP repository[1]. A variant of Faster R-CNN [26], which is pre-trained on the Visual Genome [15] dataset, is used to encode raw images into visual features. For each image, we extract 100 image regions and every region is represented as three vectors: a 2048-dimensional feature vector, a 4-dimensional positional vector of the bounding box, and a 1601-dimensional vector of class likelihood. Both the VLP encoder and the GPT-2 decoder consist of $L = 12$ transformer layers. VLP has been pre-trained on Conceptual Captions [27] and MS COCO [21]. And the Chinese GPT-2 decoder[2] has been pre-trained on the CLUECorpusSmall [35] corpus with 14GB of Chinese text data. In the adjacent layer-fusion module, we set $k = 3$, i.e., each decoder layer attends to the visual representations from 3 different encoder layers.

Models are trained using cross-entropy loss with a batch size of 16. AdamW [22] is used for optimization with $\beta_1 = 0.9$, $\beta_2 = 0.999$ and the weight decay is set to 0.01. The learning rate is first linearly warmed up from 0 to 1e-3 for the first 10% of the total steps and then undergoes a linear decay.

At the reference stage, we use the beam-search [8] algorithm for caption generation with a beam size of 5. Four standard metrics are used to evaluate the quality of the generated captions, including BLEU-4, METEOR, ROUGE-L, and CIDEr. We train each model for 30 epochs and select the checkpoint with the best validation performance for evaluation. We repeat the procedure 3 times with different seeds and report the average scores.

---

[1] https://github.com/LuoweiZhou/VLP.
[2] https://github.com/Morizeyao/GPT2-Chinese.

Table 2. Evaluation scores of different models trained on COCO-CN.

Model	BLEU-4	METEOR	ROUGE_L	CIDEr
COCO-CN-1000				
AoANet	12.0 ± 0.57	18.9 ± 0.21	38.8 ± 0.33	92.5 ± 3.45
$\mathcal{M}^2$ Transformer	12.2 ± 0.39	18.5 ± 0.33	39.2 ± 0.56	94.5 ± 1.56
Fine-Tuning	12.7 ± 0.42	18.7 ± 0.25	41.3 ± 0.24	100.3 ± 2.53
Feature-Based	14.3 ± 0.82	19.8 ± 0.71	41.6 ± 0.52	110.1 ± 5.09
TraVL	$\mathbf{15.9 \pm 0.14}$	$\mathbf{20.8 \pm 0.05}$	$\mathbf{43.7 \pm 0.40}$	$\mathbf{121.6 \pm 2.65}$
COCO-CN-2000				
AoANet	13.7 ± 0.42	18.6 ± 0.25	40.8 ± 0.46	105.3 ± 2.31
$\mathcal{M}^2$ Transformer	13.2 ± 0.12	18.8 ± 0.19	41.1 ± 0.05	107.1 ± 0.95
Fine-Tuning	15.2 ± 0.21	20.0 ± 0.14	44.2 ± 0.33	120.9 ± 1.04
Feature-Based	16.7 ± 0.37	21.3 ± 0.29	43.6 ± 0.29	126.9 ± 1.63
TraVL	$\mathbf{17.0 \pm 0.33}$	$\mathbf{21.8 \pm 0.28}$	$\mathbf{44.7 \pm 0.42}$	$\mathbf{132.3 \pm 3.11}$
COCO-CN-5000				
AoANet	15.7 ± 0.34	20.3 ± 0.25	44.1 ± 0.34	126.2 ± 2.57
$\mathcal{M}^2$ Transformer	15.1 ± 0.19	19.7 ± 0.05	43.4 ± 0.24	123.4 ± 0.80
Fine-Tuning	17.6 ± 0.25	21.4 ± 0.17	46.5 ± 0.25	148.7 ± 0.94
Feature-Based	18.4 ± 0.21	22.1 ± 0.22	45.6 ± 0.57	144.3 ± 1.22
TraVL	$\mathbf{19.6 \pm 0.11}$	$\mathbf{22.8 \pm 0.17}$	$\mathbf{47.0 \pm 0.16}$	155.6 ± 2.01

5.3   Quantitative Analysis

We establish two types of baselines to which we compare our method. The first type is the novel captioning models that score high on MS-COCO without involving VL pre-training, including AoANet [11] and $\mathcal{M}^2$ Transformer [4]. And the second type is pre-training-based approaches that transfer pre-trained VL models for cross-lingual image captioning, including the fine-tuning method and the basic feature-based approach, which are described in Sect. 3. We experiment on subsets of COCO-CN and report the results in Table 2.

Effect of VL Pre-Training. AoANet and $\mathcal{M}^2$ Transformer are inferior to the pre-training-based approaches on all data scales. This shows that the VL pre-training, although performed for a different language, can actually benefit the target-language task via knowledge transfer. Pre-trained VL models can transform image regions into representations with high-level semantics that are not specific to a certain language.

Effect of Linguistic Pre-Training. Compared to the fine-tuning approach, the feature-based approach and TraVL yield better results, especially when the training set is small. But on COCO-CN-5000, the fine-tuning approach surpasses the feature-based approach on the ROUGE_L and CIDEr metrics. We explain this result by the use of the pre-trained GPT-2 decoder. When less training data is available, the linguistic knowledge carried by GPT-2 is essential for generating fluent and coherent sentences. However, a large training set allows the model to learn sufficient linguistic knowledge during fine-tuning, making the advantage of leveraging a pre-trained language model less significant.

Effect of Joint Attention and Adjacent Layer Fusion. On all data scales, TraVL outperforms the basic feature-based approach on all evaluation metrics. The advantage of TraVL over the basic feature-based approach stems entirely from the architectural differences. For one thing, the use of joint attention contributes to better alignment of different modalities at the semantic level. For another thing, the adjacent layer-fusion mechanism allows the decoder to leverage VLP's multilayer representations in a more fine-grained manner.

5.4 Qualitative Analysis

To provide an intuitive understanding of the different methods, we present a qualitative analysis of the captions generated for a few sample images. The images are shown in Fig. 3 and the captions are presented in Table 3. For each image, we list the ground truth as well as the captions generated by three different models (i.e., TraVL, $\mathcal{M}^2$ Transformer, and the fine-tuning method) trained on COCO-CN-1000.

By comparing the generated captions with the ground truths, we identify two aspects that make TraVL better than the other two methods. First, $\mathcal{M}^2$ Transformer and the fine-tuning method sometimes suffer from a misunderstanding of the objects in the images. In the Fig. 3a, the *color kite* described by $\mathcal{M}^2$ Transformer and the *hat* described by the fine-tuning method are both misinterpretations of the objects. Secondly, TraVL captures the details more accurately. For instance, in Fig. 3b, only TraVL describes the cake as a *colorful cake*; and in Fig. 3c, only TraVL identifies the vehicle as a *double-decker bus*. The other two models either miss the attribute or completely ignore the object.

5.5 Ablation Studies

We perform analysis on ablated versions of TraVL to quantify the effect of our proposed modules. All models are trained on COCO-CN-1000 in this experiment. We compare the full-featured TraVL against the following variations:

- **Base:** Conventional cross attention is used to fuse VLP's highest-level representation into GPT-2.
- **Base + Random Init. Enc.:** The VLP encoder is randomly initialized.
- **Base + Random Init. Dec.:** The GPT-2 decoder is randomly initialized.
- **Base + Joint Attention:** Each decoder layer attends to the visual representation of its corresponding encoder layer via joint attention.
- **TraVL ($k = 5$):** Each decoder layer attends to 5 encoder layers.
- **TraVL (All):** Each decoder layer attends to all encoder layers.

And the results are presented in Table 4.

(a)   (b)   (c)

Fig. 3. Sample images from COCO-CN test set.

Table 3. Comparison of captions generated by different models trained on COCO-CN-1000. The English translations of the captions are given in parentheses.

Image	Captions
Fig. 3a	**TraVL**: 一个穿着彩色连衣的女人在草地上玩飞盘。 (A woman wearing a colorful suit is playing frisbee on the grass.) $\mathcal{M}^2$ **Transformer**: 一个女孩踩着一个彩色风筝在草地上。 (A girl is treading on a colorful kite on the grass.) **Fine-Tuning**: 一个带着帽子的女孩正在草地上。 (A girl wearing a hat is on the grass.) **GT**: 一个正在玩飞盘的女孩。 (A girl is playing frisbee.)
Fig. 3b	**TraVL**: 一个女人在公园中准备切着一个彩色的蛋糕。 (A woman in the park prepares to cut a colorful cake.) $\mathcal{M}^2$ **Transformer**: 一个女孩拿着一个彩色帽子在女人坐在蛋糕前。 (A girl is holding a colorful hat the woman sitting in front of the cake.) **Fine-Tuning**: 一个女人正在厨房里。 (A woman is in the kitchen.) **GT**: 户外一个穿着蓝色T恤的女人手里拿着一盘子五彩蛋糕。 (Outdoors a woman in a blue T-shirt is holding a plate of colorful cake.)
Fig. 3c	**TraVL**: 一辆双层公交车停在街道上。 (A double-decker bus is parking on the street.) $\mathcal{M}^2$ **Transformer**: 一条公交车停在街道上。 (A bus is parking on the street.) **Fine-Tuning**: 一辆男人正在街道上。 (A man is in the street.) **GT**: 两辆双层公共汽车行驶在城市马路上，街道两边有行人和建筑物。 (Two double-decker buses are driving on the city street, with people and buildings on both sides.)

Table 4. Ablation studies of TraVL with different variations on COCO-CN-1000.

Model	BLEU-4	METEOR	ROUGE_L	CIDEr
Base	14.3 ± 0.82	19.8 ± 0.71	41.6 ± 0.52	110.1 ± 5.09
Base + Random Init. Enc.	3.0 ± 0.57	11.3 ± 2.05	25.3 ± 1.63	15.3 ± 1.35
Base + Random Init. Dec.	11.6 ± 0.98	18.9 ± 0.51	41.3 ± 0.33	95.3 ± 4.22
Base + Joint Attention	15.1 ± 0.59	20.4 ± 0.29	42.7 ± 0.28	116.1 ± 2.08
TraVL ($k = 5$)	14.7 ± 0.25	20.4 ± 0.17	43.6 ± 0.24	118.2 ± 2.13
TraVL (All)	13.0 ± 1.29	19.2 ± 1.07	39.9 ± 0.96	93.4 ± 6.45
TraVL ($k = 3$)	$\mathbf{15.9 \pm 0.14}$	$\mathbf{20.8 \pm 0.05}$	$\mathbf{43.7 \pm 0.40}$	$\mathbf{121.6 \pm 2.65}$

Effect of Pre-Trained VLP. Compared to using the pre-trained VLP model, training the encoder from scratch substantially degrades the model performance. This demonstrates the importance of leveraging the knowledge acquired from VL pre-training. Without this knowledge, it is hard to train a powerful feature extractor from scratch with limited training data.

Effect of Pre-Trained GPT-2. Randomly initializing the GPT-2 decoder also harms the performance, showing that the captioning model can benefit from using a pre-trained language model. The linguistic pre-training on Chinese textual data allows the GPT-2 decoder to quickly adapt to the in-domain data.

Effect of Joint Attention. Compared to the base model, allowing each decoder layer to attend to the encoder layer at the same level via joint attention improves the performance by a large margin. On the one hand, aligning the encoder and decoder layers helps exploit the visual information at a more fine-grained level. On the other hand, the introduction of joint attention can compensate for the architectural mismatch and is better at aligning both modalities.

Effect of Adjacent Layer Fusion. In contrast to the intuition that fusing more encoder layers into each decoder layer can further boost the performance, the results show the opposite. Increasing k from 3 to 5 hardly does any good, and fusing all encoder layers further worsens the performance. It can be explained by the vastly increased network complexity, which not only increases computation time and memory usage but also leads to difficulties in optimization. Therefore, our choice of $k = 3$ makes a good trade-off between the richness of information and the simplicity of the network.

6   Conclusions

In this paper, we present TraVL, a framework for transferring pre-trained VL models for cross-lingual image captioning. To enforce the semantic alignment between different modalities, we propose to fuse visual and linguistic representations through joint attention instead of conventional cross attention. To fully exploit the rich hierarchy of visual information, we develop an adjacent layer-fusion mechanism that allows each decoder layer to attend to multilayer visual representations with similar semantics. Experiments on the COCO-CN dataset show that TraVL outperforms the state-of-the-art captioning models and other transfer learning methods.

Acknowledgements. This work was supported by the Key Research Program of Zhejiang Province (Grant No.2021C01109).

References

1. Anderson, P., et al.: Bottom-up and top-down attention for image captioning and visual question answering. In: CVPR, pp. 6077–6086 (2018)
2. Aneja, J., Deshpande, A., Schwing, A.G.: Convolutional image captioning. In: CVPR. pp. 5561–5570 (2018). https://doi.org/10.1109/CVPR.2018.00583
3. Chen, J., Guo, H., Yi, K., Li, B., Elhoseiny, M.: Visualgpt: Data-efficient image captioning by balancing visual input and linguistic knowledge from pretraining. CoRR abs/2102.10407 (2021)
4. Cornia, M., Stefanini, M., Baraldi, L., Cucchiara, R.: Meshed-memory transformer for image captioning. In: CVPR, pp. 10575–10584 (2020). https://doi.org/10.1109/CVPR42600.2020.01059
5. Devlin, J., Chang, M.W., Lee, K., Toutanova, K.: BERT: Pre-training of deep bidirectional transformers for language understanding. In: NAACL, pp. 4171–4186 (2019). https://doi.org/10.18653/v1/N19-1423
6. Elliott, D., Frank, S., Hasler, E.: Multi-language image description with neural sequence models. CoRR abs/1510.04709 (2015)
7. Farhadi, A., et al.: Every picture tells a story: Generating sentences from images. In: ECCV, pp. 15–29 (2010)
8. Freitag, M., Al-Onaizan, Y.: Beam search strategies for neural machine translation. In: NMT@ACL. pp. 56–60 (2017). https://doi.org/10.18653/v1/w17-3207
9. He, T., et al.: Layer-wise coordination between encoder and decoder for neural machine translation. In: NeurIPS, pp. 7955–7965 (2018)
10. Hodosh, M., Young, P., Hockenmaier, J.: Framing image description as a ranking task: data, models and evaluation metrics. JAIR **47**(1), 853–899 (2013)
11. Huang, L., Wang, W., Chen, J., Wei, X.: Attention on attention for image captioning. In: ICCV, pp. 4633–4642 (2019). https://doi.org/10.1109/ICCV.2019.00473
12. Ichikawa, K., Tamano, H.: Unsupervised qualitative scoring for binary item features. Data Sci. Eng. **5**(3), 317–330 (2020)
13. Jawahar, G., Sagot, B., Seddah, D.: What does BERT learn about the structure of language? In: ACL, pp. 3651–3657 (2019). https://doi.org/10.18653/v1/p19-1356
14. Karpathy, A., Joulin, A., Fei-Fei, L.: Deep fragment embeddings for bidirectional image sentence mapping. In: NeurIPS, pp. 1889–1897 (2014)
15. Krishna, R., et al.: Visual genome: Connecting language and vision using crowd-sourced dense image annotations. IJCV **123**(1), 32–73 (2017)
16. Kulkarni, G., et al.: Baby talk: Understanding and generating simple image descriptions. In: CVPR, pp. 1601–1608 (2011). https://doi.org/10.1109/CVPR.2011.5995466

17. Lan, W., Li, X., Dong, J.: Fluency-guided cross-lingual image captioning. In: ACM Multimedia, pp. 1549–1557 (2017). https://doi.org/10.1145/3123266.3123366
18. Li, W., et al.: UNIMO: towards unified-modal understanding and generation via cross-modal contrastive learning. In: ACL-IJCNLP, pp. 2592–2607 (2021). https://doi.org/10.18653/v1/2021.acl-long.202
19. Li, X., et al.: Coco-cn for cross-lingual image tagging, captioning, and retrieval. IEEE Multimedia **21**(9), 2347–2360 (2019). https://doi.org/10.1109/TMM.2019.2896494
20. Li, X., et al.: Oscar: Object-semantics aligned pre-training for vision-language tasks. In: ECCV, pp. 121–137 (2020)
21. Lin, T.Y., et al.: Microsoft coco: Common objects in context. In: ECCV, pp. 740–755 (2014)
22. Loshchilov, I., Hutter, F.: Decoupled weight decay regularization. In: ICLR (2019)
23. Lu, J., Batra, D., Parikh, D., Lee, S.: Vilbert: Pretraining task-agnostic visiolinguistic representations for vision-and-language tasks. In: NeurIPS, pp. 13–23 (2019)
24. Miyazaki, T., Shimizu, N.: Cross-lingual image caption generation. In: ACL, pp. 1780–1790 (2016). https://doi.org/10.18653/v1/P16-1168
25. Radford, A., Wu, J., Child, R., Luan, D., Amodei, D., Sutskever, I.: Language models are unsupervised multitask learners. Tech. rep, OpenAI (2019)
26. Ren, S., He, K., Girshick, R., Sun, J.: Faster r-cnn: Towards real-time object detection with region proposal networks. In: NeurIPS, pp. 91–99 (2015)
27. Sharma, P., Ding, N., Goodman, S., Soricut, R.: Conceptual captions: A cleaned, hypernymed, image alt-text dataset for automatic image captioning. In: ACL, pp. 2556–2565 (2018). https://doi.org/10.18653/v1/P18-1238
28. Su, W., et al.: Vl-bert: Pre-training of generic visual-linguistic representations. In: ICLR (2020)
29. Tsutsui, S., Crandall, D.J.: Using artificial tokens to control languages for multilingual image caption generation. CoRR abs/1706.06275 (2017)
30. Vaswani, A., et al.: Attention is all you need. In: NeurIPS, pp. 6000–6010 (2017)
31. Vinyals, O., Toshev, A., Bengio, S., Erhan, D.: Show and tell: A neural image caption generator. In: CVPR, pp. 3156–3164 (2015). https://doi.org/10.1109/CVPR.2015.7298935
32. Wang, B., Wang, C., Zhang, Q., Su, Y., Wang, Y., Xu, Y.: Cross-lingual image caption generation based on visual attention model. IEEE Access **8**, 104543–104554 (2020). https://doi.org/10.1109/ACCESS.2020.2999568
33. Weng, R., Yu, H., Huang, S., Cheng, S., Luo, W.: Acquiring knowledge from pre-trained model to neural machine translation. In: AAAI, pp. 9266–9273 (2020)
34. Xu, K., et al.: Show, attend and tell: Neural image caption generation with visual attention. In: ICML. vol. 37, pp. 2048–2057 (2015)
35. Xu, L., Zhang, X., Dong, Q.: Cluecorpus 2020: A large-scale chinese corpus for pre-training language model. CoRR abs/2003.01355 (2020)
36. Zhou, L., Palangi, H., Zhang, L., Hu, H., Corso, J.J., Gao, J.: Unified vision-language pre-training for image captioning and VQA. In: AAAI, pp. 13041–13049 (2020)

From Less to More: Common-Sense Semantic Perception Benefits Image Captioning

Feng Chen, Xinyi Li, Jintao Tang, Shasha Li, and Ting Wang[✉]

National University of Defense Technology, Changsha 410073, China
{chenfeng15a,tangjintao,tingwang}@nudt.edu.cn

Abstract. Most recent arts in image captioning rely solely on exploring the information contains in the image or modeling the inner-relations among visual features, which fails to generate informative captions in some cases. Part of what defines humans is the ability of common-sense reasoning behind semantic association, which is different from machines. To this end, we propose a Common-Sense Aware method (CSA) for image captioning, which capitalizes general prior knowledge to associate extra semantic information during generation to infer more informative captions. Specifically, based on ConceptNet, we extract common-sense knowledge features using pre-generated concepts to provide comprehensive associated semantic information for captioning. We conduct extensive experiments on the MS COCO dataset to demonstrate the effectiveness of CSA, results show that it furthers state-of-the-arts.

Keywords: Common-sense perception · ConceptNet · Attention mechanism · Image captioning

1 Introduction

Compared with other tasks [3, 21], image captioning is more challenging. The main challenges not only lie in generating fluent natural language sentences but also in comprehensive visual understanding (e.g., concepts, relationships) of the input image.

Inspired by the encoder-decoder framework in machine translation [20], early methods apply CNNs to encode images as vectors and RNNs to generate captions. To further focus on relevant spatial aspects (e.g., regions) of images, recent methods resort to attention mechanisms to design captioning models. Existing methods also explore the effectiveness of semantic information which is taken as extra input during generation to produce semantically consistent captions. However, today's methods [1, 5, 23, 25, 27] are good at detecting and telling explicit information, such as objects, but failing to reason. As far as we know, reasoning is a unique ability of humans to obtain comprehensive information based on the known limited information and prior knowledge.

B. Li et al. (Eds.): APWeb-WAIM 2022, LNCS 13422, pp. 356–368, 2023.
https://doi.org/10.1007/978-3-031-25198-6_27

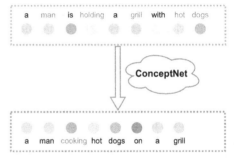

Fig. 1. An example result of our proposed method.

Recognizing important objects or concepts in an image is much easier for a machine than describing it with a comprehensive natural language sentence as humans. As the example shown in Fig. 1, the base model correctly recognizes important objects in the image and generates "a man is holding a grill with hot dogs". However, in general, humans rarely describe an image so stiffly. Given the words "grill" and "hot dogs", it's easy to reason "a man cooking hot dogs on a grill" based on common-sense knowledge. To tackle the problems of generating relatively brief captions and solely paying attention to explicit image content, knowledge-driven methods [9,25,28] are proposed, which optimize the quality of generated sentences by introducing common-sense knowledge [15,19]. These methods use detected semantic concepts(e.g., attributes) as single queries to extract associate information in external knowledge bases, and then generate captions based on the extracted information. However, they extract relative knowledge in a coarse way, while ignoring relationships between query concepts. Meanwhile, the encoding methods of extracted knowledge information are simple, which needs to be further optimized.

To effectively take advantage of common-sense knowledge and achieve semantic association, we propose a Common-Sense Aware method (CSA) for image captioning. Different from existing methods [9,25,28], CSA is novel in both knowledge extraction and caption generation. Firstly, we propose to solve the problems of inadequate caption generation and implied semantic mining from the perspective of common-sense association, and resort to ConceptNet [19] for the extension of limited semantic information, so as to expand the breadth of image semantic perception in image captioning. Specifically, in the knowledge extraction stage, this work adopts the knowledge extraction method based on path-searching [4,16] to extract common-sense paths between each pre-generated concept pair. Not only the direct connections between concepts but also multi-hops indirect connections are considered to enlarged the search scope of relevant knowledge. The final semantic features are retained through pruning, merging, and other simplified operations. Furthermore, aiming at the utilization of extracted common-sense knowledge-based semantic information, this work constructs an attention-based feature fusion layer from the perspective of multi-source information encoding. Specifically, in the caption generation stage, the feature fusion layer fuses the visual and the semantic information, which selectively pays attention to these two kinds of features when generating words.

Contributions of the proposed method are three-fold: 1) A Common-Sense Aware method (CSA) is proposed for image captioning, which takes advantage of both the visual and the semantic information based on common-sense knowledge to generate captions. 2) We propose a method based on path-searching to extract semantic information of images, which utilizes pre-generated concept pairs as query inputs. 3) To utilize the extracted semantic information during generation, we design an attention-based feature fusion layer to encode two kinds of features in a cooperative way, which deepens the visual understanding of the captioning model.

2   Related Work

Thanks to the development of deep learning related technologies and successful explorations in related fields such as machine translation, image captioning methods based on the encoder-decoder framework in early studies [8,23,26] have achieved promising progress. In general, these methods encode the input image as a single vector by a CNN in the encoding part and generate captions by an RNN in the decoding part. Attention-based methods are no longer limited by encoding the input image as a single vector but represent it as a series of vectors (such as sub-region vectors). These methods selectively pay attention to the subset related to the current word during generation and assign the importance of vectors by weights. Thus, words can be better aligned with relevant features of fine granularity of images.

In order to make the description generation model free from the shackles of brief caption generation and ignoration of explicit information, knowledge-driven methods [9,25,28] propose to introduce external knowledge to optimize the quality of the generated captions. Among them, Wu et al. [25] firstly proposes to take advantage of external knowledge to generate captions, and further solves visual question answering [2] with annotated texts in DBpedia [15]. DBpedia is a database of structured information extracted from Wikipedia. Specifically, given the image and predicted attributes, this method extracts the top five attributes to obtain the annotations of each attribute in DBpedia and then encodes the generated captions and the queried annotations to generate the answer. Zhou et al. [28] introduces common-sense knowledge based on ConceptNet [19] during generation, which comprehensively considers the visual features and knowledge features of the image to generate captions. However, this method extracts the associated terms of each detected object in ConceptNet and encodes all the associated terms as a feature representation, which lacks fine-grained selection and encoding of common-sense knowledge.

Most recently, Huang et al. [9] proposes an external knowledge-driven method to improve the performance of the captioning model. In this method, external common-sense knowledge in ConceptNet is introduced in the last step of word generation, so as to improve the generation probabilities of some words. However, the extracted knowledge semantics are not introduced into the captioning model, and the generation probabilities of some words are only changed in the last generation step, which results in inadequate use of semantic information in the model.

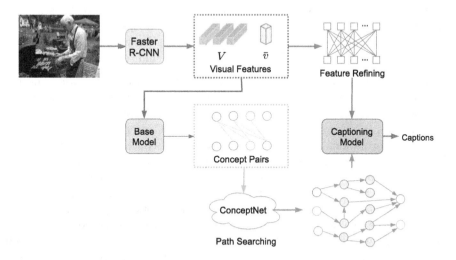

Fig. 2. An illustration of the proposed framework.

3 Methodology

From the perspective of common-sense perception, we propose a Common-Sense Aware method (CSA) for image captioning. The framework of CSA is depicted in Fig. 2, which takes advantage of extended semantic information based on external knowledge to enrich generated captions.

3.1 Feature Extraction

We use two kinds of features in CSA: sub-region visual features and common-sense knowledge-based semantic features. Specifically, the semantic features are extracted based on pre-generated concept pairs and ConceptNet, which are then further denoised and encoded.

Visual Features. Region-based visual features are extracted by an off-the-shelf Faster R-CNN [1], which is pre-trained with object detecting task on Visual Genome [14]. Specifically, These object-related sub-regions whose detection probability of each category exceeds the confidence threshold are represented as a series of vectors after an RoI pooling layer.

Common-Sense Extended Features. Taking the semantic relevance in structured knowledge base into account, we propose to optimize semantic perception of the captioning model based on a common-sense knowledge graph and limited semantic information. As far as we know, ConceptNet includes lexical and world knowledge from many different sources and contains abundant common-sense information [19]. Thus, in this paper, we obtain common-sense semantic extended information based on ConceptNet, which is then used as extra semantic guidance during generation to enrich the semantic perception of the captioning model. There are three steps to get common-sense knowledge-based features: concept pair identification, sub-graph construction, and semantic encoding.

Concept Pair Identification. Intuitively, different from semantic concepts such as objects and attributes, captions of images contain more comprehensive semantic information. Therefore, we construct concept pairs based on pre-generated captions as queries of path-searching in ConceptNet to obtain richer semantic information. The identification process of concept pairs is as follows: 1) Pre-generate captions. Given an image, we utilize Up-Down [1] model to pre-generate a caption $Y_{1:T} = \{y_1, y_2, ..., y_T\}$, where T is the number of words in Y; 2) Extract concept pairs. After filtering out stopwords in the pre-generated caption, we use the remaining words as initial concepts $C_{1:N} = \{c_1, c_2, ..., c_N\}$, where N is the number of concepts; 3) Construct concept pairs. For the ith concept in $C_{1:N}$, we pair it with the subsequent $N - 1$ ones separately and obtain concept pairs as common-sense query candidates $P = \{p_{1,2}, p_{1,3}, ..., p_{(N-1),N}\}$, where $p_{i,j} = [c_i, c_j]$, P contains $\frac{N \times (N-1)}{2}$ pairs.

Sub-graph Construction. ConceptNet represents the general knowledge, links between knowledge resources, and allows relative applications to better understand the meanings of words. In this work, we only take the English part into account, represent the common-sense knowledge as abundant triples, and further translate the common-sense knowledge-based semantic extension and sub-graph construction problem as a path-searching problem between concepts. The process of sub-graph construction has two steps: 1) Path searching. For (c_i, c_j) in P, we search paths with multi-hops between c_i and c_j in ConceptNet, and reduce noise information and computation by keeping up to 10 paths for each concept pair; 2) Sub-graph Construction. For each path in the path set G, we score every triple in G by transE [24] to reduce redundant paths and the preserved triples are constructed as the common-sense knowledge sub-graph.

Semantic Encoding. To reduce noise information and computational complexity, this work ignores relations between concept terms $E = \{e_1, e_2, ..., e_M\}$ in the common-sense sub-graph, where M is the number of extended concepts. To further introduce common-sense knowledge-based semantic information into image captioning, we use Numberbatch word vectors that trained based on ConceptNet as initial representations of extended concepts. Specifically, for $e_i \in E$, we represent it as a pre-trained vector in Numberbatch.

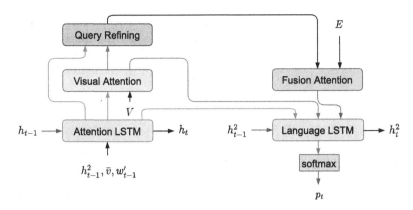

Fig. 3. An illustration of the proposed captioning model.

3.2 Common-Sense Aware Captioning Model

The common-sense aware captioning model is proposed to take advantage of both visual and common-sense knowledge-based extended semantic features for image captioning. As shown in Fig. 3, it's an attention-based model with two-layer LSTMs.

Attention LSTM. To dynamically pay attention to visual features and common-sense information, and further fuse these two kinds of features during generation, we design a visual attention layer and a fusion attention layer. At step t, an attention LSTM calculates query q_t for attention blocks based on the previous hidden state of the language LSTM h_{t-1}^2, the global image feature $\bar{v}$ which denotes the mean pooling of sub-region visual features, the previous generated word w_{t-1}', and the previous hidden state and context vector of the attention LSTM (h_{t-1}^1, c_{t-1}^1):

$$w_{t-1} = W_{em}w_{t-1}' \tag{1}$$

$$x_t = [h_{t-1}^2, \bar{v}, w_{t-1}] \tag{2}$$

$$h_t^1, c_t^1 = f_{att-lstm}(x_t, (h_{t-1}^1, c_{t-1}^1)) \tag{3}$$

$$q_t = h_t^1 \tag{4}$$

where $W_{em} \in \mathbb{R}^{|\Sigma|*D}$ is the word vectors of vocabulary, $|\Sigma|$ is the number of words, D is the dimension.

Attention Layers. Based on certain weight calculation rules, the attention layers selectively focus on visual and common-sense knowledge-based semantic features that relate to the current generated word. Specifically, it contains two parts: a visual attention layer and a fusion attention layer that focuses on multi-source feature encoding.

Before the attention layer, a feature encoding layer [10] based on the multi-head self-attention mechanism is firstly used to model the inner-relationships among visual features. Based on the current attention query vector q_t and the sub-region features V, the visual attention layer f_{V-att} calculates as follows:

$$a_i = W_l(\sigma((W_v V) \oplus (W_h q^1))) \tag{5}$$

$$\alpha_i = \frac{e^{a_i}}{\sum_{i=0}^{r} e^{a_i}} \tag{6}$$

where W_l, W_v, and W_h are learned parameters of linear transformations, $\oplus$ means element-wise add, σ represents tanh activation layer, r is the number of regions. After that, the weighted sum of V is calculated based on attention weights to obtain the visual attention result:

$$\hat{v} = \sum_{i=0}^{r} \alpha_i v_i \tag{7}$$

Not all of the extracted semantic features are related to the current word, thus the fusion attention layer selectively pays attention to these features by using visual attention result as guidance. We update attention query q_t as q_t' based on $\hat{v}$ to comprehensively consider multi-source features and mine the advantages of multi-source information. Specifically, the calculation process of fusion attention layer f_{e-att} is as follows:

$$q_t' = W_q[q_t, \hat{v}] \tag{8}$$

$$a_i = W_l'(\sigma((W_v' E) \oplus (W_h' q_t'))) \tag{9}$$

$$\alpha_i = \frac{e^{a_i}}{\sum_{i=0}^{M} e^{a_i}} \tag{10}$$

where W_q, W_l', W_v' and W_h' are learned parameters of linear transformations, M is the number of concepts in the constructed common-sense sub-graph. Then, the fusion attention result which is conditioned on the current query:

$$\hat{e} = \sum_{i=0}^{M} \alpha_i e_i \tag{11}$$

Language LSTM. q_t, $\hat{v}$ and $\hat{e}$ has encoded important information about the current word, we concatenate them as the input of language LSTM, then calculate the word distribution p_t on the pre-defined vocabulary by softmax function:

$$\tilde{x}_t = [q_t, \hat{v}, \hat{e}] \tag{12}$$

$$h_t^2, c_t^2 = f_{lan-lstm}(\tilde{x}_t, (h_{t-1}^2, c_{t-1}^2)) \tag{13}$$

$$p_t = softmax(h_t^2) \tag{14}$$

Optimization. There are two training stages for the proposed model: $-XE^*$ to minimize negative log-likelihood estimation loss, and $-RL^*$ (Reinforcement Learning-based strategy [1]) to maximize the negative expected CIDEr [22] reward. At the $-XE^*$ stage, the objective is:

$$loss = -\frac{1}{I}\sum_{i=1}^{I}\log p(T^i|V^i, E^i, \overline{v}^i; \theta) \tag{15}$$

where I is the number of images, T is the target caption, V is visual features, $\overline{v}$ is the global visual feature (mean-pooling of V), E is the extracted common-sense features, and θ is the learned parameter.

At the $-RL^*$ stage, we further employ reinforcement learning to boost the proposed model with CIDEr rewards. An update is implemented by computing the gradient of the pre-defined CIDEr reward:

$$\nabla_\theta \mathbb{E} \approx (R(c) - R(\hat{c}))\nabla_\theta log_{p_\theta}(c) \tag{16}$$

where θ is the learned parameter, R is the CIDEr reward, and $c/\hat{c}$ are random/max sampled captions by the current caption generator under the inference algorithm.

4 Experiments

4.1 Dataset and Metrics

Dataset. MS COCO dataset[1] contains $123,287$ images, in which each image has at least five captions in English. It is a standard benchmark for image captioning. We use Karpathy split [13] for training, validation, and testing.

Metrics. We use BLEU, METEOR, CIDEr, ROUGE-L, and SPICE which are publicly used metrics in image captioning evaluation[2] to report our results.

4.2 Evaluation

Comparison with State-of-the-Art Methods. For comprehensive comparisons, we show results of both the $-XE^*$ stage (Table 1) and the $-RL^*$ stage (Table 2) on Karpathy's test split of MS COCO. At the $-XE^*$ stage, the proposed model (Ours) outperforms existing knowledge-driven methods [25,28], which demonstrates its effectiveness. Meanwhile, when compared with recent state-of-the-arts [1,6,7,11,12,18,25,27], Ours achieves consistency improvements, especially on CIDEr (117.2) and SPICE (21.4). It shows that common-sense knowledge-based semantic features are beneficial to improve the quality of generated captions. At the $-RL^*$ stage, Ours outperforms existing knowledge-driven

---

[1] https://cocodataset.org/.

[2] https://github.com/tylin/coco-caption.

Table 1. Scores of methods on the public MS COCO test split in the −XE* stage. All of the scores are represented as percentages (%).

Methods	Metrics							
	B1	B2	B3	B4	METEOR	CIDEr	ROUGE-L	SPICE
CNet [28]	73.1	54.7	40.5	29.9	25.6	107.2	53.9	−
AttNet [25]	74	56	42	31	26	94	−	−
Att-CNN+LSTM [25]	74.0	56.0	42.0	31.0	26.0	94.0	−	−
LSTM-Att [27]	73.4	56.7	43.0	32.6	25.4	100.2	54.0	−
Adaptive-Att [18]	74.2	58.0	43.9	33.2	26.6	108.5	−	−
VSDA [7]	75.3	59.1	45.1	34.4	26.5	106.3	55.2	−
Stack-Cap [6]	76.2	60.4	46.4	35.2	26.5	109.1	−	−
RFNet [12]	76.4	60.4	46.6	35.8	27.4	112.5	56.5	20.5
Up-Down [1]	77.2	−	−	36.2	27.0	113.5	56.4	20.3
STMA [11]	**77.4**	**61.5**	47.6	36.5	27.4	114.4	56.8	20.5
Ours	77.1	61.3	**47.7**	**36.8**	**28.1**	**117.2**	**57.1**	**21.4**

Table 2. Scores of methods on the public MS COCO test split in the −RL* stage. All of the following models are optimized with the RL-based strategy, and scores are represented as percentages (%).

Methods	Metrics							
	B1	B2	B3	B4	METEOR	CIDEr	ROUGE-L	SPICE
IENet [9]	79.2	64.0	48.9	37.1	26.9	118.2	57.3	−
Stack-Cap [6]	78.6	62.5	47.9	36.1	27.4	120.4	56.9	20.9
RFNet [12]	79.1	63.1	48.4	36.5	27.7	121.9	57.3	21.2
Up-Down [1]	79.8	−	−	36.3	27.7	120.1	56.9	21.4
CAVP [17]	-	−	−	**38.6**	28.3	126.3	**58.5**	21.6
DeRF [5]	79.9	−	−	37.5	**28.5**	125.6	58.2	**22.3**
STMA [11]	80.2	64.4	49.7	37.7	28.2	125.9	58.1	21.9
Ours	**80.3**	**64.9**	**50.5**	38.5	**28.5**	**127.4**	**58.5**	22.0

method [9] and recent state-of-the-arts [1,5,6,11,12,17] on most metrics, e.g., obtain CIDEr score of 127.4, which further demonstrates the effectiveness of the proposed method.

Results suggest that common-sense-based semantic information of external knowledge effectively provides supplementary semantics to help the captioning model generate high-quality captions.

Ablation Studies. To figure out the contributions of different modules in the proposed model, we conduct extensive experiments and compare the following methods. **Base:** A two-layer LSTM model based on bottom-up [1], which dynamically focuses on visual features during generation. **E:** Utilizing common-sense knowledge-based semantic features; **R:** Applying visual feature refining; **C:** Feature fusion by attention result concatenation; **F:** Utilizing one attention layer to encode both visual and semantic information; **G:** The proposed feature fusion method, which contains two kinds of attention layers: visual attention and fusion attention. Ablated results are shown in Table 3.

Table 3. Results of ablation studies. All values are reported as percentage (%).

Methods	Metrics							
	B1	B2	B3	B4	METEOR	CIDEr	ROUGE-L	SPICE
Base-XE*	76.7	60.7	46.9	36.1	27.8	114.1	56.7	20.8
+E	76.7	60.7	47.0	36.1	27.8	114.6	56.8	21.0
+E+R+C	**77.2**	**61.3**	47.5	36.6	27.9	116.2	56.8	21.1
+E+R+F	**77.2**	**61.3**	47.4	36.4	27.9	116.3	56.9	21.0
+E+R+G	77.1	**61.3**	**47.7**	**36.8**	**28.1**	**117.2**	57.1	**21.4**
Base-RL*	79.6	64.0	49.1	36.9	28.0	123.2	57.8	21.4
+E	80.0	64.3	49.7	37.8	28.1	125.3	58.0	21.6
+E+R+C	80.0	64.4	49.9	38.0	28.3	126.6	58.1	21.9
+E+R+F	79.9	64.4	50.0	38.2	28.3	126.1	58.1	21.7
+E+R+G	**80.3**	**64.9**	**50.5**	**38.5**	**28.5**	**127.4**	**58.5**	**22.0**

1) Semantic association features based on common-sense knowledge. Compared with the base model (Base), +E achieves better results, especially after the RL-based training, e.g., +E exceeds the base model by 2.0 in CIDEr. It shows that +E effectively captures the relevant semantic information of the image based on common-sense knowledge. Only based on parallel attention mechanisms and a multi-source information connection layer, the performance of the model is significantly improved.

2) Fusion modeling of visual and semantic features. Compared with +E, +E+R+C performs better, e.g., boosts CIDEr/SPICE to 126.6/21.9. Specifically, +E+R+C refines visual features, encodes both the visual and the semantic features based on attention mechanisms respectively, and then concatenates them into one vector. Moreover, +E+R+G(CSA/Ours) selects knowledge-based semantic features with the guidance of visual representation and improves CIDEr/SPICE to 127.4/22.0. Results show that visual features help the proposed model (CSA) focus on relevant information more accurately, and further optimize the utilization of common-sense knowledge-based semantic information.

3) Semantic-based visual feature modeling. +E+R+F first uses common-sense knowledge-based semantic features to update visual representations based on the attention mechanism, concatenates the original visual features with the updated ones, and then generates descriptions with the same structure as Base. Compared with +E+R+C, it performs slightly better on B3/B4 and slightly worse on CIDEr/SPICE. Considering that CIDEr and SPICE focus on semantic consistency evaluation, results show that +E+R+C has higher semantic consistency.

To sum up, the above analysis shows that: 1) Compared with Base, the proposed model (CSA) relates to relevant semantic information based on common-sense knowledge, thus generating more semantically rich descriptions; 2) The internal relationship modeling of visual features deepens the visual understanding of the proposed model and improves the quality of generated descriptions;

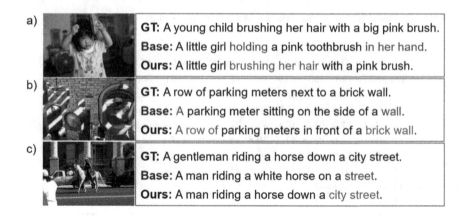

a) GT: A young child brushing her hair with a big pink brush.
Base: A little girl holding a pink toothbrush in her hand.
Ours: A little girl brushing her hair with a pink brush.

b) GT: A row of parking meters next to a brick wall.
Base: A parking meter sitting on the side of a wall.
Ours: A row of parking meters in front of a brick wall.

c) GT: A gentleman riding a horse down a city street.
Base: A man riding a white horse on a street.
Ours: A man riding a horse down a city street.

Fig. 4. Qualitative examples. We show ground truth captions (GT, annotated by humans), generated captions of the base model (Base) and the proposed CSA model (Ours).

3) The fusion modeling of the visual and semantic features further improves the captioning performance.

Case Study. Figure 4 compares generated captions of the base model (Base) and the proposed CSA model (Ours). Among them, Base accurately predicts the main content of the image, Ours takes advantage of common-sense knowledge-based semantic features, which expands the scope of semantic perception based on pre-generated concepts, and then generates better captions. For example, In Fig. 4 a), Base correctly predicts "holding *** in her hand" about the little girl but predicts "brush" as "toothbrush". In this case, Ours alleviates the problem of inaccurate target recognition and produces a more natural and reasonable description of "brushing her hair". As shown in Fig. 4 b), Ours can correct or supplement insufficient semantics. For example, "a row of parking meters" is generated from "a parking meter", and "brick wall" is generated from "wall". The final caption is of higher quality and semantically richer than that generated by Base.

The above analysis shows that: aiming at the problem of incorrect captions or insufficient semantic perception of the base model, the proposed method expands the scope of semantic perception by introducing common-sense knowledge. By paying attention to more accurate features or relevant semantic information, the proposed model generates better captions.

5   Conclusion

We propose a Common-Sense Aware method (CSA) in this paper, which takes advantage of both visual and common-sense-driven extended semantic information to generate captions. Specifically, to extract common-sense knowledge-based semantic information in ConceptNet effectively, we propose a method based on path-searching with pre-generated concept pairs as queries. Furthermore, to utilize the extracted semantic information during generation, we design an attention-based feature fusion layer to encode two kinds of image features in a cooperative way, which deepens the visual understanding of the captioning model. Experimental results demonstrate that CSA outperforms recent state-of-the-art methods.

Acknowledgement. This work was supported by the National Key Research and Development Project of China (No. 2021ZD0110700) and the National Natural Science Foundation of China (No. 62002373).

References

1. Anderson, P., et al.: Bottom-up and top-down attention for image captioning and visual question answering. In: IEEE Conference on Computer Vision and Pattern Recognition, pp. 6077–6086 (2018)
2. Antol, S., Agrawal, A., Lu, J., Mitchell, M., Parikh, D.: VQA: visual question answering. Int. J. Comput. Vis. **123**(1), 4–31 (2015)
3. Dong, G., Zhang, X., Lan, L., Wang, S., Luo, Z.: Label guided correlation hashing for large-scale cross-modal retrieval. Multimed. Tools Appl. **78**(21), 30895–30922 (2019). https://doi.org/10.1007/s11042-019-7192-5
4. Feng, Y., Chen, X., Lin, B.Y., Wang, P., Yan, J., Ren, X.: Scalable multi-hop relational reasoning for knowledge-aware question answering. In: Conference on Empirical Methods in Natural Language Processing (2020)
5. Gao, L., Fan, K., Song, J., Liu, X., Xu, X., Shen, H.T.: Deliberate attention networks for image captioning. In: AAAI Conference on Artificial Intelligence (2019)
6. Gu, J., Cai, J., Wang, G., Chen, T.: Stack-captioning: coarse-to-fine learning for image captioning. In: AAAI Conference on Artificial Intelligence (2018)
7. He, C., Hu, H.: Image captioning with visual-semantic double attention. ACM Trans. Multimed. Computi. Commun. Appl. **15**(1), 26 (2019)
8. Hodosh, M., Young, P., Hockenmaier, J.: Framing image description as a ranking task: data, models and evaluation metrics. J. Artif. Intell. Res. **47**(1), 853–899 (2015)
9. Huang, F., Li, Z., Chen, S., Zhang, C., Ma, H.: Image captioning with internal and external knowledge. In: 29th ACM International Conference on Information and Knowledge Management (2020)
10. Huang, L., Wang, W., Chen, J., Wei, X.Y.: Attention on attention for image captioning. In: IEEE International Conference on Computer Vision, pp. 4634–4643 (2019)
11. Ji, J., Xu, C., Zhang, X., Wang, B., Song, X.: Spatio-temporal memory attention for image captioning. IEEE Trans. Image Process. **29**, 7615–7628 (2020)
12. Jiang, W., Ma, L., Jiang, Y.-G., Liu, W., Zhang, T.: Recurrent fusion network for image captioning. In: Ferrari, V., Hebert, M., Sminchisescu, C., Weiss, Y. (eds.) ECCV 2018. LNCS, vol. 11206, pp. 510–526. Springer, Cham (2018). https://doi.org/10.1007/978-3-030-01216-8_31
13. Karpathy, A., Fei-Fei, L.: Deep visual-semantic alignments for generating image descriptions. In: IEEE Conference on Computer Vision and Pattern Recognition, pp. 3128–3137 (2015)

14. Krishna, R., et al.: Visual genome: connecting language and vision using crowd-sourced dense image annotations. Int. J. Comput. Vis. **123**(1), 32–73 (2017)
15. Lehmann, J., et al.: DBpedia - a large-scale, multilingual knowledge base extracted from Wikipedia. Semant. Web **6**, 167–195 (2015)
16. Lin, B.Y., Chen, X., Chen, J., Ren, X.: KagNet: knowledge-aware graph networks for commonsense reasoning. arXiv abs/1909.02151 (2019)
17. Liu, D., Zha, Z.J., Zhang, H., Zhang, Y., Wu, F.: Context-aware visual policy network for sequence-level image captioning. In: 26th ACM International Conference on Multimedia, pp. 1416–1424 (2018)
18. Lu, J., Xiong, C., Parikh, D., Socher, R.: Knowing when to look: adaptive attention via a visual sentinel for image captioning. In: IEEE Conference on Computer Vision and Pattern Recognition, vol. 6, p. 2 (2017)
19. Speer, R., Chin, J., Havasi, C.: ConceptNet 5.5: an open multilingual graph of general knowledge. In: AAAI Conference on Artificial Intelligence (2017)
20. Sutskever, I., Vinyals, O., Le, Q.V.: Sequence to sequence learning with neural networks. In: Advances in Neural Information Processing Systems, pp. 3104–3112 (2014)
21. Tan, H., Zhang, X., Lan, L., Huang, X., Luo, Z.: Nonnegative constrained graph based canonical correlation analysis for multi-view feature learning. Neural Process. Lett. **50**(2), 1215–1240 (2018). https://doi.org/10.1007/s11063-018-9904-7
22. Vedantam, R., Zitnick, C.L., Parikh, D.: CIDEr: consensus-based image description evaluation. In: IEEE Conference on Computer Vision and Pattern Recognition, pp. 4566–4575 (2015)
23. Vinyals, O., Toshev, A., Bengio, S., Erhan, D.: Show and tell: a neural image caption generator. In: IEEE Conference on Computer Vision and Pattern Recognition, pp. 3156–3164 (2015)
24. Wang, Z., Zhang, J., Feng, J., Chen, Z.: Knowledge graph embedding by translating on hyperplanes. In: AAAI Conference on Artificial Intelligence (2014)
25. Wu, Q., Shen, C., Wang, P., Dick, A., Hengel, A.V.: Image captioning and visual question answering based on attributes and external knowledge. IEEE Trans. Pattern Anal. Mach. Intell. **40**, 1367–1381 (2018)
26. Xu, K., et al.: Show, attend and tell: neural image caption generation with visual attention. Computer Science, pp. 2048–2057 (2015)
27. Yao, T., Pan, Y., Li, Y., Qiu, Z., Mei, T.: Boosting image captioning with attributes. In: IEEE International Conference on Computer Vision, pp. 22–29 (2017)
28. Zhou, Y., Sun, Y., Honavar, V.G.: Improving image captioning by leveraging knowledge graphs. In: IEEE Winter Conference on Applications of Computer Vision (WACV), pp. 283–293 (2019)

MACNet: Multi-Attention and Context Network for Polyp Segmentation

Xiuzhen Hao, Haiwei Pan$^{(\boxtimes)}$, Kejia Zhang, Chunling Chen, Xiaofei Bian, and Shuning He

Harbin Engineering University, Harbin, People's Republic of China
panhaiwei@hrbeu.edu.cn

Abstract. Colorectal cancer (CRC) is the third most common cancer worldwide. Colonoscopy is an effective technique for detecting colorectal polyps, which are closely associated with colorectal cancer. In clinical practice, segmenting polyps from colonoscopy images is of great significance, as it provides valuable information for diagnosis and surgery. Many networks have demonstrated better segmentation results. However, achieving accurate polyp segmentation remains a challenge due to the diverse in size, shape, texture and color of polyps. This paper proposes a Multi-Attention and Context Network (MACNet), which simulates the process of determining the segmentation region by clinical experts, incorporating the Balancing Attention Module (BAM), Non-local Information Statistical Attention module (Non-local), Position Rectify Module (PRM) and Focus Module (FM). BAM and PRM learn to adjust the distribution of attention in the feature map from six different perspectives: polyp region, surrounding mucosa, boundaries, channel axial, horizontal axial and vertical axial of the feature map. Non-local captures the connections between any two pixels in the feature map to supplement long-distance global dependence. FM uses context information of different scales to reason and refine the ambiguous regions in segmentation results and then achieves more accurate polyp segmentation. We evaluate the effectiveness of our network with six evaluation metrics on five polyp datasets, and it can be seen from results that our MACNet can achieve more accurate segmentation in general.

Keywords: Polyp segmentation · Semantic segmentation · Attention · Deep learning · Coloscopy

1 Introduction

Colorectal cancer is the third most common cancer worldwide, with a global mortality rate of 9.4 % and a global prevalence of 100 million in 2020 according to statistics [21]. As a result, preventing colorectal cancer has become a public

Supported by the National Natural Science Foundation of China under Grant No. 62072135 and No. 61672181.

B. Li et al. (Eds.): APWeb-WAIM 2022, LNCS 13422, pp. 369–384, 2023.
https://doi.org/10.1007/978-3-031-25198-6_28

safety issue. Studies have found that 95% of colorectal cancers are caused by the progression of adenomatous polyps in the colorectum. Therefore, early tests and diagnosis are essential to reduce the prevalence of colorectal cancer. Colonoscopy can provide colorectal polyp location and contour information to help experts remove colorectal polyps before they develop into colorectal cancer, so it is an effective colorectal cancer screening and preventive technique. Accurate polyp segmentation is of great importance in clinical practice. However, this is a challenging task due to differences in the size, shape, color and texture of polyps. In addition, boundaries between the polyps and the surrounding mucosa are unclear, which can also lead to misdiagnosis and missed tests, raising the risk of disease. Another important cause of disease is the limitations and disparities in medical resources that make screening untimely and inaccurate in many areas. Therefore, an automated and accurate polyp segmentation technique that can detect all potential polyps at an early stage is important for the prevention of colorectal cancer [13].

Early polyp segmentation used polyp appearance features such as color, texture and shape to classify them. However, these traditional methods [9,19] rely on manual labeling, which has high cost and low accuracy. In recent years, computer vision has been greatly developed. The proposal of FCNs has greatly promoted the development of image segmentation. U-Net [18] proposed a U-shaped structure network including both encoder and decoder components. U-Net has become the basic framework for many medical segmentation networks because of its simple structure and high accuracy. U-Net++ [28] combines a DenseNet-like structure based on U-Net, and these dense skip connections can improve gradient fluidity. ResUNet [27] uses residual blocks to replace the convolution operation in U-Net, making effective control over gradient disappearance and explosion through independent short-circuit connections. Such connections allow the layers behind the network to continuously learn missing information from the previous layers and reduce redundancy. After the success of the attention mechanism in natural language processing, an increasing number of people notice its application in visual tasks. The visual attention modules Squeeze and Excitation (SE) [8], Non-local Information Statistical Attention (Non-Local) [24] and Axial-Attention [7] are all plug-and-play modules that help networks focus on important regions in the feature map, construct connections between pixels and suppress unimportant regions and connections. Therefore, they have been widely used in various visual tasks. In the task of polyp segmentation, the difference in polyp size is one of the difficulties, which is also a challenge in vision tasks. Since the size of the conventional convolution kernel is fixed, the range of the receptive field corresponding to the feature map is limited, and it is difficult for the network to perform contextual reasoning through a single-scale feature to capture both small and large targets at the same time. DoubleU-Net [10] and ResUNet++ [12] use the atrous spatial pyramid pooling (ASPP) [3] to obtain multiscale features, and CaraNet [14] obtains multiscale features with the help of the channel-wise feature pyramid (CFP) [15]. These multiscale features play an important role in contextual reasoning. In addition, the proposal of dilated convolution ensures that features of different scales can be captured by adjusting the dilation rate without increasing the amount of computation and losing information in advance, which is of great help for vision tasks. Although many polyp segmentation networks have paid attention to the importance of multiscale features for context reasoning and have also added related modules to networks, the segmentation results still fall short of the target. The main reason is that indiscriminately exploring contextual features is not very helpful [16] in a small target task such as polyp segmentation, as polyps make up a small proportion of images and contextual features are dominated by surrounding mucosa that make up a large proportion of images. This is very similar to camouflage object segmentation. PFNet [16] noticed this problem and proposed a focus module (FM). This module takes the feature map of the current encoder layer and the prediction result from the upper layer as input, obtains four feature maps of different scales, uses them to infer the context information, and finds false positive distraction and false negative distraction in the prediction. Then, FM eliminates the ambiguous area by elementwise subtraction and elementwise addition to achieve more accurate segmentation. Another challenge in polyp segmentation is the unclear boundaries between polyps and the surrounding mucosa. Previous networks focused on segmenting the entire region and ignored boundary constraints, which are critical for improving segmentation performance. Pranet [6] proposed the reverse attention module (RA) to sequentially mine complementary regions and details by erasing the existing estimated polyp regions through the output of the upper layer of prediction and focusing on the background regions. ACSNet [26] designed the local context attention (LCA) module. LCA aims to combine hard sample mining when merging shallow features and pays more attention to uncertain and complex regions to achieve hierarchical feature complementation and prediction. CCBANet [17] uses the balancing attention module (BAM), which applies different attention to the foreground, boundaries and background, and provides rich local information for the feature map in the decoding stage.

This paper aims at two main difficulties in polyp segmentation, as shown in Fig. 1: (1) the size and shape of polyps in colonoscopy images are greatly different; (2) high similarity in color and texture between polyps and surrounding mucosal results in unclear boundaries. By simulating the process of determining polyps by clinicians, we propose the Multi-Attention Context Network (MACNet). First, we locate the position of polyps for coarse prediction, and then draw the outline of polyps with the help of local texture, color characters and differences between contrast pixels. MACNet contains four key modules named Position Rectify Module (PRM), Balancing Attention Module (BAM), Non-local Information Statistical Attention Module (Non-local) and Focus Module (FM). Among them, PRM supplements the position information lost due to multiple downsampling during the encoding stage and rectifies the position of polyps in the feature map. It consists of a channel axial attention module, horizontal axial attention module and vertical axial attention module in series, which can obtain long-distance relationships from different axial directions, so it can redistribute the attention of the feature map in 3D space and then reposition and make coarse predictions. BAM pays attention to the problem of unclear boundaries between polyps and the surrounding mucosa. It consists of foreground attention focusing on polyps, background attention focusing on mucosa and boundary attention. They provide rich local information for predicting segmentation. Non-local complements the global information required by the network in the decoding stage. It calculates the connection between any two pixels in the feature map, establishes long-distance dependence, and enhances the semantic representation of deep features from a global perspective. FM performs contextual inference through multiscale feature maps to reduce segmentation accuracy loss caused by large polyp size gaps. The distraction discovery module in FM uses dilated convolution of different sizes to obtain multiscale feature maps for contextual inference to find false positive distraction and false negative distraction in the prediction results from the upper layer. The distraction removal module in FM uses elementwise addition and subtraction to remove ambiguous regions from the prediction results and refines the prediction results to improve the accuracy of segmentation. MACNet has achieved high accuracy on the Kvasir-SEG, CVC-ClinicDB, CVC-ColonDB, EndoScene and ETIS datasets.

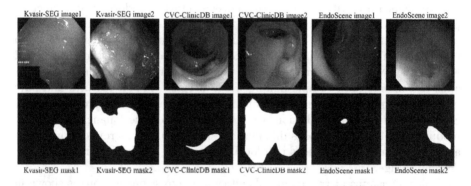

Fig. 1. Examples of images and masks for three different datasets.

2   Method

Figure 2 shows our MACNet, a classic encoder-decoder structure with five layers. Specifically, the encoder takes a colonoscopy image as input, uses ResUNet as the backbone, and obtains five feature maps $\{f_i,\ i = 1, 2, 3, 4, 5\}$ with a size of $[h/2^{k-1},\ w/2^{k-1}]$. f_5 can characterize the high-level semantic features of colonoscopy images. However, multiple downsampling operations during the encoding stage lose the position information of the polyps. We use PRM to rectify the position of polyps on three dimensions of space and then make coarse predictions. The BAM takes the output from the same encoder layer as the input, and obtains the polyps, surrounding mucous and boundary attention feature maps. The Non-local establishes the connection between any two pixels in the feature map, supplements the global information of long-distance dependence. The feature maps obtained from these two modules with rich local and global information are connected and passed in as inputs to the FM. The other two inputs of FM are the feature map f_i output from the same encoder layer and the segmentation result $pred_{i+1}$ predicted by the upper decoder layer, then use distraction discovery module to find the distraction regions in $pred_{i+1}$. The connected feature map is then refined by elementwise adding and subtracting operations to remove the distraction regions by the removal module in FM to obtain a more accurate feature map for more accurate segmentation.

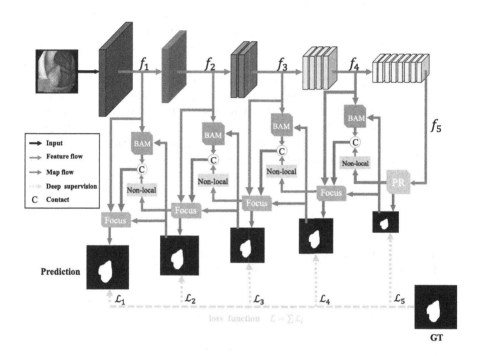

Fig. 2. Overview of the proposed MACNet.

2.1 Position Rectify Module (PRM)

We know that the feature map f_5 has rich semantic information after five encoding operations, but the location information of polyps in f_5 is also lost due to multiple downsampling operations, and directly using it for segmentation will reduce the accuracy of the results. We propose PRM, which consists of three modules: channel axial attention (SE), horizontal axial attention and vertical axial attention [7]. PRM reallocates attention of the feature map in 3D space, relocates polyp positions and makes coarse predictions.

Specifically, as shown in Fig. 3. First, PRM takes $f_5 \in R^{C \times H \times W}$ from the encoder as input, where C, H and W denote the number of channels, height and width of the feature map, respectively. PRM retains the channel direction of the feature map and uses global average pooling to compress the width and height to 1 to obtain $Z_5^c \in R^{C \times 1 \times 1}$. Then, network learns Z_5^c to obtain the attention vector $attention_5^c$ and multiplies it to f_5 to obtain a new feature map f_5', in which the position of polyps in the channel direction is rectified. As shown in formula (1):

$$
\begin{aligned}
Z_5^c &= \tfrac{1}{H \times W} \textstyle\sum_i^H \sum_j^W f_5^{i,j}, \\
attention_5^c &= \sigma \left(W_2 ReLU \left(W_1 Z_5^c \right) \right), \\
f_5' &= f_5 * F_{scale(attention_5^c)},
\end{aligned}
\tag{1}
$$

W_1 and W_2 are learnable parameters, σ is a sigmoid function, and F_{scale} extends the attention vector $attention_5^c$ to the same size as feature map f_5. Then, f_5' is passed into the horizontal axis attention module to rectify the position of polyps on the horizontal axis. First, PRM performs the 1×1 convolution operation on f_5' to obtain the query vector q, key vector k and value vector v, $q \in R^{H \times N}$, k, $v \in R^{N \times H}$, $N = C \times W$. Then, the q and k vectors are multiplied to obtain the coordination matrix $attention_5^H$, which represents the connection between the horizontal pixels in the feature map. $attention_5^H$ and v are multiplied to obtain a feature map of attention adjustment, and then f_5' is add to the result to obtain feature map f_5'', which rectified the position of polyps in the horizontal direction. As shown in formula (2).

$$
\begin{aligned}
attention_5^H &= q \times k, \\
f_5'' &= \gamma_h \left(v \times attention_5^H \right) + f_5',
\end{aligned}
\tag{2}
$$

Similarly, f_5'' is passed into the vertical axis attention module, and the position of polyps in the vertical direction is also rectified. As shown in formula (3).

$$
\begin{aligned}
attention_5^{H'} &= q' \times k', \\
f_5''' &= \gamma_v \left(v' \times attention_5^{H'} \right) + f_5'',
\end{aligned}
\tag{3}
$$

γ_h and γ_v are learnable parameters, and $\times$ is a multiplication operation. Finally, we use f_5''' to make a coarse prediction.

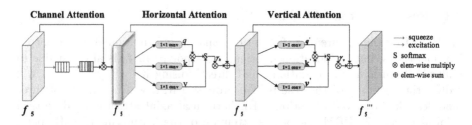

Fig. 3. Position Rectify Module.

2.2 Balancing Attention Module (BAM)

As show in Fig. 1, the polyps and surrounding mucosa are highly similar in color and texture, resulting in unclear boundaries. We pay attention to the difficulty of segmentation and integrate BAM into our MACNet. It can obtain the polyp, mucosal and boundary attention feature maps. In this way, the network can directly select the area it wants to pay attention to, avoid the interference of useless areas, and reduce the segmentation loss caused by unclear boundaries.

First, we use the prediction output from the upper decoder layer to obtain the attention scores of polyps, surrounding mucosa and boundaries, denoted as $attention_{foreground}$, $attention_{background}$ and $attention_{boundary}$, as formula (4), where $pred_{i+1}$ represents the predicted result from the upper decoder layer, and $[]_0^1$ represents limiting the value of the feature map to 0–1. Then, the attention score is multiplied with the feature map $f_i, i = 1, 2, 3, 4$ to obtain attention feature maps. After the convolution operation, we obtain a polyp feature map $f_{foreground}$, a mucosal feature map $f_{background}$ and a boundary feature map $f_{boundary}$ focusing on different regions individually. We then contact these three feature maps in the channel direction and use channelwise to redistribute channel importance so that the network can focus on the region of interest. Finally, the local attention feature map is added to the original feature map $f_i, i = 1, 2, 3, 4$ to obtain the final output of the module f_i^{BAM}. f_i^{BAM} provides the net with valuable local information during the decoding stage, greatly alleviating the problems caused by unclear boundaries.

$$
\begin{aligned}
attention_{boundary} &= 1 - \frac{|\sigma(pred_{i+1}) - 0.5|}{0.5}, \\
attention_{foreground} &= [\sigma(pred_{i+1}) - attention_{boundary}]_0^1, \\
attention_{background} &= [(1 - \sigma(pred_i + 1)) - attention_{boundary}]_0^1,
\end{aligned}
\tag{4}
$$

2.3 Non-local Information Statistical Attention Module (Non-local)

To supplement the global information required by FM in the decoding stage, we introduce a Non-local into MACNet. Non-local establishes a long-distance dependence by calculating the connections between any two pixels in the feature map, enriching the representation of deep semantic features in the feature map.

Specifically, we obtain the feature map from the upper decoder layer, represent each pixel in the feature map as a vector through a 1×1 convolution, and then calculate the connections between any two pixels in the feature map. Then, the pixel is mapped to the new feature map. As formula (5):

$$
y_i = \frac{1}{c(x)} \sum_{\forall j} f(x_i, x_j) g(x_j)
\tag{5}
$$

i and j can represent a certain spatial position of the input feature map, f is a function that calculates the similarity between any two points, g is a mapping function, and g is a 1×1 convolution. In this way, to calculate a point of the new output feature map, each pixel of the input feature map is considered, the connection between any two pixels is established, and an valuable global information feature map is obtained.

2.4 Focus Module (FM)

FM simulates the process of fine polyps outlining by clinicians. After roughly identifying the location of polyps, clinicians make a secondary judgment about the area of ambiguity by comparing the difference in color and texture between polyps and surrounding mucosa. This contextual inference learning can again refine the segmentation results. Additionally, to alleviate the segmentation problem caused by the large difference in polyp size, FM uses multiscale feature maps for context reasoning. FM also notices the small proportion of polyp pixels in the image, and the contextual features will be dominated by mucosal pixels that occupy a large proportion in the image. FM is a purposeful contextual exploration module, which consists of distraction discovery and distraction removal modules. The distraction discovery module uses multiscale features for contextual inference to find false positive and false negative regions in the prediction results from the upper layer. We call these regions false positive distraction and false negative distraction. Then, the distraction removal module uses elementwise addition and elementwise subtraction operations to remove distraction regions from the prediction results for more accurate segmentation.

Distraction Discovery. Humans rely on appearance characteristics such as texture and color to perform contextual reasoning, i.e., comparing the difference between ambiguous and confident regions to make the final decision [16]. Contextual reasoning can effectively improve the segmentation accuracy. Distraction discovery takes the output feature map $f_i, i = 1, 2, 3, 4$ from the same encoder layer and the prediction result $pred_{i+1}$ from the upper decoder layer as input and obtains false positive distraction F_{fpd} and false negative distraction F_{fnd} of $pred_{i+1}$ with the help of four contextual reasoning branches. The context reasoning consists of three layers. The first layer is used for channel reduction, the second layer uses convolution kernels of different sizes to extract local features of different scales, and the third layer uses different dilation rates for context perception. Table 1 shows the parameter settings of the four branches.

Table 1. Parameter settings of the four branches.

	Branch 1	Branch 2	Branch 3	Branch 4
Layer1	Conv 3×3	Conv 3×3	Conv 3×3	Conv 3×3
	BN+ReLU	BN+ReLU	BN+ReLU	BN+ReLU
Layer2	Conv 1×1	Conv 3×3	Conv 5×5	Conv 7×7
	BN+ReLU	BN+ReLU	BN+ReLU	BN+ReLU
Layer3	DConv 3×3	DConv 3×3	DConv 3×3	DConv 3×3
	dr $= 1$	dr $= 2$	dr $= 4$	dr $= 8$
	BN+ReLU	BN+ReLU	BN+ReLU	BN+ReLU

Conv means convolution operation, BN means batch normalization operation, ReLU means activation operation, DConv means dilated convolution, and dr means dilation rate. The outputs of all four branches are contacted in the channel dimension and fused by convolution. The ability to perceive rich context on a multiscale is obtained through the four branch networks, and the ambiguous regions in the prediction can be found.

Distraction Removal. First, we contact the output feature maps of BAM and Non-Local in the channel direction and then perform distraction removal on this contacted feature map F_h to obtain a purer feature map for polyp segmentation. An upsampling operation is performed to match the size of the distraction discovery feature map, and then the false positive area is eliminated by elementwise subtraction, and elementwise addition is used to compensate for false negative areas.

$$
\begin{aligned}
F_{up} &= U\left(CBR\left(F_h\right)\right), \\
F_r &= BR\left(F_{up} - \alpha F_{fpd}\right), \\
F_r' &= BR\left(F_r + \beta F_{fnd}\right),
\end{aligned}
\tag{6}
$$

CBR represents convolution, batch normalization and ReLU activation operations, and U represents upsampling. α and β are learnable scale parameters; Here, we set them to 1. F_r' is the feature map after removing distractions. Finally, we use F_r' to obtain more accurate predictions $pred_i, i = 1, 2, 3, 4$.

2.5 Deep Supervision

The network has a total of five predictions, one from the PRM and the other four predictions from FM. The loss of each decoder layer adopts the sum of DiceLoss and BceLoss, both of which are classical loss functions in segmentation. To match the size of the decoder predictions, we process the real labels separately $[h/2^{k-1}, w/2^{k-1}]$ and add the loss values of the five layers as the network overall loss. The formulas are as follows:

$$
l_i = l_i^{dice} + l_i^{bce}
\tag{7}
$$

$$
l_{loss} = \sum_{i=1}^{5} l_i
\tag{8}
$$

3 Experiments

3.1 Datasets

We evaluate our MACNet on five benchmark colonoscopy image datasets: ETIS [19], CVC-ClinicDB/CVC-612 [2], CVC-ColonDB [22], EndoScene [1] and Kvasir [11]. ETIS contains 196 images captured from 34 polyp video audios, and the size of the images is 1225×966. CVC-ClinicDB contains 612 images captured from 25 polyp videos, so it is also called CVC-612, and the size of the images is 384×288. The CVC-ClinincDB and ETIS datasets are provided in the 2015 MICCAI automatic polyp detection subchallenge. CVC-ColonDB contains 380 images, and the size of the images is 574×500. EndoScene is consists of CVC-300 and CVC-612, so it contains 912 images. Kvasir-SEG is the largest and most challenging dataset published recently. It contains 1000 images with large differences in polyp size. There are 700 polyp images with a large target of more than 160×160 and 48 polyp images with a small target of less than 64×64.

3.2   Metrics

We mainly use the evaluation metrics of MICCAI 2015 Challenge as the metrics to evaluate MACNet performance: Dice Score, Mean IoU (mIoU), Recall and Precision. In addition, to better demonstrate the performance of MACNet, we also use the Accuracy Rate and F2 coefficient.

$$Dice = \frac{2 \times tp}{2 \times tp + fp + fn}, \quad mIoU = \frac{tp}{tp + fp + fn}, \quad Recall = \frac{tp}{tp + fn},$$

$$Precision = \frac{tp}{tp + fp}, \quad Acc = \frac{tp + tn}{tp + tn + fp + fn}, \quad F_2 = \frac{5p \times r}{4p + r}, \tag{9}$$

Table 2. Results under the first training strategy. The **best**, <u>second</u> results are highlighted.

Dataset	Method	Dice	IoU	Rec	Pre	Acc	F_2
Kvasir-SEG	U-Net	79.94	69.35	81.51	82.91	82.17	81.79
	ResUnet++	81.33	79.27	70.64	87.74	88.40	86.86
	PraNet	89.84	83.81	**94.14**	91.12	96.53	91.93
	ACSNet	90.28	84.35	92.18	90.90	97.52	90.58
	PFNet	<u>93.35</u>	<u>87.52</u>	91.40	<u>95.38</u>	<u>97.98</u>	92.17
	CCBANet	92.59	86.21	92.21	92.98	97.43	<u>92.36</u>
	MACNet(Ours)	**94.74**	**90.01**	<u>93.85</u>	**95.65**	**98.39**	**94.21**
CVC-ClinicDB (CVC-612)	U-Net	87.62	79.47	87.32	89.99	87.36	87.84
	ResUnet++	79.55	79.62	70.22	87.85	88.30	88.66
	PraNet	94.59	90.26	95.00	94.50	<u>99.23</u>	94.90
	ACSNet	94.27	89.15	92.86	95.72	99.03	93.42
	PFNet	<u>95.67</u>	<u>91.71</u>	**97.01**	94.37	<u>99.23</u>	**96.47**
	CCBANet	95.43	91.26	94.79	**96.08**	99.22	95.05
	MACNet(Ours)	**95.88**	**92.09**	<u>95.87</u>	<u>95.89</u>	**99.28**	<u>95.88</u>
CVC-EndoSceneStill	U-Net	65.87	54.08	76.75	69.39	76.75	75.16
	ResUnet++	51.09	42.74	78.27	47.57	78.28	69.32
	PraNet	83.62	**76.55**	**88.33**	87.18	96.60	**88.10**
	ACSNet	84.78	73.58	<u>79.37</u>	90.97	97.37	81.45
	PFNet	<u>84.97</u>	73.87	78.14	93.11	97.44	80.74
	CCBANet	**85.79**	<u>75.12</u>	79.29	**93.45**	**97.57**	<u>81.77</u>
	MACNet(Ours)	84.88	73.73	77.81	<u>93.37</u>	<u>97.44</u>	80.49

3.3 Experiment Results

We use two mainstream training strategies to train the network to verify the performance of our network.

The first is to train and test separately according to CCBANet [17]. We divide the three datasets of Kvasir-SEG, CVC-Clinic DB and CVC-ColonDB according to 8:1:1. Table 2 shows the experimental results under the first training strategy.

The second strategy is to mix the Kvasir-SEG and CVC-ClinicDB datasets according to Pranet [6]. The network takes 80% of the mixed dataset as the training set and takes the remaining 20%, CVC-ColonDB, test set CVC-300 of EndoScene and ETIS datasets as the test sets to verify the effectiveness of the network. Table 3 shows the experimental results under the second training strategy.

Table 3. Results under the second training strategy. The **best**, second results are highlighted.

Method	Kvasir-SEG		CVC-ClinicDB		CVC-300		ETIS		CVC-ColonDB	
	mDice	mIoU	mDice	mIoU	mDice	mIoU	mDice	mIoU	mDice	mIoU
PraNet	89.8	84.0	89.9	84.9	87.1	79.7	62.8	56.7	70.9	64.0
SANet [25]	90.4	84.7	91.6	85.9	88.8	81.5	75.0	65.4	75.3	67.0
CaraNet	91.8	**86.5**	93.6	88.7	90.3	83.8	74.7	67.2	**77.3**	**68.9**
PFNet	92.2	85.5	94.6	89.8	**92.1**	**85.4**	82.2	69.8	75.9	61.2
MACNet	**92.3**	85.8	**95.2**	**90.8**	91.8	84.8	**82.3**	69.9	73.9	58.6

3.4 Experimental Parameters

The implementation of the network uses the PyTorch framework, and RTX3090 is used for training and testing. In the training phase, batch is set to 8, and input images are cropped to 256×256. The other parameters are the same as those in CCBANet. We adopt the Adam optimization algorithm with a momentum beta1 of 0.9, a momentum beta2 of 0.999, and a weight decay of 1×10^{-5} to optimize all parameters of the network. The initial learning rate is 0.001 and adjusted with $lr = init\_lr \times (1 - \frac{epoch}{nEpoch})^{power}$, where nEpoch is 200 and power is 0.9.

Table 4. ISIC experimental results. The **best** results are highlighted.

Method	Dice	IoU	Recall	Precision
U-Net	67.40	54.90	70.80	–
Deeplabv3+(Xception) [4]	87.72	81.28	86.81	**92.72**
Deeplabv3+(Mobilenet) [4]	87.81	82.36	88.30	92.44
MSRF-Net [20]	88.13	83.25	89.03	92.67
MACNet(Ours)	**91.42**	**84.19**	**92.29**	90.50

3.5   Learning Ability and Generalization Capability

Regarding the learning ability, CVC-ClinicDB and Kvasir-SEG, which are involved in training, are better than the other networks in various evaluation metrics under the two training strategies. In addition to the polyp dataset, we also verify the learning ability of MACNet on the ISIC2018 [5, 23] skin cancer dataset, and it can be seen from Table 4 that various metrics are also better than most networks.

We verified the generalization capability of MACNet on three datasets (CVC-ColonDB, CVC-300 and ETIS) that are not involved in the training. Except for CVC-ColonDB, the evaluation metrics on the other two datasets were higher than the other networks.

3.6   Ablation Study

We adopt ablation studies to verify the effectiveness of the introduced two modules FM and PRM, and the baseline is the CCBANet with the CCM removed. The effect of FM and PRM on network performance improvement can be seen in Table 5 and Fig. 4.

Table 5. Ablation study for MACNet on the Kvasir-SEG dataset.

Setting	Dice	IoU	Rec	Pre	Acc	F2
Backbone	92.85	86.65	89.40	96.57	97.87	90.75
Backbone+PRM	93.21	87.28	90.79	95.76	97.95	91.74
Bcakbone+FM	93.84	88.40	92.78	94.93	98.12	93.20
Backbone+PRM+FM	94.74	90.01	93.85	95.65	98.39	94.21

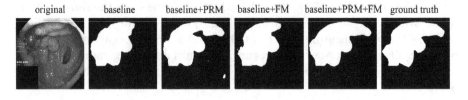

original   baseline   baseline+PRM   baseline+FM   baseline+PRM+FM   ground truth

Fig. 4. Qualitative results visualization of ablation experiments.

4   Experimental Result Analysis

As shown in Table 2 and Table 3, compared with other models, MACNet has improved performance, which is related to the multi-perspective processing of information and the increase of model parameters.

Specifically, the processing of information is carried out from two perspectives: 1) Span range of information: feature maps with rich local and global information are essential. To obtain local information, we retain the BAM in CCBANet, which implements separate attention for foreground, boundary and background, and then learns to fuse these attention feature maps to obtain local feature maps. To get global information, we refer the Non-local to MACNet. Non-local mainly captures long-distance dependencies by calculating the dependencies between any two pixels in the feature map. 2) Information determination process: We simulate the process of experts to determine the polyps, first use PRM to determine the position of the polyps, make a rough prediction, and then use the FM to correct the false positive area and false negative area to achieve accurate segmentation. Figure 5-1 shows that the position prediction results is inaccurate. Figure 5-2 is more accurate in predicting the position of polyps after adding PM, but there are still a small number of false negative areas, which is caused by the unclear boundary. We add FM to MACNet. As shown in Fig. 5-4, the segmentation result is indeed optimized in the difficult segmentation area at the edge. And in Fig. 5-3, we only added FM, although the model is accurate in the segmentation of difficult-to-segment areas, but there are a large range of false negative areas. So the idea of determining the position of the polyp first and then performing the fine segmentation is very necessary.

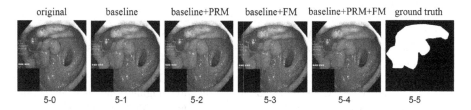

Fig. 5. Qualitative results visualization on colonoscopy image of ablation experiments (best viewed in color).

The number of parameters of our model is 39,249,755, which is an increase of 7,674,491 parameters compared to our baseline model CCBANet. After analysis, this is mainly the amount of parameters extracted from multi-scale features in the decoding stage. The decoding module of CCBANet uses a single-scale convolution kernel with small parameters, but this is difficult to pay attention to polyps with large targets and small targets at the same time. We notice this problem and use four parallel convolution kernels of different scales to extract multi-scale features, which improves the model effect. Also the increased parameter quantity increases the upper limit of the model capability. It is very friendly to big data sets. Table 2, on the kvasir-SEG dataset of 1000 images, various indicators have been significantly improved, among which the Dice coefficient is increased by 2.5, the Mean IoU is increased by 3.8, the Recall is increased by 1.64, the Precision is increased by 2.67, and the Accuracy is increased by 0.96, F2 increased by 1.85. But we can also observe that the CVC-EndoSceneStill doesn't work well, mainly because this dataset is composed of two small datasets, CVC-

300 and CVC-612, and the domain gap between the two datasets is obvious. Therefore, the increase of the number of model parameters are acceptable.

5 Conclusion

In this paper, our proposed MACNet network exhibits superior performance on polyp segmentation. MACNet uses PRM to rectify the position of polyps in the feature map for coarse segmentation prediction, BAM and Non-local provide FM local informative feature maps and global informative feature maps, and then MACNet uses FM to find false positive distraction and false negative distraction that cause low accuracy and removes ambiguous regions by elementwise subtraction and elementwise addition to achieve finer segmentation. In the future, we will continue to improve the performance of MACNet and explore applications in polyp video segmentation.

Acknowledgements. The work was supported by the National Natural Science Foundation of China under Grant No. 62072135 and No. 61672181.

References

1. Vázquez, D., et al.: A benchmark for endoluminal scene segmentation of colonoscopy images. J. Healthc. Eng. **2017**, 9 (2017). Article ID 4037190. https://doi.org/10.1155/2017/4037190
2. Bernal, J., Sánchez, F.J., Fernández-Esparrach, G., Gil, D., Rodríguez, C., Vilariño, F.: WM-DOVA maps for accurate polyp highlighting in colonoscopy: validation vs. saliency maps from physicians. Comput. Med. Imaging Graph. **43**, 99–111 (2015)
3. Chen, L.C., Papandreou, G., Schroff, F., Adam, H.: Rethinking atrous convolution for semantic image segmentation. arXiv preprint arXiv:1706.05587 (2017)
4. Chen, L.-C., Zhu, Y., Papandreou, G., Schroff, F., Adam, H.: Encoder-decoder with atrous separable convolution for semantic image segmentation. In: Ferrari, V., Hebert, M., Sminchisescu, C., Weiss, Y. (eds.) ECCV 2018. LNCS, vol. 11211, pp. 833–851. Springer, Cham (2018). https://doi.org/10.1007/978-3-030-01234-2_49
5. Codella, N.C., et al.: Skin lesion analysis toward melanoma detection: a challenge at the 2017 international symposium on biomedical imaging (ISBI), hosted by the international skin imaging collaboration (ISIC). In: 2018 IEEE 15th International Symposium on Biomedical Imaging (ISBI 2018), pp. 168–172. IEEE (2018)
6. Fan, D.-P., et al.: PraNet: parallel reverse attention network for polyp segmentation. In: Martel, A.L., et al. (eds.) MICCAI 2020. LNCS, vol. 12266, pp. 263–273. Springer, Cham (2020). https://doi.org/10.1007/978-3-030-59725-2_26
7. Ho, J., Kalchbrenner, N., Weissenborn, D., Salimans, T.: Axial attention in multi-dimensional transformers. arXiv preprint arXiv:1912.12180 (2019)
8. Hu, J., Shen, L., Sun, G.: Squeeze-and-excitation networks. In: Proceedings of the IEEE Conference on Computer Vision and Pattern Recognition, pp. 7132–7141 (2018)
9. Iwahori, Y., et al.: Automatic polyp detection in endoscope images using a hessian filter. In: MVA, pp. 21–24 (2013)

10. Jha, D., Riegler, M.A., Johansen, D., Halvorsen, P., Johansen, H.D.: DoubleU-Net: a deep convolutional neural network for medical image segmentation. In: 2020 IEEE 33rd International Symposium on Computer-Based Medical Systems (CBMS), pp. 558–564. IEEE (2020)

11. Jha, D., et al.: Kvasir-SEG: a segmented polyp dataset. In: Ro, Y.M., et al. (eds.) MMM 2020. LNCS, vol. 11962, pp. 451–462. Springer, Cham (2020). https://doi.org/10.1007/978-3-030-37734-2_37

12. Jha, D., et al.: ResUNet++: an advanced architecture for medical image segmentation. In: 2019 IEEE International Symposium on Multimedia (ISM), pp. 225–2255. IEEE (2019)

13. Jia, X., Xing, X., Yuan, Y., Xing, L., Meng, M.Q.H.: Wireless capsule endoscopy: a new tool for cancer screening in the colon with deep-learning-based polyp recognition. Proc. IEEE **108**(1), 178–197 (2019)

14. Lou, A., Guan, S., Loew, M.: CaraNet: context axial reverse attention network for segmentation of small medical objects. arXiv preprint arXiv:2108.07368 (2021)

15. Lou, A., Loew, M.: CFPNet: channel-wise feature pyramid for real-time semantic segmentation. In: 2021 IEEE International Conference on Image Processing (ICIP), pp. 1894–1898. IEEE (2021)

16. Mei, H., Ji, G.P., Wei, Z., Yang, X., Wei, X., Fan, D.P.: Camouflaged object segmentation with distraction mining. In: Proceedings of the IEEE/CVF Conference on Computer Vision and Pattern Recognition, pp. 8772–8781 (2021)

17. Nguyen, T.-C., Nguyen, T.-P., Diep, G.-H., Tran-Dinh, A.-H., Nguyen, T.V., Tran, M.-T.: CCBANet: cascading context and balancing attention for polyp segmentation. In: de Bruijne, M., et al. (eds.) MICCAI 2021. LNCS, vol. 12901, pp. 633–643. Springer, Cham (2021). https://doi.org/10.1007/978-3-030-87193-2_60

18. Ronneberger, O., Fischer, P., Brox, T.: U-net: convolutional networks for biomedical image segmentation. In: Navab, N., Hornegger, J., Wells, W.M., Frangi, A.F. (eds.) MICCAI 2015. LNCS, vol. 9351, pp. 234–241. Springer, Cham (2015). https://doi.org/10.1007/978-3-319-24574-4_28

19. Silva, J., Histace, A., Romain, O., Dray, X., Granado, B.: Toward embedded detection of polyps in WCE images for early diagnosis of colorectal cancer. Int. J. Comput. Assist. Radiol. Surg. **9**(2), 283–293 (2014)

20. Srivastava, A., et al.: MSRF-Net: a multi-scale residual fusion network for biomedical image segmentation. arXiv preprint arXiv:2105.07451 (2021)

21. Sung, H., et al.: Global cancer statistics 2020: GLOBOCAN estimates of incidence and mortality worldwide for 36 cancers in 185 countries. CA Cancer J. Clin. **71**(3), 209–249 (2021)

22. Tajbakhsh, N., Gurudu, S.R., Liang, J.: Automated polyp detection in colonoscopy videos using shape and context information. IEEE Trans. Med. Imaging **35**(2), 630–644 (2015)

23. Tschandl, P., Rosendahl, C., Kittler, H.: The ham10000 dataset, a large collection of multi-source dermatoscopic images of common pigmented skin lesions. Sci. Data **5**(1), 1–9 (2018)

24. Wang, X., Girshick, R., Gupta, A., He, K.: Non-local neural networks. In: Proceedings of the IEEE Conference on Computer Vision and Pattern Recognition, pp. 7794–7803 (2018)

25. Wei, J., Hu, Y., Zhang, R., Li, Z., Zhou, S.K., Cui, S.: Shallow attention network for polyp segmentation. In: de Bruijne, M., et al. (eds.) MICCAI 2021. LNCS, vol. 12901, pp. 699–708. Springer, Cham (2021). https://doi.org/10.1007/978-3-030-87193-2_66

26. Zhang, R., Li, G., Li, Z., Cui, S., Qian, D., Yu, Y.: Adaptive context selection for polyp segmentation. In: Martel, A.L., et al. (eds.) MICCAI 2020. LNCS, vol. 12266, pp. 253–262. Springer, Cham (2020). https://doi.org/10.1007/978-3-030-59725-2_25

27. Zhang, Z., Liu, Q., Wang, Y.: Road extraction by deep residual U-net. IEEE Geosci. Remote Sens. Lett. **15**(5), 749–753 (2018)

28. Zhou, Z., Rahman Siddiquee, M.M., Tajbakhsh, N., Liang, J.: UNet++: a nested u-net architecture for medical image segmentation. In: Stoyanov, D., et al. (eds.) DLMIA/ML-CDS 2018. LNCS, vol. 11045, pp. 3–11. Springer, Cham (2018). https://doi.org/10.1007/978-3-030-00889-5_1

Transformer-Based Representation Learning on Temporal Heterogeneous Graphs

Longhai Li[1], Lei Duan[1,3(✉)], Junchen Wang[1], Guicai Xie[1], Chengxin He[1], Zihao Chen[1], and Song Deng[2]

[1] School of Computer Science, Sichuan University, Chengdu, China
{lilonghai,wangjunchen,guicaixie,hechengxin,chenzihao}@stu.scu.edu.cn,
leiduan@scu.edu.cn
[2] Institute of Advanced Technology, Nanjing University of Posts
and Telecommunications, Nanjing, China
dengsong@njupt.edu.cn
[3] Med-X Center for Informatics, Sichuan University, Chengdu, China

Abstract. Temporal heterogeneous graphs can model lots of complex systems in the real world, such as social networks and e-commerce applications, which are naturally time-varying and heterogeneous. As most existing graph representation learning methods cannot efficiently handle both of these characteristics, we propose a Transformer-like representation learning model, named THAN, to learn low-dimensional node embeddings preserving the topological structure features, heterogeneous semantics, and dynamic evolutionary patterns of temporal heterogeneous graphs simultaneously. Specifically, THAN first samples heterogeneous neighbors with temporal constraints and projects node features into the same vector space, then encodes time information and aggregates the neighborhood influence in different weights via type-aware self-attention. Experiments on three real-world datasets demonstrate that THAN outperforms the state-of-the-arts in terms of effectiveness with respect to the temporal link prediction task.

Keywords: Temporal heterogeneous graphs · Graph neural networks · Graph representation learning · Transformer

1 Introduction

Graph representation learning, as an important task in machine learning, has significant practical value in areas such as social networks and recommendation systems. Existing graph representation learning methods usually take static graphs

This work was supported in part by the National Key Research and Development Program of China (2018YFB0704301-1), the National Natural Science Foundation of China (61972268), the Med-X Center for Informatics Funding Project (YGJC001).

B. Li et al. (Eds.): APWeb-WAIM 2022, LNCS 13422, pp. 385–400, 2023.
https://doi.org/10.1007/978-3-031-25198-6_29

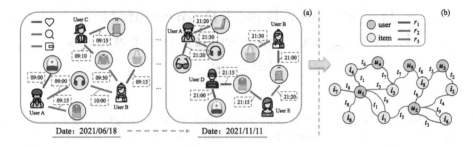

Fig. 1. A toy example of the temporal heterogeneous graph from a user-item interactions network. (a) User-item interactions network; (b) Temporal heterogeneous graph.

as input to obtain low-dimensional embeddings by encoding local non-Euclidean structures, and have achieved extensive excellent performance in downstream tasks such as link prediction [14,21] and node classification [15,25].

However, most graphs in the real world are naturally heterogeneous and dynamic, which cannot be accurately represented by static homogeneous graphs. Taking the example of a user-item interaction network in e-commerce scenarios [23], illustrated in Fig. 1(a), there are two types of nodes (*user* and *item*) and three types of interactions (*browse*, *favorite*, and *buy*). Additionally, each interaction is associated with a continuous timestamp to indicate when it occurred. In this paper, we define such interaction sequences between different types of nodes as temporal heterogeneous graphs (THG). It is of great significance to learn representations of THG with dynamic and heterogeneous characteristics for modeling real-world complex systems.

In the case of the user-item interactions network shown in Fig. 1(a), THG representation learning has the following challenges compared to static homogeneous graph representation learning:

- **(C1)** *How to model the heterogeneity?* The nodes and edges in THG are of various types and have rich semantics, making it difficult to obtain sufficient heterogeneous information just by encoding local graph structure.
- **(C2)** *How to model the continuous dynamics?* The edges in the THG are time-informed and time-dependent, i.e., each event occurs with a timestamp and current event may affect the occurrence of future events. For instance, there might be causal relationships between the interaction of searching for headphones on 18 June and the interaction of purchasing headphones on 11 November by user A. Therefore, both reasonably efficient methods of converting temporal information into dynamic features and temporal constraints are needed to avoid violating the temporal causality between interactions.
- **(C3)** *How to deal with new nodes?* The dynamics of the THG imply that new nodes will emerge in the future (e.g., users D and E are two new nodes that appeared on 11 November compared to 18 June), which makes it necessary to construct an inductive modeling approach.

As for the heterogeneity, earlier methods [2,5,31] preserve heterogeneous information by designing semantic meta-paths to generate heterogeneous sequences, and recent studies [8,21,26,33] aggregate information from heterogeneous neighborhood by extending the message-passing process of graph neural networks (GNNs). Concerning dynamics, it is general to split temporal graphs into several static snapshots (i.e. discrete-time dynamic graph, DTDG [13]) and use RNNs or attention to capture the evolutionary patterns between snapshots [3,18,20,30]. Although these methods can learn graph dynamics of the THG to some extent, the temporal information within the same snapshot is usually ignored, and the scale of snapshots needs to be predetermined in advance. Recently, researchers have proposed continuous-time dynamic graph (CTDG [13]) approaches [9,11,16,27,29] to capture dynamics via passing information between different interactions, or using continuous-time functions to generate temporal embedding. In regard to the new nodes, inductive graph representation learning methods [7,25,27,29] recognize structural features of node neighborhood by learning trainable aggregation functions, so that rapidly generate node embeddings in new subgraphs. Plenty of studies have attempted to solve the above challenges, nevertheless, few approaches can address them at the same time.

In this paper, we propose a novel Temporal Heterogeneous Graph Attention Network (THAN), which is a continuous-time THG representation learning method with Transformer-like attention architecture. To handle **C1**, we design a time-aware heterogeneous graph encoder to aggregate information from different types of neighbors. To handle **C2**, THAN samples temporally constrained neighbors from historical heterogeneous events, converts dynamic features into time embeddings by a time encoder, and incorporates them into the information propagation procedure. To handle **C3**, THAN is designed as a message-passing model based on neighbor sampling to ensure that the entire learning process does not introduce global priori information. The main contributions of our work are summarized as follows:

- We propose an inductive continuous-time THG representation learning method, which can capture both heterogeneous information and dynamic features.
- We introduce the transfer matrix and self-attention mechanism to implement the information aggregation of heterogeneous neighbors.
- We conduct experiments on three public datasets and the results demonstrate the superior performance of THAN over state-of-the-art baselines on the task of temporal link prediction.

2 Related Work

Our work is related to representation learning on static graphs, temporal graphs (i.e. dynamic graphs), and self-attention mechanism on graphs.

Representation Learning on Static Graphs. Graph representation learning produces low-dimensional embeddings by modeling the topology and node attribute information. Early methods [2,5,6,19] generate sequences of nodes by random walks and then learn node co-occurrences to obtain representations. Luo et al. [17] defines ripple distance to optimize the walking procedure. In order to integrate rich node features while learning network structure information, the GNN-based approaches [7,8,15,21,25,26,33] update node embeddings by aggregating neighborhood influence and propagating information across a multi-layer network to capture the high-order patterns. Focus on dealing with heterogeneity, meta-paths and heterogeneous attention are the two most used strategies. However, these methods cannot deal with temporal evolutionary patterns.

Representation Learning on Temporal Graphs. According to the way temporal graphs are constructed, the temporal graph representation learning methods can be divided into two categories: *discrete-time methods*, which describes the temporal graph as an ordered list of graph snapshots; *continuous-time methods*, which treats the temporal graph as an event stream with timestamps.

For the former, EvolveGCN [18] uses GCN to encode static graph structure and evolves the parameters of GCN by RNN. DySAT [20] uses structural attention to aggregate information from different neighbors in each snapshot and uses temporal attention to capture evolution over multiple snapshots. DHNE [31] performs random walks in THG under the guidance of meta-paths and proposes the dynamic heterogeneous skip-gram model to learn the node embeddings. DyHATR [30] adopts hierarchical attention to learn heterogeneous information and applies RNNs with temporal attention to capture dependencies among snapshots. HTGNN [3] jointly models heterogeneous spatial and temporal dependencies through intra-relational, inter-relational, and cross-temporal aggregation. Although the discrete-time methods succeed in learning the dynamic patterns of the temporal graphs, they ignore the time information within the same snapshot and lead to a weakened connection between graph snapshots.

Recent studies [9,11,16,22,27,29] have shown the superior performance of continuous-time methods in dealing with temporal graphs. JODIE [16] uses RNNs to propagate information in interactions and update node representations smoothly at different timesteps. TGAT [29] is designed as a GAT-like neural network, which propagates node information by sampling and aggregating historical neighbors, and learns high-order patterns by stacking multiple layers. CAW-N [27] proposes Causal Anonymous Walks (CAWs) to inductively represent a temporal graph and uses RNN to encode the walk sequences. These methods make full use of temporal information and model the evolution of the graph without taking into account the heterogeneity. THINE [9] and HPGE [11] combine heterogeneous attention and Hawkes process to model graph heterogeneity and dynamics but do not consider the edge attributes.

Self-attention Mechanism. Transformer [24] proposed by Vaswani et al. for machine translation has achieved great success in NLP and CV tasks, which has recently been attempted for graph representation learning. For example, Graphormer [32] generalizes positional encoding to the graph domain and uses scaled dot-product attention for message passing. Transformer relies on the self-attention mechanism to learn contextual information for sequences. A scaled dot-product attention layer can be defined as:

$$Attn(\mathbf{Q}, \mathbf{K}, \mathbf{V}) = softmax(\frac{\mathbf{Q}\mathbf{K}^{\top}}{\sqrt{d}})\mathbf{V} \tag{1}$$

where $\mathbf{Q}$ denotes the 'queries', $\mathbf{K}$ the 'keys' and $\mathbf{V}$ the 'values'. They are the projections of the input $\mathbf{Z}$ on the matrices W_Q, W_K and W_V, where $\mathbf{Z}$ contains the node embeddings and their positional embeddings.

3   Preliminaries

In this section, we introduce the definition of temporal heterogeneous graphs and the problem of temporal heterogeneous graph representation learning.

Definition 1. *Temporal Heterogeneous Graph.* *A temporal heterogeneous graph is $\mathcal{G} = (V, E, T, \phi, \varphi)$, where V denotes the set of nodes corresponding to a node type mapping function $\phi : V \to A$, E denotes the temporal events (i.e. edges) corresponding to an event type mapping function $\varphi : E \to R$, and T denotes the set of timestamps. A and R are node type and event type sets, respectively, and $|A| + |R| > 2$. Notice that, event $e = (u, v, t, r, \chi)$ means that there is an edge from u to v at time t, where $r = \varphi(e)$ denotes the event type and χ denotes the edge feature.*

For instance, a temporal heterogeneous graph about user-item interactions in Fig. 1(b) consists of 13 nodes, 17 events (a smaller subscript of t indicates an earlier event), 2 types of nodes, and 3 types of events. Specifically, $V = \{u_1, ..., i_8\}$, $E = \{(u_1, i_1, t_1, r_2), ..., (u_5, i_5, t_8, r_3)\}$, $A = \{user, item\}$, $R = \{r_1, r_2, r_3\}$, $\phi(u) = user$ and $\phi(i) = item$.

For any node pair (u, v), a *temporal causal path* is a set of events consisting of u as the source node of the start event and v as the target node of the terminal event. Therefore, the *temporal shortest path distance* $d_t(u, v)$ is defined as the minimum length of the temporal causal path from u to v with all events on the path occurring no later than t. Denote V_t as the set of nodes that appear up to time t, and for each node $v \in V_t$, define its k-hop *temporal neighbors* as:

$$\mathcal{N}_t^k(v) = \{u : d_t(u, v) \le k, u \in V_t\} \tag{2}$$

For node v, we define its k-hop *temporal neighborhood* as $G_t^k(v)$, which is a subset of the temporal heterogeneous graph $\mathcal{G}$ and can be induced by $\mathcal{N}_t^k(v)$. $G_t^k(v)$ contains the source node v and its neighbors $\mathcal{N}_t^k(v)$, events between the nodes, and timestamps of these temporal events. The final representation of node v will generate relying on $G_t^k(v)$. Notice that, we use $\mathcal{N}_t(v)$ and $G_t(v)$ to simplify the representation of $\mathcal{N}_t^1(v)$ and $G_t^1(v)$ in this paper, respectively.

Definition 2. *Temporal Heterogeneous Graph Representation Learning.* *Given a temporal heterogeneous graph $\mathcal{G}$ and the node features X, it aims to learn a mapping function $\mathcal{F} : \mathcal{F}(\mathcal{G}, X) \to \mathbb{R}^{|V| \times d}$, where $|V|$ is the node size and d is the dimension of embeddings, $d \ll |V|$. This function maps nodes to low-dimensional vector space while preserving temporal, structural, and semantic information.*

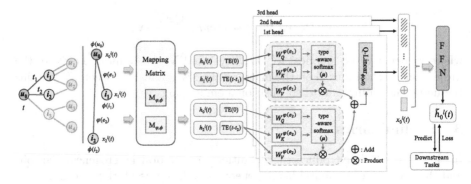

Fig. 2. The architecture of the l-th THAN layer for node u_0 at time t.

4   The Proposed Model

In this section, we present a Transformer-like attention architecture named THAN. It uses mapping matrices to project node embeddings into the same vector space, then passes neighborhood information by dot-product attention corresponding to different event types. Similar to GAT [25], THAN can be thought of as a local aggregation operator that captures higher-order information by stacking multiple THAN layers. Fig. 2 shows the architecture of the l-th THAN layer, which has three components: *temporal heterogeneous neighbor sampling*, *dynamic embedding mapping* and *temporal heterogeneous graph attention layer*. After encoding, we design a heterogeneous graph decoder for the temporal link prediction task, which receives the node representations from THAN as inputs.

4.1   Temporal Heterogeneous Neighbor Sampling

For the purpose of improving the induction and generalization performance of the model, THAN does not select all but a certain number of neighbors from the temporal neighbors as input. Given a node v_0 and time t, sample N neighbors from its 1-hop temporal neighbors $\mathcal{N}_t(v_0)$, denoted as $\{v_1, ..., v_N\}$.

We discuss two neighbor sampling strategies: *uniform random sampling*, where all temporal neighbors are randomly selected with equal probability; *top-N recent sampling*, where the time difference with the source node is calculated and sorted in ascending order, then select the top N neighbors. Intuitively, recent interactions reflect the node's current state better than distant interactions and have a greater influence on future events. On the contrary, the distant interactions may introduce noise. Therefore, we use the *top-N recent sampling* strategy to sample neighbors.

In the temporal heterogeneous graph, the number of different-typed events varies greatly, which can easily lead to an unbalanced distribution of the types of sampled neighbors. To avoid sampling bias as far as possible, THAN limits the number of samples of each event type to no more than M. Assuming that the total number of event types related to the source node is γ ($\gamma \leq |R|$), the total number of sampled neighbors N satisfies $N \leq \gamma * M$.

4.2 Dynamic Embedding Mapping

For different nodes, TGAT [29] assumes that they are in the same feature distribution and share parameter matrices, which does not hold in heterogeneous graphs. Furthermore, in the real world, there might exist multiple types of edges between two nodes, so we must consider both node and edge types when propagating node information.

Inspired by TransD [10], THAN uses mapping matrices to project node features from the node-type space to the event-type space, which uses projection vectors to reduce model parameters and avoid matrix multiplication calculations than directly parameterizing the matrices. Given an event $e = (u, v, t)$ with its meta relation $\langle \phi(u), \varphi(e), \phi(v) \rangle$ [8], define the mapping matrices as:

$$\mathbf{M}_{eu} = \mathbf{e}_{\varphi(e)} \mathbf{n}_{\phi(u)}^{\top} + \mathbf{I}^{d \times d} \tag{3}$$

$$\mathbf{M}_{ev} = \mathbf{e}_{\varphi(e)} \mathbf{n}_{\phi(v)}^{\top} + \mathbf{I}^{d \times d} \tag{4}$$

where $\mathbf{e}$ and $\mathbf{n}$ denote the projection vectors of event types and node types, both of which are trainable. The projected node embedding are:

$$\mathbf{h}_u(t) = \mathbf{M}_{eu} \mathbf{x}_u(t) = \mathbf{n}_{\phi(u)}^{\top} \mathbf{x}_u(t) \mathbf{e}_{\varphi(e)} + \mathbf{x}_u(t) \tag{5}$$

$$\mathbf{h}_v(t) = \mathbf{M}_{ev} \mathbf{x}_v(t) = \mathbf{n}_{\phi(v)}^{\top} \mathbf{x}_v(t) \mathbf{e}_{\varphi(e)} + \mathbf{x}_v(t) \tag{6}$$

where $\mathbf{x}_u(t)$ and $\mathbf{x}_v(t)$ are the input embeddings of node u and v, respectively.

4.3 Temporal Heterogeneous Graph Attention Layer

Different events in a temporal heterogeneous graph may have different features, for example, in a question answering network, an answer interaction can be regarded as an event, and its features can be determined by the content. To enable event features to be propagated when aggregating information, THAN adds them to the node embeddings followed by a normalization layer (e.g., LayerNorm [1]). The event features will be resized to the same dimension as the node embeddings, and the output is:

$$\mathbf{z}_i(t_i) = \text{LayerNorm}(\mathbf{h}_i^l(t_i) + \chi_{0,i}(t_i)) \tag{7}$$

where i indicates the i-th neighbor, $\chi_{0,i}(t_i)$ denotes the feature of event between node v_0 and v_i at time t_i, and set $\chi_{0,0}(t)$ as zero vector for the source node.

Transformer [24] uses positional encoding to model relative position relationships, thus solving the problem that the attention mechanism cannot capture the sequential relationships between entities. In temporal graphs, a functional time encoder [12, 28] is usually used to map the time interval between nodes into a d_T-dimensional vector in place of positional encoding. THAN uses a Bochner-type functional time encoding [28, 29] as:

$$\text{TE}(t) = \sqrt{\frac{1}{d_T}} [\cos(\omega_1 t), \sin(\omega_1 t), ..., \cos(\omega_{d_T} t), \sin(\omega_{d_T} t)] \tag{8}$$

where $\{\omega_i\}$s are learnable parameters. We merge the time embeddings with the node representations to obtain the node-temporal feature matrices as:

$$\mathbf{Zs}(t) = [\mathbf{z}_0^{e_1}(t)\|\text{TE}(0), ..., \mathbf{z}_0^{e_N}(t)\|\text{TE}(0)]^\top \tag{9}$$

$$\mathbf{Zn}(t) = [\mathbf{z}_1(t_1)\|\text{TE}(t - t_1), ..., \mathbf{z}_N(t_N)\|\text{TE}(t - t_N)]^\top \tag{10}$$

where $\mathbf{z}_0^{e_i}$ and $\mathbf{z}_i$ denote the mapped embeddings of the source node v_0 and its neighbor v_i corresponding to event e_i, respectively, and $\|$ denotes the 'concatenate' operation. $\mathbf{Zs}$ and $\mathbf{Zn}$ are forwarded to three different linear projections to obtain the 'query', 'key', and 'value':

$$\mathbf{Q} = \mathbf{Zs}(t)W_Q^{\varphi(e_i)} \tag{11}$$

$$\mathbf{K} = \mathbf{Zn}(t)W_K^{\varphi(e_i)} \tag{12}$$

$$\mathbf{V} = \mathbf{Zn}(t)W_V^{\varphi(e_i)} \tag{13}$$

where e_i denotes the event between v_0 and v_i, $W_Q^{\varphi(e_i)}$, $W_K^{\varphi(e_i)}$, and $W_V^{\varphi(e_i)} \in \mathbb{R}^{(d+d_T)\times d}$ denote the projection matrices. Due to the edge heterogeneity, the projection matrices cannot be shared directly, thus we use matrices of different types to distinguish different events while capturing the semantics of events. The attention weight α_i is given by:

$$\alpha_i = \frac{\mathbf{Q}_i\mathbf{K}_i^\top}{\sum_{j=1}^N \mathbf{Q}_j\mathbf{K}_j^\top} \cdot \frac{\mu_{\phi(v_0),\varphi(e_i)}}{\sqrt{d}} \tag{14}$$

and it reveals how v_i attends to the feature of v_0 through event e_i. In addition, not all types of events have the same contribution to the source node, so we set a learnable tensor $\mu \in \mathbb{R}^{|A|\times|R|}$ to adaptively adjust the scale of attention to different-typed events.

The self-attention aggregates the features of temporal neighbors and obtains the hidden representation for node v_i as $\alpha_i\mathbf{V}_i$, which can capture both node features and topological information. The next step is to map the representations back to the type-specific distribution of node v_0 so that they can be fused with the features of node v_0. We use a linear projection named Q-Linear to do this and the updated neighborhood representation is:

$$\mathbf{s}(t) = \sum_{i=1}^N \text{Q-Linear}_{\phi(v_0)}(\alpha_i\mathbf{V}_i) \tag{15}$$

To combine neighborhood representation with the source node feature, we concatenate and pass them to a feed-forward neural network just as in TGAT [29]:

$$\tilde{\mathbf{h}}_0^l(t) = \text{FFN}(\mathbf{s}(t)\|\mathbf{x}_0^l(t)) \equiv \text{ReLU}([\mathbf{s}(t)\|\mathbf{x}_0^l(t)]W_0^l + b_0^l)W_1^l + b_1^l \tag{16}$$

Multi-head attention can effectively improve the model performance and stability, and THAN can be easily extended to support a multi-head setup. Assuming the self-attention outputs from P different heads, i.e. $s^i \equiv Attn^i(\mathbf{Q}, \mathbf{K}, \mathbf{V})$,

$i = 1, ..., P$. We first concatenate the P neighborhood representations with the source node feature and then carry out the same procedure in Eq. 16 as:

$$\tilde{\mathbf{h}}_0^l(t) = \text{FFN}(\mathbf{s}^1(t)\|...\|\mathbf{s}^P(t)\|\mathbf{x}_0^l(t)) \tag{17}$$

where $\tilde{\mathbf{h}}_0^l(t) \in \mathbb{R}^d$ is the final output representation for node v_0 at time t, and it can be used for link prediction task with an encoder-decoder framework.

4.4 Heterogeneous Graph Decoder

Heterogeneous graph decoder aims to reconstruct heterogeneous edges of the graph relying on the node representations, in other words, it scores edge triples through a function $\mathcal{H} : \mathbb{R}^d \times \mathbb{R}^{d_r} \times \mathbb{R}^d \rightarrow \mathbb{R}$, where d_r denotes the dimension of edge type embeddings. We compute node representations through a l-layer THAN encoder and use a feed-forward neural network as the scoring function, thus an event (u, v, t, r) is scored as:

$$\mathcal{H}(u, v, t, r) = \text{FFN}(\tilde{\mathbf{h}}_u^l(t)\|\mathbf{r}_r\|\tilde{\mathbf{h}}_v^l(t)) \tag{18}$$

where u, v denotes the source and target node, r denotes the edge type and $\mathbf{r} \in \mathbb{R}^{d_r}$ is edge type embedding.

As in previous work [21,29], we train the model with negative sampling. For each observed example, we change the target node to construct a new event that does not actually exist as a negative sample, so we have the same number of positive and negative samples. We optimize the cross-entropy loss as:

$$\mathcal{L} = \frac{1}{|\varepsilon|} \sum_{(u,v,t,r,y)\in\varepsilon} -y \log \sigma(\mathcal{H}(u,v,t,r)) - (1-y) \log(1-\sigma(\mathcal{H}(u,v,t,r))) + \lambda\|\theta\|_2^2 \tag{19}$$

where ε denotes the total set of positive and negative triples, σ denotes the logistic sigmoid function, y denotes the sample label and takes the value of 1 for positive samples and 0 for negative samples, θ denotes the model parameters and λ controls the L2 regularization.

5   Experiments

In this section, we present the details of experiments including experimental settings and results. Firstly, we introduce the dataset, baselines, and parameter settings. The performance comparisons are then demonstrated in detail. Finally, we test the inductive capability of our proposed model.

5.1   Experimental Settings

Datasets. We evaluate our model on three public datasets: Movielens, Twitter, and MathOverflow. The statistics of these datasets are listed in Table 1.

Table 1. Statistics of the three public datasets.

Dataset	Node types	#Nodes	Event types	#Events	Time span
Movielens	User	943	5	100,000	7 months
	Movie	1,682			
Twitter	User	304,691	3	563,069	188 days
MathOverflow	User	24,818	3	506,550	2,350 days

- **Movielens**[1] is a dataset of user ratings of movies at different times collected from the MovieLens website. We select two types of nodes: user and movie. Regarding different ratings of movies as different types of events, a total of 5 types of events are obtained.
- **Twitter**[2] collects public data on three types of relationships (retweet, reply, and mention) between users from the US social network Twitter.
- **MathOverflow**[3] is from MathOverflow, a question and answer site for professional mathematicians. There are three relationships between users in this dataset: a user answered or commented on another user's question, and a user commented on an answer.

Baselines. To demonstrate the effectiveness, we compare THAN with ten popular graph representation learning methods, which can be divided into three groups: static graph embedding (DeepWalk [19], metapath2vec [2], GraphSAGE [7], RGCN [21], HGT [8]), discrete-time dynamic graph embedding (DySAT [20], DHNE [31], DyHATR [30]), and continuous-time dynamic graph embedding (TGAT [29], HPGE [11]). We use the implementations of static graph embedding methods provided in the PyTorch Geometric (PyG) package [4], and for other baselines, use the code submitted by the authors on GitHub. Besides, We ignore the heterogeneity for homogeneous methods and ignore the temporal information for static methods. For fairness, the same decoder declared in Sect. 4.4 is used for the downstream temporal link prediction task.

- **DeepWalk** and **metapath2vec:** They are random walk-based network embedding methods designed for static graphs.
- **GraphSAGE:** A GNN model for homogeneous graphs that updates the node representation by sampling from neighborhood.
- **RGCN** and **HGT:** They are two static heterogeneous GNN methods, where the former maintains a unique linear projection weight for each edge type while the latter uses mutual attention based on meta relations to perform message passing on heterogeneous graphs.
- **DySAT:** A discrete-time temporal graph embedding method and we split graph snapshots with the guidance in the paper.

---

[1] https://grouplens.org/datasets/movielens/100k.
[2] http://snap.stanford.edu/data/higgs-twitter.html.
[3] http://snap.stanford.edu/data/sx-mathoverflow.html.

Table 2. Overall performance comparison on link prediction task. All results are converted to a percentage by multiplying by 100, and the best result is bolded.

Dataset	Movielens		Twitter		MathOverflow	
Model	AUC	AP	AUC	AP	AUC	AP
DeepWalk	67.35(0.3)	71.26(0.3)	57.73(0.7)	63.63(0.9)	63.73(0.2)	73.47(0.4)
metapath2vec	68.43(0.2)	71.82(0.2)	66.29(0.3)	74.67(0.2)	72.59(0.9)	81.13(1.1)
GraphSAGE	72.34(0.4)	75.94(0.4)	76.88(3.2)	85.1(2.1)	83.48(2.4)	89.20(3.4)
RGCN	69.49(0.4)	76.51(0.5)	84.18(0.6)	91.41(0.8)	84.02(0.2)	92.91(0.2)
HGT	73.44(1.1)	80.01(0.7)	88.54(0.5)	93.06(0.3)	86.53(1.5)	**93.88(1.4)**
DySAT	73.13(0.4)	72.1(0.3)	83.03(0.3)	86.89(0.2)	83.12(0.3)	85.84(0.1)
DHNE	59.78(0.1)	59.00(0.2)	55.37(0.2)	57.66(0.2)	58.06(0.4)	59.31(0.3)
DyHATR	80.21(0.7)	77.54(1.3)	79.73(0.1)	81.78(0.4)	75.22(0.1)	78.21(0.2)
TGAT	82.00(0.4)	79.46(0.4)	89.55(0.3)	90.43(0.2)	82.23(0.6)	83.25(0.6)
HPGE	85.25(0.1)	82.16(0.2)	73.55(0.1)	73.91(0.1)	81.12(0.2)	82.61(0.2)
THAN	**88.63(0.1)**	**86.77(0.2)**	**91.84(0.2)**	**93.43(0.2)**	**90.33(0.1)**	90.62(0.2)

- **DHNE** and **DyHATR:** They are two discrete-time THG embedding methods. DHNE performs meta path-based random walk between historical snapshots and the current snapshot. DyHATR uses hierarchical attention to learn heterogeneous information and incorporates RNNs with temporal attention to capture evolutionary patterns.
- **TGAT:** A continuous-time temporal graph embedding method that aggregates historical neighbors by self-attention to obtain node representations.
- **HPGE:** A continuous-time THG embedding method that integrates the Hawkes process into graph embedding to capture the excitation of historical heterogeneous events to current events.

Parameter Settings. THAN was implemented in PyTorch. We split the training and test set as 8:2 according to time order. For a fair comparison, we use the default parameter settings of the baselines and set the embedding (i.e. node output embeddings, time embeddings, and event type embeddings) dimension d as 32, regularization weight λ as 0.01, and dropout rate as 0.1. We employ Adam as the optimizer with a learning rate of 0.001. We randomly initialize the node vector if the dataset does not provide node features, and similarly, initialize the event features as zero vectors. For DeepWalk, metapath2vec, GraphSAGE, RGCN, and HGT, we set the max training epochs as 500 and use an early stopping strategy with the patience of 50. For DySAT, DHNE, and DyHATR, we split datasets into 10 snapshots. For our THAN, we set the event embedding dimension as 16, the number of layers as 2, attention heads as 4, epochs as 20 (30 for Movielens), learning rate as 0.001 (0.0001 for Twitter), batch size as 800 (500 for Movielens), and the number of samples for each type of neighbors as 10 (8 for Movielens). The implementation of THAN is publicly available[4].

---

[4] https://github.com/moli-L/THAN.

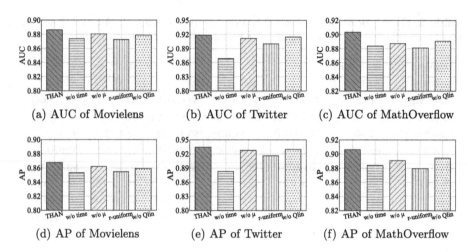

Fig. 3. Ablation study of THAN.

5.2 Effectiveness Analysis

We conduct the temporal link prediction task to verify the effectiveness, which asks if a type-r edge exists between two nodes at time t. We run all methods five times on three datasets and evaluate the average AUC (Area under the receiver operating characteristic curve) and AP (Average precision score) scores. The overall results are shown in Table 2.

Obviously, THAN achieves the state-of-the-art performance in AUC metric on all three datasets. Although THAN does not outperform all other methods in AP metric, it also has a considerable performance (i.e. AP score achieves the SOTA result on Movielens and Twitter datasets and over 0.9 on MathOverflow dataset). Besides, the GNN-based approaches achieve better performance than the random walk-based approaches since they capture much more useful information about the graph structure. HGT and GCN perform better than GraphSAGE which indicates that integrating semantics can benefit graph representation learning. DySAT and DyHATR obtain performance improvements due to considering the changes of graph structure over time. In addition, TGAT, HPGE, and our THAN perform better than DySAT and DyHATR, this phenomenon shows that it is important to make full use of temporal information compared with simply preserving evolving structures between snapshots.

5.3 Ablation Study

To demonstrate the effectiveness of each component in THAN, we conduct ablation experiments by removing/replacing a specific component at a time. We rename them as: (1) THAN w/o time: remove time embeddings; (2) THAN w/o μ: remove event type attention weight; (3) THAN w/o Qlin: remove linear projection Q-Linear; (4) THAN r-uniform: use uniform random sampling strategy.

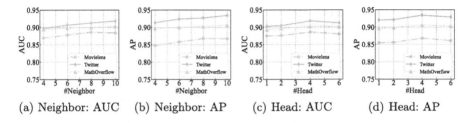

(a) Neighbor: AUC   (b) Neighbor: AP   (c) Head: AUC   (d) Head: AP

Fig. 4. Sensitivity analysis on the number of neighbor samples and attention heads.

Table 3. Results of inductive learning task.

Dataset	Movielens		Twitter		MathOverflow	
Model	AUC	AP	AUC	AP	AUC	AP
TGAT	78.35(0.4)	76.97(0.3)	85.87(0.3)	88.61(0.3)	74.14(0.5)	75.96(0.4)
THAN	82.71(0.2)	80.67(0.2)	88.41(0.2)	90.69(0.3)	80.93(0.3)	80.52(0.3)

We report the results of the ablation study in Fig. 3, from which we have the following observations: (1) THAN outperforms the others with components removed in all metrics; (2) Time embedding plays an important role in temporal graph representation learning; (3) Setting different attention weights for different edge types helps to learn heterogeneous semantic information; (4) More recent neighbors are more useful for extracting temporal evolutionary patterns and better reflect the current state of the source node; (5) It makes sense to keep the same feature space to fuse features from different nodes. Besides, it is noteworthy that removing the Q-Linear component did not have a significant impact on model performance on the Twitter and MathOverflow datasets, that is because both these datasets have only one type of node, and there is no need to consider the consistency of feature distribution across different types of nodes.

5.4 Parameter Sensitivity

To investigate the robustness of THAN and find the most suitable hyperparameters, we analyzed the effect of the number of neighbor samples and attention heads on three datasets shown in Fig. 4. For fairness, we select the number of neighbor samples from $\{4, 6, 8, 10\}$, the number of attention heads from $\{1, 2, 4, 6\}$, and the rest of the parameters remain the same as the experimental settings in Sect. 5.1.

On the one hand, Fig. 4(a) and (b) can lead to the following conclusion: the scores of AUC and AP improve as the number of neighbor samples increases, but on the Movielens dataset there is a decreasing trend instead, which may be caused by the dense connections between nodes. Sampling more neighbors may introduce more noise, resulting in smooth node representations. On the other hand, Fig. 4(c) and (d) show that the number of attention heads affects the performance of the model. Multi-head attention helps to obtain different aspect representations from different subspaces, thus enhancing the expressiveness.

5.5 Inductive Capability Analysis

We further discuss the inductive performance of THAN with the same settings as TGAT, i.e., mask 10% of the nodes from the training set and predict the existence of future events containing these masked nodes. In this paper, we choose TGAT as the comparison model, which is proposed as an inductive representation learning method on temporal graphs, and its inductive capability is demonstrated experimentally. Experiments were conducted on three datasets and the results are shown in Table 3. Intuitively, THAN outperformed the TGAT in two metrics on all datasets, which demonstrates the inductive capability of THAN.

6   Conclusion

Existing graph representation learning methods cannot well capture the information of temporal heterogeneous graphs. This paper proposes the THAN, which is a continuous-time temporal heterogeneous graph representation learning method. THAN uses transfer matrices to map different-typed nodes to the same feature space and aggregates neighborhood information based on the type-aware self-attention mechanism. To efficiently utilize temporal information, THAN uses a functional time encoder to generate time embeddings which are naturally integrated into the neighbor aggregation process. THAN is an inductive message-passing model based on historical neighbor sampling that not only captures temporal evolutionary patterns but also efficiently extracts topological features. Experimental results on three public datasets demonstrate that THAN outperforms the baselines on the temporal link prediction task. In the future, we plan to improve the computational efficiency of the model to deal with large-scale temporal heterogeneous graphs.

References

1. Ba, L.J., Kiros, J.R., Hinton, G.E.: Layer normalization. CoRR abs/1607.06450 (2016)
2. Dong, Y., Chawla, N.V., Swami, A.: metapath2vec: scalable representation learning for heterogeneous networks. In: SIGKDD, pp. 135–144 (2017)
3. Fan, Y., Ju, M., Zhang, C., Zhao, L., Ye, Y.: Heterogeneous temporal graph neural network. CoRR abs/2110.13889 (2021)
4. Fey, M., Lenssen, J.E.: Fast graph representation learning with PyTorch geometric. CoRR abs/1903.02428 (2019)
5. Fu, T., Lee, W., Lei, Z.: HIN2Vec: explore meta-paths in heterogeneous information networks for representation learning. In: CIKM, pp. 1797–1806 (2017)
6. Grover, A., Leskovec, J.: node2vec: scalable feature learning for networks. In: SIGKDD, pp. 855–864 (2016)

7. Hamilton, W.L., Ying, Z., Leskovec, J.: Inductive representation learning on large graphs. In: NIPS, pp. 1024–1034 (2017)
8. Hu, Z., Dong, Y., Wang, K., Sun, Y.: Heterogeneous graph transformer. In: WWW, pp. 2704–2710 (2020)
9. Huang, H., Shi, R., Zhou, W., Wang, X., Jin, H., Fu, X.: Temporal heterogeneous information network embedding. In: IJCAI, pp. 1470–1476 (2021)
10. Ji, G., He, S., Xu, L., Liu, K., Zhao, J.: Knowledge graph embedding via dynamic mapping matrix. In: ACL, pp. 687–696 (2015)
11. Ji, Y., Jia, T., Fang, Y., Shi, C.: Dynamic heterogeneous graph embedding via heterogeneous Hawkes process. In: Oliver, N., Pérez-Cruz, F., Kramer, S., Read, J., Lozano, J.A. (eds.) ECML PKDD 2021. LNCS (LNAI), vol. 12975, pp. 388–403. Springer, Cham (2021). https://doi.org/10.1007/978-3-030-86486-6_24
12. Kazemi, S.M., et al.: Time2Vec: learning a vector representation of time. CoRR abs/1907.05321 (2019)
13. Kazemi, S.M., et al.: Representation learning for dynamic graphs: a survey. J. Mach. Learn. Res. **21**, 70:1–70:73 (2020)
14. Kipf, T.N., Welling, M.: Variational graph auto-encoders. CoRR abs/1611.07308 (2016)
15. Kipf, T.N., Welling, M.: Semi-supervised classification with graph convolutional networks. In: ICLR (2017)
16. Kumar, S., Zhang, X., Leskovec, J.: Predicting dynamic embedding trajectory in temporal interaction networks. In: SIGKDD, pp. 1269–1278 (2019)
17. Luo, J., Xiao, S., Jiang, S., Gao, H., Xiao, Y.: ripple2vec: node embedding with ripple distance of structures. Data Sci. Eng. **7**, 156–174 (2022). https://doi.org/10.1007/s41019-022-00184-6
18. Pareja, A., et al.: EvolveGCN: evolving graph convolutional networks for dynamic graphs. In: AAAI, pp. 5363–5370 (2020)
19. Perozzi, B., Al-Rfou, R., Skiena, S.: DeepWalk: online learning of social representations. In: SIGKDD, pp. 701–710 (2014)
20. Sankar, A., Wu, Y., Gou, L., Zhang, W., Yang, H.: DySAT: Deep neural representation learning on dynamic graphs via self-attention networks. In: WSDM, pp. 519–527 (2020)
21. Schlichtkrull, M., Kipf, T.N., Bloem, P., van den Berg, R., Titov, I., Welling, M.: Modeling relational data with graph convolutional networks. In: Gangemi, A., et al. (eds.) ESWC 2018. LNCS, vol. 10843, pp. 593–607. Springer, Cham (2018). https://doi.org/10.1007/978-3-319-93417-4_38
22. Trivedi, R., Farajtabar, M., Biswal, P., Zha, H.: DyRep: learning representations over dynamic graphs. In: ICLR (2019)
23. Tuteja, S., Kumar, R.: A unification of heterogeneous data sources into a graph model in e-commerce. Data Sci. Eng. **7**, 57–70 (2022). https://doi.org/10.1007/s41019-021-00174-0
24. Vaswani, A., et al.: Attention is all you need. In: NIPS, pp. 5998–6008 (2017)
25. Velickovic, P., Cucurull, G., Casanova, A., Romero, A., Liò, P., Bengio, Y.: Graph attention networks. In: ICLR (2018)
26. Wang, X., et al.: Heterogeneous graph attention network. In: WWW, pp. 2022–2032 (2019)
27. Wang, Y., Chang, Y., Liu, Y., Leskovec, J., Li, P.: Inductive representation learning in temporal networks via causal anonymous walks. In: ICLR (2021)
28. Xu, D., Ruan, C., Körpeoglu, E., Kumar, S., Achan, K.: Self-attention with functional time representation learning. In: NIPS, pp. 15889–15899 (2019)

29. Xu, D., Ruan, C., Körpeoglu, E., Kumar, S., Achan, K.: Inductive representation learning on temporal graphs. In: ICLR (2020)
30. Xue, H., Yang, L., Jiang, W., Wei, Y., Hu, Y., Lin, Yu.: Modeling dynamic heterogeneous network for link prediction using hierarchical attention with temporal RNN. In: Hutter, F., Kersting, K., Lijffijt, J., Valera, I. (eds.) ECML PKDD 2020. LNCS (LNAI), vol. 12457, pp. 282–298. Springer, Cham (2021). https://doi.org/10.1007/978-3-030-67658-2_17
31. Yin, Y., Ji, L., Zhang, J., Pei, Y.: DHNE: network representation learning method for dynamic heterogeneous networks. IEEE Access 7, 134782–134792 (2019)
32. Ying, C., et al.: Do transformers really perform bad for graph representation? CoRR abs/2106.05234 (2021)
33. Zhao, J., Wang, X., Shi, C., Hu, B., Song, G., Ye, Y.: Heterogeneous graph structure learning for graph neural networks. In: AAAI, pp. 4697–4705 (2021)

SCBERT: Single Channel BERT for Chinese Spelling Correction

Hong Gao[1,2(✉)], Xuezhen Tu[1], and Donghai Guan[1]

[1] Nanjing University of Aeronautics and Astronautics, Nanjing, China
Gaoh1803@126.com
[2] ZTE Corporation, Shenzhen, China

Abstract. Chinese spelling correction (CSC) and BERT pre-training task can both be regarded as text denoising. In this work, to further narrow the gap between the pre-training and CSC tasks, we present a Single Channel **BERT** (SCBERT) which incorporates semantics, pinyin and glyph of typos to provide effective spelling correction. In model pre-training, we introduce fuzzy pinyin and glyph of Chinese characters and adjust mask strategies to restore the pinyin or glyph information of the "[MASK]" token under certain probabilities. Therefore, we can mask out the char channel of the typo and only provide its pinyin or glyph information in order to reduce the input noise when using our models, as the char information of typos in CSC is a kind of noise. Moreover, we apply synonym replacement and sentence reordering for paraphrasing to improve the accuracy of the correction step. We conduct experiments using widely accepted benchmarks. Our method outperforms state-of-the-art approaches under zero-shot learning condition and achieves competitive results when fine-tuning.

Keywords: Chinese spelling correction · Model pre-training · Zero-shot learning

1 Introduction

Chinese spelling correction (CSC), which aims to detect and correct spelling errors in texts, is highly similar to pre-training tasks of Pre-trained Language Models (PLMs). Thus, it is feasible and rewarding to train a PLM for CSC task. Chinese spelling correction is an important task in NLP. It is essential for many natural language applications, including optical character recognition [1] (OCR), automatic speech recognition [2] (ASR) and search engine [3]. For the correction of spelling errors, a general background knowledge of language is required. Usually, PLMs help achieve the point. In this paper, we consider Chinese spelling error correction at character-level.

In recent years, large-scale pre-trained language models have been extensively studied and become the fundamental backbone for various Natural Language Processing (NLP) tasks. An established paradigm in the NLP field is to pre-train a model on a large amount of texts and then fine-tune it afterward. The similarity between the pre-training task and the downstream task affects the performance of fine-tuning.

B. Li et al. (Eds.): APWeb-WAIM 2022, LNCS 13422, pp. 401–414, 2023.
https://doi.org/10.1007/978-3-031-25198-6_30

In Chinese, spelling errors are mainly caused by misusing phonetically or visually similar characters [4]. Correcting errors using semantics is a common practice. However, it can be challenging to determine the right word without the information from pinyin and glyph. When attempting to correct spelling errors, it is usual to consider the glyph and phonetic aspects of the typo. Evidently, the meaning of the typo is the noise for CSC and the pinyin and glyph of the character offer clues for error correction.

Human-level language comprehension is a necessity for correcting the misspelled words. Therefore, PLMs are nowadays used to obtain State-of-the-Art (SOTA) results. Notably, the pre-training tasks of PLMs are very similar to the text error correction task. For instance, among the three mask strategies of BERT [5], 80% of the selected tokens are masked out via "[MASK]" tokens, 10% of the selected tokens are replaced with a random tokens and 10% of the selected tokens are stay unchanged. The first two pre-training tasks can be regarded as spelling correction tasks. Therefore, BERT is an out-of-the-box solution for CSC.

To implement existing PLMs in CSC task, such as BERT, either the original char text is provided for prediction, or the candidate typos are replaced with "[MASK]" tokens. Both methods have disadvantages. It is inevitable that the first approach will introduce noise about wrong characters. Though the second approach alleviates the weakness, important pinyin and glyph information are lost. Solution to this problem is to make the model accept single channel inputs, that is, the model has ability to function well when only pinyin or glyph information of character are provided. According to our knowledge, there is currently no prior works that comply with the requirement.

To address the above issues, we propose SCBERT, a **S**ingle **C**hannel **BERT** for Chinese spelling correction, which not only has the capability of providing dynamic word vectors as other PLMs, but also brings a new pattern of spelling errors correction via single channel input mechanism. We further improve accuracy of CSC by using synonym replacement and sentence reordering for paraphrasing before the correction step.

In summary, the paper contributes the followings:

- We propose a BERT based SCBERT and present the datasets, model hyperparameters and training strategies for effective pre-training task. The essential components of SCBERT include single channel input mechanism and fuzzy information modeling.
- We propose a novel single channel input mechanism which jointly learns the semantics, pinyin and glyph of Chinese characters while reducing the noise caused by typos in the CSC task.
- We utilize fuzzy pinyin and glyph of Chinese characters to enables SCBERT to model the similarities in pinyin and glyph.
- Finally, We introduce paraphrasing methods before the correction step to improve accuracy even further.

2   Related Work

2.1   PLMs in NLP

The pre-training/fine-tuning paradigm reigns in NLP tasks. GPT [6], BERT, XLNet [7] and BART [8] have brought significant performance gains and several downstream tasks have benefited from their utilization. In general, the primary technical innovation of PLMs is the application of Transformer, a popular and scalable attention model. GPT is an autoencoding model, which guesses the next token after reading all the previous ones. BERT is pre-trained by corrupting the input tokens in some way and then attempting to reconstruct the original sentence. BART keeps both the encoder and the decoder of the original Transformer and is a typical Seq2Seq model. Although all those models can be fine-tuned and used to accomplish a variety of specific downstream tasks, their areas of expertise also differ due to the differences in their pre-training tasks. Due to this, it can be beneficial to use a pre-trained model that aligns with the downstream task in order to enhance performance. Most of the pre-training tasks can be adapted to the CSC task with a little modification. It is still possible to improve the performance of PLM applied to CSC by adjusting the pre-training task, such as SpellBERT [9] and Soft-Masked BERT [10].

2.2 Pinyin and Glyph Information Modeling

Incorporating pinyin and glyph information for Chinese NLP tasks [10] has gained increased attention in recent times. For character feature extraction, UMRC [11] employs a specific CNN structure and uses image classification as an auxiliary objective to mitigate the influence of images in a very small number of instances. ChineseBert [12] uses the glyph and pinyin information of Chinese characters to enhance its capability of capturing context semantics from surface character forms as well as disambiguating polyphonic Chinese characters. PLOME [13] jointly learns semantics and misspelled knowledge thanks to the confusion set based masking strategy. These methods show that adding more training information, such as pinyin and glyph, will improve model performance.

2.3 Chinese Spelling Correction

Spelling error correction can be challenging since it essentially requires human-level language understanding skills to achieve a satisfactory result. In some cases, researchers took advantage of unsupervised approaches, usually using a confusion set to identify correct candidates and employing language models to determine which of them should be selected. In addition, Hybrid [14] is based on a BiLSTM model trained on a generated dataset. FASPell [15] adopts a Seq2Seq model for CSC employing BERT as a denoising autoencoder and a decoder. Confusionset [16] is a Seq2Seq model consisting of both a pointer network and a copy mechanism. Overall, after the pre-training models appear, all SOTA models in the CSC task will be required to use them. Therefore, it is extremely important to apply an appropriate pre-training model to correct spelling errors.

3 Our Approach

In this section, we demonstrate details regarding dataset processing, pinyin-glyph feature construction, and SCBERT pre-training tasks. We modify the pre-training task to make it more appropriate to cope with fuzzy pinyin and glyph data. The character char, pinyin, and glyph of Chinese token will be embedded separately within the input layer of the model, as shown in Fig. 1. The mask strategy will be adapted to allow the pre-trained model to accept single channel inputs.

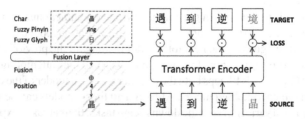

Fig. 1. A forward propagation example of SCBERT. **Left:** This component illustrates the overall architecture of the proposed embedding layer. Input characters are represented by concatenated char, pinyin and glyph vectors. **Right:** This component makes prediction for input sentence and calculates loss. We propose the single channel input mechanism in this component to narrow gap between pre-training task and CSC task.

3.1 Fuzzy Pinyin and Glyph

Spelling errors in Chinese can be mainly categorized into two types: phonological errors and visual errors, caused by the misuse of both phonologically similar and visually similar characters. Hence, phonetic (pinyin) and glyph information play crucial roles in CSC.

Table 1. Fuzzy pinyin and fuzzy glyph examples

Char	Precise Pinyin	Fuzzy Pinyin	Fuzzy Glyph
房	Fang	Huang	方
陈	Chen	Cheng	东
盛	Sheng	Sheng	成
了	Le	Ne	了
思	Si	Shi	田

Chinese pinyin, a standard system of romanized spelling using the Latin alphabet and diacritic, represents the pronunciation of characters. In this paper, we use pypinyin[1] to obtain the character-pinyin mapping, which is capable of handling polyphonic words. Similar pinyin representations are pronounced similarly. Considering that we use pinyin information more often than char information when communicating orally, modeling of pinyin is important for CSC of speech-related tasks, such as ASR text error correction. Unlike word vectors, similar pinyin does not have semantically similar representations. In this way, existing pre-training tasks have difficulty capturing pinyin similarity. To circumvent this problem, On the one hand, we model pinyin at the character level instead of considering pinyin as a whole, and on the other hand, we propose to replace accurate pinyin with fuzzy pinyin.

Table 1 illustrates some fuzzy pinyin examples. Specifically, there are two steps. (1) To eliminate the diacritic of each pinyin. (2) According to the rules of mispronunciation in

---

[1] https://github.com/mozillazg/python-pinyin.

everyday spoken language, we convert pinyin in advance to make similar pronunciation have same pinyin representation.

Chinese characters are pictograms, as they developed from the graphical regularity of the transactions. Therefore, visually similar words tend to have common parts. In analyzing glyph information, some existing pre-training models use different strokes by disassembling Chinese characters to represent glyph. Some researchers converted characters into pixels according to specific fonts. Nevertheless, text error correction does not require precise glyph, for exact glyph of typos would bring noise during the correction process. In this paper, we use ChaiZi Database[2] to split the character into parts and select the main part (non-radical part) as the fuzzy glyph. No splitting is conducted for characters with less than five strokes. See Table 1 for specific examples of fuzzy glyph.

In practice, we first tokenize each sentence, and calculate the fuzzy pinyin and fuzzy glyph information of each token. Then use a separate network layer for three type embeddings. Inspired by Chinesebert [12], we map fuzzy glyph to font images for modeling.

3.2 SCBERT Pre-training Task

BERT is a model with the strong expressiveness that based on the Transformer's structure. The purpose of the BERT pre-training tasks is to construct a self-supervised task using a plain text corpus, so that the machine learning model can learn the semantic representation of the text. The similarity between pre-training task and downstream tasks will influence the effect of fine-tuning. When pre-training SCBERT, we mask out 15% of the words in the input following BERT pre-training routine and then only the masked words are to predicted. In this work, we make the following improvements to the original BERT pre-training task.

Combination of WWM and CM. Chinese Whole Word Mask (WWM) is different from English WWM. Chinese WWM first segments the sentences, and then masks the characters within these segments simultaneously. In this paper, Chinese word segmentation is based on jieba[3]. On the contrary, Char Mask (CM) treats Chinese characters as independent segments. WWM needs higher-level semantic understanding and is more challenging than CM. However, in CSC tasks, it is not recommended to use only WWM [17], because the CM mask is indispensable for correcting errors of a single character, which is also the most frequent type of spelling errors. Likewise, WWM has a great influence on the error correction of two or more consecutive typos. Therefore, We apply WWM 20% of the time and CM 80% of the time. We use the dynamic masking strategy, which means different input masks are fed to the model on every single epoch.

MLM Loss. We pre-train the model using Masked Language Model (MLM) loss without Next Sentence Prediction (NSP) loss, since it has been proved to offer no benefits for improving downstream performances [18] and we also consider the fact that the CSC

---

[2] https://github.com/howl-anderson/hanzi_chaizi.

[3] Https://github.com/fxsjy/jieba.

task will not be conducted on sentences from different documents. Additionally, other pre-training tasks such as Sentence Order Prediction (SOP) [19] tasks are not obviously related to text error correction tasks. Their effectiveness will be assessed in future work.

Table 2. Single channel masking strategy

Masking Type	Sub Type	Example	Probability
"[MASK]" replacement	-all	**Char:** 一 [MASK] 风 顺 **Pinyin:** Yi [MASK] Hong Shun **Glyph:** 一 [MASK] 风 页	7.2% (0.15*0.8*0.6)
	-char-glyph	**Char:** 一 [MASK] 风 顺 **Pinyin:** Yi Hang Hong Shun **Glyph:** 一 [MASK] 风 页	3.6% (0.15*0.8*0.3)
	-char-pinyin	**Char:** 一 [MASK] 风 顺 **Pinyin:** Yi [MASK] Hong Shun **Glyph:** 一 凡 风 页	1.2% (0.15*0.8*0.1)
Random replacement	-	**Char:** 一 徼 风 顺 **Pinyin:** Yi Jing Hong Shun **Glyph:** 一 敬 风 页	1.5% (0.15*0.1)
Unchanged	-	**Char:** 一 帆 风 顺 **Pinyin:** Yi Hang Hong Shun **Glyph:** 一 凡 风 页	1.5% (0.15*0.1)

Single Channel Input. The single channel input mechanism is to reduce the noise of the input. Table 2 illustrates this with a specific example. When tokens are selected to be replaced by "[MASK]", one or two channels are reserved without masking for pinyin or/and glyph channels. In other words, unlike other PLMs models that modeling and glyph, our model does not always mask out the input char, pinyin and glyph channels simultaneously. The single-channel mask is a key feature of our PLMs, as un-masked char channels will lead to errors and the model will also tend to predict the original wrong characters. Nevertheless, if all channels are been masked out, the semantically appropriate answer given by PLMs may not be accurate. We have included the modeling channels of pinyin and glyph for Chinese characters in SCBERT, which allows for a single channel "[MASK]" token to be kept of the single channel input ensures that as much information as possible is provided for text error correction with-out introducing noise. Additionally, there is a possibility that the pinyin and glyph channels will be reserved at the same time, which means the pinyin and glyph channels of typos can be used in correction step. SCBERT can be used as an ordinary BERT in downstream applications. On top of that, if we use the same single channel input of pre-training task in the CSC tasks, we can better utilize the knowledge in SCBERT. The specific mask ratio for single channel input is provided in Table 2.

3.3   Model Setup

Generally speaking, training hyperparameters have a great influence on the performance for deep models. In this paper, we pre-train models of two structure, $SCBERT_{tiny}$ and $SCBERT_{base}$, which is composed of 4/12 transformer layers respectively, with input dimensionality of 312/768. The model structure refers to the RoBERTa model. The embedding layer is modified to accept three-channel input, as shown in Fig. 1. Both of $SCBERT_{tiny}$ and $SCBERT_{base}$ model is initialized with random parameters. We alse train a $SCBERT_{base}$, whose Transformer layers are initialized with Roberta-zh[4].

Table 3. Hyperparameters for pre-training $SCBERT_{base}$ and $SCBERT_{tiny}$.

Hyperparameter	$SCBERT_{base}$	$SCBERT_{tiny}$
Number of Layers	12	4
Hidden size	768	312
Optimizer	LAMB	LAMB
Lr Scheduler	OneCycleLR	OneCycleLR
Warmup Percentage	0.025	0.01
Batch Size	3072	3072
Anneal Strategy	Linear	Linear
Weight Decay	0.01	0.01
Max Steps	$800k_{phase1} + 200k_{phase2}$	$400k_{phase1} + 100k_{phase2}$

To further reduce the size of the tiny model, we replaced the 23236 large dictionary with the 8035 small dictionary published in ChineseCLUE [20]. Since text error correction does not require long-term attention dependency, we set the *max_len* of position embedding to 128, and correspondingly increase the training batch size to ensure the total number of tokens in each batch. To speed up pre-training task in our experiments, similar to BERT, we first train 800k steps on 64 sequence length with batch size 4k. We then train 200k on 128 sequence length with batch size 3k to make the model more suitable for tasks with longer sequence lengths. Each sample is packed with full sentence sampled contiguously from one or more documents.

We chose LAMB [21] instead of Adam as the optimizer, which can guarantee the convergence in case of large batch sizes. We have trained the random initialized models with maximum learning rate 5e−4, OneCycleLR [22] scheduler and warmup of 20k steps. We set maximum learning rate 1e−4 for Roberta-zh initialized model. The specific parameter list is shown in Table 3. Our training strategy included DDP, mixed precision [23], and gradient accumulation. Training was done on 8 × 32 GB Nvidia V100 GPUs.

---

[4] Https://github.com/brightmart/roberta_zh.

3.4 SCBERT for Chinese Spelling Correction

In this section, we demonstrate the application of SCBERT to the CSC task. The model can be directly applied to downstream tasks after pre-training, in the same manner as BERT. Moreover, if integrated with the single channel input mechanism and paraphrasing technique, our models can significantly enhance the accuracy of text error correction on both fine-tuning condition and zero-shot learning condition. The results of the experiments and ablation study are described in the next section.

Paraphrasing. We develop a way of paraphrasing to better retrieve the information from our PLMs. As illustrated in Fig. 2, for the sentence "我这个里拜有很多事" (I have a lot to do this week.), we first figure out "里拜" may be typos in detection step. Then we paraphrase the remaining correct part of sentence, for example, replacing the "这个" (this) with "那个" (that) to obtain another sentence "我那个里拜有很多事" (I have a lot to do that week.). Obviously, the correction result "礼拜" should make the paraphrasing sentence fluent and plausible as well. We also change the sequence of text when it encompasses sub-sentences.

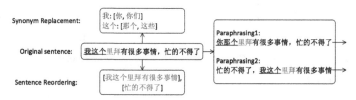

Fig. 2. Paraphrasing for correction. Large chunks of text will split into small segment to make sure each text sample has no more than 3 sub-sentences.

Prediction Inference. We recommend using SCBERT in conjunction with the single channel inputs and paraphrasing techniques for both fine-tuning and zero-shot learning conditions. As illustrated in Fig. 3, the CSC process should be split into detection and correction steps firstly and then we utilize paraphrasing and single channel input mechanism before the correction step. In the detection step, for sentence "遇到逆晶" (Encounter retrograde), tokens that are inconsistent with the original text or tokens whose softmax probability falls below the threshold (0.6) are considered as candidate typos, which is "晶" in this case. Next, based on the results of the detection step, by synonym replacement, we generate additional equivalent sample "遭遇逆晶" and for each sample, we apply single channel input to the candidate typo ("晶") to reduce the noise brought by its char channel. Accordingly, one sample is augmented to 4 samples before the correction step. Finally, all samples will be predicted by SCBERT, and then the softmax values of candidate typo will be calculated and averaged to obtain the final error correction result. Following the above steps, the error sample "遇到逆晶" will be corrected to "遇到逆境" (Encounter adversity).

The core idea of our methodology is to adjust the BERT pre-training task and the downstream CSC task in order to make them more similar so as to enhance the effectiveness of error correction in fine-tuning and zero-shot learning conditions. To this end,

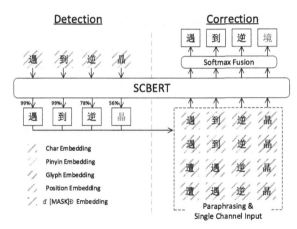

Fig. 3. An application of SCBERT in CSC. There are two steps to the scheme: typo detection step and error correction step. At the detection step, tokens with a prediction probability lower than threshold or tokens that are inconsistent with the original text are placed into the typo candidate set. At the correction step, the single-channel input and paraphrasing pipeline are applied to correct the detected typos.

we propose single channel input mechanism and apply them in both pre-training task and CSC task.

For the improvement of the pre-training task, in addition to the single channel input method, fuzzy pinyin and fuzzy glyph are used for modeling the similarity of pinyin and glyph. For CSC tasks, as depicted in Fig. 3, the strategy of utilizing single channel input before the correction step is combined with paraphrasing technique. Single channel input method is to reduce the input noise of the model when correcting the already positioned typos, and paraphrasing is used to better stimulating PLM capability.

4 Experiments

4.1 Dataset

Pre-training Data. We use zhwiki-20220101[5] as the pre-training corpus, which consists of 1.25 million Chinese pages. We also collect THUCNews and SinaNews as extended data. Total dataset contains 6.5GB uncompressed text. We split those pages and articles into sentences and obtain 56 million sentences.

Fine-tuning Data. Following previous work, we evaluate our pre-trained models on on the most widely used benchmark dataset SIGHAN13[6], SIGHAN14 and SIGHAN15. The fine-tuning dataset contains 10K annotated samples. OpenCC[7] is used to convert traditional characters into simplified characters.

---

[5] Https://dumps.wikimedia.org/zhwiki/.

[6] Http://nlp.ee.ncu.edu.tw/resource/ncu_nlplab_csc.zip.

[7] https://pypi.org/project/OpenCC/.

Evaluation Data. We evaluate the proposed model using the latest SIGHAN test dataset, which includes 1100 texts and 461 types of errors.

4.2 Baselines

For comparison, We use the following three pretrained language models as the baselines, which are all capable of zero-shot inference. We report the results of these PLMs based on the same fine-tuning and few-shot learning condition. **BERT** is an original but highly competitive PLM. **ChineseBert** and **PLOME** are variants of BERT, both capable of modeling pinyin and glyph. PLOME is a PLM trained for CSC and jointly considering the target pronunciation and character distributions, whereas ChineseBert is a more universal PLM. For a fair comparison, base structure is chosen for each baseline model.

4.3 Results

Table 4 illustrates the experimental results under fine-tuning and zero-shot learning conditions and our models have produced outstanding results.

The fine-tuning procedure is implemented to adjust the PLM's output according to the data distribution of downstream tasks. In case of fine-tuning, the detection F1 scores of *BERT-Finetune* and *ChineseBert-Finetune* are lower because their pre-training tasks are not adapted for CSC. The F1 values of the other two models are at the same level. Given that the amount of SCBERT pre-training data is only one third of that of PLOME, our model has already achieved SOTA results. For correction, $SCBERT_{base}$-*Finetune* achieved 1.7% and 1.3% absolute improvements over *ChineseBert-Finetune* and *PLOME-Finetune* respectively. Therefore, Our model can provide a robust word vector, which is a prerequisite for downstream work. Additionally, the single channel input and paraphrasing pipeline used between the detection step and the correction step achieved the expected effect as well.

We further explore the results of model fine-tuning using part of training data. Upon beginning with zero-shot learning condition, comparison experiments are performed for every 10% increase in training data. The correction F1 scores in results are presented in Fig. 4. We find that our model outperforms other models at different scales of the training set. In the case of few-shot learning (10% or 20% of the total), our model has obvious advantages, which achieves 1% improvements over the other three counterpart models. This implies that SCBERT obtains better representations of pinyin and glyph during the pre-training stage.

Without fine tuning, that is, under zero-shot learning condition, our model surpasses the other models with remarkable gains in the correction step. As shown in second group in Table 4, $SCBERT_{base}$-*pretrain* has a higher error correction F1 score than others by 1.5%. Especially, $SCBERT_{base}$-*pretrain* has achieved SOTA results in both detection and correction step. Our model is, therefore, an exceptional out-of-the-box CSC model. We believe the performance of CSC models in zero-shot learning condition is very important. Because a robust labeled dataset for CSC should be updated and extended timely and regularly to adapt to the current expressions, emerging words and new entities. The maintenance costs may be hard to bear. However, PLMs can utilize the content of the

Table 4. The performances of our approach and baseline models. Rows 1 to 4 list the results for the models fine-tuned on SIGHAN train data (10k). Rows 5 to 8 list results on zero-shot learning condition. Each experiment is run four times and the average metrics are reported.

Category	Method	Detection (%)			Correction (%)		
		P	R	F	P	R	F
SIGHAN	*BERT-Finetune*	88.4	86.2	87.3	89.9	78.6	83.9
	ChineseBert-Finetune	91.6	84.2	87.7	93.2	79.9	86.0
	PLOME-Finetune	**92.3**	85.5	**88.7**	93.5	80.4	86.4
	SCBERT$_{base}$-Finetune	90.3	**86.7**	88.5	**94.2**	**81.8**	**87.7**
Zero-Shot	*BERT-pretrain*	64.2	63.3	63.7	83.6	60.2	70.0
	ChineseBert-pretrain	**68.7**	64.5	66.5	83.7	70.5	76.5
	PLOME-pretrain	66.4	71.3	68.8	83.2	72.5	77.5
	SCBERT$_{base}$-pretrain	67.2	**72.0**	**69.5**	**85.4**	**73.5**	**79.0**

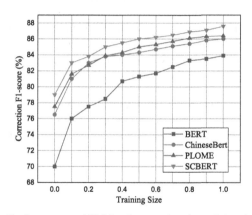

Fig. 4. Performances of PLMs when varying the training data size.

latest textual information without labeling, which makes zero-shot learning with PLMs a promising solution for CSC in the big data era.

4.4 Ablation Study

Model Capacity. As can be seen from the results in the first and last rows of the Table 5, *SCBERT$_{tiny}$* obtains only 10% less in the correction step F1 score than *SCBERT$_{base}$*, however the size of the model is much smaller. Thus, Our tiny model has a great deal of potential in resource-limited deployment scenarios.

Single Channel Input. When compared with normal input scenarios, single channel input is far better than normal input. For *SCBERT$_{base}$-no-sc*, we do not restore the pinyin or glyph channel of inputs but keep the paraphrasing procedure. *SCBERT$_{base}$*

Table 5. The performances of our PLMs.

Category	Method	Detection (%)			Correction (%)		
		P	R	F	P	R	F
Zero-Shot	$SCBERT_{base}$	67.2	72.0	69.5	85.4	73.5	79.0
	$SCBERT_{base}$-no-sc	67.2	72.0	69.5	83.6	71.5	77.1
	$SCBERT_{base}$-no-pa	67.2	72.0	69.5	84.8	72.7	78.3
	$SCBERT_{base}$-rand	65.5	69.7	67.5	84.1	70.8	76.9
	$SCBERT_{tiny}$	60.6	64.5	62.5	72	65.8	68.8

scores 2 points higher than $SCBERT_{base}$-no-sc, in terms of correction F1 score. The result in Table 5 illustrates that the single channel input can effectively reduce the noise of the input for the model and providing the pinyin and glyph information for CSC.

Paraphrasing. The proposed synonym replacement and sentence reordering can improve the accuracy of the correction step. Results of $SCBERT_{base}$-no-pa is obtained without paraphrasing before the correction step, which leads to a drop by 0.7 point in F1 score. While our paraphrasing methods are still relatively primitive, it nonetheless demonstrates the effectiveness of exploration in this direction. In future research, we will explore the effects of paraphrasing using the latest prompt techniques [24].

Initialization Strategy. The initialization strategy impacts the final convergence of the model, affecting performance in both detection step and correction step, compared to $SCBERT_{base}$-rand, $SCBERT_{base}$ presents an overwhelming boost, demonstrating that Roberta-zh's rich semantic information can enhance the model generalizability. We believe more training steps and more training data are required to close the gap when pre-training language models from scratch.

5   Conclusion

In this paper, we present a novel pre-trained language model for Chinese spelling correction task. To the best of our knowledge, SCBERT is the first task-specific language model considering the input noise introduced by typos. We propose the concept of the single channel input mechanism and evaluate its implementation in both PLM pre-training task and CSC task. Combined with the proposed paraphrasing technique, we further improve the accuracy of correction step. Besides, our work is a successful practice of model pre-training and potentially useful for similar projects. Experimental results in fine-tuning condition and zero-shot learning condition illustrate that SCBERT outperforms the state-of-the-art methods utilizing BERT. In the future, we plan to extend SCBERT to other related tasks such as grammar error correction and explore potential applications combining with prompt techniques.

References

1. Islam, N., Islam, Z., Noor, N.: A survey on optical character recognition system. arXiv preprint arXiv:1710.05703 (2017)
2. Benzeghiba, M., et al.: Automatic speech recognition and speech variability: a review. Speech Commun. **49**, 763–786 (2007)
3. Guo, J., Sainath, T.N., Weiss, R.J.: A spelling correction model for end-to-end speech recognition. In: ICASSP 2019-2019 IEEE International Conference on Acoustics, Speech and Signal Processing (ICASSP), pp. 5651–5655. IEEE (2019)
4. Liu, C.-L., Lai, M.-H., Chuang, Y.-H., Lee, C.-Y.: Visually and phonologically similar characters in incorrect simplified chinese words. In: Coling 2010: Posters, pp. 739–747 (2010)
5. Devlin, J., Chang, M.-W., Lee, K., Toutanova, K.: Bert: pre-training of deep bidirectional transformers for language understanding. arXiv preprint arXiv:1810.04805 (2018)
6. Radford, A., Narasimhan, K., Salimans, T., Sutskever, I.: Improving language understanding by generative pre-training (2018)
7. Yang, Z., Dai, Z., Yang, Y., Carbonell, J., Salakhutdinov, R.R., Le, Q.V.: Xlnet: generalized autoregressive pretraining for language understanding. Adv. Neural Inf. Process. Syst. **32** (2019)
8. Lewis, M., et al.: Bart: denoising sequence-to-sequence pre-training for natural language generation, translation, and comprehension. arXiv preprint arXiv:1910.13461 (2019)
9. Ji, T., Yan, H., Qiu, X.: SpellBERT: a lightweight pretrained model for chinese spelling check. In: Proceedings of the 2021 Conference on Empirical Methods in Natural Language Processing, pp. 3544–3551 (2021)
10. Zhang, A., Li, B., Wang, W., Wan, S: A Novel Text Classification Model Combining Deep Active Learning with BERT. CMC-COMPUTERS MATERIALS & CONTINUA
11. Li, X., Feng, J., Meng, Y., Han, Q., Wu, F., Li, J.: A unified MRC framework for named entity recognition. arXiv preprint arXiv:1910.11476 (2019)
12. Sun, Z., et al.: Chinesebert: Chinese pretraining enhanced by glyph and pinyin information. arXiv preprint arXiv:2106.16038 (2021)
13. Liu, S., Yang, T., Yue, T., Zhang, F., Wang, D.: PLOME: pre-training with misspelled knowledge for Chinese spelling correction. In: Proceedings of the 59th Annual Meeting of the Association for Computational Linguistics and the 11th International Joint Conference on Natural Language Processing, vol. 1, Long Papers, pp. 2991–3000 (2019)
14. Wang, D., Song, Y., Li, J., Han, J., Zhang, H.: A hybrid approach to automatic corpus generation for Chinese spelling check. In: Proceedings of the 2018 Conference on Empirical Methods in Natural Language Processing, pp. 2517–2527 (2018)
15. Hong, Y., Yu, X., He, N., Liu, N., Liu, J.: FASPell: A fast, adaptable, simple, powerful Chinese spell checker based on DAE-decoder paradigm. In: Proceedings of the 5th Workshop on Noisy User-generated Text (W-NUT 2019), pp. 160–169 (2019)
16. Wang, D., Tay, Y., Zhong, L.: Confusionset-guided pointer networks for chinese spelling check. In: Proceedings of the 57th Annual Meeting of the Association for Computational Linguistics, pp. 5780–5785 (2018)
17. Dai, Y., et al.: Is Whole Word Masking Always Better for Chinese BERT?: probing on Chinese Grammatical Error Correction. arXiv preprint arXiv:2203.00286 (2022)
18. Liu, Y., et al.: Roberta: A robustly optimized bert pretraining approach. arXiv preprint arXiv:1907.11692 (2019)
19. Lan, Z., Chen, M., Goodman, S., Gimpel, K., Sharma, P., Soricut, R.: Albert: a lite bert for self-supervised learning of language representations. arXiv preprint arXiv:1909.11942 (2019)
20. Xu, L., Zhang, X., Dong, Q.: CLUECorpus2020: a large-scale Chinese corpus for pre-training language model. arXiv preprint arXiv:2003.01355 (2020)

21. You, Y., et al.: Large batch optimization for deep learning: Training bert in 76 min. arXiv preprint arXiv:1904.00962 (2019)
22. Smith, L.N., Topin, N.: Super-convergence: Very fast training of neural networks using large learning rates. In: Artificial intelligence and machine learning for multi-domain operations applications, p. 1100612. International Society for Optics and Photonics (2020)
23. Micikevicius, P., et al.: Mixed precision training. arXiv preprint arXiv:1710.03740 (2017)
24. Liu, P., Yuan, W., Fu, J., Jiang, Z., Hayashi, H., Neubig, G.: Pre-train, prompt, and predict: a systematic survey of prompting methods in natural language processing. arXiv preprint arXiv:2107.13586 (2021)

Improving Robustness of Medical Image Diagnosis System by Using Multi-loss Hybrid Adversarial Function with Heuristic Projection

Chufan Cheng[(✉)] and Fang Chen[(✉)]

Department of Computer Science and Engineering, Nanjing University of Aeronautics
and Astronautics, Nanjing, China
{trayvon,chenfang}@nuaa.edu.cn

Abstract. Medical image diagnosis system by using deep neural networks (DNN) can improve the sensitivity and speed of interpretation of chest CT for COVID-19 screening. However, DNN based medical image diagnosis is known to be influenced by the adversarial perturbations. In order to improve the robustness of medical image diagnosis system, this paper proposes an adversarial attack training method by using multi-loss hybrid adversarial function with heuristic projection. Firstly, the effective adversarial attacks which contain the noise style that can puzzle the network are created with a multi-loss hybrid adversarial function (MLAdv). Then, instead of adding these adversarial attacks to the training data directly, we consider the similarity between the original samples and adversarial attacks by using an adjacent loss during the training process, which can improve the robustness and the generalization of the network for unanticipated noise perturbations. Experiments are finished on COVID-19 dataset. The average attack success rate of this method for three DNN based medical image diagnosis systems is 63.9%, indicating that the created adversarial attack has strong attack transferability and can puzzle the network effectively. In addition, with the adversarial attack training, the augmented networks by using adversarial attacks can improve the diagnosis accuracy by 4.75%. Therefore, the augmented network based on MLAdv adversarial attacks can improve the robustness of medical image diagnosis system.

Keywords: Medical image diagnosis system · Adversarial attack · Multi-loss hybrid function · Heuristic projection · Attack transferability

1 Introduction

Machine learning methods, particularly deep neural networks (DNN), have been considered to be among the most effective methods for medical image analysis. For example, deep convolutional neural networks (CNN) are used to classify chest X-ray images and diagnose pneumonia and COVID-19. However, clean images can be altered with imperceptible perturbations (called adversarial noise) to generate adversarial examples, and such adversarial samples can puzzle CNN classifiers and make incorrect predictions with high confidence. Adversarial attacks have also disturbed the CNN based medical image diagnosis system [1], confirming the high sensitivity of CNN diagnostic systems to adversarial noise. Therefore, there is a high demand for improving the robustness of intelligent diagnostic systems.

Here, taken the COVID-19 diagnosis system as an example, the robustness of this system can be improved by effective adversarial training. Specifically, as shown in Fig. 1(a), the COVID-19 dataset usually has a long-tailed distribution of classification results, under-represented early data for COVID-19 samples and very rapid strain variation making the ReLU network tend to form unbounded decision boundaries, which leaves the network at risk of activation by arbitrary noise [2]. We propose multi-loss function to efficiently generate adversarial samples (as shown in Fig. 1(b)), and with adjacent loss based adversarial training, the activated boundaries can converge to the original boundaries of the samples, thus the robustness is improved.

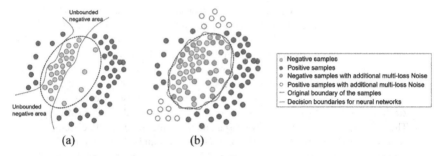

Fig. 1. A conceptual illustration of the proposed training scheme.

2   Related Works

By providing the system with the ability to defend against adversarial attacks, the robustness of the system can be improved. Several defense methods have been proposed. In this section, we briefly review two lines of previous work on defense methods, including adversarial training and denoising methods.

Adversarial Training. An intuitive idea of adversarial training is to ensure that the convolutional layer in the classifier converts all neighboring adversarial samples around each clean original sample to the same points in the semantic feature space as the clean image sample. The popular adversarial training which adds adversarial examples to the training set, can be considered a simplified implementation of this idea. Some researchers have proven that adversarial training and its variants can improve the defense of a system by adding one or more adversarial examples to the training data during classifier training [3–5].

Denoising Method. Denoising based robustness improvement method is to remove the adversarial samples from the dataset by using denoising methods. In general, denoising approaches typically perform preprocessing autoencoders on images with a specific type, aiming to eliminate potential adversarial noise before feeding the image into a classifier

[6, 7]. However, denoising approaches usually reduce the accuracy when classifying the clean original images [6, 7]. For solving this problem, the improved denoising method is proposed, which is to train a distillation network that can improve the defense by effectively expanding the gap between class distributions in the high-level semantic feature space [8].

In this work, we propose a new plug-and-play mechanism to defend against adversarial attacks, thus improving the robustness of the system. Firstly, we introduce a realistic adversarial data generation approach (MLAdv) by using multi-loss hybrid adversarial artifacts. Compared with existing methods, our adversarial samples have stronger attack mobility and appear legitimate to a human observer. Instead of directly putting the adversarial samples into the model for learning, we use an adjacent loss approach to consider the relationship between the adversarial samples and the original samples during the adversarial training. Experimental results validated that our approach can effectively improve the robustness of the model without relying on large amounts of data to train the network.

3 Methods

3.1 Overview

In this paper, we propose a multi-loss hybrid adversarial camouflage (MLAdv) to generate effective adversarial attacks which contain the noise style that can puzzle the network. And we use an adjacent loss approach to retrain the model with the relationship of adversarial samples and the original samples.

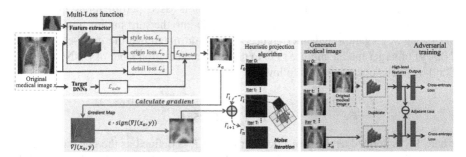

Fig. 2. Overview of the proposed approach.

Problem Formulation
Given an original medical image x with output label y, our MLAdv method finds the effective adversarial sample x_a by solving the following optimization problems:

$$minimize \|x - x_a\|_p + \mathcal{L}_{hybrid}(x, x_a)$$
$$s.t. \ x_a \in [0, 255], D(x, x_a) \leq \varepsilon, \tag{1}$$

where $\mathcal{L}_{hybrid}$ is the multi-loss function, ε means the threshold. With these adversarial samples, the model is retrained to improve the robustness by considering the relationship of adversarial samples and the original samples.

Solutions

Figure 2 shows an overview of our proposed model robustness improvement approach. User defines the clean source image, expected target style. Our proposed MLAdv then generates the adversarial sample with the desired style in the desired region. Firstly, we extract to high-dimensional features by feature extractor and optimize $\mathcal{L}_{hybrid}$ with multi-loss method to generate the adversarial sample x_a. Then, we introduce the heuristic projection-based perturbations iteration to improve the transferability of adversarial sample. We compute the target models' gradient $\nabla J(x_a, y)$ and make gradient map, where $J(x_a, y)$ is the cross-entropy loss. We further add a small perturbation $\varepsilon \cdot sign(\nabla J(x_a, y))$ to the image of each iteration to achieve perturbations iteration. Finally, with these adversarial samples, the model is retrained to improve the robustness by considering the relationship of adversarial samples and the original samples.

Firstly, we develop a multi-loss hybrid function $\mathcal{L}_{hybrid}$ about the style, origin, and detail to craft medical adversarial samples into realistic looking styles. The final multi-loss function is a combination of the adversarial loss $\mathcal{L}_{adv}$, the style loss $\mathcal{L}_s$ for the style generation, the origin loss $\mathcal{L}_o$ to preserve the original features of the source image, and the detail loss $\mathcal{L}_d$ to ensure that the adversarial example is smooth in detail. The definition is as follows:

$$\mathcal{L}_{\text{hybrid}} = (\mathcal{L}_s + \mathcal{L}_o + \mathcal{L}_d) + \tau \cdot \mathcal{L}_{\text{adv}}$$
$$s.t.\ x_a \in [0, 255], D(x, x_a) \le \varepsilon \qquad (2)$$

Secondly, we further use a heuristic projection-based perturbations iteration to project the added noise in the adversarial example into the surrounding area, to ensure the strong aggregation and interference. We should maximize the cross-entropy loss $J(x_t^a, y)$ of target model and use $f(x)$ to denote the prediction label of DNNs to ensure $f(x_t^a) \ne y$. The operation is as follows:

$$x_a' = \text{Clip}(x_a + \varepsilon \cdot sign(\nabla J(x_a, y)) + \gamma \cdot sign(W_o * P))$$
$$s.t.\ x_a, x_a' \in [0, 255] \qquad (3)$$

where W_o is defined as the projection kernel and P is defined as the generated perturbation. We use a heuristic projection algorithm to project those noises that exceed the threshold ε into the surrounding area. This is based on the assumption that pixels that are more likely to exceed the ε threshold will have a higher probability of being in the highlighted areas of the class activation map (CAM) visualization area [11], and that we naturally expand the aggregation of perturbations in these areas by using this projection method.

Finally, instead of adding these adversarial attacks to the training data directly, we consider the similarity between the original samples and adversarial attacks by using an adjacent loss during the training process, which can improve the robustness and the generalization of the network for unanticipated noise perturbations.

3.2 Multi-loss Function for Generating Adversarial Samples

The final multi-loss function is a combination of an adversarial loss $\mathcal{L}_{adv}$ for adversarial strength, a style loss $\mathcal{L}_s$ for style generation, an origin loss $\mathcal{L}_o$ to preserve the content of the original image and a detail loss $\mathcal{L}_d$ to ensure that the adversarial example is smooth in detail.

Style Loss

Image concealment is defined as $||x - x_a||_p$, where $|| \cdot ||_p$ usually uses L_p norm, where L_2 and L_∞ are typically used. For our proposed image generation, style similarity is defined by the style metric between the adversarial example and the style reference image x^s. The style distance between two images can be defined by their differences in the style representation, as follows:

$$\mathcal{L}_s = \sum_{l \in P_l} \left\| \mathcal{G}\left(\tilde{D}_l\left(x^s\right)\right) - \mathcal{G}\left(\tilde{D}_l\left(x_a\right)\right) \right\|_2^2, \tag{4}$$

where $\tilde{D}$ is a feature extractor of deep neural network (DNN), and $\mathcal{G}$ is a Gram matrix [9] of deep features extracted at a set of style layers of $\tilde{D}$. As different style can be learned at different layers, we use all the convolution layers of the network as the style layer.

Origin Loss

The above pattern loss can be used to generate an adversarial image in the reference style, but the content of the adversarial image may be very different from that of the original image. The content of the original image can be saved by the loss of content preservation, as follows:

$$\mathcal{L}_o = \sum_{l \in O_l} \left\| \tilde{D}_l\left(x\right) - \tilde{D}_l\left(x_a\right) \right\|_2^2, \tag{5}$$

where D_l is the set of content layers used to extract the content representation. This is to ensure that the adversarial image has very similar content to the original image in the depth representation space. We use the deeper layer of the feature extractor network as the content layers.

Detail Loss

By reducing the change between adjacent pixels and ensure that the picture is smooth in detail. For adversarial image x_a, the detail loss is defined as:

$$\mathcal{L}_d = \sum \sqrt{\left(x_{ai,j} - x_{i+1,j}\right)^2 + \left(x_{ai,j} - x_{i,j+1}\right)^2}, \tag{6}$$

where $x_{ai,j}$ are pixels at coordinates (i, j) of image x_a. Intuitively, this will encourage the image to have local patches with low variance. Because Sharif et al. [10] pointed out in the paper that the smooth term is useful to improve the robustness of adversarial examples in physical environment.

Adversarial Loss

For adversarial loss $\mathcal{L}_{adv}$, we use the following cross-entropy loss:

$$\mathcal{L}_{adv} = \begin{cases} log\left(p_y(x_a)\right), & untargeted\ category, \\ -log\left(p_{yadv}(x_a)\right) + log\left(p_y(x_a)\right), & targeted\ category \end{cases},\quad (7)$$

where $p_{yadv}()$ is the probability output (SoftMax on Logits) of target model F.

3.3 Heuristic Projection Based Perturbations Iteration

Smooth patches are the basic components of natural and medical images [12], and distinguishable areas are usually concentrated in some of these patches. However, as can be seen in Fig. 3, DNN generally focus on discriminative regions (has been circled in the CAM in Fig. 3), which usually contain clustered pixels instead of scattered ones. Besides, Li *et al.* [13] have demonstrated that regionally homogeneous perturbations are strong in attacking defense models, which is especially helpful to learn transferable

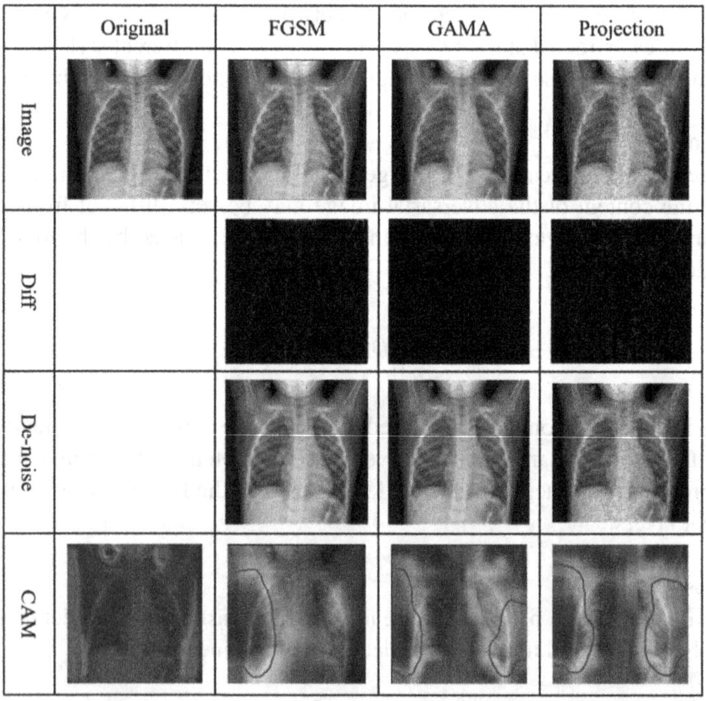

Fig. 3. We show the adversarial examples generated by FGSM, GAMA and our method (Projection based method) for ResNet50 model respectively. The maximum perturbation ε is limited to 10, and the maximum number of iterations is limited to 20. **Top row:** the adversarial image. **Second row:** the adversarial noise. **Third row:** the image after denoising with BM3D. **Botton row:** the Gradient-weighted Class Activation Mapping (Grad-CAM) image. Our MLadv projection method can generate adversarial noise which has the same clustering property as the activation map and also well covers the different discriminative regions.

adversarial examples in the black-box setting. For this reason, we believe that noises perturbations with the characteristic of aggregation in these discriminative regions are more likely to attack successfully because they perturb more significant information. For example, as shown in Fig. 3, we find that the more migratory FGSM (Fast Gradient Sign Method) algorithm has some aggregation characteristics in the noise visualization results by comparing GAMA (Guided Adversarial Margin Attack) and FGSM.

Although FGSM reflects some aggregation, it still has sparse characteristics. From the CAM image, we can see that the discriminative regions tend to be clustered in a few specific parts (the third row in Fig. 3). To solve this problem, after the multi-loss function based adversarial samples generation, we introduce the heuristic projection-based perturbations iteration to enhance the aggregation characteristics of noise perturbations. From the CAM image in Fig. 3, we can see that our noise also has strong aggregation compared to FGSM and GAMA, which is more easily to puzzle the network.

We propose a heuristic projection-based perturbations iteration and the inspiration comes from Rosen Project Gradient Method [14], which method project the gradient direction when the iteration point is on the edge of the feasible region in order to ensure the iteration point remains within the feasible region after updating. However, performing this method is a little complex and needs additional computational cost. Hence, we project those noises that exceed the threshold due to the amplification step to the surrounding area, which results in noises with stronger aggregation. We argue that the part of the noise vector which is easier to break ε-ball limitation has a higher probability of being in the highlighted area of discriminative regions. Our strategy can simply reuse the noise to increase the degree of aggregation in these regions without additional huge computational costs.

We need to get the noise after the iteration, and if $L_\infty - norm$ of x_a exceeds the threshold ε, we cut out the perturbation by:

$$P = clip\left(|x_a| - \varepsilon, 0, \infty\right) \cdot sign(x_a) \tag{8}$$

The final adversarial image with perturbations iteration is then defined as:

$$x_a' = \text{Clip}(x_a + \varepsilon \cdot sign(\nabla J(x_a, \ y)) + \gamma \cdot sign(W_o * P)), \tag{9}$$

where W_o is a special uniform project kernel of size $w \times w$; and $sign(W_o * P)$ is the feasible direction of the cutting perturbation. In this paper, we simply define W_o, as follows:

$$W_o[i, j] = \begin{cases} 0, & i = w/2, j = w/2. \\ 1/\left(w^2 - 1\right), & else. \end{cases} \tag{10}$$

We also test other types of kernels, such as Gaussian kernels. However, the experimental results show that there is no significant difference (only $\sim$1%). Moreover, the uniform kernel does not require additional parameters. Therefore, we finally chose it. As shown in the third row of Fig. 3, compared with the FGSM and GAMA methods, the noise perturbations iteration using the heuristic projection method is difficult to be removed by the BM3D denoising approach. Therefore, heuristic projection-based perturbations iteration can ensure the noises with stronger aggregation, that is not easily removed by the denoising algorithm.

In summary, we provide a multi-loss function to generate adversarial samples close to the decision boundary and ensure that the difference between the adversarial and original samples is not detectable to the naked eye. Then a heuristic projection-based perturbations iteration is used to make the generated noise aggregated and cover multiple discriminative regions to acquire the effective adversarial attack samples. The following figure gives the pseudo-code flow of this method.

Algorithm 1 : Multi-Loss Hybrid Adversarial Function and Heuristic Projection Algorithm to Generate Adversarial Examples

---

Input: The cross-entropy loss function J of target model; iterations T; L_∞ constraint; project kernel W_o; project factor γ; a clean image x (Normalized to [-1,1]) and the corresponding groud-truth label y;

Output: The adversarial example x'_a;

1: **for** $t = 0 \to T$ **do**
2:     Calculate $L_s = \sum_{l \in P_l} \|\mathcal{G}(\widetilde{D}_l(x^s)) - \mathcal{G}(\widetilde{D}_l(x_a))\|_2^2$
3:     Calculate $L_o = \sum_{l \in O_l} \|\widetilde{D}_l(x) - \widetilde{D}_l(x_a)\|_2^2$
4:     Calculate $L_d = \sum \sqrt{(x_{a\,i,j} - x_{i+1,j})^2 + (x_{a\,i,j} - x_{i,j+1})^2}$
5:     Calculate L_{adv} use the cross-entropy loss in equation 7
6:     minimize $(L_s + L_o + L_d + \tau \cdot L_{adv})$ to genarate x_a
7: **end for**
8: Initialize cumulative amplification noise $\mathcal{T}_0$ and cut noise P to 0;
9: $x'_a{}^0 = x_a$
10: **for** $t = 0 \to T$ **do**
11:     Calculate the gradient $\nabla_x J(x'_a{}^t, y)$;
12:     $\mathcal{T}_{t+1} = \mathcal{T}_t + \epsilon \cdot sign(\nabla_x J(x'_a{}^t, y))$;
13:     **if** $\|\mathcal{T}_{t+1}\|_\infty >= \epsilon$ **then**
14:         $P = clip(|\mathcal{T}_{t+1}| - \epsilon, 0, \infty) \cdot sign(\mathcal{T}_{t+1})$
15:         $\mathcal{T}_{t+1} = \mathcal{T}_{t+1} + \gamma \cdot sign(W_o * P)$;
16:     **else**
17:         $P = 0$;
18:     **end if**
19:     $x'_a{}^{t+1} = Clip_{x,\epsilon}\{x'_a{}^t + \epsilon \cdot sign(\nabla_x J(x'_a{}^t, y) + \gamma \cdot sign(W_o * C))\}$;
20:     $x'_a{}^{t+1} = clip(x'_a{}^{t+1}, -1, 1)$;
21: **end for**
22: **return** $x'_a = x'_a{}^T$

3.4   Retrain Model with Adjacent Loss

The adversarial samples all fall in a small neighborhood of the corresponding clean image in the image space, so a classifier additionally trained with the generated adversarial samples in the neighborhood of each training image should be more robust to adversarial attacks, in the sense that all the adversarial samples around the clean image will be projected to the same point in the semantic feature space. In practice, however, it is not feasible to collect all the adversarial samples. By trying to project samples with transferable attacks as a small subset of general adversarial images in the neighborhood to the same point in the semantic feature space as the clean images, because we believe that samples with transferability are more representative samples. In this case, the adversarial samples are more likely to be identified as the same class of clean images, thus improving the robustness of the classifier.

Formally, for the i-th clean image x^i in the original training set, the adversarial samples of the clean image are denoted by $x_a^{i'}$, and $f(x^i)$ and $f(x_a^{i'})$ are the corresponding feature vectors generated by the output of the final convolution layer in the classifier. The objective of transforming the neighboring adversarial samples to the same point in the semantic feature space can then be formulated as an optimization problem, i.e., training the classifier (see Fig. 2) such that the following loss function L_a (called adjacent loss) is minimized.

$$L_a = \frac{1}{N} \sum_{i=0}^{N} \left\| f\left(x^i\right) - f\left(x_a^{i'}\right) \right\| \tag{11}$$

A more robust classifier can be obtained by training the classifier simultaneously with the constraint that the clean and adversarial samples are similar to each other in the semantic feature space, i.e., by minimizing the loss function L,

$$L = L_c + \lambda_a L_a \tag{12}$$

Here L_c denotes the cross-entropy loss of the classifier itself to improve its classification performance on both clean and adversarial images, and λ_a is the hyperparameter to control the relative weight of the loss term L_a. Next, we retrain the model with this method to enhance the generalization of the model.

4   Experiments and Results

In this study, the efficiency of the proposed adversarial image generation method MLAdv was evaluated through an image classification task on COVID-19.

Medical Image Classification Task On COVID-19 Dataset. The public COVID-19 dataset was used to classify covid-19, normal and pneumothorax patient cases. In our experiment, there were 1200 COVID-19 positive images, 1341 normal images and 1345 viral pneuomonia images. The data with COVID-19 cases was collected from different publicly accessible dataset, online sources and published papers [15–17]. Normal and Viral Pneumonia data were collected from the Chest X-Ray Images (pneumonia) database [18].

During the experiment, we first give the implementation details of the experimental setup (Sect. 4.1). Then, we verify the high aggressiveness of our adversarial samples in white-box mode and the attack transferability in black-box mode in comparison with several state-of-the-art attack methods (Sect. 4.2). After that, we use ablation experiments to verify the effectiveness of the multi-loss function for generating adversarial samples (Sect. 4.3). Finally, we verify that retraining the model with adjacent loss by using our adversarial samples can improve the robustness (Sect. 4.4).

4.1 Implementation Details

In order to evaluate how the adversarial attack training method can improve the medical image diagnosis system, the standard medical diagnosis model ResNet50 was trained and tested on two types of datasets: 1) original COVID-19 dataset; 2) the adversarial samples (Several state-of-the-art adversarial sample generation methods are compared, including FGSM, GAMA, RayS (Ray Searching attack) and our method).

The ResNet50 image classification model was implemented based on TensorFlow in GPU. We set the base learning rate to 0.001. The Batch size was set to 8, and the epoch was 30 during training. We adopted the Stochastic Gradient Descent (SGD) optimizer where momentum and weight decay rate were set to 0.9 and 0.0001 respectively. In this experiment, the original data set x was divided into two parts x_{train} and x_{val} with the ratio of 0.85/0.15. In addition, the adversarial image dataset x' were also divided into x'_{train} and x'_{val} with the same ratio of 0.85/0.15. The used original image dataset was x_{train}, and the adversarial image dataset was the combination of x_{train} and x'_{train}. The models were retrained with adjacent loss by using the adversarial samples (Table 1).

4.2 Verify the Attack Transferability of Adversarial Samples

We compare our method (multi-Loss with heuristic projection) with FGSM, GAMA, and Rays to verify the attack transferability of our method. In Table 2, the top row models were substitute models. We use ResNet50 to generate adversarial examples by FGSM, GAMA, Rays and our method. Then we attack ResNet101, Vgg16 and ResNet152 with generated adversarial examples to verify the attack transferability of different attack methods. As shown in Table 2, our Multi-Loss can improve the attack success rate by 28.9% on average compared to other attack methods, and 35.1% when we attacked the ResNet 101. This was because our perturbation has the property of aggregation, and the attack was more transferable.

Table 1. The success rate (%) of non-targeted attacks.

Use ResNet50 to generate adversarial examples					
Attacks	White-box	Black-box			
	ResNet 50	ResNet 101	Vgg16	ResNet 152	Average
FGSM	80.9	38	33.1	33.9	35
GAMA	100	29.6	19.4	20.3	23.1
RayS	99.8	54.1	43.5	50.9	49.5
Our	100	73.1	51.2	67.4	63.9

4.3 Influence of Multi-loss Function

In this section, we will verify that the multi-loss function can achieve adversarial samples with strong imperceptibility and can improve the classification performance by using these adversarial samples. We set up ablation experiments, subtracting a portion of the loss for comparison. Figure 2 illustrated three sets of adversarial examples with or without two optional loss terms (origin loss $\mathcal{L}_o$ and detail loss $\mathcal{L}_d$). When incorporating one loss term, it is directly added to the final object by following Eq. (2). It can be observed that the content loss term can help preserve the original content, while smoothness term loss can help produce smooth object surfaces. Therefore, as shown in Fig. 2(d), the multi-loss function was effective to generate adversarial examples that appear legitimate to human observers. In addition, by adding the heuristic projection-based perturbations iteration, the generated noise was highly aggregated and difficult to be removed (as shown in Fig. 2(e)). Table 2 shows image classification performance comparison by using the adversarial samples which were generated by using different loss function, indicating that the multi-loss function produces more effective adversarial samples (Fig. 4).

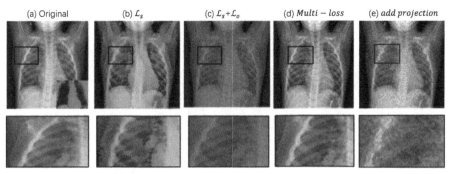

Fig. 4. Visual comparison. The results of generated adversarial examples by using different loss terms.

Table 2. Quantitative comparison. Image classification performance comparison. Evaluation using accuracy, specificity, sensitivity, and precision. The comparison of classification accuracy using partial and full loss and after adding the noise projection algorithm into the source image dataset.

Method	Accuracy	Recall	Precision	F1_score
Source	0.9166	0.9166	0.9267	0.9157
Only $\mathcal{L}_s$	0.9440	0.9440	0.9455	0.9332
$\mathcal{L}_s + \mathcal{L}_o$	0.9465	0.9465	0.9477	0.9464
Multi-loss	0.9615	0.9615	0.9620	0.9615

4.4 Verify the Robustness Enhancement

In this section, we verify that retraining the model using our generated adversarial samples can improve the robustness. On the one hand, we evaluate the robustness enhancement from the classification accuracy after model retraining. Figure 5 and Table 3 represented the comparison of classification performance with different adversarial data generation methods. For the medical classification task on COVID-19, the model which used an adversarial image dataset for training achieved better performance than the model using original data for training. Compared with other methods, our method achieves the best results in the improvement of classification accuracy. The classification accuracy was increased from 93.56% to 97.43% by using adversarial examples. Our experimental results confirmed the proposed method to be effective, and capable of improving the generalization and accuracy of the classification model.

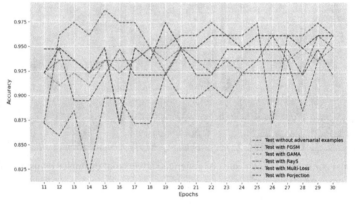

Fig. 5. Image classification accuracy comparison. Several state-of-the-art adversarial sample generation methods are compared, including FGSM, GAMA, RayS, and our methods.

Table 3. Evaluation Image classification performance with different methods, using accuracy, specificity, sensitivity, and precision.

Method	Accuracy	Recall	Precision	F1_score
Source	0.9166	0.9166	0.9267	0.9157
FGSM	0.9589	0.9589	0.9597	0.9590
GAMA	0.9294	0.9294	0.9309	0.9288
RayS	0.9474	0.9474	0.9486	0.9473
Multi-loss	0.9615	0.9615	0.9620	0.9615
Projection	0.9641	0.9641	0.9643	0.9641

On the other hand, we compare the data distribution after model retraining by using our generated adversarial samples method, in order to verify that our method can improve the robustness. In this experiment, the Receiver operating characteristic (ROC) curve comparison and feature layer presentation under t-Stochastic Neighbor Embedding representation were used. ROC curves and their AUCs (Area Under the Curve) is also used to evaluate the performance of classification performance. We use trained models to predict test inputs and generate ROC data, and the ResNet50 trained on the combined image set significantly outperforms the model trained on the original image set.

We use micro and macro averaging to evaluate the overall performance across all classes.

$$precision = PRE = \frac{TP}{TP+FP} \tag{13}$$

In "micro averaging", we'd calculate the performance from the individual true positives, true negatives, false positives, and false negatives of the k-class model (k = 3 in this case):

$$PRE_{micro} = \frac{TP_1+\cdots+TP_k}{TP_1+\cdots+TP_k+FP_1+\cdots+FP_k} \tag{14}$$

And in macro-averaging, we average the performances of each individual class by follow:

$$PRE_{macro} = \frac{PRE_1+\cdots+PRE_k}{k} \tag{15}$$

Figure 6 shows the ROC curves of ResNet50 trained on the original and combined sets. Our combined set improving the generalization and accuracy of classification model because adversarial images make the area under the curve significantly larger for class 2(COVID class). The performance of model trained on combined set outperforms original set with an AUC of 1 for PRE_{micro} and 1 for PRE_{macro}.

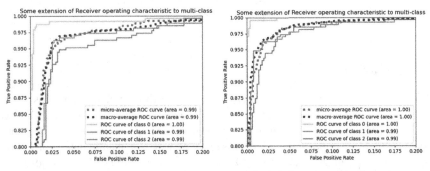

Fig. 6. The performance of ResNet50 trained on the original (left) and combined sets (source and adversarial dataset with MLAdv method) (right) in terms of ROC curves and AUC values. Classification accuracy of the ResNet50 is improved by combing the original dataset with adversarial images with MLAdv method.

We plot the t-Stochastic Neighbor Embedding representation (t-SNE) from ResNet50 for the clean and adversarial examples (MLAdv) in Fig. 7 for the classification task to further illustrate this difference. Contrasting Fig. 7 (Above) with Fig. 7 (Below), we clearly observe that the adversarial examples were embedded closer to the clean data, and adversarial data allows the dataset to produce clearer boundaries. Ultimately, adversarial examples enable the COVID-19 category to be clearly separated out, completely excluding normal and pneumothorax samples. The result of the experiment supports our hypothesis that adversarial examples can improve the robustness of the network and the activated boundaries can converge to the original boundaries of the samples.

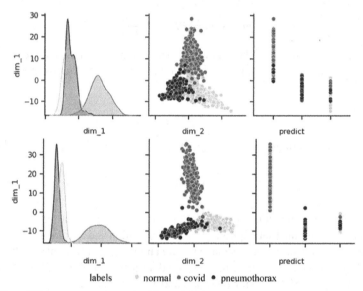

Fig. 7. The t-SNE represents the embedding of 3 classes (purple, pink and black) from covid, normal and pneumothorax. The adversarial examples are embedded closer to the clean data, and adversarial data allows the dataset to produce clearer boundaries.

5   Conclusion

This paper proposes an adversarial attack training method by using multi-loss hybrid adversarial function with heuristic projection. Then we consider the similarity between the original samples and adversarial attacks by using an adjacent loss during the training process, which can improve the robustness of the network for unanticipated noise perturbations. The experiments on COVID-19 classification tasks validated that the adversarial image examples generated by MLAdv can improve the accuracy of disease diagnosis by 4.75%. Therefore, the augmented network based on MLAdv adversarial attacks can improve the robustness and generalization ability of medical image diagnosis system.

Acknowledgement. This work is supported in part by the National Nature Science Foundation of China grants (U20A20389, 61901214).

References

1. Paschali, M., Conjeti, S., Navarro, F., Navab, N.: Generalizability vs. robustness: investigating medical imaging networks using adversarial examples. In: Frangi, A.F., Schnabel, J.A., Davatzikos, C., Alberola-López, C., Fichtinger, G. (eds.) MICCAI 2018. LNCS, vol. 11070, pp. 493–501. Springer, Cham (2018). https://doi.org/10.1007/978-3-030-00928-1_56
2. Hein, M., Andriushchenko, M., Bitterwolf, J.: Why ReLU networks yield high-confidence predictions far away from the training data and how to mitigate the problem. In: Proceedings of the IEEE/CVF Conference on Computer Vision and Pattern Recognition, pp. 41–50 (2019)
3. Goodfellow, I.J., Shlens, J., Szegedy, C.: Explaining and harnessing adversarial examples. In: ICLR (2015)
4. Kurakin, A., Goodfellow, I.J., Bengio, S.: Adversarial machine learning at scale. CoRR: abs/1611.01236 (2016)
5. Tramèr, F., Kurakin, A., Papernot, N., Goodfellow, I., Boneh, D., McDaniel, P.: Ensemble adversarial training: attacks and defenses. arXiv:1705.07204 (2017)
6. Akhtar, N., Liu, J., Mian, A.: Defense against universal adversarial perturbations. In: CVPR, pp. 3389–3398 (2018)
7. Sarki, R., Ahmed, K.: Wang, H: Image preprocessing in classification and identification of diabetic eye diseases. Data Sci. Eng. 6(4), 455–471 (2021)
8. Papernot, N., McDaniel, P., Wu, X., Jha, S., Swami, A.: Distillation as a defense to adversarial perturbations against deep neural networks. In: IEEE Symposium on Security and Privacy, pp. 582–597 (2016)
9. Sarki, R., Ahmed, K., Wang, H., Zhang, Y., Ma, J., Wang, K.: Image preprocessing in classification and identification of diabetic eye diseases. Data Sci. Eng. 6(4), 455–471 (2021). https://doi.org/10.1007/s41019-021-00167-z
10. Sharif, M., Bhagavatula, S., Bauer, L., et al.: Accessorize to a crime: real and stealthy attacks on state-of-the-art face recognition. In: Proceedings of the 2016 ACM SIGSAC Conference on Computer and Communications Security, pp. 1528–1540 (2016)
11. Selvaraju, R.R., Cogswell, M., Das, A., et al.: Grad-CAM: visual explanations from deep networks via gradient-based localization. In: Proceedings of the IEEE international conference on computer vision, pp. 618–626 (2017)
12. Mahendran, A., Vedaldi A.: Understanding deep image representations by inverting them. In: Proceedings of the IEEE Conference on Computer Vision and Pattern Recognition, pp. 5188–5196 (2015)
13. Li, Y., Bai, S., Xie, C., Liao, Z., Shen, X., Yuille, A.: Regional homogeneity: towards learning transferable universal adversarial perturbations against defenses. In: Vedaldi, A., Bischof, H., Brox, T., Frahm, J.-M. (eds.) ECCV 2020. LNCS, vol. 12356, pp. 795–813. Springer, Cham (2020). https://doi.org/10.1007/978-3-030-58621-8_46
14. Rosen, J.B.: The gradient projection method for nonlinear programming. Part I. Linear constraints. J. Soc. Ind. Appl. Math. 8(1), 181–217 (1960)
15. https://github.com/armiro/COVID-CXNet
16. https://github.com/ml-workgroup/covid-19-image-repository/tree/master/png
17. https://sirm.org/category/senza-categoria/covid-19/
18. https://www.kaggle.com/paultimothymooney/chest-xray-pneumonia

SSCG: Spatial Subcluster Clustering Method by Grid-Connection

Yihang Zhang[1], Xuming Han[2], Limin Wang[3(✉)], Weitong Chen[4], and Linliang Guo[5]

[1] School of Computer Science and Engineering, Changchun University of Technology, Changchun, China
[2] College of Information Science and Technology, Jinan University, Guangzhou, China
[3] School of Information, Guangdong University of Finance and Economics, Guangzhou, China
20211016@gdufe.edu.cn
[4] University of Adelaide, Adelaide, SA, Australia
[5] School of Mathematics and Statistics, Changchun University of Technology, Changchun, China

Abstract. Existing clustering methods rely on prior knowledge of the data set to cluster it, so the quality of the clustering effect is entirely dependent on the user's familiarity with the data set. Furthermore, when extracting the information from a data set, existing clustering algorithms frequently ignore the geometric distribution of data, making it difficult to identify data objects in their entirety and detect local spatial structures. To address these issues, this paper proposes a spatial subcluster clustering method by grid-connection, which automatically obtains subclusters by iterative local labeling without requiring a priori knowledge of the data set and efficiently extracts correlations between data by establishing relationships between subclusters by grid-connecting. Experiments are conducted to validate the proposed algorithm against existing state-of-the-art algorithms on 9 synthetic and 4 real data sets. The results show SSCG can efficiently utilize the information on the grid space without relying on a priori knowledge, and the overall performance is better than the existing advanced algorithms.

Keywords: Spatial clustering · Clustering quality · Relabeling · Grid-connection

1 Introduction

Clustering is a typical unsupervised learning method, which can discover intrinsic connections between data on unlabeled data sets [19]. The primary goal of clustering is to divide the data points into clusters according to the similarity so that the similarity of points in the same cluster is as high as possible, and the

---

Y. Zhang and X. Han—Equally contributed to this work

B. Li et al. (Eds.): APWeb-WAIM 2022, LNCS 13422, pp. 430–443, 2023.
https://doi.org/10.1007/978-3-031-25198-6_32

similarity of points in different clusters is as low as possible [14]. Although there have been many studies on clustering methods, there is still no general solution. Two key problems need to be solved:

(1) Over-reliance on a priori knowledge of the data set;
    Most existing clustering methods use the priori knowledge of the data set as a parameter [1,11,16], and the well-known k-means clustering method [10] uses the number of clusters as a priori parameter of the algorithm. However, this kind of priori knowledge conflicts with the definition of unsupervised learning [6] to a certain extent, and the requirement for a priori knowledge makes these algorithms perform poorly in practical clustering tasks.
(2) Less attention is paid to the geometric distribution information of data objects, which resulting in the spatial information of data sets is insufficient utilization.

Clustering essentially constructs the concept of similarity [5,20,21], and most clustering methods strive to capture the similarity by calculating the distance between data objects [13] in the local area. However, this measurement method only focuses on the local spatial information of data objects and ignores their global spatial information in the data set. As a result, these algorithms are ineffective in clustering complex shapes.

The above problems limit the further development of clustering algorithms. For the problems of existing methods, this paper proposes a grid-connected spatial subcluster clustering method (SSCG). Different from other methods, SSCG automatically obtains subclusters through iterative local labeling, which can efficiently generate subclusters without a priori knowledge of the data set. The method also establishes the relationship between subclusters by grid connection, and efficiently extracts the correlation between the data to form the final result. Experiments show that this method can well balance high-density and low-density clusters, and the complex-shaped data sets perform extremely well with lower time complexity.

Our contributions are as follows:

- A density-guided local iterative labeled subcluster generation method is designed to perform iterative local labeling according to the grid density. The method does not require a priori knowledge of the data set and takes well into account regions of different densities and works well for clustering complex shapes. This subcluster generation method has not been done by others within our knowledge.
- A subcluster merging method via grid connectivity is proposed, where the similarity between subclusters is calculated based on the grid path connectivity between subclusters, and the subclusters are merged based on that similarity. This method focuses more on the geometric distribution of the data and can efficiently use the clustering space information for similarity statistics and identification. This merging method has not been done by others within our knowledge.

2   Related Work

2.1   Identify the Main Part of Clusters

Clustering algorithms' first task is to identify the subject area of each cluster. A popular method for grid clustering is to separate dense grids according to a global density threshold, with neighboring dense grids considered as belonging to the same cluster and non-adjacent dense grids considered as not belonging to the same cluster [1, 20]. This identification method ignores the imbalance of the density distribution of the data set. To solve this problem, Bo Wu et al. [18] proposed DGB algorithm, which sets a global percentage parameter and enhances the performance of the algorithm on unbalanced data sets by computing density thresholds for each cluster individually to separate the subjects of each cluster, but the method requires setting parameters such as the number of clusters, edge thresholds, and grid size simultaneously, increasing the need for a priori knowledge and significantly increases the running time due to its topological neighborhood diffusion. Brown et al. [4] proposed the Fast Density-Grid Based Clustering Method, which finds the grids with locally maximum density by density histogram and connects these grids with their nearby low-density grids. In this way, the cluster centers and the labels of each grid are obtained. This method has low time complexity and can also obtain the center region of each cluster accurately, but it requires the number of clusters in the data set and often misclassifies the cluster edge regions. New shifting grid clustering algorithm [9] was proposed by Eden et al. to solve the single grid by moving the grid limitations and reducing the algorithm's need for prior knowledge but also increases the time complexity. The existing methods for identifying the class of cluster subjects mainly focus on using density thresholds to identify dense grids, this method fully considers the nature of density on the grid structure. But the method requires some priori knowledge of the data set and works poorly for imbalanced data sets, and has the problem of unreasonable edge assignment. Different from existing algorithms, we design a density-guided local iterative labeled subcluster generation method, which avoids the problem of prior knowledge parameters in the form of nonparametric iterative initialization, and not only enhances the algorithm's ability to recognize each density region but also reduces the number of parameters.

2.2   Merge Subclusters

An important step in merging subclusters is to determine the similarity between each subcluster. Many inter-cluster similarity measures have been proposed. MKCE algorithm [3] connected the centers of subclusters and counts the points on the path as a density-based inter-subcluster similarity measure. The metric is more reliable, but because it uses full connectivity and borrows the normalized spectral clustering algorithm, its time complexity is high, and it ignores the merging problem of complex manifolds, underutilizes the clustering space, and is not reliable for imbalanced data sets. RSEC algorithm [15] redesigned the inter-cluster similarity association matrix, which enhances the reliability of the co-association matrix, but it requires multiple executions of the clustering algorithm on a data set, has low utilization of the location information of subclusters, and has high time complexity. To address the problem that there is no subcluster similarity measure designed for grid structure, in this paper, we propose a subcluster merging method through grid linkage with full consideration of the features of the grid structure. The method fully takes into account the structural information of the grid space, can efficiently use the grid space and is more accurate in measuring the similarity between subclusters.

3   Methodology

Before presenting the algorithm further, we must first consider the basic definition of SSCG. We define the set of data objects in the L-dimensional data set as $\Pi = \{\pi_i\}_{i=1}^{N}$, which contains N sample points π. The clustering space is mapped to a grid structure $G = \{g_i\}_{i=1}^{M}$ consisting of M standard grids g. The dimensionality of the clustering space $D = \{d_i\}_{i=1}^{L}$. The set of labels on the lattice structure is $L = \{label(g_i)\}_{i=1}^{M}$. For the relevant base definitions that appear later, we state here uniformly.

3.1   Preliminaries

Before presenting the algorithm further, we must first consider the basic definition of SSCG. We define the set of data objects in the data set as $\Pi = \{\pi_i\}_{i=1}^{N}$, which contains N sample points π. The clustering space is mapped to a grid structure $G = \{g_i\}_{i=1}^{M}$ consisting of M standard grids g. The dimensionality of the clustering space $D = \{d_i\}_{i=1}^{D}$. For the relevant base definitions that appear later, we state uniformly here.

Definition 1. *(Density of grids) The density of grid g_i is defined as the number of data objects within it, as shown in Eq. (1)*

$$\rho_{g_i} = \underset{\pi \in g_i}{num}(\pi) \tag{1}$$

In Eq. (1), $\rho = \{\rho_{g_i}\}_{i=1}^{M}$ is the density set of M grids, num is a function that counts the number of p.

Definition 2. *(Grid distance) The distance between grid g_i and g_j in the n-th dimension is defined as Eq. (2)*

$$Dis_{d_n}(g_i, g_j) = \frac{Dc_{d_n}(g_i, g_j)}{l} \tag{2}$$

where $Dc_{d_n}(g_i, g_j)$ is the distance between the centers of grid g_i and g_j in the n-th dimension. l is the width of the standard grid on the n-th dimension.

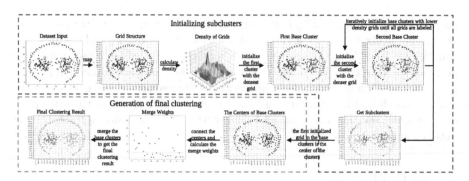

Fig. 1. Graphical illustration of principle demonstration of SSCG.

Definition 3. *(Set of grids with density ρ_i) The set G_ρ of all standard grids in the grid structure whose density is ρ is defined as Eq.*

$$G_{\rho_i} = \{g|num(g) = \rho\} \qquad (3)$$

Definition 4. *(Set of g_i's neighbors)A point's nearest neighbor is the point whose distance from the point is less than or equal to 1 in all dimensions, defined as Eq.*

$$Nen(g_i) = \{g_j|Dis_d(g_i, g_j) <= 1 \cap g_j \neq g_i, d = 1, 2, ..., D\} \qquad (4)$$

3.2 Initializing Subclusters

First, the SSCG algorithm maps the clustering space onto a standard grid structure. This part is specified by dividing each dimension on the clustering space with equal length l, which in turn obtains a grid structure G on the clustering space, as shown in Grid Structure in Fig. 1. Each dimensional edge length of each grid within this structure is equal, and this grid is called **standard grid**. The standard grid avoids the adverse effect of different scales in each dimension on accuracy. Each data object point within the clustering space is attributed to the standard grid within this structure. Based on the distribution of the grid density value ρ_g in the space, the algorithm can distinguish the core part of a cluster from its edges, and the intersection region between two clusters.

Unlike other methods, the dimensions of the cluster space are not normalized in SSCG, which is to make the spatial information of each dimension within the cluster space as complete as possible. The density value of each grid is obtained by performing Eq. (1) for each grid. It is important to note that the computation of this step is highly parallel.

It should be noted that after this step all empty meshes are discarded and only meshes with density greater than or equal to zero are kept in memory. The meshes operated on below are non-empty, which allows us to avoid the memory overflow problem that occurs with mesh clustering in high-dimensional spaces.

After obtaining the standard grid structure and the density of each grid, we can cluster on the grid structure to obtain subclusters. We designed **density-guided local iterative labeled subcluster generation method(DGL)** based on the features of the standard grid structure.

DGL: First, find the grid with the highest density in grid structure and label them as the first cluster and pre-label their neighborhood grid; Second, find grids with lower density in grid structure, label them with their pre-label if it already has pre-label, and pre-label their neighborhood as their pre-lable, and if not pre-labeled, then treat them as the center of a new subcluster and label them with their new label; Third, repeat the second step until all the lattices are labeled. For a detailed description of this method see Algorithm 1.

Algorithm 1. Density-guided local iterative labeled subcluster generation method (DGL) and Edge grid reassignment (EGR)

Input: ρ(density set for individual grids);
Output: L(lebels of base clusters) and C(centers of base clusters).
 1: Initialization: l=0, $C = \emptyset$;
 2: **MGB:**
 3: **for** ρ_i in ρ: **do**
 4:     **for** g_j in G_{ρ_i}, and $Label(g_j)$ does not exists **do**
 5:         **if** $preLabel(g_j)$ dose not exists **then**
 6:             Label g_j with l and pre-label its neighborhood as l;
 7:             Regard g_j is the center grid of a new subcluster, C.append(g_j) and $l + +$;
 8:         **else**
 9:             Label g_j with its pre-label and pre-label its neighborhood as the pre-lable of g_j;
10:         **end for**
11:     **end if**
12: **end for**
13: **EGR:**
14: **for** Edge grid g_k in Eg_i **do**
15:     get the maximal number of data points with the same cluster label $label_m$ in g_k and $Nen(g_k)$;
16:     $label(g_k) = label_m$;
17: **end for**
18: **return** L and C.

After DGL, we get the labels of the initial subcluster. To make the edge delineation of the subclusters more reasonable and improve the confidence of the subclusters, **edge grid reassignment(EGR)** is proposed to reassign the edges of the subclusters. **Edge grid** is defined as a grid adjacent to other marked grids as follows.

$$Eg = \{g_j | label(g_j) \neq label(Nen(g_j))\} \tag{5}$$

EGR: First, the labels and densities in both the edge grid and its neighbors are counted, and the labels are reassignment based on the statistical information; Then, the above operation is performed for all edge grids. The edge grid reassignment makes the edge assignment of each subcluster more reasonable and is beneficial to improve the reliability of the initialized generated subclusters. For a detailed description of the method, see Alg. 1.

3.3   Construction of Cluster Relations and Generation of Final Clustering

After obtaining the final subcluster partition, we need to merge the subclusters to obtain the final result. **Subcluster merging method by path-finding(SMM)** is proposed to merge the subclusters to obtain the final result. Before detailing its method, we need to make some definitions.

Let the clustering centers of the adjacent clusters Bc_i and Bc_j be C_i and C_j, respectively, then the parameter of the Beam Path Search Algorithm, Beam width is the number of neighborhood points of a point, it is related to the dimensionality of the data set. The objective function for optimization within each time step is shown in Eq. (6)

$$PathWeight(G_C, G_T) = Dis(G_C, C_j) + Dis(G_C, G_T) \qquad (6)$$

In Eq. (6), G_C is the current grid, G_T is the target grid, and Dis refers to the grid-based Manhattan distance. At each time step, Beam Path Search Algorithm tries to find the target grid in the neighborhood of the current grid that makes the objective function value, i.e., $PathWeight$ in Eq. (6), the smallest, and takes this optimal target grid as the new current grid. The algorithm records all current grids passed as connected paths until the current grid coincides with the end grid.

After getting the connected paths, we can calculate the merge weights of Bc_i and Bc_j based on the information on the paths, which is defined as shown in Eq. (7)

$$Mew(Bc_i, Bc_j) = len(Path_{ij}) * std(Path_{ij}) \qquad (7)$$

where $len(Path_{ij})$ is the length of the path $Path_{ij}$, and $std(Path_{ij})$ is the standard deviation of the grid density on the path $Path_{ij}$.

SMM: First, find all base clusters that are neighbors of each other, connect their clustering centers using the beam path search algorithm, and record the paths; second, the merge weights are computed from the statistical information on the paths using Eq. (7); third, compute the merging threshold using the merging parameter γ and perform merging; fourth, the labels of the grid are mapped to the data objects within the grid to obtain the final clustering results. The SMM is detailed in Algorithm 2.

Algorithm 2. Subcluster merging method by path-finding(SMM)

Input: L(lebels of base clusters), C(centers of base clusters) and γ(merge parameter);
Output: L_F(final labels for individual data points).

1: **for** Bc_i and Bc_j in set of pairwise neighbors **do**
2:    Connect the centers C_i and C_j of base clusters Bc_i and Bc_j using the Beam Path Search Algorithm, and record the path $Path_{ij}$;
3:    Calculate the length $len(Path_{ij})$ of $Path_{ij}$ and variance $std(Path_{ij})$ of all grid densities on $Path_{ij}$;
4:    The merge weight between Bc_i and Bc_j is calculated by Eq. (7);
5: **end for**
6: Merge all pairwise clusters with weights below $\gamma\%$ of the set of combined weights;
7: Map the labels on the grid to the data objects in the grid to get the final label set L_F of the data objects;
8: **return** L_F.

---

4   Experience Results and Analysis

4.1   Data Sets and Compared Methods

In this paper, we selected 9 well-known synthetic data sets containing several clusters of complex shapes with different densities and non-spherical surfaces, separating these clusters would be a challenging clustering task. Furthermore, we also selected 5 well-known UCI real data sets to verify the algorithm's ability to handle real data, including clustering tasks for two large-scale data sets: (1) Wireless sensor network information clustering for predicting the pattern of user movements in real-world office environments from time-series, the applied data set is Indoor User Movement Prediction from RSS Data Set (RSS) [2]; (2) Urban road accidents data clustering, which is used to delineate different geographic areas based on traffic accident data, the applied data set is urbanGB, a huge real data set including more than 300,000 data objects. Table 1 shows the details of the test data sets.

On these data sets, we compare SSCG with the k-means-based clustering ensemble algorithm RSEC [15], the grid clustering algorithm MSGC [7], the DGB [18] algorithm, and the CLIQUE [1] algorithms by quantitatively evaluating and contrasting effectiveness and time costs. We introduce three widely used external criteria ARI [8], NMI [17] and FMI [12] to measure the similarity between clustering results and true labels on the data set, and compute the ARI, NMI, and FMI of the clustering results using Python's Sklearn module. The larger the ARI, NMI, and FMI, the better the clustering performance. All the experiments are implemented based on the same hardware: Windows 10 64 bit operating system with Intel Core i5 (I5-9300) @2.4 GHz 8.00 GB memory. SSCG and CLIQUE are run under a Python3.60 programming environment, the rest of the algorithm is run in MATLAB (2019b). For each algorithm compared, we adjusted according to the parameter settings suggested by the author, and recorded the best performance of the algorithm under each parameter, and the parts that did not converge and could not be run were represented by the '-' flag.

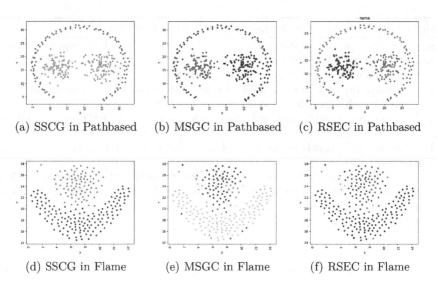

(a) SSCG in Pathbased   (b) MSGC in Pathbased   (c) RSEC in Pathbased

(d) SSCG in Flame   (e) MSGC in Flame   (f) RSEC in Flame

Fig. 2. Comparison of clustering results of SSCG, MSGC, and RSEC on two synthetic data sets.

Table 1. Description of data sets: Number of data objects (Num), Number of dimensions (Dim), Number of clusters (Clu).

Data set	Num	Dim	Clu
Path-based	300	2	3
Flame	240	2	3
Aggreation	788	2	7
Jain	373	2	2
Sym	350	2	3
Spiral	994	2	3
VaryDensity	135	2	3
R15	600	2	15
D31	3100	2	31
Seed	210	4	3
Iris	150	4	3
RSS	13197	4	2
UrbanGB	360177	2	470

4.2   Experimental Results

The clustering results of SSCG, MSGC, and RSEC in Path-based data set and Flame data set are shown in Fig. 2. And the ARI, NMI, and FMI of the clustering results of each algorithm are shown in Table 2, Table 3 and Table 4. SSCG obtained the best results in second-best on all data sets except some data sets such as VaryDensity and Jain. After comparison, it can be seen that SSCG is superior to existing algorithms in both accuracy and adaptability. Regarding the run time, as displayed in Table 4, SSCG is slower than CLIQUE and RSEC in some synthetic data sets, but SSCG is the fastest or second fastest on low-dimensional real data sets and most synthetic data sets, especially on the large data set UrbanGB, the running time is extremely short.

4.3   Complexity Analysis

In this section, we will analyze the time complexity of SSCG and assume the number of grids is G. The time complexity of SSCG mainly depends on the following parts: (1) Mapping the original data to the grid structure, due to the highly parallel algorithm, the time complexity of this part is $O(1)$; (2) Initialization Base cluster, the time complexity of this part is $O(G\sum_{g_i \in G}^{G} num(Nen(g_i)))$, because of the proposed mechanism of skipping labeled grids and $num(Nen(g_i)) < G$, the time complexity of this part of the algorithm is less than $O(G^2)$; (3) Connect the cluster centers and merge the clusters, the time complexity of this part is $O(len(Path))$, $len(Path)$ is the grid number of all connection paths, $O(len(Path)) < G$. To sum up, the total time complexity of SSCG is $O(G\sum_{g_i \in G}^{G} num(Nen(g_i)))$.

Table 2. Comparison of the performance of SSCG with other four algorithms, based on ARI, on 13 data sets, the best results are boldfaced.

Data set	SSCG	MSGC	DGB	CLIQUE	RSEC
Path-based	**0.9600**	0.6062	0.9162	0.6746	0.7020
Flame	**0.9833**	0.7515	0.9170	0.4390	0.9441
Aggreation	**1.0000**	0.8085	0.8085	0.8689	0.9725
Jain	0.9887	0.9193	0.9971	0.9024	**1.0000**
Sym	**0.8510**	0.7802	0.7606	0.7764	0.7429
Spiral	**1.0000**	**1.0000**	0.5127	0.9149	0.7502
VaryDensity	0.8946	0.5096	**0.9066**	0.7435	0.8137
R15	**0.9857**	0.9039	0.9302	0.9429	0.7744
D31	**0.9352**	0.6321	0.7364	0.7369	–
Seed	**0.6791**	–	–	0.5727	0.6279
Iris	**0.8842**	–	–	0.5681	0.5486
RSS	**0.0104**	–	–	0.0020	–
UrbanGB	**0.8732**	–	–	0.0515	–

Table 3. Comparison of the performance of SSCG with other four algorithms, based on NMI, on 13 data sets, the best results are boldfaced.

Data set	SSCG	MSGC	DGB	CLIQUE	RSEC
Path-based	**0.9482**	0.6171	0.8752	0.6729	0.7408
Flame	**0.9348**	0.6280	0.8299	0.4355	0.9100
Aggreation	**1.0000**	0.8881	0.9333	0.8799	0.9743
Jain	0.9713	0.7227	0.9871	0.7393	**1.0000**
Sym	0.8374	0.6919	0.7689	0.7969	0.7765
Spiral	**1.0000**	**1.0000**	0.6705	0.8837	0.7990
VaryDensity	**0.8761**	0.5118	0.8474	0.7101	0.7999
R15	**0.9885**	0.9291	0.9383	0.9550	0.9306
D31	**0.9540**	0.8453	0.8577	0.8433	–
Seed	**0.6735**	–	–	0.5478	0.6556
Iris	**0.8500**	–	–	0.7316	0.6696
RSS	**0.0281**	–	–	0.0011	–
UrbanGB	**0.8967**	–	–	0.3313	–

Table 4. Comparison of the performance of SSCG with other four algorithms, based on FMI, on 13 data sets, the best results are boldfaced.

Data set	SSCG	MSGC	DGB	CLIQUE	RSEC
Path-based	**0.9733**	0.7255	0.9437	0.7785	0.8019
Flame	**0.9867**	0.8750	0.9610	0.7215	0.9817
Aggreation	**1.0000**	0.8650	0.9856	0.8799	0.9802
Jain	0.9956	0.9626	0.9989	0.9607	**1.0000**
Sym	**0.9010**	0.7585	0.8453	0.8567	0.8376
Spiral	**1.0000**	**1.0000**	0.7159	0.9564	0.8763
VaryDensity	**0.9295**	0.6578	0.8810	0.8283	0.8660
R15	**0.9866**	0.9133	0.9351	0.9471	0.8001
D31	**0.9373**	0.8453	0.7451	0.7464	–
Seed	**0.7912**	–	–	0.7009	0.7583
Iris	**0.9216**	–	–	0.7715	0.7451
RSS	0.5047	–	–	**0.7104**	–
UrbanGB	**0.8829**	–	–	0.3019	–

Table 5. The running time of SSCG and the other four algorithms on 13 data sets, in seconds, the fastest results are boldfaced.

Data set	SSCG	MSGC	DGB	CLIQUE	RSEC
Path-based	0.0408	0.8876	0.1289	**0.0351**	3.3802
Flame	**0.0360**	0.8099	0.1221	0.0393	2.9640
Aggregation	0.0988	1.8806	0.0839	**0.0423**	13.2845
Jain	**0.0420**	1.0287	0.1346	0.0440	5.4360
Sym	0.1738	0.9290	0.0840	**0.0447**	4.2603
Spiral	0.2156	0.9676	0.1852	**0.0485**	22.7274
VaryDensity	**0.0214**	0.6715	0.1072	0.0239	2.0152
R15	**0.0051**	1.3837	0.2800	0.0797	7.0452
D31	**0.0358**	5.3094	1.2738	0.3610	–
Seed	6.9074	–	–	**5.2672**	6.8541
Iris	0.8902	–	–	**0.4115**	11.6130
RSS	9.7260	–	–	**8.9627**	–
UrbanGB	**5.9348**	–	–	14.22442	–

5   Conclusion and Outlook

In this paper, we propose a spatial subcluster clustering method by grid-connection (SSCG), which automatically obtains subclusters by iterative local labeling without pre-set density thresholds, reduces the number of parameters to be adjusted during clustering, and takes good care of regions with different densities within the data set. In addition, SSCG establishes the relationship between subclusters by connecting the neighboring subclusters and derives a similar relationship between each cluster according to the linkage path, which extracts the correlation between data efficiently and makes better use of spatial information.

We compare the SSCG algorithm with several similar advanced algorithms on synthetic and real data sets to verify the generalizability and effectiveness of the SSCG method. The experimental results show that SSCG has high accuracy and robustness for data of different shapes and densities, performs well on data sets of different shapes, and outperforms other advanced algorithms in general.

Due to the nature of the grid structure, SSCG still has some limitations, such as: (1) depends on the choice of parameters; (2) poor performance for ultra-high dimensional data. In future research, we will investigate clustering algorithms that are more insensitive to parameters. In addition, we plan to propose a clustering integration algorithm for high-dimensional data.

References

1. Agrawal, R., Gehrke, J., Gunopulos, D., Raghavan, P.: Automatic subspace clustering of high dimensional data for data mining applications. In: Proceedings of the 1998 ACM SIGMOD international conference on Management of data, pp. 94–105 (1998)
2. Bacciu, D., Barsocchi, P., Chessa, S., Gallicchio, C., Micheli, A.: An experimental characterization of reservoir computing in ambient assisted living applications. Neural Comput. Appl. **24**(6), 1451–1464 (2014)
3. Bai, L., Liang, J., Cao, F.: A multiple k-means clustering ensemble algorithm to find nonlinearly separable clusters. Inform. Fusion **61**, 36–47 (2020)
4. Brown, D., Japa, A., Shi, Y.: A fast density-grid based clustering method. In: 2019 IEEE 9th Annual Computing and Communication Workshop and Conference (CCWC), pp. 0048–0054. IEEE (2019)
5. Du, G., Zhou, L., Yang, Y., Lü, K., Wang, L.: Deep multiple auto-encoder-based multi-view clustering. Data Sci. Eng. **6**(3), 323–338 (2021)
6. Ghahramani, Z.: Unsupervised learning. In: Bousquet, O., von Luxburg, U., Rätsch, G. (eds.) ML -2003. LNCS (LNAI), vol. 3176, pp. 72–112. Springer, Heidelberg (2004). https://doi.org/10.1007/978-3-540-28650-9_5
7. Gui, Z., Peng, D., Wu, H., Long, X.: Msgc: multi-scale grid clustering by fusing analytical granularity and visual cognition for detecting hierarchical spatial patterns. Future Gener. Comput. Syst. **112**, 1038–1056 (2020)
8. Li, H., Liu, X., Li, T., Gan, R.: A novel density-based clustering algorithm using nearest neighbor graph. Patt. Recogn. **102**, 107206 (2020)
9. Ma, E.W., Chow, T.W.: A new shifting grid clustering algorithm. Patt. Recogn. **37**(3), 503–514 (2004)
10. MacQueen, J., et al.: Some methods for classification and analysis of multivariate observations. In: Proceedings of the Fifth Berkeley Symposium on Mathematical Statistics and Probability. vol. 1, pp. 281–297. Oakland, CA, USA (1967)
11. Mautz, D., Plant, C., Böhm, C.: Deepect: the deep embedded cluster tree. Data Sci. Eng. **5**(4), 419–432 (2020)
12. Powers, D.M.: Evaluation: from precision, recall and f-measure to roc, informedness, markedness and correlation. arXiv preprint arXiv:2010.16061 (2020)
13. Sarfraz, S., Sharma, V., Stiefelhagen, R.: Efficient parameter-free clustering using first neighbor relations. In: Proceedings of the IEEE/CVF Conference on Computer Vision and Pattern Recognition, pp. 8934–8943 (2019)
14. Sarle, W.S.: Algorithms for clustering data (1990)
15. Tao, Z., Liu, H., Li, S., Fu, Y.: Robust spectral ensemble clustering. In: Proceedings of the 25th ACM International on Conference on Information and Knowledge Management, pp. 367–376 (2016)
16. Tu, L., Chen, Y.: Stream data clustering based on grid density and attraction. ACM Trans. Knowl. Disc. Data (TKDD) **3**(3), 1–27 (2009)
17. Vinh, N.X., Epps, J., Bailey, J.: Information theoretic measures for clusterings comparison: Variants, properties, normalization and correction for chance. J. Mach. Learn. Res. **11**, 2837–2854 (2010)
18. Wu, B., Wilamowski, B.M.: A fast density and grid based clustering method for data with arbitrary shapes and noise. IEEE Trans. Indust. Inform. **13**(4), 1620–1628 (2016)
19. Yue, L., Zuo, W., Peng, T., Wang, Y., Han, X.: A fuzzy document clustering approach based on domain-specified ontology. Data Knowl. Eng. **100**, 148–166 (2015)

20. Zarikas, V., Poulopoulos, S.G., Gareiou, Z., Zervas, E.: Clustering analysis of countries using the covid-19 cases dataset. Data Brief **31**, 105787 (2020)
21. Zhu, Q., Pei, J., Liu, X., Zhou, Z.: Analyzing commercial aircraft fuel consumption during descent: a case study using an improved k-means clustering algorithm. J. Cleaner Prod. **223**, 869–882 (2019)

Shap-PreBiNT: A Sentiment Analysis Model Based on Optimized Transformer

Kejun Zhang[1], Liwen Feng[1]([✉]), and Xinying Yu[2]

[1] Beijing Electronic Science and Technology Institute, Beijing, China
fengliwen101@126.com
[2] Beijing University of Posts and Telecommunications, Beijing, China

Abstract. Confront the problems of the semantic representation, classification accuracy and efficiency of text sentiment analysis methods need to be optimized, this paper proposes an optimized text sentiment classification model (Shap-PreBiNT) based on Transformer. In the text embedding stage, a Shap-Word model is devised to improve the ability of text semantic representation. The Shapley-value method is introduced to calculate the contribution weight of words in sentences, which fused with Word2Vec word vector. In the feature extraction stage, a bidirectional normalization layer is designed to regulate the feature distribution from multi-dimension. In the stage of network structure optimization, a pre-normalization structure is adopted to stabilize the gradient norm and accelerate the convergence rate. The experimental results demonstrate that the proposed model has a better performance by comparing with other related models. On the IMDB English dataset, the classification accuracy and F1-score reach 94.87 and 94.83%, which are 1.48% and 1.47% higher than Transformer. On the ChnSentiCorp Chinese dataset, our model achieves the highest accuracy and F1-Score, which are 91.82% and 91.66%, respectively, and increased by 2.43% and 2.51%.

Keywords: Text sentiment classification · Shapley-value · Word vector optimization · Gradient normalization

1 Introduction

Text sentiment analysis is a research hotspot in natural language processing in recent years. Its task is to analyze, process and induce texts with personal subjectivity, and it is the division of authors' tendencies, opinions and attitudes [12]. Sentiment analysis is widely used in topics inference, public opinion monitoring, information prediction and other aspects [1]. It provides important decision-making basis for governments, enterprises and other institutions by identifying sentiment trends and change rules. In fact, the application of text sentiment analysis has spread from computer science to management science and social science. Research on sentiment analysis will not only improve natural language processing, but also help in information-related fields such as marketing, finance, political science, and history.

© The Author(s), under exclusive license to Springer Nature Switzerland AG 2023
B. Li et al. (Eds.): APWeb-WAIM 2022, LNCS 13422, pp. 444–458, 2023.
https://doi.org/10.1007/978-3-031-25198-6_33

With the support of large-scale corpus, feed forward neural network structure based on deep learning technology has become an effective research method in the field of sentiment analysis. In 2017, Transformer was born, which relies on multi-head self-attention to realize the encoder. It has become a new milestone in the research field of NLP and entered a new stage in various application fields including sentiment analysis. Although Transformer is outstanding in text sentiment analysis tasks, it still has some problems. Firstly, the network structure using self-attention is not sequential and the word order relative position information is lost. Secondly, the multi-attention with residual connections and the structure of the feed forward neural network may also lead to instability in the initial training stage of the model. Thirdly, it has the problem of insufficient semantic representation, and optimization of classification accuracy and efficiency is also a challenge.

To solve the above problems, we proposes an optimized text sentiment classification model Shap-PreBiNT, which has made improvements in word embedding, gradient normalization and network structure. The main contributions of this paper can be summarized as follows:

- Shap-Word model. We design a word vector model combining Shapley-value method with Word2Vec, which use Shapley algorithm in game theory to calculate the contribution weight of words in the whole text. This model can not only represent the local relationships between words, but also pay attention to the whole information of the text.
- Bidirectional normalization. We design a normalization layer consists of a batch normalization sub-layer and a layer normalization sub-layer. The former normalizes data distribution for a single statement from the vertical dimension, and the later normalizes data distribution for different features from the horizontal dimension. This method can avoid the problems of gradient disappearing or gradient explosion and accelerate the training convergence speed.
- Pre-normalization network structure. The gradient normalization is placed before the multi-head attention with residual connection and the feed forward neural network. So that the gradient value do not increase with the deepening of the network layers, which can reduce the parameter sensitivity in the pretraining stage and make the gradient norm of each layer remain stable.

The rest of this paper is organized as follows: Section 2 briefly introduces the relevant work of the text sentiment classification; Secition 3 describes the model structure of Shap-PreBiNT and its algorithm principle in details; Sect. 4 is the comparison of the experimental model and results from analysis; At last, we draw a conclusion in Sect. 5.

2   Related Work

Deep learning-based methods can actively learn the underlying semantic features of texts. It shows unique advantages in large-scale corpus and has good flexibility and robustness. In early deep learning classification models, convolutional neural network (CNN) or recurrent neural network (RNN) structures are used for feature extraction. The multi-layer perceptron has good learning ability for complex, high and nonlinear syntactic information. Kim [11] proposed TextCNN model to complete text feature extraction through convolution window sliding. Wang et al. [23] proposed a capsule network model combining CNN and Bidirectional Gated Recurrent unit (Bi-GRU), and added dynamic routing strategy to improve the semantic understanding of the model. The Google Mind team [15] first used the attention mechanism on RNN to classify image. Shen [17] proposed a neural network model combining RNN and ONLSTM. The introduction of long short term memory network [8] (LSTM) and gate recurrent unit [4] (GRU) can alleviate the gradient instability in training to a certain extent. However, it still cannot solve the problem of low efficiency in large-scale computation of RNN structure. As for CNN, it is unable to capture long-distance features due to the limitation of the convolution window size. So it can only optimize the model by superimposing convolution layer depth and adding auxiliary networks [9].

Attention mechanism introduced from the field of image processing has been widely used in deep learning models in recent years [2]. It is worth noting that Transformer [21], which does not rely on traditional neural network structure and only adopts multi-head attention to build the whole encoder, has excellent performance in various sequence modeling. The model generates the semantic representation by calculating different positions in the correlation sequence, and has the advantages of capturing long distance dependency of text and paralleling computation [22]. Google Brain combined the advantages of RNN sequence modeling and multi-attention, and proposed the Transformer-XL model [5], which adopted fragment level cyclic recursion mechanism and relative positional embedding method to improve the long-distance dependence information capability of the model.

3   Shap-PreBiNT Model

In order to improve the shortcomings of the original Transformer, the Shap-PreBiNT model proposed in this paper aims to provide Transformer with high quality word embedding as input, improve its gradient normalization and optimize the network structure to achieve better performance in sentiment classification tasks. The architecture of the model is displayed in Fig. 1. It consists of three parts: Shap-Word word vector model, positional embedding and encoder based on bidirectional normalization with the pre-normalization structure.

The Shap-Word model completes the text embedding and the idea is to fuse the vector matrix generated by Shapley and Word2vec. The positional embedding module utilizes the sinusoidal absolute positional embedding to record the position information [24]. The encoder based on bidirectional normalization mainly includes four modules: 1) The multi-head attention module divides data into multiple heads that pay attention to information in different spaces, which can enhance the ability of capturing feature information. 2) The feed forward neural network module consists of two fully connected layers, using the

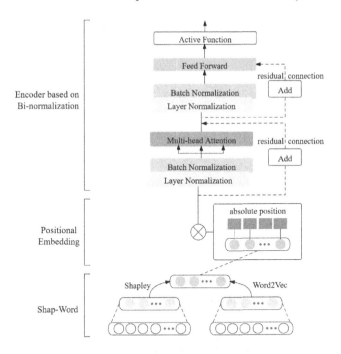

Fig. 1. The architecture of Shap-PreBiNT

ReLU activation function. 3) The Add module adds the output of the positional embedding module and the output of the multi-head attention module, which can reduce the loss of information transmission between the network layers and prevent the problem of gradient disappearing. 4) The bidirectional normalization module is composed of a batch normalization sub-layer and a layer normalization sub-layer, which normalizes the data distribution from the horizontal and vertical aspects of the feature matrix respectively, so as to optimize the feature extraction process and improve the convergence efficiency.

3.1 Shap-Word Model

As a basic task of sentiment analysis, the essence of word vector technology is to transform natural language text into dense vector representations and facilitate the mining of features between words and sentences.

Word2Vec [14] is a word vector model to measure the similarity between words. Although it can complete the semantic embedding of words, it contains limited information and can only perform limited tasks such as semantic similarity calculation, while ignoring the global statistical information of words.

Shapley-value method [10] is a mathematical method to solve the problem of n individuals cooperative game in game theory. Each person in the cooperation can reasonably distribute the benefits according to his or her contribution. In the field of machine learning, this method is generally used to interpret the output of the model.

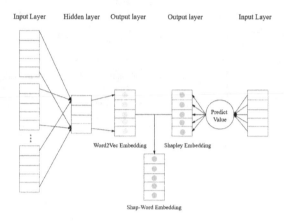

Fig. 2. The architecture of Shap-Word model

In order to improve the expression effect of word vector, this paper proposes a fused word vector model Shap-Word, as shown in Fig. 2. It uses Shapley value algorithm to calculate the contribution weight matrix of words in the whole text, and integrates it with Word2Vec word vector matrix to fully capture global statistics and local context relevance while keeping a small training cost.

Firstly, the Word2Vec model in CBOW mode is used to generate the initial vector matrix of the text. In the first step, the context of a word in the text is expressed as a sequence $X = [x_1, x_2, \ldots, x_c]^T$, where c represents the number of words and x_i represents the i-th word; The second step is to multiply the sequence X by the input weight matrix W, add and average the obtained vectors as the hidden layer vector h_i, and then multiply by the output weight matrix W'. The third step is to adjust the probability distribution using the activation function.

Secondly, use the Shapley-value method to capture the contribution value of words and generate the vector matrix of text. The text is represented as a sequence $T = [t_1, t_2, \ldots, t_c]^T$. If there is a real-valued function $v(S)$ means the sum of the contribution values of each word corresponding to any subset S and if the following two conditions are met:

The real-valued function of the empty set is 0:$v(\emptyset) = 0$;

The real-valued function of the union of two disjoint subsets is greater than or equal to the sum of the real-valued functions of the two subsets:

$$v(S_i \cup S_j) = v(S_i) + v(S_j), S_i \cap S_j = \emptyset \tag{1}$$

Then, $[T, v]$ is the multi-player cooperative game and v is the characteristic function of the game. The marginal contribution of word i to the classification accuracy in the word subset S is:

$$\Delta_i S = v(S \cup i) - v(S) \tag{2}$$

And the Shapley value of feature i is:

$$\gamma_i(v) = \frac{1}{n!} \sum_{\pi \in \Pi} \Delta_i(S_i(\pi)) \tag{3}$$

where Π is the set of feature subsets, $S_i(\pi)$ is all of the feature subsets in Π that contain feature i.

Finally, the vector matrix generated by Word2vec is fused with the Shapley-value vector matrix to generate a new vector matrix. The fusion mode is described by the following formula:

$$Z_i = (X_i + T_i) \in R^m \tag{4}$$

3.2   Multi-head Attention

Multi-head attention mechanism is the key link of feature extraction in Transformer [20]. Taking the fusion vector matrix output by Shap-Word model as input, it is divided into query vector Q, key vector K and value vector V. The query vector Q and key vector K are calculated through the dot product similarity function to get the weight of the corresponding value vector V. Then the weighted sum of the value vectors is obtained, as shown in Fig. 3.

The single self-attention cannot obtain the feature information of multiple sub-spaces, while the multi-head self-attention calculates the semantic representation of sequence by associating different positions in the sequence in different sub-spaces. The text sequence is divided into h subspace, and a linear mapping is performed in the unit of the subspace, that is, the attention function is performed for h times in parallel. Then, the h outputs obtained are spliced together and the attention function is executed again to obtain the final attention matrix, as shown in Formula (5–6).

$$MultiHead(Q, K, V) = Concat(head_i)W \tag{5}$$

$$head_i = Attention(QW_i^Q, KW_i^K, VW_i^V) \tag{6}$$

where W, QW_i^W, KW_i^K, VW_i^V is the parameter matrix.

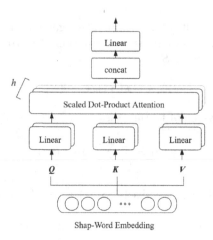

Fig. 3. The architecture of multi-head attention

3.3 Bidirectional Normalization

In the field of machine learning, normalization plays a key role in the performance of neural network model. In order to ensure good convergence of model, normalization makes the characteristics of each dimension of input have similar distribution before the relative importance of each dimension is clear.

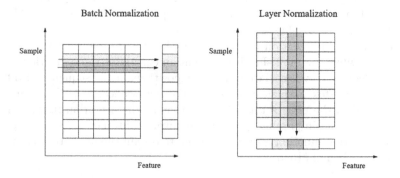

Fig. 4. The conceptual structure batch normalization and layer normalization

The original Transformer has some problems such as hyperparameter sensitivity and slow convergence in the optimization process. The improvement of Shap-PreBiNT is to design a bidirectional normalization module with sub-layers of batch normalization and layer normalization. Batch normalization normalizes different features in the same batch from longitudinal dimension, which reduces the dependence of gradient on parameters, improves network generalization ability and speeds up training convergence speed. Layer normalization normalizes different time steps from the transverse dimension, so that the gradient norm of each hidden layer keeps stable distribution to avoids the problems of gradient disappearance and gradient explosion. Figure 4 depicts the conceptual structure of batch and layer normalization.

Batch normalization takes a batch as the basic unit, calculates its mean and standard deviation, appends additional normalization operations on the hidden layer of the neural network. It distributes the activation function within the linear interval, so as to increase the step length of search and speed up the convergence.

Algorithm1: Algorithm of batch normalization

Input: Values of x over a mini-batch: $B = [x_1, x_2, ..., x_m]$
Parameters to be learned: γ, β
Output: $y_i = BN_{\gamma,\beta}(x_i)$

$\mu_B \leftarrow \frac{1}{m} \sum_{i=1}^{m} x_i$
$\sigma_B^2 \leftarrow \frac{1}{m} \sum_{i=1}^{m} (x_i - \mu_B)^2$
$\hat{x}_i \leftarrow \frac{x_i - \mu_B}{\sqrt{\sigma_B^2 + \epsilon}}$
$y_i \leftarrow \gamma \hat{x}_i + \beta = BN_{\gamma,\beta}(x_i)$

Layer normalization comprehensively considers the input of all feature dimensions of a layer, calculates the average mean and variance of the layer, and then normalizes the input with the same normalized operation according to the data distribution of different feature specifications.

Algorithm2: Algorithm of layer normalization

Input: Values of x over a feature: $L = [x_1, x_2, ..., x_m]$
Parameters to be learned: g, b
Output: $y_i = LN_{g,b}(x_i)$

$\mu_L \leftarrow \frac{1}{m} \sum_{i=1}^{m} x_i$
$\sigma_L^2 \leftarrow \frac{1}{m} \sum_{i=1}^{m} (x_i - \mu_L)^2$
$\hat{x}_i \leftarrow \frac{x_i - \mu_L}{\sqrt{\sigma_L^2 + \epsilon}}$
$y_i \leftarrow g \hat{x}_i + b = LN_{g,b}(x_i)$

3.4 Pre-normalization Structure

The original Transformer uses layer normalization and is designed behind two modules: the multi-attention with residual connections and the feed forward neural network (FFN) [7]. This structure is often called post-normalization and achieves state-of-the-art performance in many machine learning tasks, including language modeling and machine translation. However, it requires more complex parameter optimization than other sequential sequence models in training. The core reason is that the input scale of each module is independent of the number of network layers if the layer normalization is set after each module. This kind of constantly changing distribution among different layers will inevitably affect the convergence effect of the model, leading to the instability of the training process in the initial stage.

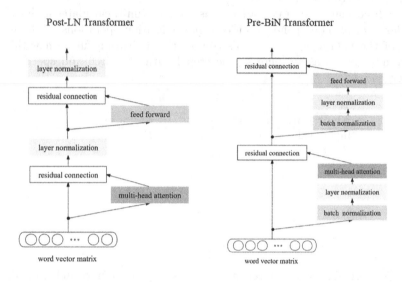

Fig. 5. The comparison of normalization structure

To address this problem, the Shap-PreBiNT model proposed in this paper adopts the pre-normalization structure, which designs the bidirectional normalization layer before the multi-head attention layer and feed forward neural network layer, as shown in Fig. 5. The vector matrix output by Shap-Word model first goes through bidirectional normalization layer, so that the vector matrix has a more normalized distribution, and the gradient value does not expand with the deepening of network layer. Then the optimized vector matrix is transferred to the multi-head attention layer or the feed forward neural network layer, so as to avoid the over-fitting phenomenon in the initial stage and maintain the stability of the model.

4   Experiments

4.1   Experimental Settings and Datasets

The experimental environment and configuration of this paper: Windows 10 operating system, Intel Core I7 processor, 16G memory, Python 3.7 programming language, PyCharm 64 integrated development environment, Keras model framework.

An English dataset and a Chinese dataset are used in the experiments. (1) The IMDB Internet English movie review dataset contains 50000 labeled review data, which are divided into positive and negative sentiment categories. The training set contains 25000 reviews, the validation set contains 2000 reviews, and the test set contains 2000 reviews. (2) ChnSentiCorp [19] is a dataset of Chinese hotel reviews organized by Dr. Tan, including 6000 tagged comment data, the training set contains 4000 comments, the validation set contains 1000 comments, and the test set contains 1000 comments. The division of the two datasets is shown in Table 1.

Table 1. Datasets

Datasets	Training set	Test set	Validation set	Vocabulary
IMDB	25000	2000	2000	56273
ChnSentiCorp	4000	1000	1000	22052

4.2 Performance of Shap-Word Model

In order to verify the effectiveness of Shap-Word model, we set the word vector model of represented statically, such as Word2vec (CBOW) and Glove, as contrast models respectively. And we alse set the word vector model of represented dynamically ELMo as a contrast model. The experiments carry out the sentiment classification task in four ChnSentiCorp datasets: htl-2000, htl-4000, htl-6000 and htl-10000.

In the experiment, the dimension of word vector is set to 100, the size of input window is 5, and the number of iterations is 10. The training sample size of the classifier is set as 64, and the learning rate is 0.01. According to tag attributes, the loss function is *binary_Crossentropy*.

Table 2. Accuracy under different word vector models on ChnSentiCorp

Model	Dataset			
	htl-2000	*htl-4000*	*htl-6000*	*htl-10000*
Word2Vec	90.77	90.5	89.39	89.57
Glove	89.79	90.44	90.32	90.56
ELMo	**92.77**	91.82	91.05	**91.23**
Shap-Word	91.38	**91.87**	**91.8**	90.84

As can be seen from Table 2, on htl-6000 dataset, when using Shap-Word vector as input, the classification accuracy is improved by 2.41% compared with Word2vec, 1.48% compared with Glove, 0.75% compared with ELMo. In other dataset experiments, the performance of the Shap-Word model is also significantly better than that of the statically represented word vector, which proves the effectiveness of the proposed Shap-Word model.

4.3   Performance of the Improved Normalization Structure

In order to investigate the influence of bidirectional normalization on model performance, this paper adopts layer normalization and bidirectional normalization proposed in this paper to conduct experimental comparison on ChnSentiCorp Chinese dataset(htl-6000) and IMDB English dataset.

Figure 6 shows the variation trend of model classification accuracy with the number of iterations on ChnSentiCorp Chinese dataset. It can be seen from Fig. 6 that when a single layer normalization is adopted and the number of iterations is 28, the classification accuracy of the model tends to be stable, with the highest accuracy of 0.904. When the bidirectional normalization is adopted and the number of iterations is 25, the classification accuracy of the model tends to be stable, and the highest accuracy is 0.909. The experimental results show that the bidirectional normalization normalizes the feature distribution from longitudinal and transverse dimensions, thus speeding up the convergence of the model.

Figure 7 shows the variation trend of model classification accuracy with the number of iterations on IMDB English dataset. It can be seen from Fig. 7 that when a single layer normalization is adopted and the number of iterations is 28, the classification accuracy of the model tends to be stable, with the highest accuracy of 0.934. When the bidirectional normalization is adopted and the number of iterations is 26, the classification accuracy of the model tends to be stable, and the highest accuracy is 0.943.

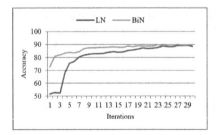 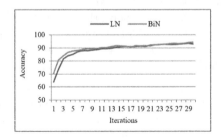

Fig. 6. The influence of different normalization on model performance (on ChnSentiCorp)

Fig. 7. The influence of different normalization on model performance(on IMDB)

Table 3. Accuracy under different normalization structure on ChnSentiCorp

Model	Dataset			
	htl-2000	*htl-4000*	*htl-6000*	*htl-10000*
Post-LN Transformer	90.77	90.5	89.39	89.57
Post-BiN Transformer	91.05	90.82	89.6	90.04
Pre-BiN Transformer	91.29	90.91	89.55	90.16

In order to further explore the impact of the bidirectional normalization and the pre-structure on the model performance, ablation experiments are conducted on IMDB datasets. Compared with the Post-LN Transformer and Post-BiN Transformer, the classification accuracy of the model is improved, which shows that the construction of bidirectional normalization in the encoder effectively improves the classification performance of the model normalizes the data distribution from different dimensions. Compared with Post-BiN Transformer and Pre-BiN Transformer, the classification accuracy of the model is is basically unchanged, indicating that the pre-structure has no obvious impact on the classification performance of the model, and its role is to maintain the stability of the model.

4.4 Model Comparisons

In order to verify the classification performance of the model, this section compares the model we proposed with other related shallow learning models and deep learning models on the IMDB English dataset and ChnSentiCorp Chinese dataset. The comparison models are as follows:

SVM [16]: A traditional machine learning algorithm named support vector machine.

CNN [11]: A single convolutional neural network.

CNN+Attention [6]: A hybrid network model which combines convolutional neural network with attention mechanism.

BiLSTM+Attention [3]: A hybrid network model combines BiLSTM with attention mechanisms to capture emotional features over long distances.

Transformer [21]: A network model based on multi-head attention, which is composed of positional embedding, feed forward neural network and normalization layer, can realize efficient parallel computation.

Transformer-XL [5]: On the basis of Transformer, fragment level cyclic recursion mechanism and relative position embedding mechanism are added.

Table 4. Comparison of model classification results on ChnSentiCorp

Model	Acc	F1	Recall	Precision
SVM	81.63	82.17	82.4	83.1
CNN	88.17	89.22	89.16	89.47
CNN+Attention	88.32	87.19	87.13	87.27
BiLSTM+Attention	88.66	88.06	87.98	88.14
Transformer-base	89.39	89.15	91.42	87.81
Transformer-XL	90.78	90.48	93.55	89.71
Shap-PostBiNT	91.3	91.25	93.38	89.3
Shap-PreBiNT	**91.82**	**91.66**	**93.54**	**90.42**

In the ChnSentiCorp dataset experiment, Word2Vec is used to initialize the text as a vector, and the dimension is set as 100. In Transformer architecture, multiple head number is set to 4, dropout rate is set to 0.5, and activation function is set to ReLU function. Table 4 shows the comparison of accuracy, F1-score, recall and precision with other learning models. As can be seen from the table, the accuracy rate of Shap-PreBiNT is 91.82% and the F1-score is 91.66%. The best accuracy and F1-score are achieved.

In the IMDB dataset experiment, the data is imported through Keras framework, and the initialization vector is generated by sequence filling and word embedding. The dimension is set as 100. In Transformer architecture, multiple head number is set to 4, dropout_rate is set to 0.5, and the activation function is set to ReLU. According to the Table 5, the accuracy of Shap-PreBiNT is 94.87%, which is 1.48% higher than Transformer. And the F1-score is 94.83%, 1.47% higher than Transformer.

Table 5. Comparison of model classification results on IMDB

Model	Acc	F1	Recall	Precision
SVM	88.64	88.24	91.68	85.95
CNN	90.94	90.57	93.68	88.78
CNN+Attention	91.29	90.33	93.61	89.22
BiLSTM+Attention	91.61	91.67	**94.91**	89.19
Transformer-base	93.49	93.46	93.37	93.58
Transformer-XL	93.6	93.78	93.42	93.75
Shap-PostBiNT	94.02	94.01	93.93	93.89
Shap-PreBiNT	**94.87**	**94.83**	94.62	**95.14**

4.5 Effects of Different Activation Functions on Model Performance

To explore the effect of activation function on model performance of Shap-PreBiNT, Sigmoid [18], ReLU [13], TanH [18] activation functions are used to test the classification accuracy on IMDB and ChnSentiCorp datasets separately and compare them with different models. As can be seen from Table 6, the classification effect of Shap-PreBiNT is superior to other models in ReLU function.

Table 6. Accuracy under different activation functions on ChnSentiCorp

Model	ChnSentiCorp			IMDB		
	Sigmoid	ReLU	TanH	Sigmoid	ReLU	TanH
CNN+Attention	88.73	88.32	87.73	91.66	91.29	91.03
Transformer-base	89.85	89.39	90.07	**92.87**	93.49	–
Shap-PreBiNT	**90.9**	**91.82**	**91.1**	91.87	**94.87**	**94.58**

5   Conclusion

This paper proposed a sentiment classification model Shap-PreBiNT on Chinese and English comment datasets, and improved the problems of original Transformer in text vectorized representation, semantic information extraction and network structure. In the stage of word embedding, a Shap-Word model was proposed. The Word2Vec generates the initial vector matrix of the text to indicate the importance of the words, and the Shapley-value method generates the contribution weight matrix to indicate the importance of the words. This model can not only fully express the local connections between words, but also pay attention to the global information of text. In the stage of semantic information extraction, a bidirectional normalization layer combining batch normalization and layer normalization was proposed to regulate data distribution from multi-dimensional dimensions, which overcome the problem of gradient disappearance or gradient explosion in the training process. In the stage of network structure optimization, the pre-normalization structure was promotes the model convergence efficiency. Experimental results show that our model keeps the efficiency and stability of the training process on both Chinese and English datasets, and achieves higher classification accuracy.

References

1. Alashri, S., Kandala, S.S., Bajaj, V., Ravi, R., Desouza, K.C.: An analysis of sentiments on facebook during the 2016 u.s. presidential election. In: 2016 IEEE/ACM International Conference on Advances in Social Networks Analysis and Mining (ASONAM) (2016)
2. Bahdanau, D., Cho, K., Bengio, Y.: Neural machine translation by jointly learning to align and translate. arXiv preprint arXiv:1409.0473 (2014)
3. Chen, P., Sun, Z., Bing, L., Yang, W.: Recurrent attention network on memory for aspect sentiment analysis. In: Proceedings of the 2017 Conference on Empirical Methods in Natural Language Processing, pp. 452–461 (2017)
4. Cho, K., et al.: Learning phrase representations using rnn encoder-decoder for statistical machine translation. arXiv preprint arXiv:1406.1078 (2014)
5. Dai, Z., Yang, Z., Yang, Y., Carbonell, J., Salakhutdinov, R.: Transformer-xl: Attentive language models beyond a fixed-length context (2019)
6. Dey, A.: Attention based lstm cnn framework for sentiment extraction from bengali texts. In: 2020 11th International Conference on Electrical and Computer Engineering (ICECE), pp. 226–229. IEEE (2020)
7. Dong, Y., Cordonnier, J.B., Loukas, A.: Attention is not all you need: Pure attention loses rank doubly exponentially with depth. arXiv preprint arXiv:2103.03404 (2021)
8. Hochreiter, S., Schmidhuber, J.: Long short-term memory. Neural Comput. **9**(8), 1735–1780 (1997)
9. Johnson, R., Zhang, T.: Deep pyramid convolutional neural networks for text categorization. In: Proceedings of the 55th Annual Meeting of the Association for Computational Linguistics (Volume 1: Long Papers), pp. 562–570 (2017)

10. Keinan, A., Sandbank, B., Hilgetag, C.C., Meilijson, I., Ruppin, E.: Fair attribution of functional contribution in artificial and biological networks. Neural Comput. **16**(9), 1887–1915 (2004)
11. Kim, Y.: Convolutional neural networks for sentence classification. Eprint Arxiv (2014)
12. Liu, B.: Sentiment Analysis: Mining Opinions, Sentiments, and Emotions. Mining Opinions, Sentiments, and Emotions, Sentiment Analysis (2020)
13. Maas, A., Daly, R.E., Pham, P.T., Huang, D., Ng, A.Y., Potts, C.: Learning word vectors for sentiment analysis. In: Proceedings of the 49th Annual Meeting of the Association for Computational Linguistics: Human Language Technologies, pp. 142–150 (2011)
14. Mikolov, T., Chen, K., Corrado, G., Dean, J.: Efficient estimation of word representations in vector space. arXiv preprint arXiv:1301.3781 (2013)
15. Mnih, V., Heess, N., Graves, A., et al.: Recurrent models of visual attention. In: Advances in Neural Information Processing Systems, pp. 2204–2212 (2014)
16. Mullen, T., Collier, N.: Sentiment analysis using support vector machines with diverse information sources. In: Proceedings of the 2004 Conference on Empirical Methods in Natural Language Processing, EMNLP 2004, A meeting of SIGDAT, a Special Interest Group of the ACL, held in conjunction with ACL 2004, 25–26 July 2004, Barcelona, Spain (2004)
17. Shen, Y., Tan, S., Sordoni, A., Courville, A.: Ordered neurons: Integrating tree structures into recurrent neural networks (2018)
18. Shen, Y., Tan, S., Sordoni, A., Courville, A.: Ordered neurons: Integrating tree structures into recurrent neural networks. arXiv preprint arXiv:1810.09536 (2018)
19. Tan, S., Zhang, J.: An empirical study of sentiment analysis for Chinese documents. Expert Syst. Appl. **34**(4), 2622–2629 (2008)
20. Tay, Y., Dehghani, M., Bahri, D., Metzler, D.: Efficient transformers: A survey. arXiv preprint arXiv:2009.06732 (2020)
21. Vaswani, A., et al.: Attention is all you need. In: Advances in Neural Information Processing Systems, pp. 5998–6008 (2017)
22. Wang, W., Li, B., Feng, D., Zhang, A., Wan, S.: The ol-dawe model: tweet polarity sentiment analysis with data augmentation. IEEE Access **8**, 40118–40128 (2020)
23. Wang, Y., Sun, A., Han, J., Liu, Y., Zhu, X.: Sentiment analysis by capsules. In: Proceedings of the 2018 World Wide Web Conference, pp. 1165–1174 (2018)

Universum-Inspired Supervised Contrastive Learning

Aiyang Han and Songcan Chen[✉]

Nanjing University of Aeronautics and Astronautics, Nanjing, China
{aiyangh,s.chen}@nuaa.edu.cn

Abstract. Mixup is an efficient data augmentation method which generates additional samples through respective convex combinations of original data points and labels. Although being theoretically dependent on data properties, Mixup is reported to perform well as a regularizer and calibrator contributing reliable robustness and generalization to neural network training. In this paper, inspired by Universum Learning which uses out-of-class samples to assist the target tasks, we investigate Mixup from a largely under-explored perspective - the potential to generate in-domain samples that belong to none of the target classes, that is, *universum*. We find that in the framework of supervised contrastive learning, universum-style Mixup produces surprisingly high-quality hard negatives, greatly relieving the need for a large batch size in contrastive learning. With these findings, we propose **Uni**versum-inspired **Con**trastive learning (UniCon), which incorporates Mixup strategy to generate universum data as *g-negatives* and pushes them apart from anchor samples of the target classes. Our approach not only improves Mixup with hard labels, but also innovates a novel measure to generate universum data. With a linear classifier on the learned representations, on Resnet-50, our method achieves 81.68% top-1 accuracy on CIFAR-100, surpassing the state of art by a significant margin of 5% with a much smaller batch size.

Keywords: Mixup · Contrastive learning · Supervised learning · Universum

1 Introduction

As a strong augmentation technique in supervised learning, Mixup has empirically and theoretically been proved to boost the performance of neural networks with its regularization power [2,23,35]. Despite its reliable performance, Mixup is also reported to strengthen deep models with better calibration [30], robustness [11,36] and generalization [36], thus being widely used in adversarial training [23], domain adaptation [10], imbalance problems [8] and so on. However, as Mixup-style training depends heavily on data properties [7], on certain cases, chances are that traditional Mixup labels cannot correctly describe the augmented data. These labels, when taken as the ground truth, may provide unreliable supervision for learners.

B. Li et al. (Eds.): APWeb-WAIM 2022, LNCS 13422, pp. 459–473, 2023.
https://doi.org/10.1007/978-3-031-25198-6_34

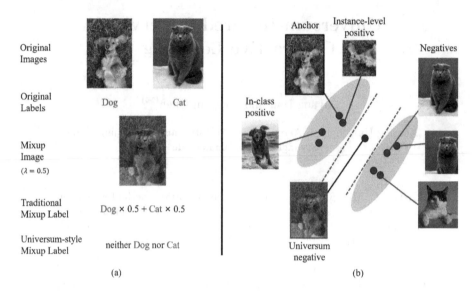

Fig. 1. The intuition behind our model. **(a)**: When processing Mixup labels, traditional method uses the mixture of original labels, but universum-style method regards Mixup data points as belonging to neither of the original classes, thus assigning the new points to a generalized negative class which is compulsorily limited to some desired region. **(b)**: In the framework of supervised contrastive learning, universum-style Mixup images can serve as negative samples for all anchor samples of the target classes. By pushing these g-negatives (universum data) apart from other data points, the model can better separate images from different classes.

Universum learning allows us to see Mixup in a new light. Introduced by [3,31], universum is referred to as in-domain samples that belong to none of the target classes in classification. In universum learning, usually a new dataset of universum is introduced to assist classification of the target dataset (e.g. hand-written letters are introduced to help classify hand-written digits) [5,6,26]. Although universum data cannot be assigned to the classes in question, they still can be constructed into a regularization term so as to improve the model performance with their domain knowledge and negativity [3]. From the perspective of universum learning, here comes a natural question: instead of using the linear interpolations of original labels, why don't we assign Mixup samples to a generalized negative class? Just as humans may perceive, if an animal is half dog and half cat, it is actually of neither species. As is shown in Fig. 1(a), universum-style Mixup regards the Mixup image as neither dog nor cat, but rather a *universum* data point. With this approach, models can be free from the concern of unreliable ground truth labels in Mixup. What's more, the combination of universum learning and Mixup also introduces a new way to acquire universum data, which extends universum learning to fully-supervised setting. Compared with foreign samples such as hand-written letters in the classification of hand-written digits, universum data produced by Mixup are semantically closer to target data, which may provide better regularization effects in training.

Recently, contrastive learning has greatly boosted deep learning via pulling together positive sample pairs and separating negative pairs in the embedding space [1,4,14,20,34]. Early contrastive models only take augmentations of the same image as positive pairs, while treating all other sample pairs as negative pairs [4,14]. Specially, SupCon model extended contrastive learning to the fully-supervised setting by including samples from the same class into positives for each anchor sample [20].

Although contrastive learning and Mixup both improve the performance of supervised learning, the combination of the two can be especially difficult due to their opposite ways of organizing data. While Mixup softly assigns augmented data to multiple classes [32], contrastive learning requires hard labels to compute the contrastive loss. A few attempts have been made to conjoin contrastive learning and Mixup either by designing a Mixup version of InfoNCE loss [21] or by using the naïve addition of the InfoNCE loss and the Mixup-style cross entropy loss [17]. A better exploration might be MoCHi [19], which applies Mixup only to the hard negatives in the memory bank so as to acquire more and harder negatives. However, these methods pay more attention to softening the contrastive learning rather than innovating Mixup strategy, ignoring the innate potential of Mixup to produce negative samples.

In this paper, inspired by universum learning, we introduce a novel measure to combine contrastive learning and Mixup with the simple idea that *Mixup samples could be hard negatives*. Following the framework of supervised contrastive learning, we go a step further to include Mixup images into the contrastive loss by viewing them as *g-negatives* - universum data which are negative to the global dataset - in contrast with traditional negatives that are negative for a limited group of anchor samples. As is shown in Fig. 1(b), we incorporates Mixup to generate g-negatives and pushes them apart from anchor samples of the target classes. For each anchor sample, a contrast sample is chosen from other classes to synthesize a universum data point, which helps establish clearer margins among different instances as well as different classes. Since traditional Mixup strategy that samples the Mixup parameter from Beta distribution [35] may generate samples semantically close to a target class, we fix the Mixup parameter to a constant, thereby driving the synthesized universum data out of the regions of target classes in the data space. Although the idea is simple, there is no prior knowledge on how to contrast these universum negatives with anchor samples. We design a small-scale contrast loss and a large-scale contrast loss, and empirically show that the large-scale contrast loss achieves better performance on datasets. Please note that here "large-scale" and "small-scale" refer to the scale of contrast rather than the training batch size. Indeed, our training batch size is much smaller than that of other contrastive models [4,20]. Our use of universum data spares us the efforts for hard negative mining, as Mixup samples naturally become hard negatives with their visual ambiguity.

Our work provides an effective method for fully-supervised learning and can be applied to other contrastive learning methods in need of large amounts of negatives. We validate the performance of UniCon on a range of datasets. On ResNet-50 [15], UniCon achieves 81.68% top-1 accuracy on CIFAR-100 and 97.23% on CIFAR-10 [22], which surpasses the state of art [20] by 1.23% and 5.18% respectively. Our main contributions are as follows:

- We investigate Mixup from the perspective of universum learning, thus unearthing Mixup's potential of generating samples that lie in the same domain as the target data yet belong to none of the target classes. We dig out Mixup as a novel measure to acquire universum data from a fully supervised dataset.
- We introduce **Uni**versum-inspired **Con**trastive learning (UniCon), which incorporates Mixup to generate universum data as g-negatives and pushes them apart from anchor samples of the target classes. Different from other contrastive models where the negativity of samples varies with anchors, g-negatives in our model are negative to the global dataset. To our best knowledge, this is the first time that Mixup is used to produce a generalized negative class.
- We find that in the framework of supervised contrastive learning, Mixup samples can work surprisingly good as hard negatives.
- We show that our model can achieve outstanding performance on a range of datasets with a relatively small-scale neural network as well as a smaller batch size.

2   Related Works

In this section, we will give a brief introduction of Mixup, universum learning and contrastive learning, as well as their relation to our method.

2.1   Mixup

Since Mixup was proposed by [35], it has been widely accepted as an effective and efficient measure for deep training [2,23]. Despite Mixup's outstanding performance, recently the foundations of Mixup have also been scrutinized in theory. [2] theoretically proves that Mixup is a strong regularizer and equals to a standard empirical risk minimization estimator in the face of noises. [30] focuses on Mixup's effects of improving calibration and predictive uncertainty. [36] gives a theoretical explanation on how Mixup contributes to robustness and generalization of deep models. While Mixup is empirically and theoretically proved a reliable method, [7] demonstrates its data dependency by computing a closed form for the Mixup-optimal classification, and thereby providing a failure case of Mixup. This failure case indicates that Mixup could also be misleading as the synthesized data points are still softly connected with the original labels. Our method intends to disconnect the Mixup data from all known classes so that the additional domain knowledge could be learned without misleading information.

2.2   Universum Learning

Universum was introduced by Vapnik as "an alternative capacity concept to the large margin approach", which indicates a group of samples that cannot be assigned to any target class in classification [31]. Universum learning is mostly explored as a new research scenario where a relevant dataset is introduced to assist the tasks on the target dataset. [3] has theoretically proved that the use of universum data could benefit Support Vector Machines (SVM) with regularization effects. Inspired by universum learning, our model, instead of importing a dataset, generates a group of universum samples from the target dataset to assist classification.

2.3   Contrastive Learning

Contrastive learning learns deep representations through contrasting positive sample pairs against negative ones. The definition of positive and negative pairs varies with different contrastive models. SimCLR [4] and MoCo [14] only admit augmentations of the same image as positive pairs, while cluster-based methods like SupCon [20] and SwAV [1] also give in-class positives a pass. While classical contrastive models use the InfoNCE loss [25], more contrastive losses have flourished. For example, Barlow Twins [34] aims to reduce data redundancy with a cross-correlation matrix, while BYOL [13] strengthens the consistency among views by predicting the second view from the first one.

Several attempts have been made to construct Mixup-style contrastive models [17,19,21,29]. Mixco [21] pulls Mixup data towards their original images in a Mixup way, while MoCHi [19] uses Mixup only on the hard negatives to capture the hardest negatives. In the unsupervised setting, Un-Mix [29] only mixes the images, closing the distance among the Mixup image and an augmented version of Mixup data. Different from them, UniCon does not combine Mixup and contrastive learning in a naïve way. Instead, we delve into the nature of hard negatives, adopting Mixup as a way of hard negative generation. In this way, we not only train a more effective model, but also relieve the need for a large batch size in contrastive learning as is shown in the latter experimental results.

3   Method

This section begins with a brief introduction of self-supervised and supervised contrastive losses, after which we present universum-style Mixup method. Then, with the g-negatives produced by Mixup, small-scale and large-scale UniCon losses are proposed, while the latter is empirically proved to be a better one.

Following the framework of [20], our approach is in nature a representation learning method. A deep encoder f is adopted to learn the representations of target samples through minimizing a proposed loss. With N being the batch size, each data point and its label are denoted by x_k and y_k ($k = 1, 2, .., N$), while the corresponding augmented sample and its label is denoted by $\widetilde{x}_k$ and $\widetilde{y}_k$

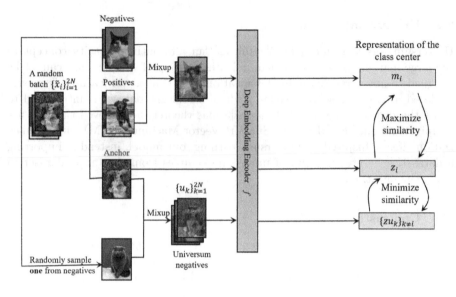

Fig. 2. An overview of UniCon.

$(k = 1, 2, .., 2N)$. Note that $\tilde{x}_{2k-1}$ and $\tilde{x}_{2k}$ are two transformed augmentations derived from x_k, thus $\tilde{y}_{2k-1} = \tilde{y}_{2k} = y_k$. Since most of our operations are performed on the augmented set, we will refer to this set of $2N$ samples as "a training batch" in the following part. The framework of UniCon is depicted in Fig. 2.

3.1   Contrastive Loss

Our proposed method is based on contrastive learning. As the most used contrastive loss, InfoNCE loss [25] draws positive pairs close to each other while separating the negative ones. InfoNCE loss is defined in this form:

$$L_{contrast} = -\frac{1}{2N} \sum_{i=1}^{2N} log \frac{exp(z_i \cdot z_{p(i)}/\tau)}{\sum_{k \neq i} exp(z_i \cdot z_k/\tau)}, \tag{1}$$

where $z_i = f(\tilde{x}_i)$ represents the normalized deep embedding for each data point, τ is a temperature parameter, and $p(i)$ indicates a *positive* for anchor i while the rest indices are *negatives*.

Considering that Eq. 1 does not encode the label information, SupCon loss [20] involves in-class samples into the positives:

$$L_{sup} = \sum_{i=1}^{2N} \frac{-1}{|D_i|} \sum_{d \in D_i} log \frac{exp(z_i \cdot z_d/\tau)}{\sum_{k \neq i} exp(z_i \cdot z_k/\tau)}, \tag{2}$$

where $D_i \equiv \{k|k \in \{1, 2, .., 2N\}, k \neq i, \tilde{y}_k = \tilde{y}_i\}$ is a set of indices that refer to samples in the same class with i, and $|D_i|$ denotes the capacity of the set.

Both two losses pay limited attention to negative pairs, simply recycling the non-positive sample pairs.

3.2   Universum-Style Mixup

Motivated by universum learning, universum-style Mixup intends to provide a set of additional negatives to boost the performance of contrastive learning. It is assumed that by rejecting visual ambiguity, classes can be better separated with margins among them. Just like traditional Mixup method, Universum-style Mixup convexly combine each anchor sample $\widetilde{x}_i$ in a training batch, and its out-of-class negative $\widetilde{x}_{q(i)}$ to generate a universum negative u_i. Different from traditional Mixup strategy, in our approach the Mixup parameter λ is set to a certain number rather than randomly sampled from Beta distribution. Moreover, for each anchor sample in the training batch, an image is randomly chosen from the samples that do not belong to the same class with the anchor, after which the random image and the anchor are mixed to generate a universum data point. By doing so, we minimize the possibility of the universum data falling into the regions of target classes in the data space, thereby ensuring the negativity of g-negatives. The universum is acquired through the following process:

$$u_i = \lambda \cdot \widetilde{x}_i + (1 - \lambda) \cdot \widetilde{x}_{q(i)}, \quad i = 1, 2, .., 2N, \tag{3}$$

where q(i) is randomly chosen from $\cup_{k \neq i} D_k$ and λ is the Mixup parameter. In the remainder of this paper, u_i will be referred to as a "g-negative" and $\widetilde{x}_i$ will be referred to as its "anchor". Please note that universum-style Mixup does not mix the labels, and therefore the synthesized samples should belong to, if any, a generalized negative class. By doing so, our method completely drops the effect of label smoothing in Mixup [2], in return earning a group of samples with hard labels.

3.3   Universum-Inspired Supervised Contrastive Learning

In this paper, our approach introduces a set of universum data $u_k{}_{k=1}^{2N}$ (which has been elaborated in Eq. 3) into the contrastive loss. The normalized encoded representation of u_l is denoted as $zu_k = f(u_k)$. As Fig. 2 shows, our proposed method intends to draw anchor samples close to the center of their class while pushing them from negatives. Here two solutions (L_{add} and L_{UniCon}) are presented in the following parts.

Universum Data as Additional Negatives. A straightforward way of combining supervised contrastive learning and Mixup-induced universum is to use universum data as additional negatives.

$$L_{add} = \sum_{i=1}^{2N} \frac{-1}{|D_i|} \sum_{d \in D_i} log \frac{exp(z_i \cdot z_d / \tau)}{\sum_{k \neq i} exp(z_i \cdot z_k / \tau) + \sum_{k=1}^{2N} exp(z_i \cdot zu_k / \tau)} \tag{4}$$

L_{add} generally adopts the original form of Eq. 2, yet further contrasting anchor samples with universum negatives. This loss function aims to use large amounts of universum negatives to alleviate the need for large amounts of negative samples in contrastive learning [4, 20]. However, experiments show that this loss function is not very effective on CIFAR-100 dataset. To justify such a result, here are two possible causes. On the one hand, it is deduced that the problem of "manifold intrusion" in Mixup (e.g. an image of number "1" and image of number "4" are mixed into a image that somewhat looks like number "4") may also appear in our universum-style Mixup, leading to poor results. On the other hand, L_{add} may overemphasize negatives, which produces undesirable disequilibrium.

An Entirely Universum-Based Method. Here is the main loss function we use in this paper. This strategy is entirely based on universum data, both for contrast with negatives and derivation of class centers in the embedding space.

$$L_{UniCon} = \sum_{i=1}^{2N} \frac{-1}{|D_i|} \sum_{d \in D_i} log \frac{exp(z_i \cdot zu_d/\tau)}{\sum_{k \neq i} exp(z_i \cdot zu_k/\tau)} \tag{5}$$

To differentiate Eq. 5 from a naïve combination of Mixup-induced universum and Eq. 2, L_{UniCon} is further derived into the following form.

$$L_{UniCon} = \sum_{i=1}^{2N} \frac{-1}{|D_i|} \left[\frac{z_i}{\tau} \cdot \sum_{d \in D_i} zu_d - \sum_{d \in D_i} log \sum_{k \neq i} exp(z_i \cdot zu_k/\tau) \right] \tag{6}$$

$$= \sum_{i=1}^{2N} \left[-\frac{z_i}{\tau} \cdot m_i + log \sum_{k \neq i} exp(z_i \cdot zu_k/\tau) \right] \tag{7}$$

where $m_i = (\sum_{d \in D_i} zu_d)/|D_i|$ is the mean of the representations of universum data points around the cluster of z_i. According to Eq. 7, L_{UniCon} does not simply push together universum data derived from samples of the same class, but rather drives in-class data points close to m_i. As Fig. 3 illustrates, chances are that the mean of universum data points could better represent the center of a class while these data points themselves loosely surround the in-class space as out-of-class negatives. In this way, universum data are also utilized for derivation of class prototypes, thereby relieving manifold intrusion mentioned above. Meanwhile, L_{UniCon} only adopts universum data as negatives, dropping out negatives in the conventional sense, which further improves model robustness. Still, it should be admitted that this strategy is coarse and primary, yet the experimental results show that it is especially effective.

Table 4 empirically demonstrates that L_{UniCon} works better than L_{add}. The performance of L_{add} is even worse than the loss without the extra universum negatives, which implies that an entirely universum-based framework is crucial for utilizing the universum data. Based on these findings, our method generalizes better to the test set for the following reasons:

Noise Injection. In the aforementioned situation, our method injects noises to the training data (e.g. anchors in class "4" regard number "4" synthesized by "1" and "4" as a negative sample). On the one hand, such technique is widely used in adversarial training as well as contrastive learning to learn a more robust model [12, 27, 28]. On the other hand, since Mixup-induced universum are used in both contrast with negatives and class centers, these two kinds of contrast are in a restrictive relation with each other. Noises in universum negatives can help derive a more accurate class center, and vice versa.

A Different Approach of Contrast. Our method does not directly contrast anchors with conventional out-of-class negatives in [20]. However,

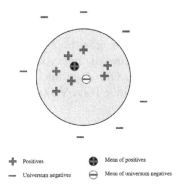

+ Positives	⬤ Mean of positives
— Universum negatives	⊖ Mean of universum negatives

Fig. 3. The illustration of using the mean of universum negatives to represent a class. Here the universum negatives are synthesized from the positives of the class in question. Since these universum data are expected to be distributed in the margin space around the in-class space, their mean may better describe the class especially when the positive samples are not evenly distributed in a minibatch.

UniCon still uses universum data as negatives, which differentiates itself from absolutely contrast-free methods like [13]. By contrasting with universum negatives and benefiting from their data diversity, UniCon not only avoids contrastive models' dependency on large batch sizes, but also allows a balanced network design easier to optimize.

4 Experiments

4.1 Setup

We evaluate our model on several widely used benchmarks including CIFAR-10, CIFAR-100 [22] and TinyImageNet [24]. Detailed information of dataset settings can be viewed in Table. 2. Here input size refers to the transformed size of neural network input. Without special statement, the encoder network is trained for 1000 epochs with a batch size of 256. As for hyperparameters, temperature τ and Mixup parameter λ are respectively fixed to 0.1 and 0.5. We set the learning rate to 0.05 with 10 epochs of warm-up. As the purpose of this paper is to show how universum improves contrastive learning rather than to explore the effects of different augmentation techniques on our model, we empirically use a set of augmentations that was chosen by [20] through AutoAugment [9]. In the evaluation period, a classifier of batch size 512 is trained for 100 epochs with the deep representations extracted by the encoder while the encoder itself is frozen. On both stages, we use SGD optimizer with cosine annealing for weight decay. All datasets are split into the training set and test set according to their official division. We do not use a validation set because the performance of our model is promising without the need of hyperparameter adjustment. Indeed, we empirically chose the hyperparameters which, as Sect. 4.3 shows, may not be the optimal ones. Our experiments are implemented in PyTorch framework on at most four Nvidia Tesla V100 GPUs in an online computing center.

Table 1. Top-1 classification accuracy (in percentage %) on various datasets. We compare our model (UniCon) with a deep classifier using cross-entropy loss, SimCLR [4], and SupCon [20]. For fairness, we use the baseline numbers published by [20] on CIFAR-10 and CIFAR-100 since we did not achieve higher results in our own re-implementation. For TinyImageNet, we implement experiments to acquire the baseline results. We use **bold** to indicate the best results, and underline the second best ones. Also please note that the batch size of our model is only 256, which is much smaller than that of the baseline models.

Method	Architecture	Batch size	CIFAR-10	CIFAR-100	TinyImageNet
Cross-Entropy	ResNet-50	1024	95.0	75.3	58.3
SimCLR [4]	ResNet-50	1024	93.6	70.7	34.6
SupCon [20]	ResNet-50	1024	96.0	76.5	50.4
UniCon(ours)	ResNet-18	*256*	96.2	78.9	58.4
	ResNet-50	*256*	**97.2**	**81.7**	**65.6**

4.2 Classification Accuracy

We compare UniCon with a cross-entropy classifier, Sim-CLR [4], and SupCon [20] on their top-1 accuracy on CIFAR-10, CIFAR-100 and TinyImageNet. Although these methods have all be proposed for a few years, so far they are the mainstream methods for fully-supervised learning. Follow-up methods either focus on a specific application scenario or adapt the aforementioned models to other settings, failing to propose a better model on fully-supervised learning. Therefore, we still adopt these three old but effective models as our baselines. As is shown in Table. 1, UniCon outperforms other models on all datasets, while adopting smaller batch sizes and encoder backbones. Our model achieves 97.2%, and 81.7% on CIFAR-10 and CIFAR-100, respectively, which surpasses the state of art by a significant margin of 1.2% and 5.2% with only one fourth the batch size. Even with a backbone of ResNet-18 and batch size 256, UniCon outperforms its counterparts with ResNet-50 and batch size 1024. UniCon also achieves 65.6% top-1 accuracy on TinyImageNet. Please note that we input images as 32 × 32 patches, which is

Table 2. Dataset settings.

Dataset	Images	Classes	Input size
CIFAR-10	60,000	10	32 × 32
CIFAR-100	60,000	100	32 × 32
TinyImageNet	100,000	200	32 × 32

way smaller than the input sizes (e.g. 224×224) of other models [21] that report better performance of cross-entropy classifiers. In this sense, our performance gain over the cross-entropy classifier is also significant on TinyImageNet.

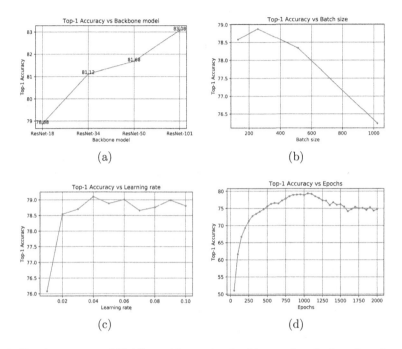

(a)

(b)

(c)

(d)

Fig. 4. Top-1 accuracy of UniCon with varying backbones, batch sizes, learning rates and pretraining epochs. The experiments are conducted on CIFAR-100, and except for the backbone analysis, Resnet-18 is adopted for model encoders.

4.3   Hyper-parameter Analysis

Figure 4 illustrates UniCon's stability to different hyper-parameters on CIFAR-100. We modify the backbone networks, batch sizes, learning rates and training epochs one at a time to observe whether our model is sensitive to the punctuation of hyper-parameters. Generally speaking, UniCon shows promising performance even in the worst situation.

We evaluate our model with a backbone of Resnet-18, Resnet-34, Resnet-50, and Resnet-101, respectively. In the aspect of model sizes, a deeper network would always improve the performance. Specially, UniCon achieves 83.08% on CIFAR-101. We deduce that stronger networks like PreAct ResNet [16], WideResNet [33] and DenseNet [18] can further boost the performance of our model, which is beyond the scope of this paper.

It is worth noting that our model may not always perform better with a larger batch size, as its top-1 accuracy on batch size 512 and 1024 is lower than that on batch size 256. Since a lot of papers have shown that large batch sizes benefit the training of contrastive models [4,14,20], such results can be intriguing. We conjecture that our model, with additional hard negatives generated by Mixup, is a beneficiary of frequent gradient descents. For training epochs of a fixed number, large batches inevitably lead to a decline in optimization times, thereby resulting into worse performance. It is necessary to make a trade-off between large batch sizes and optimization frequencies. As is shown in Fig. 4(b), we find that 256 is the optimal batch size for most cases.

Figure 4(d) shows the convergence of UniCon for 2000 epochs. Since cosine annealing we use for learning rate decay is sensitive with different training epochs, for reproductivity we divide the training period into first 1000 epochs and second 1000 epochs, each with a complete process of cosine annealing.

Table 3. CIFAR-100 classification accuracy for different Mixup Settings. We either set λ to a constant or sampled λ from Beta distribution.

λ	Top-1 accuracy
0.3	74.50
0.4	76.32
0.5	**78.88**
0.6	77.39
0.7	75.22
$\lambda \sim Beta(0.5,\ 0.5)$	77.05

4.4   Mixup Strategies

We test different strategies of choosing λ in Mixup. We either fix λ to 0.3, 0.4, 0.5, 0.6 and 0.7, respectively, or assume that λ is a random number subject to Beta distribution, following [35]. As Table. 3 demonstrates, the model achieves best performance when two images are equally mixed to produce a universum negative. This result is in line with our intuition that the Mixup image is farthest from its original images in semantics when two images make equal contributions to their mixture.

4.5   Ablation Study

To further understand the effectiveness of each designed component of our model, an ablation study is conducted. We examine what data are regarded as negatives for contrast and whether universum data are used for class center derivation for each loss. As Table 4 demonstrates, it is crucial that universum data are utilized to derive the class centers. The use of additional universum negatives does harm to the model performance, while the mere use of universum for class center derivation will result into deteriorated performance. However, when universum negatives are used in combination with universum-derived class centers, the model acquires the best performance.

Table 4. Ablation study. The loss functions are examined on what data are regarded as negatives for contrast and whether universum data are used for class center derivation.

Loss function	Negatives for contrast		Class center derivation	Top-1 Accuracy
	Universum	Out-of-class	from universum data	
L_{UniCon}	✓	✗	✓	78.88
	✓	✓	✓	78.58
	✗	✓	✓	2.4
L_{sup}	✗	✓	✗	70.46
L_{add}	✓	✓	✗	66.69

5   Conclusion

This paper explores Mixup from the perspective of Universum Learning, thus proposing to assign synthesized samples into a generalized negative class in the framework of supervised contrastive learning. Our model achieves state-of-the-art performance on CIFAR-10, CIFAR-100 and TinyImageNet. The results of our experiments reveal the potential of Mixup to generate hard negative samples, which may open a new window for further studies.

References

1. Caron, M., Misra, I., Mairal, J., Goyal, P., Bojanowski, P., Joulin, A.: Unsupervised learning of visual features by contrasting cluster assignments. Adv. Neural. Inf. Process. Syst. **33**, 9912–9924 (2020)
2. Carratino, L., Cissé, M., Jenatton, R., Vert, J.P.: On mixup regularization. arXiv preprint arXiv:2006.06049 (2020)
3. Chapelle, O., Agarwal, A., Sinz, F., Schölkopf, B.: An analysis of inference with the universum. In: Advances in Neural Information Processing Systems, vol. 20 (2007)
4. Chen, T., Kornblith, S., Norouzi, M., Hinton, G.: A simple framework for contrastive learning of visual representations. In: International Conference on Machine Learning, pp. 1597–1607. PMLR (2020)
5. Chen, X., Chen, S., Xue, H.: Universum linear discriminant analysis. Electron. Lett. **48**(22), 1407–1409 (2012)
6. Cherkassky, V., Dhar, S., Dai, W.: Practical conditions for effectiveness of the universum learning. IEEE Trans. Neural Netw. **22**(8), 1241–1255 (2011)
7. Chidambaram, M., Wang, X., Hu, Y., Wu, C., Ge, R.: Towards understanding the data dependency of mixup-style training. arXiv preprint arXiv:2110.07647 (2021)
8. Chou, H.-P., Chang, S.-C., Pan, J.-Y., Wei, W., Juan, D.-C.: Remix: rebalanced Mixup. In: Bartoli, A., Fusiello, A. (eds.) ECCV 2020. LNCS, vol. 12540, pp. 95–110. Springer, Cham (2020). https://doi.org/10.1007/978-3-030-65414-6_9
9. Cubuk, E.D., Zoph, B., Mane, D., Vasudevan, V., Le, Q.V.: Autoaugment: Learning augmentation strategies from data. In: Proceedings of the IEEE/CVF Conference on Computer Vision and Pattern Recognition, pp. 113–123 (2019)

10. Davoudian, A., Chen, L., Tu, H., Liu, M.: A workload-adaptive streaming parti-tioner for distributed graph stores. Data Sci. Eng. **6**(2), 163–179 (2021)
11. Erichson, N.B., Lim, S.H., Utrera, F., Xu, W., Cao, Z., Mahoney, M.W.: Noisymix: Boosting robustness by combining data augmentations, stability training, and noise injections. arXiv preprint arXiv:2202.01263 (2022)
12. Goodfellow, I.J., Shlens, J., Szegedy, C.: Explaining and harnessing adversarial examples. arXiv preprint arXiv:1412.6572 (2014)
13. Grill, J.B., et al.: Bootstrap your own latent-a new approach to self-supervised learning. Adv. Neural. Inf. Process. Syst. **33**, 21271–21284 (2020)
14. He, K., Fan, H., Wu, Y., Xie, S., Girshick, R.: Momentum contrast for unsupervised visual representation learning. In: Proceedings of the IEEE/CVF Conference on Computer Vision and Pattern Recognition, pp. 9729–9738 (2020)
15. He, K., Zhang, X., Ren, S., Sun, J.: Deep residual learning for image recognition. In: Proceedings of the IEEE Conference on Computer Vision and Pattern Recognition, pp. 770–778 (2016)
16. He, K., Zhang, X., Ren, S., Sun, J.: Identity mappings in deep residual networks. In: Leibe, B., Matas, J., Sebe, N., Welling, M. (eds.) ECCV 2016. LNCS, vol. 9908, pp. 630–645. Springer, Cham (2016). https://doi.org/10.1007/978-3-319-46493-0_38
17. Hou, J., Xu, J., Feng, R., Zhang, Y., Shan, F., Shi, W.: Cmc-cov19d: Contrastive mixup classification for covid-19 diagnosis. In: Proceedings of the IEEE/CVF Inter-national Conference on Computer Vision, pp. 454–461 (2021)
18. Huang, G., Liu, Z., Van Der Maaten, L., Weinberger, K.Q.: Densely connected convolutional networks. In: Proceedings of the IEEE Conference on Computer Vision and Pattern Recognition, pp. 4700–4708 (2017)
19. Kalantidis, Y., Sariyildiz, M.B., Pion, N., Weinzaepfel, P., Larlus, D.: Hard negative mixing for contrastive learning. Adv. Neural. Inf. Process. Syst. **33**, 21798–21809 (2020)
20. Khosla, P., et al.: Supervised contrastive learning. Adv. Neural. Inf. Process. Syst. **33**, 18661–18673 (2020)
21. Kim, S., Lee, G., Bae, S., Yun, S.Y.: Mixco: Mix-up contrastive learning for visual representation. arXiv preprint arXiv:2010.06300 (2020)
22. Krizhevsky, A., Hinton, G., et al.: Learning multiple layers of features from tiny images (2009)
23. Lamb, A., Verma, V., Kannala, J., Bengio, Y.: Interpolated adversarial training: Achieving robust neural networks without sacrificing too much accuracy. In: Pro-ceedings of the 12th ACM Workshop on Artificial Intelligence and Security, pp. 95–103 (2019)
24. Le, Y., Yang, X.: Tiny imagenet visual recognition challenge. CS 231N **7**(7), 3 (2015)
25. Van den Oord, A., Li, Y., Vinyals, O., et al.: Representation learning with con-trastive predictive coding. arXiv preprint arXiv:1807.03748 **2**(3), 4 (2018)
26. Qi, Z., Tian, Y., Shi, Y.: Twin support vector machine with universum data. Neural Netw. **36**, 112–119 (2012)
27. Shafahi, A., et al.: Adversarial training for free! In: Advances in Neural Information Processing Systems, vol. 32 (2019)
28. Shen, Y., Shen, Z., Wang, M., Qin, J., Torr, P., Shao, L.: You never cluster alone. In: Advances in Neural Information Processing Systems, vol. 34 (2021)
29. Shen, Z., Liu, Z., Liu, Z., Savvides, M., Darrell, T., Xing, E.: Un-mix: Rethinking image mixtures for unsupervised visual representation learning. arXiv preprint arXiv:2003.05438 (2020)

30. Thulasidasan, S., Chennupati, G., Bilmes, J.A., Bhattacharya, T., Michalak, S.: On mixup training: Improved calibration and predictive uncertainty for deep neural networks. In: Advances in Neural Information Processing Systems, vol. 32 (2019)
31. Weston, J., Collobert, R., Sinz, F., Bottou, L., Vapnik, V.: Inference with the universum. In: Proceedings of the 23rd International Conference on Machine Learning, pp. 1009–1016 (2006)
32. Yun, S., Han, D., Oh, S.J., Chun, S., Choe, J., Yoo, Y.: Cutmix: Regularization strategy to train strong classifiers with localizable features. In: Proceedings of the IEEE/CVF International Conference on Computer Vision, pp. 6023–6032 (2019)
33. Zagoruyko, S., Komodakis, N.: Wide residual networks. arXiv preprint arXiv:1605.07146 (2016)
34. Zbontar, J., Jing, L., Misra, I., LeCun, Y., Deny, S.: Barlow twins: Self-supervised learning via redundancy reduction. In: International Conference on Machine Learning, pp. 12310–12320. PMLR (2021)
35. Zhang, H., Cisse, M., Dauphin, Y.N., Lopez-Paz, D.: mixup: Beyond empirical risk minimization. arXiv preprint arXiv:1710.09412 (2017)
36. Zhang, L., Deng, Z., Kawaguchi, K., Ghorbani, A., Zou, J.: How does mixup help with robustness and generalization? arXiv preprint arXiv:2010.04819 (2020)

A Task-Aware Attention-Based Method for Improved Meta-Learning

Yue Zhang$^{(\boxtimes)}$, Xinxing Yang, Feng Zhu, Yalin Zhang, Meng Li, Qitao Shi,
Longfei Li, and Jun Zhou

Ant Group, Hangzhou, China
{yue.zy,xinxing.yangxx,zhufeng.zhu,lyn.zyl,lm168260,qitao.sqt,
longyao.llf,jun.zhoujun}@antgroup.com

Abstract. Based on massive data, deep neural networks have been proven to have a powerful learning capability of non-linear relationships. However, training deep neural networks on limited samples is still challenging, which may lead to the over-fitting problem. To alleviate this problem, meta-learning was proposed to train a model that can rapidly adapt to a new task with only a few related examples. However, existing meta-learning approaches tend to ignore the domain gap between different tasks. For a specific task, some of the features are unrelated or even disruptive, which may cause damage to the effectiveness of meta-learning. To address this issue, in this paper, we propose a novel attention-based method that can skip the useless features and highlight the task-specific information. We design two simple but effective attention modules, which take task representation as input and produce attention weights for features from two different perspectives. Experiments conducted on four benchmarks validate that our method outperforms state-of-the-art methods, and the main idea can be applied to various existing meta-learning models.

Keywords: Meta-learning · Attention mechanism · Few-shot classification · Cold-start recommendation

1 Introduction

The data sparsity problem may seriously damage the effect of deep learning models. Meta-learning is proposed to tackle this issue [3,8,19,20,23,24], which aims to prevent over-fitting and improve generalization by learning from a collection of tasks. Specifically, meta-learning trains the model with several learning tasks sampled from the original dataset, so as to learn the general knowledge across different tasks. With the learned general knowledge, the model can rapidly adapt to new tasks with a limited number of labeled data.

It is realized that not all of the features in different tasks play equally important roles. However, most existing meta-learning models ignore this kind of feature difference. Take the few-shot image classification as an example, suppose

B. Li et al. (Eds.): APWeb-WAIM 2022, LNCS 13422, pp. 474–482, 2023.
https://doi.org/10.1007/978-3-031-25198-6_35

we have such two tasks: 1) discerning "apple" versus "pear", 2) discerning "pear" versus "lemon". Intuitively, each few-shot classification task requires a different weight allocation of the feature dimensions. For the first task, we only need to focus on the color, which is the significant difference between apple and pear. While for the second task, the shape is the crucial difference between pear and lemon. Thus the model should pay more attention to this feature dimension and reduce interference from other dimensions such as color. In particular, there is another issue of exiting methods for few-shot image classification. Not all of the spatial parts of an intermediate feature map contribute equally to a specific task. Suppose we have two meta-learning tasks: 1) comparing "dog" and "cat", 2) comparing "cat" and "tiger". We would intuitively focus on the positions of the ears, nose, and other organs when discerning between cats and dogs. When comparing cat and tiger, we may pay more attention to the position of fur, whose pattern is the key difference between cat and tiger. Recently, some works [2,7,10,14] have tried to solve part of these problems by designing a task-aware model architecture. TADAM [14] proposes a solution from the perspective of metric, which uses an adaptive metric to learn a task-dependent metric space. CTM [10] uses a concentrator and a projector to find the relevant feature by traversing across and within classes. Most recently, [12] provides an idea to randomly prune the learned features from the pre-trained stage. Nevertheless, none of these methods have simultaneously overcome the two issues mentioned above.

To address the two challenges of existing meta-learning, in this work, we propose a task-aware attention-based method to help the model avoid interference from irrelevant information given specific tasks. To summarize, our main contributions are three folds:

- We propose a novel task-aware method for meta-learning, which strengthens the ability of the model to extract and utilize task-specific information by reducing interference from irrelevant features.
- We design two types of embedding attention module, Dimension Attention Module (DAM) and Region Attention Module (RAM), to pinpoint the task-specific information from two different perspectives.
- We apply the proposed modules to both metric-based and optimization-based meta-learning methods, and evaluate their performance on four meta-learning benchmarks. Compared with competitive baseline approaches, our method achieves great improvement.

2   Related Work

Meta-Learning. The main idea of meta-learning is to learn a model that can quickly generalize to new tasks with limited labeled examples (learning to learn). [23] introduced an essential concept called episode training to mimic the test scenario based on meta-learning. In episode training, a support set and a query set are sampled from the training data to build a learning task as an input data item. The support set is used for producing the task-oriented model, and the query set is used to test whether the model has learned the essential information of the learning task. Most existing meta-learning approaches can be classified into two main types: metric-based and optimization-based meta-learning. The metric-based approach aims to learn a comparison model which can classify an unseen test sample with a small number of labeled instances by determining their similarity. In previous studies [19,20,23], it is usually used to perform the few-shot classification tasks. The optimized-based method targets learning to fine-tune the initial model to adapt to the new tasks. One effective idea aims to learn good parameter initialization so that the predictor suitable for a new task can be produced with limited labeled samples and a few gradient update steps. Examples include MAML [24] and its derivative works [3,8,13].

Attention Mechanism. The attention mechanism is firstly introduced in [1], which is designed to provide more precise alignment for each position in machine translation. Recently, the attention mechanism is widely used in various fields [4,18,21] to highlight important local features to extract more discriminative information. Our approach utilizes the strength of the attention mechanism to emphasize the task-related information and exclude the irritated information by designing two task-aware soft attention modules.

3 Proposed Method

3.1 Overall Procedure

Our overall framework is illustrated in Fig. 1. All the samples in the support and query set are first embedded into feature representations by $\phi_x = f_\theta(\mathbf{x})$. $f(\cdot)$ is an embedding function. A task-level embedding presentation τ is then calculated from the support samples. The proposed Task-Aware Attention Module (TAAM) takes τ as input and produces an attention assignation function $\mathbf{\Omega}(\cdot)$, making the learned feature representations task-related. In this way, the proposed model can adaptively reduce the weights of irrelated feature information and improve the weights of crucial information for the specific meta-learning task, like the process of human recognition. After that, with the modified feature representations $\psi_\mathbf{x} = \mathbf{\Omega}_\tau(\phi_\mathbf{x})$, the label of query instance x_{query} can be predicted by prediction function $\mathbf{P}(\cdot,\cdot)$.

$$\hat{\mathbf{y}}_{\text{query}} = \mathbf{P}\left(\psi_{\mathbf{x}_{\text{query}}}, \{\psi_\mathbf{x}, \forall(\mathbf{x},\mathbf{y}) \in \mathcal{D}_{\text{support}}\}\right) \tag{1}$$

Our model is applicable to various types of feature embedding model $f(\cdot)$ and prediction function $\mathbf{P}(\cdot,\cdot)$. The feature embedding model $f(\cdot)$ and the attention assignation function $\mathbf{\Omega}_\tau(\cdot)$ are optimized with different learning rates over tasks sampled from training set $\mathbf{D}$.

3.2 Task-Aware Attention Module (TAAM)

Dimension Attention Module (DAM). As each dimension of a feature representation is considered as a feature detector [25], our Dimension Attention

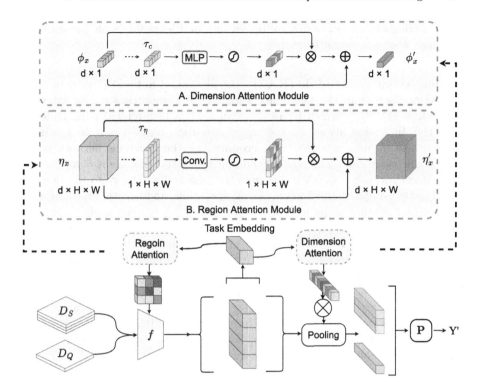

Fig. 1. An overview of the proposed TAAM applied to existing meta-learning model and the details of Dimension Attention Module and Region Attention Module.

Module focuses on which feature dimensions are helpful given an random given task. As illustrated in Fig. 1 (A), for this module, given an N-way K-shot learning task, a task embedding τ_c is computed as the mean of every centroids of N categories: $\tau_c = \frac{1}{N}\sum_i \mathbf{c}_i$, where $\mathbf{c}$ is the mean of all K feature representations in a class: $\mathbf{c} = \frac{1}{K}\sum_k \phi_k$. Next, given an feature representation $\phi_x \in \mathbb{R}^{d\times1}$ and a task embedding $\tau_c \in \mathbf{R}^{d\times1}$, we compute a task-related dimension attention map $\mathbf{M_d} \in \mathbf{R}^{d\times1}$ by forwarding the task embedding to a multi-layer perceptron (MLP) with one hidden layer. The hidden activation is set to $\mathbb{R}^{d/16\times1}$ in order to balance performance and efficiency. Then a softmax function is employed to compute the attention values given the output of the MLP network. After that, we randomly prune some of the attention dimensions using a dropout layer with the pruning rate r. In brief, the task-related dimension attention is computed as: $\mathbf{M_d}(\tau_c) = Dropout(Softmax(MLP(\tau_c)))$. Then we apply the dimension attention map to the feature representation in a residual way. We multiply the dimension attention map by a learnable scale parameter λ and perform an element-wise multiplication operation with the feature representation. The overall attention process can be summarized as:

$$\phi'_x = \mathbf{\Omega}_d(\phi_x) = \phi_x + \lambda\mathbf{M_d}(\tau_c) \otimes \phi_x \tag{2}$$

where $\otimes$ denotes element-wise multiplication, and learnable scale parameter λ gradually learns a weight from 0.

Region Attention Module (RAM). As illustrated in Fig. 1 (B), by utilizing the information from the task embedding, RAM computes a region attention map $\mathbf{M_r} \in \mathbf{R}^{1 \times H \times W}$ which focuses on 'where' is a task-related part. Different from the dimension attention, the region attention plays a part in the intermediate feature map $\eta \in \mathbf{R}^{d \times H \times W}$ produced by the backbone network before the pooling operation. The task embedding τ_η for this module is computed as: $\tau_\eta = MaxPool(\frac{1}{N} \sum_n \bar{\eta}_n)$. We compress the task embedding τ_η to the shape of $1 \times H \times W$ using a channel max-pooling operation, and $\bar{\eta}$ is the mean of all K intermediate feature maps η in one class: $\bar{\eta} = \frac{1}{K} \sum_k \eta_k$. We apply a standard convolution layer to the task representation to produce a two-dimensional region attention map $\mathbf{M_r} \in \mathbf{R}^{1 \times H \times W}$. Similar to the dimension attention, we compute the attention value using a softmax function, followed by a random dropout layer for avoiding over-fitting. In a word, the region attention is computed as: $\mathbf{M_r}(\tau_\eta) = Dropout(Softmax(Conv(\tau_\eta)))$. Then the region attention is applied back to the intermediate feature map η, in a residual way. A learnable scale parameter μ is multiplied by the region attention to adjust its influence on the original model. Finally, we perform a element-wise multiplication between the region attention and the intermediate feature map:

$$\eta_x' = \mathbf{\Omega}_r(\eta_x) = \eta_x + \mu \mathbf{M_r}(\tau_\eta) \otimes \eta_x \tag{3}$$

where $\otimes$ denotes element-wise multiplication and μ is initialized as 0 and gradually learns to assign more weight during training.

4   Experiments

4.1   Experimental Datasets

We evaluate the proposed model in two typical meta-learning problems: few-shot classification and cold-start recommendation. We use four real-world benchmark datasets: *mini*ImageNet [23], *tiered*ImageNet [16], MovieLens 1M [6], and Taobao Display Ad Click[1]. The first two datasets are used for the few-shot classification, and the other two are for the cold-start recommendation.

4.2   Performance Comparison and Analysis

Result 1: Comparison with State-of-the-Art. The results of our method and recent SOTA methods on *mini*ImageNet and *tiered*ImageNet benchmarks are summarized in Table 1. It can be seen that our method significantly outperforms other methods, not limited to the metric-based methods but also compared

---

[1] https://tianchi.aliyun.com/dataset/dataDetail?dataId=56.

with the optimization-based methods. In particular, compared with other relevant task-aware methods like TADAM [14], CAN [7], CTM [10], and ATL-Net [2], our method achieves a better performance. The performance improvement proves that the proposed method can extract and utilize task-specific information in a better way.

Table 1. Comparison with the state-of-the-art methods on *mini*ImageNet and *tiered*ImageNet. The first block of methods are optimization-based, and the second block are metric-based.

Method	miniImageNet		tieredImageNet	
	1-shot	5-shot	1-shot	5-shot
MAML [24]	48.7	63.11	51.67	70.30
Meta-SGD [11]	54.24	70.86	62.95	79.34
PFA* [15]	59.60	73.74	-	-
LEO*† [17]	61.76	77.59	66.33	81.44
MetaOptNet-SVM† [9]	64.09	80.00	65.81	81.75
TADAM [14]	58.50	76.70	-	-
SimpleShot* [22]	62.85	80.02	69.09	84.58
CAN [7]	63.85	79.44	69.89	84.23
wDAE-GNN*† [5]	62.96	78.85	-	-
CTM* [10]	64.12	80.51	68.41	84.28
ATL-Net [2]	54.30	73.22	-	-
TAAMs (ours)	**65.66**	**81.66**	**70.01**	**84.81**

Methods with * utilize a deeper backbone network.
Methods with † are trained on training set and validation set.
− means not reported in original paper.

Result 2: TAAMs with Existing Meta-Learning Methods. To validate the effectiveness of TAAMs, we applied them to three existing metric-based methods [19,20,23] and compared their performance with that of the original models. Table 2 reports the performance improvement after incorporating TAAMs into these three methods on *mini*ImageNet and *tiered*ImageNet. After applying the proposed modules, the improvement in performance is significant. We also apply the proposed module to an optimization-based meta-learning model and compare it with another MAML-based method named MeLU [8] on two datasets. As reported in Table 3, both of the meta-learning methods achieve significant improvement over the traditional methods. Our approach showed better performance than the other MAML-based method with a relative improvement of 9.19% on MovieLens 1M and 7.25% Taobao Display Ad Click dataset. The experiment results validate that our approach can improve the learning-to-learn performance of both metric-based and optimization-based meta-learning models.

Result 3: Ablation Study. We test the two proposed modules' performance in few-shot image classification with the *mini*ImageNet dataset. As reported in the second row and third row of Table 4, both of the two modules achieve significant improvement over the baseline method, and the DAM performs much better than RAM. The result reported in the last row shows that when the two modules are combined, the model achieves the best performance.

Table 2. Performance improvement after incorporating TAAMs into existing metric-based methods on two benchmarks. The evaluation metric is accuracy.

Method	miniImageNet		tieredImageNet	
	1-shot	5-shot	1-shot	5-shot
Matching Net	64.45	71.45	56.29	70.30
w/ TAAM(ours)	**65.24**(+1.21%)	**72.23**(+1.09%)	**57.33**(+1.85%)	**70.35**(+0.07%)
Prototypical Net	62.29	80.53	68.23	84.03
w/ TAAM(ours)	**65.66**(+5.41%)	**81.66**(+1.40%)	**70.01**(+2.61%)	**84.81**(+0.93%)
Relation Net	57.03	69.71	60.66	76.73
w/ TAAM(ours)	**58.09**(+1.86%)	**70.87**(+1.66%)	**61.21**(+0.91%)	**77.45**(+0.94%)

Table 3. Performance improvement after incorporating TAAMs into existing optimization-based methods on two benchmarks. The evaluation metric is AUC.

Method	Movielens 1M	Taobao Display Ad Click
DeepFM	0.5896	0.4998
WDL	0.5884	0.5082
MeLU	0.6607	0.5323
w/ TAAM(ours)	**0.7214**(+9.19%)	**0.5709**(+7.25%)

Table 4. Ablation study results on *mini*ImageNet in 5-way 1-shot settings.

Method	Accuracy	Improvement	Relative improvement
Baseline	62.29	0.00	0.00%
DAM only	65.12	+2.83	+4.54%
RAM only	63.05	+0.76	+1.22%
Full model (ours)	**65.66**	**+3.37**	**+5.41%**

5   Conclusion

In this work, we proposed a novel task-aware attention model for meta-learning, which adjusts feature representation to be more task-related and discriminative. Specifically, we introduce Dimension Attention Module and Region Attention

Module to capture the task information and compute weights for feature dimensions and feature map regions, respectively. Our approach achieves outstanding performance consistently on four benchmarks. It has a highly competitive performance compared with the state-of-the-art. Moreover, the proposed method is a general module that can be applied to various existing meta-learning models.

References

1. Bahdanau, D., Cho, K., Bengio, Y.: Neural machine translation by jointly learning to align and translate. In: Bengio, Y., LeCun, Y. (eds.) ICLR 2015 (2015)
2. Dong, C., Li, W., Huo, J., Gu, Z., Gao, Y.: Learning task-aware local representations for few-shot learning. In: Bessiere, C. (ed.) IJCAI, pp. 716–722 (2020)
3. Dong, M., Yuan, F., Yao, L., Xu, X., Zhu, L.: MAMO: memory-augmented meta-optimization for cold-start recommendation. In: KDD 2020. ACM (2020)
4. Fu, J., et al.: Dual attention network for scene segmentation. In: CVPR 2019, pp. 3146–3154 (2019)
5. Gidaris, S., Komodakis, N.: Generating classification weights with GNN denoising autoencoders for few-shot learning. In: CVPR 2019, pp. 21–30 (2019)
6. Harper, F.M., Konstan, J.A.: The movielens datasets: history and context. ACM Trans. Interact. Intell. Syst. 5(4), 19:1–19:19 (2016)
7. Hou, R., Chang, H., Ma, B., Shan, S., Chen, X.: Cross attention network for few-shot classification. In: NeurIPS 2019, pp. 4005–4016 (2019)
8. Lee, H., I.m, J., Jang, S., Cho, H., Chung, S.: MELU: meta-learned user preference estimator for cold-start recommendation. In: SIGKDD 2019, pp. 1073–1082 (2019)
9. Lee, K., Maji, S., Ravichandran, A., Soatto, S.: Meta-learning with differentiable convex optimization. In: CVPR 2019, pp. 10657–10665 (2019)
10. Li, H., Eigen, D., Dodge, S., Zeiler, M., Wang, X.: Finding task-relevant features for few-shot learning by category traversal. In: CVPR 2019, pp. 1–10 (2019)
11. Li, Z., Zhou, F., Chen, F., Li, H.: Meta-SGD: learning to learn quickly for few shot learning. CoRR abs/1707.09835 (2017)
12. Liu, C., Xu, C., Wang, Y., Zhang, L., Fu, Y.: An embarrassingly simple baseline to one-shot learning. In: CVPR Workshops 2020, pp. 4005–4009. IEEE (2020)
13. Lu, Y., Fang, Y., Shi, C.: Meta-learning on heterogeneous information networks for cold-start recommendation. In: KDD 2020, pp. 1563–1573. ACM (2020)
14. Oreshkin, B.N., López, P.R., Lacoste, A.: TADAM: task dependent adaptive metric for improved few-shot learning. In: NeurIPS 2018, pp. 719–729 (2018)
15. Qiao, S., Liu, C., Shen, W., Yuille, A.L.: Few-shot image recognition by predicting parameters from activations. In: CVPR 2018, pp. 7229–7238 (2018)
16. Ren, M., et al.: Meta-learning for semi-supervised few-shot classification. In: ICLR 2018 (2018)
17. Rusu, A.A., et al.: Meta-learning with latent embedding optimization. In: ICLR 2019 (2019)
18. Shen, T., Zhou, T., Long, G., Jiang, J., Pan, S., Zhang, C.: DiSAN: directional self-attention network for RNN/CNN-free language understanding. In: AAAI 2018 (2018)
19. Snell, J., Swersky, K., Zemel, R.S.: Prototypical networks for few-shot learning. In: NeurIPS 2017, pp. 4077–4087 (2017)
20. Sung, F., Yang, Y., Zhang, L., Xiang, T., Torr, P.H.S., Hospedales, T.M.: Learning to compare: relation network for few-shot learning. In: CVPR 2018 (2018)

21. Vaswani, A., et al.: Attention is all you need. In: NeurIPS (2017)
22. Wang, Y., Chao, W., Weinberger, K.Q., van der Maaten, L.: Simpleshot: evisiting Nearest-neighbor classification For Few-shot Learning. CoRR (2019)
23. Wu, J., Li, B., Ji, Y., Tian, J., Xiang, Y.: Text-enhanced knowledge graph representation model in hyperbolic space. In: Li, B., et al. (eds.) ADMA 2022. LNCS (LNAI), vol. 13088, pp. 137–149. Springer, Cham (2022). https://doi.org/10.1007/978-3-030-95408-6_11
24. Zang, Y., et al.: GISDCN: a graph-based interpolation sequential recommender with deformable convolutional network. In: DASFAA, pp. 289–297. Springer (2022). https://doi.org/10.1007/978-3-031-00126-0_21
25. Zeiler, M.D., Fergus, R.: Visualizing and understanding convolutional networks. In: Fleet, D., Pajdla, T., Schiele, B., Tuytelaars, T. (eds.) ECCV 2014. LNCS, vol. 8689, pp. 818–833. Springer, Cham (2014). https://doi.org/10.1007/978-3-319-10590-1_53

One-Stage Deep Channels Attention Network for Remote Sensing Images Object Detection

Jinyun Tang, Wenzhen Zhang, Guixian Zhang, Rongjiao Liang,
and Guangquan Lu[✉]

Guangxi Key Lab of Multi-Source Information Mining and Security,
Guangxi Normal University, Guilin 541004, China
{tangjinyun,zgxcs}@stu.gxnu.edu.cn, jianjiu17@outlook.com,
lugq@mailbox.gxnu.edu.cn

Abstract. Although existing remote sensing image object detection methods have made significant evolution in deep learning, they did not fully consider the problem of features loss caused by the correspondingly different importance of different channels of feature maps in the convolution pooling. Therefore, a one-stage deep channels attention network for remote sensing images object detection was proposed. First, through a multi-scale feature representation of the Single Shot MultiBox Detector (SSD) Network, the model can combine semantic information with detailed features to better integrate feature layers with different resolutions. Second, for each additional feature extraction layer, the squeeze and excitation (SE) module is introduced, which adaptively re-calibrates the interdependencies between deep channels, then they achieve the response of channel properties in order to learn more efficient feature information. According to experimental results on the RSOD dataset and NWPU VHR-10 dataset, the models proposed in this paper all realize advanced results and achieve state-of-the-art technical performance.

Keywords: Remote sensing images · Muti-scale feature · Semantic information · Deep channels attention

1 Introduction

Remote sensing images object detection is considered one of the most critical research areas that play a massive role in both military and civil fields. Remote sensing images are bird-eye-view usually. One of the disadvantages of such images is the uneven distribution of the target. Other the size of the object is generally tiny and occupies fewer pixels in the whole image, so it is challenging to distinguish between the background and objects. How to improve precision is a problem that has been studied.

To the best of our knowledge, deep convolutional neural networks (CNNs)have been great succeeded in images processing domains [9]. More people are using one-stage algorithms to reduce time complexity for object detection while achieving high accuracy. The SSD algorithm has the characteristics

B. Li et al. (Eds.): APWeb-WAIM 2022, LNCS 13422, pp. 483–491, 2023.
https://doi.org/10.1007/978-3-031-25198-6_36

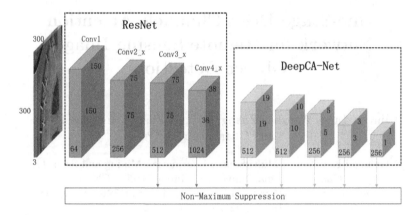

Fig. 1. The overall architecture of our framework. The whole network structure consists of improved ResNet and additional convolutional feature extraction layers which is propossed DeepCA-Net layers.

of multi-scale feature fusion, end-to-end training improves the feature expression power and dramatically reduces the computational complexity. To obtain adaptively re-calibration and realize the response of the interdependence of deep channel features. The SE module is introduced to each of the additional convolutional feature extraction layers. We model the feature correlation between the deep channels while obtaining the global receptive fields and highlight high-level semantic features between deep channels. This paper proposed a one-stage DeepCA-Net for remote sensing images object detection. Our contribution is mainly in three aspects:

- This article proposes a new object detection network for remote sensing images, named DeepCA-Net. We introduce the SE module at each additional convolutional feature extraction layer of the model, strengthening the ability to extract sufficient feature information to highlight high-level semantic features between deep channels.
- Aiming at the defect of small objects semantic feature extraction, we improve the backbone network by adding an additional feature layer of a scale to the feature extraction layer, which increases the diversity of small object semantic feature and enhances the overall detection performance.
- Extensive comparison and ablation experiments were performed on the RSOD dataset and NWPU VHR-10 dataset. The results show that compared with other advanced algorithms, the DeepCA-Net achieves maximum results on both data sets.

2   Methods

Due to the low precision for remote sensing images, this paper proposed model uses the SSD as the base framework with improved ResNet-50 as the backbone.

In addition, we utilize the SE module to enable the model to capture useful feature information better, while a generalized IoU is utilized as a distance assessment indicator.

2.1 Deep Channels Attention Network

This paper uses SSD [7] as the base framework, and the popular benchmark model ResNet-50 [3] are utilized as the backbone and improves it.

SSD has multi-scale characteristics, which can conduct targeted detection of different object sizes. In this work, the scaling sizes of 75×75, 38×38 is set as the first two feature extraction layers, then removed all layers after the original backbone Conv4_x layer, adding five additional convolutional feature layers, setting five different scale sizes of 19×19, 10×10, 5×5, 3×3, 1×1 for high-level semantic information extraction, as shown in Fig. 1.

Aiming at the deficiency of small objects semantic feature extraction, the first residual structure of Conv3_x is modified by us to change the stride of the convolutional kernel on shortcut connection from 2 to 1 and the output of Conv3_x to $75 \times 75 \times 512$. The improved Conv3_x layer, which is the first feature extraction layer, increases the diversity of small object semantic features (shown by Fig. 2). The second feature extraction layer is the Conv4_x layer, which output is $38 \times 38 \times 1024$.

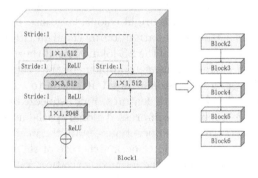

Fig. 2. We modify the convolution kernel step size of the shortcut connection in the first residual structure of Conv3_x from 2 to 1, and the output of Conv3_x becomes $75 \times 75 \times 512$.

To highlight the feature correlation between deep channels, the SE module is introduced into the additional convolutional feature extraction layer, which framework, as shown in Fig. 3. First, we reduce the dimension using a convolutional kernel for 1×1. The stride is set to 1 to ensure no dimensions change to the feature layer. Then, we add BN (Batch Normalization) to normalize the output characteristic layer to accelerate the convergence of the network. The activation function is LeakyReLU, which prevents problems that fail to learn practical features because partial weights cannot be continuously updated. Finally, a $3 \times$

3 convolutional kernel is used to regulate the number of channels in ascending dimensions. The stride is set to 2 to reduce the output feature size. In the above operation cycles five times, this model achieves the results of the multi-scale output feature layer, obtains rich feature information, and improves the detection performance.

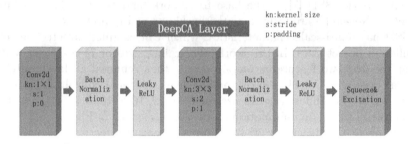

Fig. 3. The first deep channels attention layer structure. This structure add the Batch Normalization layer, the LeakyReLU activation function and the SE module to each additional convolutional feature extraction layer.

2.2 Squeeze and Excitation

In the field of humans visual, due to the visual attention response to the target, the first to notice is the target with bright colour, large size, etc. We can get more effective feature information if the attention mechanism [10] is applied to computer vision. This paper introduces the SE [5] module at the end of the additional convolutional feature extraction layer to implement the deep channels attention mechanism, enabling more precise and rapid attention to adequate information when extracting high-level semantic features. First, squeeze operations were performed global average pooling of compressed global spatial information to obtain channel features. Then, excitation operations obtain the weights for each channel in the feature map through the bottleneck structure of the two fully connected layers. Finally, the information of the learned activation was scaled to the basic features by using scale operations to obtain the output of the SE block and the weighted feature map as input to the following layer network.

2.3 Generalized Intersection over Union

Because of the current CNN detection method, there is a mismatch between classification and location reliability, and we use the GIoU [11] metric proposed in 2019. The specific calculation process of GIoU is as follows: For any A box, B box, find a minimum box C box that can wrap them. Step 1: calculate the C box area minus the area of the intersection of the A box and the B box; Step 2: The ratio of the final area to the C box area in the previous step was calculated; Step 3: Calculate the IoU values for the A and B boxes; Step 4: The

IoU value obtained in step 3 minus the ratio obtained in step 2 finally in GIoU. The formula for the GIoU is shown in (1).

$$GIoU = \frac{area(B_G \cap B_P)}{area(B_G \cup B_P)} - \frac{|C \backslash (A \cup B)|}{|C|} \tag{1}$$

where B_G represents the ground truth bounding box, and B_P represents the prediction box.

3 Experiments Results and Analysis

This article performs extensive experiments on the RSOD dataset and the NWPU VHR-10 dataset to evaluate the performance of the DeepCA-Net. And compared with other advanced technologies, ablation experiments are implemented on the DeepCA-Net.

3.1 Experimental Setting

Data Sets. The RSOD dataset [8] is a public object detection dataset, which was downloaded, collected, and annotated by Wuhan University for remote sensing image object detection. The dataset contains four categories: aircraft, oiltank, playground, and overpass, which have 976 pictures in total and labeled in the format of the Pascal VOC dataset. To reasonably allocate the dataset, we assigned 90% as the training/validation set and 10% as the test set.

The NWPU VHR-10 is a public object detection dataset for the study, collected and annotated by the Northwestern University of Technology. The NWPU VHR-10 contains ten classes: airplane, ship, storage tank, baseball diamond, tennis court, basketball court, ground track field, harbour, bridge, and vehicle. It consists of 715 high spatial resolution colour images and 85 ultra-high spatial resolution generalized sharpening Color Infrared (CIR) images [2]. To obtain experimental results of the same properties, we assigned the same assignment benchmark as the RSOD.

In the RSOD, due to too few data samples in each category, the trained models have problems such as poor generalization ability. Therefore, We performed the flip scale of the RSOD appropriately and adjusted for colour and saturation.

Implementation Details. We set the experimental parameters uniformly in the experiments on the two datasets. The weight decay set is 0.0005, the momentum set 0.9, and the optimizer adopts the stochastic gradient descent (SGD).

All experiments were performed on an NVIDIA RTX 2080Ti GPU Ubuntu server, implementing the DeepCA-Net using PyTorch, Cuda10.2.

Evaluation Metrics. We used mean Average Precision [4] as performance evaluation criteria. In practice, AP values are the area under the curve plotted using combinations of different precision and recall points.

The average AP of all target types in the dataset can be denoted as mAP, which represents the average accuracy of object detection for an object.

3.2   Comparison with Other Advances Models

To verify that the DeepCA-Net is significantly improved in remote sensing images, compared it on the RSOD and the NWPU VHR-10 with other algorithms.

We refer all the contrast data on the RSOD in [6,16], as shown in Table 1, which includes algorithms such as CF2PN, BDFFDN, SSD, YOLOv3. According to Table 1, the DeepCA-Net achieves 95.59% mAP on the RSOD, and improves 4.18%, 1.98%, and 4.32% compared to BDFFDN, CF2PN, and YOLOv3 respectively. The CF2PN is better than this paper in aircraft, which is adopted focal loss to balance positive and negative samples from background. In Table 2, we cite the advanced methods in [1,12–16] for experimental comparison of the NWPU VHR-10, they were increased by 9.26%, 4.5%, and 2.55%, when compared to BDFFDN, R2CNN, and RFN, respectively.

Table 1. Comparing the results of proposed model with other algorithms on the RSOD dataset.

Method	Aircraft	Oil tank	Overpass	Playground	MAP
RetinaNet	75.01	99.23	54.68	94.66	80.90
YOLOV4-tiny	66.47	99.42	80.68	99.31	86.47
SSD	57.05	98.89	93.51	100.00	87.36
M2Det	80.99	**99.98**	79.10	100.00	90.02
YOLOv3	84.80	99.10	81.20	100.00	91.27
BDFFDN	90.81	90.73	84.12	100.00	91.41
CF2PN	**95.52**	99.42	83.82	95.68	93.61
Proposed	85.97	98.80	**97.60**	**100.00**	**95.59**

3.3   Ablation Study

To validate the DeepCA-Net, we performed ablation studies using the uniform settings for the RSOD and the NWPU VHR-10, the first two rows are the experimental results of Original-Net, and the last three rows are the experimental results of DeepCA-Net proposed in this paper, as shown in Table 3. First, the LeakyReLU activation function was verified. Compared to the ReLU activation function, LeakyReLU does not always output 0 when the input value is negative. This ensures that the neurons can update the parameters in time. The results show that 87.41% and 92.34% of the mAP were obtained, respectively. Then, the IoU was replaced by the GIoU, and we found the mAP improved again, achieving 89.62% and 94.06% mAP, respectively. Finally, the DeepCA layer is an additional convolutional feature extract layer to better capture the semantic information in the deep layers. The results show the importance of re-calibrating the feature of deep channels, significantly improving the DeepCA-Net performance. The average accuracy of 95.59% and 96.25% was obtained on the RSOD and the NWPU VHR-10, respectively.

Table 2. Comparing the results of proposed model with other algorithms on the NWPU VHR-10 dataset.

Method	BDFFDN	RDAS512	R^2CNN	NL-TLFPN	HSP	RFN	Proposed
Airplane	99.02	99.60	**100.00**	**100.00**	99.79	98.00	99.88
Ship	78.89	85.50	89.41	89.47	92.45	89.70	**99.04**
Storage tank	90.67	89.00	97.22	90.91	96.96	83.60	**100.00**
Baseball diamond	90.68	95.00	97.00	96.80	**98.55**	98.20	98.06
Tennis court	90.91	89.60	83.15	96.65	90.37	93.30	**97.70**
Basketball court	81.50	94.80	87.54	99.19	91.48	**100.00**	89.94
Ground track field	100.00	95.30	99.17	**100.00**	99.04	96.70	99.59
Habor	90.70	82.60	99.40	90.09	88.90	100.00	**100.00**
Bridge	86.23	77.20	75.51	79.05	87.14	85.10	**90.70**
Vehicle	81.30	86.50	90.10	90.16	89.07	**92.40**	87.59
MAP	88.99	89.50	91.75	93.23	93.38	93.70	**96.25**

Table 3. Ablation experiments for the RSOD dataset and the NWPU VHR-10 dataset.

Model	SGD	LeakyReLU	GIoU	DeepCA	RSOD	NWPU
Original-Net	✓				86.49	90.74
	✓	✓			87.41	92.34
DeepCA-Net	✓	✓	✓		89.62	94.06
	✓	✓		✓	91.45	94.11
	✓	✓	✓	✓	95.59	96.25

4   Conclusions

This paper proposes a one-stage DeepCA-Net for remote sensing images object detection, aiming to solve the problem of low accuracy of remote sensing image detection. The model introduces the SE module in the additional convolutional feature extraction layer. This method automatically obtains the feature correlation between the deep channels through learning and improves the sensitivity to the features in the deep channels. An additional scale feature extraction layer is added to enrich the semantic features of small objects and improve the overall detection performance to extract the semantic features of small objects better. We perform ablation experiments on the RSOD and the NWPU VHR-10, and is also compared with other algorithms. According experimental results show that the DeepCA-Net model achieves the current other advanced model's performance.

Acknowledgements. This work is partially supported by the Project of Guangxi Science and Technology (GuiKeAD20159041), the Research Fund of Guangxi Key Lab of Multi-source Information Mining & Security (No.20-A-01–01, No.20-A-01–02, MIMS21-M-01, MIMS20-M-01, MIMS20-04) and the Innovation Project of Guangxi

Graduate Education (YCSW2022124); the Guangxi "Bagui" Teams for Innovation and Research, China, the Guangxi Collaborative Innovation Center of Multi-Source Information Integration and Intelligent Processing.

References

1. Chen, S., Zhan, R., Zhang, J.: Geospatial object detection in remote sensing imagery based on multiscale single-shot detector with activated semantics. Remote Sens. **10**(6), 820 (2018)
2. Cheng, G., Han, J., Zhou, P., Guo, L.: Multi-class geospatial object detection and geographic image classification based on collection of part detectors. ISPRS J. Photogrammetry Remote Sens. **98**, 119–132 (2014)
3. He, K., Zhang, X., Ren, S., Sun, J.: Deep residual learning for image recognition. In: Proceedings of the IEEE Conference on Computer Vision and Pattern Recognition, pp. 770–778 (2016)
4. Henderson, P., Ferrari, V.: End-to-End Training of Object Class Detectors for Mean Average Precision. In: Lai, S.-H., Lepetit, V., Nishino, K., Sato, Y. (eds.) ACCV 2016. LNCS, vol. 10115, pp. 198–213. Springer, Cham (2017). https://doi.org/10.1007/978-3-319-54193-8_13
5. Hu, J., Shen, L., Sun, G.: Squeeze-and-excitation networks. In: Proceedings of the IEEE Conference on Computer Vision and Pattern Recognition, pp. 7132–7141 (2018)
6. Huang, W., Li, G., Chen, Q., Ju, M., Qu, J.: Cf2pn: a cross-scale feature fusion pyramid network based remote sensing target detection. Remote Sens. **13**(5), 847 (2021)
7. Liu, W., et al.: SSD: single shot multibox detector. In: Leibe, B., Matas, J., Sebe, N., Welling, M. (eds.) ECCV 2016. LNCS, vol. 9905, pp. 21–37. Springer, Cham (2016). https://doi.org/10.1007/978-3-319-46448-0_2
8. Long, Y., Gong, Y., Xiao, Z., Liu, Q.: Accurate object localization in remote sensing images based on convolutional neural networks. IEEE Trans. Geosci. Remote Sens. **55**(5), 2486–2498 (2017)
9. Lu, G., Gan, J., Yin, J., Luo, Z., Li, B., Zhao, X.: Multi-task learning using a hybrid representation for text classification. Neural Comput. Appl. **32**(11), 6467–6480 (2020)
10. Lu, G., Li, J., Wei, J.: Aspect sentiment analysis with heterogeneous graph neural networks. Inf. Proces. Manage. **59**(4), 102953 (2022)
11. Rezatofighi, H., Tsoi, N., Gwak, J., Sadeghian, A., Reid, I., Savarese, S.: Generalized intersection over union: a metric and a loss for bounding box regression. In: Proceedings of the IEEE/CVF Conference on Computer Vision and Pattern Recognition, pp. 658–666 (2019)
12. Xu, C., Li, C., Cui, Z., Zhang, T., Yang, J.: Hierarchical semantic propagation for object detection in remote sensing imagery. IEEE Trans. Geosci. Remote Sens. **58**(6), 4353–4364 (2020)
13. Yang, X., et al.: R2CNN++: multi-dimensional attention based rotation invariant detector with robust anchor strategy. **2**, 7 (2018) arXiv preprint arXiv:1811.07126
14. Zhang, W., Jiao, L., Li, Y., Huang, Z., Wang, H.: Laplacian feature pyramid network for object detection in VHR optical remote sensing images. In: IEEE Trans. Geosci. Remote Sens. (2021)

15. Zhou, K., Zhang, Z., Gao, C., Liu, J.: Rotated feature network for multiorientation object detection of remote-sensing images. IEEE Geosci. Remote Sens. Lett. **18**(1), 33–37 (2020)
16. Zhou, L., et al.: Aircraft detection for remote sensing image based on bidirectional and dense feature fusion. Comput. Intell. Neurosci. **2021**, 14 (2021)

Multi-objective Global Path Planning for UAV-assisted Sensor Data Collection Using DRL and Transformer

Rongtao Zhang[1,2(✉)], Jie Hao[1,2], Ran Wang[1], Hai Deng[1], and Hui Wang[1]

[1] College of Computer Science and Technology, Nanjing University of Aeronautics and Astronautics, Nanjing, China
{zrt1997,haojie,wangran,denghai,wanghui9727}@nuaa.edu.cn
[2] Collaborative Innovation Center of Novel Software Technology and Industrialization, Nanjing, China

Abstract. Unmanned-Aerial-Vehicles' (UAVs) inherent features such as high dynamicity, quick deployment, and line of sight communication have motivated the research of UAV-assisted IoT networks. In such networks, one critical issue is path planing scheduling, which unfortunately is a complex multi-objective optimization problem (MOP). Although there exist extensive traditional MOP algorithms, the efficiency is unacceptable due to the resource constrains and they are unscalable for dynamic scenarios. In order to achieve a more efficient yet scalable multi-objective path planing algorithm, we innovatively propose a framework integrating deep reinforcement learning (DRL) and transformer. We firstly decompose the MOP problem into a series of sequencing subproblems with weighted objectives, and then we present a modified transformer network to solve each sequencing subproblem and further a DRL algorithm to facilitate the subproblem network training. Experimental results demonstrate that the proposed algorithm is superior to NSGA-II, MOEA/D and pointer network in terms of robustness, convergence, diversity of solutions, and temporal complexity.

Keywords: Transformer · Multi objective optimization · Deep reinforcement learning

1 Introduction

A UAV assisted IoT-sensor network has attracted extensive attentions recently. In a data collection application, UAV needs to fly to the target IoT sensor nodes (SNs) and hover over them while collecting data in sequence. Therefore, the global path planning (GPP), involving the scheduling on the visiting sequence, the hovering time and the data transmission amount of each target SN, is critical [4].

This work is supported in part by the National Key R&D Program of China under Grant 2019YFB2102000, and in part by the Collaborative Innovation Center of Novel Software Technology and Industrialization.

B. Li et al. (Eds.): APWeb-WAIM 2022, LNCS 13422, pp. 492–500, 2023.
https://doi.org/10.1007/978-3-031-25198-6_37

In fact, GPP usually needs to achieve several conflicting performance metrics. However, the existing MOP algorithms are unsatisfactory in terms of efficiency and scalability. In this paper, we propose an efficient yet scalable GPP algorithm, which can solve the MOP problem elegantly by combining deep reinforcement learning (DRL) with transformer. The main contributions of our work are as follows:

- We propose a framework combining DRL with transformer for solving the multi-objective GPP problem, which utilizes a neighborhood-based parameter transfer strategy to speed up the training process. To the best of our knowledge, it is the first time to use DRL and transformer to solve multi-objective GPP problem.
- We conduct extensive experiments and demonstrate that the proposed algorithm has superior convergence, diversity, and time complexity compared with the traditional MOP algorithms and pointer network. When the locations of SNs change, traditional MOP algorithms have to compute from scratch due to the variable distance matrix, which is unacceptable in practice. Our algorithm is robust to the dynamicity and can be applied to a new IoT network without effort.

2   Related Work

Most of literatures focusing on GPP for UAV-assisted IoT networks [2,4,7] are single objective optimization problems (SOP), which can be divided into three categories according to the optimization objectives: energy consumption minimization, data throughput maximization, and latency minimization. Although this works solve SOP efficiently, they are not suitable to solve the MOP for GPP.

Traditional MOP algorithms such as NSGA-II and MOEA/D have proven to be effective in dealing with MOP. However, they are not suitable for dynamic scenarios and still have room for effect improvement due to the limitation of distance matrix.

In recent years, DRL shows its potential in solving MOPs. An end-to-end framework using DRL has been demonstrated to be superior regarding generalization ability, convergence speed, and solution quality [3,6,8]. It is also capable to integrate pointer network based subproblem solvers. However, the performance is limited by the pointer network and the training speed is relatively slow. As Kool et al. [1] have verified a transformer model based on multi-head attention mechanism that is able to fasten the training speed in a SOP, we are motivated to propose an efficient MOP algorithm for GPP integrating DRL and transformer.

3   System Model

Considering the network scenario shown in Fig. 1, massive battery powered SNs are randomly distributed in a two-dimensional plane and a UAV fles over the target area to collect the data from SNs and return to the data center finally. We assume a clustering routing protocol is adopted and cluster head nodes (CHNs) are elected accordingly [5]. In this paper, we assume that the data amount of each CHN must exceed a given threshold value to ensure the data quality requirement.

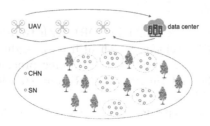

Fig. 1. Network scenario.

Let $C = \{c_1, c_2, \cdots, c_n\}$ represent the CHNs and c_0 represents the data center. The visiting sequence and data amount to be collected are expressed as $\boldsymbol{\rho} = [c_0, \rho_1, \rho_2, \cdots, \rho_n, c_0]^T$ and $\boldsymbol{D} = [d_1^{rec}, d_2^{rec}, \cdots]^T$, respectively, where ρ_k represents the k_{th} visiting target, which is a CHN or the data center, d_k^{rec} denotes the data collection amount of the k_{th} visiting CHN ρ_k. For the GPP, we have the following constraints: $\rho_i \neq \rho_j$, if $i \neq j$ and $d_k \leq d_k^{rec} \leq d_k^{max}$, which means that a CHN can only be visited once and the data collection amount of ρ_k must be beyond a threshold d_k and below its data storage capacity d_k^{max}.

The first objective of GPP is to maximize the data collection amount of all CHNs, which is formulated as

$$\min_{D} \; f_1 = -\sum_{k=1}^{n} log_{d_k^{max}}\left(\frac{\kappa + d_k^{rec}}{\kappa}\right), \tag{1}$$

where κ is a positive number to ensure that the antilogarithm is greater than 1. The second objective is to minimize the total waiting time of all CHNs, consists of the travailing time of the UAV and the data collection time (hovering time over all CHNs). Thus, the second objective is formulated as :

$$\min_{\rho, D} \; f_2 = \sum_{k=1}^{n} (t_k^{tra} + t_k^{col}), \tag{2}$$

where t_k^{tra} and t_k^{col} are the traveling time and data collection time of UAV when visiting the k_{th} CHN, respectively. Here comes the specific definitions: $t_k^{tra} = \sum_{j=0}^{k} \frac{dist(\rho_j, \rho_{j+1})}{v}$, $t_k^{col} = \sum_{j=1}^{k} \frac{d_j^{rec}}{\epsilon}$, where $dist(\rho_j, \rho_{j+1})$ represents the distance between ρ_j and ρ_{j+1}, v and ϵ refer to the speed of a UAV and data transmission rate between CHNs and the UAV respectively.

The aim of our MOP is to determine the visiting sequence and actual data collection amount of each CHN while maximizing the data collection amount and minimizing the waiting time of all CHNs simultaneously:

$$\min_{\rho,D} \quad \boldsymbol{L} = [f_1, f_2] \tag{3}$$

4 Optimization Algorithm

4.1 General Framework

We define a set of uniformly distributed weight vectors $\boldsymbol{\lambda}^0, \boldsymbol{\lambda}^1, \cdots, \boldsymbol{\lambda}^M$, where $\boldsymbol{\lambda}^i = (\lambda_1^i, \lambda_2^i)$, and M is the number of subproblems. The original MOP is decomposed into $M + 1$ scalar optimization subproblems,

$$\min_{\rho,D} \quad \boldsymbol{f}^i = \lambda_1^i f_1 + \lambda_2^i f_2$$

$$s.t. \quad (3), (3) \tag{4}$$

where $\lambda_1^i = 1 - \frac{1}{M}i$, and $\lambda_2^i = 1 - \lambda_1^i$. We adopt a DRL framework and model each subproblem as a neural network and train the networks cooperatively to reduce the computational complexity.

Specifically, we denote the parameters of the i_{th} neural network $\boldsymbol{f}^i$ as $\boldsymbol{P}^i$. Two neighboring subproblems, say $\boldsymbol{f}^{i-1}$ and $\boldsymbol{f}^i$, should have similar solutions as their weight vectors are numerically close. Thus, $\boldsymbol{P}^{i-1}$ can be utilized to obtain $\boldsymbol{P}^i$ and accelerate the overall model training.

4.2 Subproblem Solution

A modified transformer network is utilized to solve the subproblems, and a greedy rollout baseline algorithm [1] is used to train this network.

First, we introduce the input form of the subproblem $\boldsymbol{X} = \{\boldsymbol{x}^1, \cdots, \boldsymbol{x}^n\}$, where $\boldsymbol{x}^i = (a^i, b^i, d^i)$ is a tuple composed of the geographic abscissa, ordinate, and minimum transmitted data amount of the i_{th} CHN. Then, the output of the subproblem model is the permutation of CHNs, i.e., $\boldsymbol{Y} = \{\rho_t, t = 1, 2, \cdots, n\}$. Y is calculated by decomposing the sequence using the chain rule:

$$P(Y|X) = \prod_{t=1}^{n-1} P(\rho_{t+1}|\rho_1, \cdots, \rho_t). \tag{5}$$

A modified transformer network is utilized to model (5), which consists of an encoder and a decoder. The structure of the encoder is shown in Fig. 2. From the 3-dimensional input features $\boldsymbol{X}$, the encoder computes the initial d_h-dimensional embedding $\boldsymbol{H} = \{\boldsymbol{h}_1, \cdots, \boldsymbol{h}_n\} = W^{emb}\boldsymbol{X} + b^{emb}$ through a linear projection, where W^{emb} and b^{emb} are trainable parameters. Then N attention layers are utilized to process the embedding $\boldsymbol{H}$. Each attention layer is consisted of two sublayers : a multi-head attention (MHA) layer with M heads and a node-wise fully connected feed-forward (FF) layer. In addition, batch normalization (BN) and residual (RES) connect are used in each sublayer to speed up training and mitigate information loss. We denote the outcome of the i_{th} attention layer as $A^i = \{\alpha_1^i, \cdots, \alpha_n^i\}$, $i \in \{1, \cdots, N\}$. The encoder also computes an aggregated vector $\boldsymbol{avg}$ according to the last attention layer's outcome $A^N : \boldsymbol{avg} = \frac{1}{n} \sum_{i=1}^{n} \alpha_i^N$.

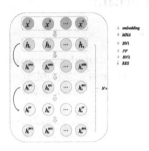

Fig. 2. The structure of the encoder.

We calculate the probability of visiting CNHs $\boldsymbol{pro}$ using single-head attention mechanism at decoding steps. We augment h_c to represent the decoding context : $h_c = [\boldsymbol{avg}, previous, rest]$. Here $[\cdot, \cdot, \cdot]$ is horizontal concatenation operator. *previous* is the information of the last visited CHN processed by the encoder, which is one of the $A^N = \{\alpha_1^N, \cdots, \alpha_n^N\}$. *rest* is the remaining storage capacity of the UAV. Here we compute a new decoding context h_c' using M-head MHA at each decoding step:

$$Q' = W^{Q'} h_c \qquad K' = W^{K'} A^N \qquad V' = W^{V'} A^N \tag{6}$$

$$temp = \begin{cases} -\infty & rest \leq d^{rec} \\ \frac{Q' K'^T}{\sqrt{d_k}} & rest > d^{rec} \end{cases} \tag{7}$$

$$Attention(Q', K', V') = softmax(temp)V'$$

$$head_i = Attention(Q' W_i^{Q'}, K' W_i^{K'}, V' W_i^{V'}) \tag{8}$$

$$h_c' = Concat(head_1, \cdots, head_n)W^{O'}$$

where $W^{Q'}$, $W^{K'}$, $W^{V'}$, $W_i^{Q'}$, $W_i^{K'}$, $W_i^{V'}$ and $W^{O'}$ are trainable parameters. d^{rec} is the data collection amount of all CHNs and its details will be given in Sect. 4.3. Then we compute the possibilities of visiting the next CHN $\boldsymbol{pro}$ using single-head attention mechanism:

$$Q^{pro} = W^{Q^{pro}} h'_c, \qquad K^{pro} = W^{K^{pro}} A^N, \tag{9}$$

$$temp^{pro} = \begin{cases} -\infty & rest \leq d^{rec} \\ C \cdot tanh(\frac{Q^{pro} K^{proT}}{\sqrt{d_k}}) & rest > d^{rec}, \end{cases} \tag{10}$$

$$\boldsymbol{pro} = [pro_1, \cdots, pro_n] = softmax(temp^{pro}), \tag{11}$$

where Q^{pro} and K^{pro} are trainable parameters.

4.3 Determination of Data Collection Amount

Given next visiting target ρ_i, its appropriate data transmission amount is determined by optimizing:

$$\max \quad w_1 \cdot log_{d_i^{max}}(\frac{\kappa + d_i^{rec}}{\kappa}) - w_2 \cdot \frac{l \cdot d_i^{rec}}{\epsilon}, \tag{12}$$

$$s.t. \qquad d_i \leq d_i^{rec} \leq d_i^{max}$$

where w_1 and w_2 refer to the weight of the data collection benefit and waiting time respectively, l is the number of remaining CHNs, that is $l = n - i + 1$. Accordingly the solution of Eq.(12) can be obtained as follows:

$$d_i^{rec} = \begin{cases} d_i & s_i < d_i \\ s_i & d_i \leq s_i < d_i^{max} \\ d_i^{max} & d_i^{max} \leq s_i \end{cases} \tag{13}$$

4.4 Training Method

Here we define the loss function according to (3): $\mathcal{L} = L(Y)$, where $Y = Trans_\theta(\boldsymbol{X})$. We optimize $\mathcal{L}$ by gradient descent : $\nabla\mathcal{L} = \mathbb{E}[(L(Y) - L(BL(\boldsymbol{X}))) \nabla_\theta \log Trans_\theta(\boldsymbol{X})]$, where BL is a rollout baseline network that measures the effectiveness of the modified transformer network. Algorithm 1 presents the training procedure. First, we initialize θ and θ^{BL} randomly in line 1, which are the parameters of transformer network and baseline network. Then, two visiting sequences Y_i and Y_i^{BL} are obtained by using our modified transformer network and baseline networks respectively in line 4 and 5. Gradient information can be calculated from two permutations in line 6 and the parameters of transformer network can be updated in line 7. As for baseline network, the parameters are updated only if the improvement of target network is significant according to a paired t-test.

Algorithm 1. REINFORCE with Rollout Baseline

Input: number of epochs E, batch numbers of each epoch T, batch size B, significance
        γ
1: Init θ, θ^{BL} ← $Random\_Initialize$
2: **for** epoch $=$ ← $1, \cdots, E$ **do**
3:     **for** i $=$ ← $1, \cdots,$ T **do**
4:         Y_i ← Sample(X_i, θ)
5:         Y_i^{BL} ← Greedy(X_i, θ^{BL})
6:         $\nabla\mathcal{L}$ ← $(L(\pi_i) - L(\pi_i^{BL}))\, \nabla_\theta \log Trans_\theta(\pi_i)$
7:         θ ← $Adam(\theta, \nabla\mathcal{L})$
8:     **end for**
9:     **if** OneSidedPairedTTest(θ, θ^{BL}) $< \gamma$ **then**
10:        θ^{BL} ← θ
11:    **end if**
12: **end for**

5   Results and Discussion

We first compare the PF (Pareto Front) obtained by our algorithm, traditional
MOP algorithms and pointer network with 20 and 30 CHNs. As shown in Fig. 3,
all of the compared algorithms show a great ability of convergence. It is observed
obviously that our method can obtain less f_2 for varying f_1, that is it can collect
more data with varying total waiting time.

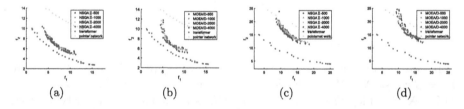

(a)   (b)   (c)   (d)

Fig. 3. PF from modified transformer, traditional MOP algorithms and pointer network. (a) and (b) show the case of 20 CHNs (c) and (d) show the case of 30 CHNs.

We further compare the running time and HV values of the algorithms. As
shown in Table 1, our modified transformer network performs best. In addition,
due to the limitation of distance matrix, NSGA-II and MOEA/D have to execute
every time once the optimization problem changes. In contrast, our DRL and
transformer method is robust to the problem dynamics.

Table 1. Comparison of algorithm execution time and HV values

	20 CHNS		30 CHNS	
	Time/s	HV	Time/s	HV
NSGAII-500	5.75	62.51	5.98	55.93
NSGAII-1000	11.86	66.03	10.99	62.60
NSGAII-2000	20.95	67.20	21.14	59.09
NSGAII-4000	41.92	68.61	44.07	73.30
MOEA/D-500	24.04	54.02	24.13	33.97
MOEA/D-1000	48.25	57.27	49.64	37.76
MOEA/D-2000	99.72	60.09	103.40	57.74
MOEA/D-4000	207.94	63.64	211.14	60.78
Pointer network	5.28	10.43	**7.68**	10.02
Ours	**2.65**	**94.08**	8.38	**220.00**

6 Conclusions

In this paper, we present a novel algorithm to solve the multi-objective global path planning for UAV-assisted IoT, where the data collection benefit is maximized and the waiting time of CHNs is minimized. MOP based on DRL is still in its infancy. Our work, as we believe, shall contribute useful insights helping the development and application of UAVs and MOP based on DRL.

References

1. Kool, W., Hoof, H.V., Welling, M.: Attention, learn to solve routing problems! In: 7th International Conference on Learning Representations, ICLR 2019, New Orleans, LA, USA, May 6–9, 2019. OpenReview.net, pp. 1–25 (2019)
2. Li, J., et al.: Joint optimization on trajectory, altitude, velocity, and link scheduling for minimum mission time in UAV-aided data collection. IEEE Internet of Things J. **7**(2), 1464–1475 (2019)
3. Li, K., Zhang, T., Wang, R.: Deep reinforcement learning for multiobjective optimization. IEEE Trans. Cybern. **51**(6), 3103–3114 (2020)
4. Samir, M., Sharafeddine, S., Assi, C.M., Nguyen, T.M., Ghrayeb, A.: UAV trajectory planning for data collection from time-constrained IoT devices. IEEE Trans. Wirel. Commun. **19**(1), 34–46 (2019)
5. Singh, J., Yadav, S.S., Kanungo, V., Pal, V., et al.: A node overhaul scheme for energy efficient clustering in wireless sensor networks. IEEE Sens. Lett. **5**(4), 1–4 (2021)
6. Wang, H., Wang, R., Xu, H., Kun, Z., Yi, C., Niyato, D.: Multi-objective mobile charging scheduling on the internet of electric vehicles: a DRL approach. In: 2021 IEEE Global Communications Conference (GLOBECOM), pp. 01–06. IEEE (2021)

7. Zhan, C., Zeng, Y.: Aerial-ground cost tradeoff for multi-UAV-enabled data collection in wireless sensor networks. IEEE Trans. Commun. **68**(3), 1937–1950 (2019)
8. Zhu, B., Bedeer, E., Nguyen, H.H., Barton, R., Henry, J.: UAV trajectory planning in wireless sensor networks for energy consumption minimization by deep reinforcement learning. IEEE Trans. Veh. Technol. **70**(9), 9540–9554 (2021)

SAC-PER: A Navigation Method Based on Deep Reinforcement Learning Under Uncertain Environments

Xinmeng Wang, Lisong Wang[⊠], Shifan Shen, and Lingling Hu

College of Computer Science and Technology,
Nanjing University of Aeronautics and Astronautics, Nanjing 211106, China
{Wxinmeng,wangls,shenshifan,linglinghu}@nuaa.edu.cn

Abstract. In real scenarios, robots usually face dynamically changing environments, and traditional navigation methods require a predefined high-precision map, which limits the achievability of navigation in dynamic and uncertain environments. To solve this problem, this paper uses a Partially Observable Markov Decision Process (POMDP) to model the uncertain navigation planning problem and proposes a soft actor-critic with prioritized experience replay (SAC-PER) method based on multi-sensor perception to achieve efficient navigation. The method uses multi-source information fusion for environment perception and Deep Reinforcement Learning (DRL) for continuous control of navigation. The multi-source SAC-PER method can effectively avoid obstacles and enable robots to perform navigation tasks autonomously in uncertain environments without building high-precision maps. We evaluate the proposed method using Robot Operating System (ROS) and Gazebo simulator. The results demonstrate that the SAC-PER method has high efficiency and robustness in different environments, and shows good generalization ability.

Keywords: Uncertain environments · Multi-sensor data · POMDP model · Deep reinforcement learning · Navigation and obstacle avoidance

1 Introduction

The intelligent autonomy of mobile robots enables them to be used in more complex or dangerous environments to replace human tasks, such as ground exploration [9], and disaster rescue [5]. With the expansion of the application range, many uncertainties have been added to the operating environment, which makes the field of motion planning and navigation extremely challenging. Therefore, navigation control in unknown environments, especially in dynamic uncertain environments, has been a research hotspot in the field of mobile robots [2,4].

© The Author(s), under exclusive license to Springer Nature Switzerland AG 2023
B. Li et al. (Eds.): APWeb-WAIM 2022, LNCS 13422, pp. 501–510, 2023.
https://doi.org/10.1007/978-3-031-25198-6_38

Traditional navigation technologies mostly make path planning through complete map information [6]. In an unknown environment, traditional algorithms can only use the knowledge stored in the map and need to manually adjust the parameters, which is difficult to generalize to more complex scenarios [11]. The current navigation research shows that although traditional navigation technologies have made great progress, they still cannot deal with environmental uncertainty, which limits the generalization ability in complex environments.

To solve the above problems, many scholars apply reinforcement learning to navigation systems to control mobile robots [3,12]. However, the classical reinforcement learning algorithm has the disadvantages of slow convergence speed and manual extraction of data features in complex and dynamic environments. To make up for these defects, Google's AI team DeepMind proposes a deep reinforcement learning algorithm, which combines the perception ability of deep learning with the autonomous decision-making ability of reinforcement learning.

To solve the motion planning problem of robots in unknown and uncertain environments and further optimize navigation efficiency, this paper carries out research work from two aspects of perception and decision-making.

- Environmental perception: A multi-sensor perception method is proposed that fuses precise data from laser measurement with complementary information from visual images. This method can extract richer environmental features, enhance the observability of the environment, and make navigation in uncertain environments more reliable and robust.
- Decision making: We model the uncertain navigation problem as a POMDP model and propose a deep reinforcement learning SAC-PER algorithm to solve the navigation model. Our method takes perceptual signals as input and motion control as output, avoiding complex dynamics theory. It can deal with the change of uncertain dynamic environments and achieve collision-free navigation safely with low cost and computation.

2   Related Work

There are two main decision-making methods for navigation, one is the classic map-based navigation, and the other is the map-less navigation. According to environmental information obtained by robots, map-based navigation methods are divided into global path planning and local path planning, such as the D* algorithm, Rapidly Exploring Random Tree (RRT) [1], Probabilistic Roadmap Method (PRM) [10], and so on, which are relatively easy to implement. However, the environmental information perceived by the robot is limited, and the environmental information is unpredictable and uncertain [8]. The above traditional navigation methods are overly dependent on environmental maps and easily affected by environmental changes, which makes it difficult to ensure the safety and stability of navigation.

Traditional navigation methods need to prepare prior knowledge of environments in advance and have poor flexibility. Therefore, map-less navigation based on deep reinforcement learning (DRL) is being actively researched. It is better than traditional methods and has better adaptability to high-dimensional dynamic unknown environments. Lei Tai et al. [7] propose a map-less method based on DRL, which takes laser results and target position as input, and takes control commands as output. The trained planner can be applied to complex environments. Although the DRL-based method is a hot topic in the field of navigation, most research works do not consider dynamic obstacle avoidance.

3   Uncertain POMDP Model and SAC-PER Approach

Considering the dynamic properties and uncertainties of the environment, we design the POMDP process to model the navigation problem, and use the deep-RL strategy to convert the multi-source fusion information into robot actions.

3.1   Designed Uncertain Model for Navigation Problem

The movement of obstacles will lead to environmental uncertainty. To find an effective path, we design the dynamic navigation problem as a POMDP model:

$$tuple = (S, A, P, R, \Omega, O, \gamma) \tag{1}$$

where S is the state space; A is the action space; P is the state transition function; R is the environmental reward; Ω is the observation space, where the observations represent obstacles information; O is the observation transfer function. Below are descriptions of the important components of the POMDP model.

State Space S: Through the analysis of the odometer sensor information, the state space is the set of all possible states of the mobile robot:

$$s_i = (position.x, position.y, yaw) \quad i \in (0, n) \tag{2}$$

Observation Space Ω: The observation space at time t is composed of multi-sensor information, the current speed of the robot, and the target position:

$$O_t = [O_t^{percep}, O_t^v, O_t^{goal}] \tag{3}$$

where $O_t^{percep} = [O_t^{laser}, O_t^{camera}]$, which represents obstacle distance samples observed by multiple sensors. It consists of two parts, O_t^{laser} represents laser measurement samples, and O_t^{camera} represents the 'Pseudo-laser' distance data converted from a depth image. O_t^v represents the speed of the robot, including the linear and angular velocity, and O_t^{goal} represents the relative target position.

Action Space A: An action space is a set of velocity instructions. The motion of the robot consists of linear and angular velocities, namely $a_t = [v_t, w_t]$.

Model Goal: The goal of the POMDP model is to maximize the cumulative discounted reward R_t. γ is a discount factor for future rewards, which represents uncertainty by discounting future rewards.($\gamma \in (0,1)$).

$$R_t = \sum_{i=t}^{\infty} \gamma^{i-t} r_i \qquad (4)$$

Reward Value Mechanism: We combine sparse and dense rewards to design the mechanism, which reduces the number of interactions, accelerates learning process, and improves the utilization of samples.

$$r(s_t, a_t, s(t+1)) = \begin{cases} r_{arrive} & \text{\textit{arrive to the goal}} \\ r_{collision} & \text{\textit{collisions occur}} \\ r_{distance} : \mu(d_t - d_{t+1}) & \text{\textit{close to the goal}} \end{cases} \qquad (5)$$

Model Uncertainty Analysis: (1) State uncertainty analysis: Since the environment changes at any time, even if the agent takes deterministic actions, the new state may be uncertain. POMDP model uses the state transition matrix P to represent the new uncertain state.

(2) Environmental observation uncertainty analysis: The observation space Ω and the observation transfer function O are results of uncertain environmental observation. Ω contains observation values, and O is calculated from states and actions, which can effectively deal with environmental uncertainty.

Figure 1 is the framework diagram of the navigation method proposed, which mainly includes three parts: sensor data processing, POMDP dynamic navigation model, and model solution based on SAC-PER learning algorithm.

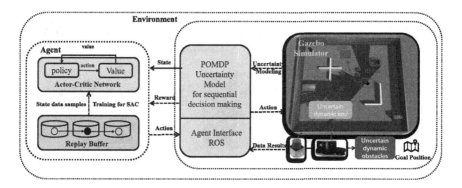

Fig. 1. Structure diagram of map-less navigation approach based on DRL.

3.2 Environmental Obstacle Perception

Algorithm 1. Multi-sensor data processing

Input: Laser distance and camera image data (sensors readings)
Output: Distance O_t from obstacles observed by multiple-sensors
1: Initialize T_{final}, Vector R, R_1, L[n], V[n]
2: **while** $Time < T_{final}$ **do**
3:     R ← laser scan ranges received at time t
4:     **for** $i = 1$ to $length(R)$ **do**
5:         **if** R(i) == NaN **then**
6:             $R(i) \leftarrow R_{max}$
7:         **else**R(i) == NULL
8:             $R(i) \leftarrow 0$
9:     Convert RGB to Depth Map D_t via Depth Estimation Model
10:     (Pseudo-Laser) R_1 ← Compress the D_t into Pseudo-Laser Data
11:     **for** $i = 1$ to n **do**
12:         L ← take n feature samples from R
13:         V ← take n feature samples from R_1
14:         L ← L/max(L)// Normalized laser feature samples to the (0,1) range
15:         V ← V/max(V)// Normalized camera feature samples to the (0,1) range
16:     O_t ← concat distance information L from laser and V from camera

---

To collect environmental information accurately, we use the fusion data from the laser and the camera as perception information for navigation. For laser data, image data has a lot of redundant information, which leads to difficulty in convergence. To solve this problem, we restore the depth information from the camera data and convert it into 'pseudo-laser' data. This not only retains the advantages of simple and easy transfer from the laser but also retains important information from images. The process of multi-sensor perceiving environmental information and data processing is shown in Fig. 2 and Algorithm 1.

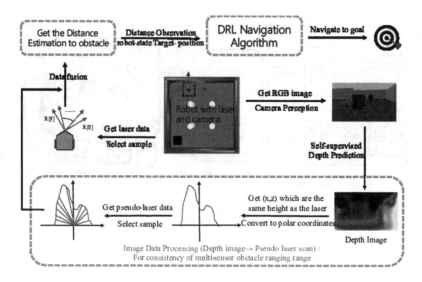

Fig. 2. Structure diagram of multi-sensor perception module.

3.3 Navigation and Obstacle Avoidance Strategy: SAC-PER

Our navigation strategy SAC-PER does not require prior knowledge of the environment. The strategy obtains the optimal objective function through maximum entropy. Compared with on-policy algorithms, SAC-PER can reuse past experience and improve the utilization efficiency of training samples. Compared with deterministic strategies, SAC-PER adopts a stochastic strategy which improves the sensitivity of hyper-parameters, and poor convergence.

Algorithm 2. SAC with Prioritized Experience Replay (SAC-PER)

Input: Initial state s_t
Output: Action a_t and DRL_trained_models
1: Initialize network parameters, weights and an experience buffer D
2: **for** each episode **do**
3:     **for** each environment step **do**
4:         Observe state s_t and select $a_t \sim \pi_\phi(a_t|s_t)$// get action using ActorNet
5:         Execute a_t and observe immediate reward r and new state s'
6:         Calculate the sampling priority of each transition (s,a,r,s')
7:         Add transition $(s, a, r, s', P(i))$ to experience buffer D
8:     **for** each gradient step **do**
9:         Sample a mini-batch of N transitions according to the priority P(i)
10:        Calculate the TD-error of each sample and update it to buffer D
11:        Update network weights and each transition' priority
12:        Update soft Q-value parameters and soft V-value parameters
13:        Update policy network parameters by sampling strategy gradient
14:        Update target networks' parameters

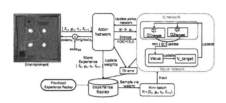 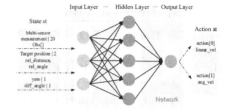

Fig. 3. SAC-PER algorithm structure.

Fig. 4. SAC-PER network input and output.

The SAC-PER algorithm is shown in Fig. 3 and Algorithm 2. The SAC-PER network' input and output are shown in Fig. 4. To speed up the convergence, we integrate the Prioritized Experience Replay mechanism (PER) into the SAC network to improve the algorithm. The method uses metrics to measure sample value, prioritizes each sample based on its value, and finally non-uniformly samples empirical data in order of their priorities, which can quickly search for high-priority experience samples.

4   Experiments and Performance Evaluation

4.1   Experiment Setup

The experimental platform is based on ROS-melodic. We use Gazebo as a simulator and use TensorFlow to implement the SAC-PER algorithm and neural network model. Table 1 gives the optimal parameters obtained by experiments. We design four environments in Gazebo for training and testing (see Fig. 5).

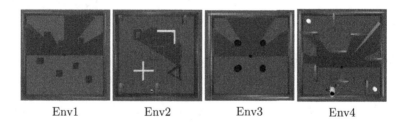

Env1   Env2   Env3   Env4

Fig. 5. Gazebo simulation scenarios.

Table 1. Hyper-parameters

Parameter	Value	Parameter	Value	Reward	Value	Action	Value
$\gamma : discount factor$	0.99	threshold_arrive	0.25	$r_{reached}$	100	V_{max}	0.25
Replay memory size	1×10^5	threshold_collision	0.15	$r_{collision}$	−100	W_{max}	1
Batch size	256	Max episode	1000				

4.2   Experimental Results

Training Results. DDPG and PPO are widely used algorithms and have achieved good results in tasks of robots. Therefore, we choose them to verify the efficiency of the SAC-PER algorithm. We train the model in Env1 (see Fig. 6). According to Fig. 6(a)-(d), the robot has many random behaviors at the beginning of training. As the number of training episodes increases, the robot learns the environment and the navigation ability faster by getting more positive rewards. The training results show our method tends to be stable after about 400 episodes, and the average reward is better than the DDPG and PPO algorithms. Figure 6(e) shows that our algorithm is better than the other in the accuracy of obstacle avoidance. To test the influence of the multi-sensor perception algorithm in Sect. 3.2, we conduct a comparison experiment. As shown in Fig. 6(f), the SAC-PER training models of different sensors all begin to converge after about 500 episodes. And the average reward of the multi-sensor perception is better than that of the single-sensor perception.

Testing and Analysis. As is shown in Fig. 7, there are a few timeouts or collisions at first. After 60 episodes, the robot can successfully avoid obstacles, and the success rate reaches 95%. As shown in Fig. 8, the average navigation time is lower than comparison algorithms, which can complete the navigation faster. To further evaluate the generalization ability of the navigation model, we conduct some experiments (see Table 2). We can find that when the environment becomes complex, the success rate of PPO and DDPG decreases greatly. And our algorithm still maintains stable efficiency and has a better avoidance effect.

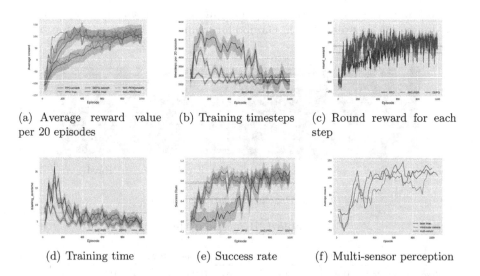

(a) Average reward value per 20 episodes

(b) Training timesteps

(c) Round reward for each step

(d) Training time

(e) Success rate

(f) Multi-sensor perception

Fig. 6. Performance comparison results on Env1 for 1000 training episodes.

Table 2. Success rate and average path length results

Metric	Method	Env1	Env2	Env3	Env4
Success rate (%)	PPO	0.87	0.75	0.80	0.73
	DDPG	0.90	0.74	0.78	0.6
	SAC-PER (Ours)	0.93	0.88	0.91	0.85
AVG path length (m)	PPO	2.26	4.24	3.68	8.85
	DDPG	2.03	5.45	4.56	10.47
	SAC-PER (Ours)	1.59	3.75	3.51	6.52

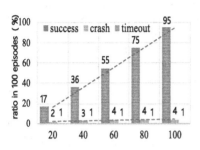

Fig. 7. SAC-PER test results for 100 trajectories in Env1.

Fig. 8. Navigation time.

5 Conclusion

To solve the poor robustness of traditional navigation in uncertain environments, our method uses sensor perception and the POMDP uncertainty model to solve the navigation process. To improve the efficiency of obstacle avoidance, we propose a multi-sensor fusion dynamic perception method and the SAC-PER algorithm to learn the navigation strategy. Future research work is to consider complex outdoor environments to improve the network structure.

Acknowledgment. This work is supported by Civil Aircraft Special Scientific Research Project under grant number MJ-2018-S-29.

References

1. Chen, J., Hu, K., Li, Y.: Research on UAV multi-point navigation algorithm based on mb-rrt*. Comput. Sci. **45**(6A), 85–90 (2018)
2. Duong, T., Das, N., Yip, M.: Autonomous navigation in unknown environments using sparse kernel-based occupancy mapping. In: 2020 IEEE International Conference on Robotics and Automation (ICRA), pp. 9666–9672 (2020)
3. Gavrilov, A.V., Lenskiy, A.: Mobile robot navigation using reinforcement learning based on neural network with short term memory. In: Advanced Intelligent Computing, pp. 210–217 (2012)

4. Mohanan, M., Salgoankar, A.: A survey of robotic motion planning in dynamic environments. Robot. Auton. Syst. **100**, 171–185 (2018)
5. Sarkar, M., Yan, X., Erol, B.A.: A novel search and survey technique for unmanned aerial systems in detecting and estimating the area for wildfires. Robot. Auton. Syst. **145**, 103848 (2021)
6. Shade, R., Newman, P.: Choosing where to go: Complete 3d exploration with stereo. In: 2011 IEEE International Conference on Robotics and Automation, pp. 2806–2811 (2011)
7. Tai, L., Paolo, G., Liu, M.: Virtual-to-real deep reinforcement learning: Continuous control of mobile robots for mapless navigation. In: 2017 International Conference on Intelligent Robots and Systems (IROS), pp. 31–36 (2017)
8. Thrun, S.: Probabilistic robotics. Commun. ACM **45**, 52–57 (2002)
9. Toan, N.D., Woo, K.G.: Mapless navigation with deep reinforcement learning based on the convolutional proximal policy optimization network. In: IEEE International Conference on Big Data and Smart Computing, pp. 298–301 (2021)
10. Yan, F., Zhuang, Y., Xiao, J.: 3d prm based real-time path planning for uav in complex environment. In: 2012 IEEE International Conference on Robotics and Biomimetics (ROBIO), pp. 1135–1140 (2012)
11. Zhou, L., Koppel, D.: Lidar slam with plane adjustment for indoor environment. IEEE Robot. Automat. Lett. **6**(4), 7073–7080 (2021)
12. Zuo, B., Chen, J., Wang, L., Wang, Y.: A reinforcement learning based robotic navigation system. In: 2014 IEEE International Conference on Systems, Man, and Cybernetics (SMC), pp. 3452–3457 (2014)

MAFT: An Image Super-Resolution Method Based on Mixed Attention and Feature Transfer

Xin Liu[1], Jing Li[1][(✉)], Yuanning Cui[2], Wei Zhu[3], and Luhong Qian[3]

[1] College of Computer Science and Technology/College of Artificial Intelligence,
Nanjing University of Aeronautics and Astronautics, Nanjing 211106, China
{liuxin123,lijing}@nuaa.edu.cn
[2] Nanjing University, Nanjing 210023, China
[3] Kunshan Huaheng Welding Co., Ltd., Suzhou 215300, China

Abstract. Reference-based image super-resolution methods, which enhance the restoration of a low-resolution (LR) images by introducing an additional high-resolution (HR) reference image, have made rapid and remarkable progress in the field of image super-resolution in recent years. Most of the existing methods use an implicit correspondence matching approach to transfer HR features from the reference image (Ref) to the LR image. However, these methods lack the further judgment and processing of the HR features from Ref, which limits them in challenging cases. In this paper, We propose an image super-resolution method based on mixed attention and feature transfer (MAFT). First, we obtain the deep features of the LR and Ref images through the encoder network, then extract the transferred features from Ref through the attention network, and perform adaptive optimization processing on the features, and finally fuse the transferred features with LR features to achieve a high-quality image reconstruction. The quantitative and qualitative experiments on these benchmarks, i.e., CUFED5, Urban100 and Manga109, show that MAFT outperforms the state-of-the-art baselines with significant improvements.

Keywords: Computer vision · Machine learning · Super-resolution · Attention mechanism

1 Introduction

Image super-resolution (SR) is a fundamental computer vision task that aims to recover natural high-frequency details from a given low-resolution image. The study of image super-resolution is usually divided into two types: single-image super-resolution (SISR) and reference-based image super-resolution (RefSR). SISR methods relies primarily on the prior knowledge learned by the model to recover the image, However, due to the inherent lack of information between LR and HR images, these classical SISR methods [3,6] often results in blurry effects or visual artifacts.

B. Li et al. (Eds.): APWeb-WAIM 2022, LNCS 13422, pp. 511–519, 2023.
https://doi.org/10.1007/978-3-031-25198-6_39

In this paper, we explore the RefSR method, which additionally introduces HR images as Ref and provides finer details to the LR image by transferring the texture features of Ref to achieve good reconstruction performance. State-of-the-art Approaches [2,5] find the deep feature correspondence between LR and Ref images by an implicit correspondence matching approach to transfer more accurate texture features. However, due to the huge difference between LR and Ref images, some noise information that remains in such texture features will have negatively affect the subsequent fusion with low-resolution image features, network convergence and final results.

To address the above problems, we propose a RefSR method called MAFT, which implements a filtering mechanism oriented to the feature transfer process by combining multiple attention structures, and can effectively distinguish important information from noisy information in the process of transfering features, and enhance the learning of important information and suppress the propagation of noisy information.

2   MAFT Method

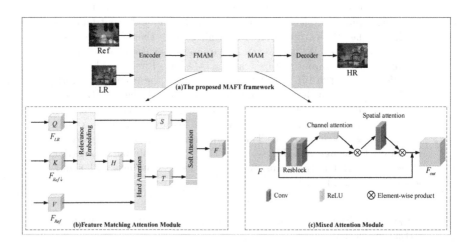

Fig. 1. (a) Framework of the proposed MAFT, which consists of an encoder, Feature Matching Attention Modules (FMAM), Mixed Attention Modules (MAM) and an decoder. (b) Structure of Feature Matching Attention Modules (FMAM), which is used to search for available textures in Ref features. (c) Structure of Feature Matching Attention Modules (MAM), which is used to adapt to strengthen the important information and weaken the noise information in the feature.

As shown in Fig. 1, our framework mainly consists of an encoder network, a decoder network, and an attention layer, where the encoder network is used for feature extraction to extract deep feature information from LR and Ref images, and the attention layer consists of a feature matching attention module and a mixed attention module. The feature matching attention module obtains the correlation between LR features and Ref features and uses the correlation to find the matching features from the high-resolution Ref features, while the mixed attention module further adaptively adjusts the matched features to obtain the final transferred features, and through this coarse-to-fine feature selection mode, the useful feature information on the reference image is accurately mined, and finally the decoder network combines the transferred features with the original LR features to achieve high-quality image recovery.

2.1 Feature Matching Attention Module

The structure of the feature matching attention module is shown in Fig. 1(b). LR denotes the input low-resolution image, Ref denotes the high-resolution image referenced by this low-resolution image, Q and V are the feature maps obtained by passing the LR and Ref through the encoder respectively, and K is the feature map obtained by passing the Ref through a 4-fold factor downsampling first and then through the encoder. The correlation between LR and Ref is first obtained by Q and K, and then the obtained correlation is applied to V to obtain accurate texture transfer information. Details will be discussed below.

Relevance Embedding. The relevance between LR and Ref images is embedded by estimating the similarity between Q and K. First, We unfold both Q and K into patches, denoted as q_i ($i \in [1, H_{LR} \times W_{LR}]$) and k_j ($j \in [1, H_{Ref} \times W_{Ref}]$), with each position corresponding to one patch, and each patch consists of a tensor composed of that position and eight surrounding positions. Then for each patch q_i in Q and k_j in K, we calculate the relevance $r_{i,j}$ between these two patches by normalized inner product:

$$r_{i,j} = \left\langle \frac{q_i}{\|q_i\|}, \frac{k_j}{\|k_j\|} \right\rangle \tag{1}$$

The relevance is further used to obtain the hard-attention map and the soft-attention map.

Hard-Attention. Find the most relevant position in V for each position in Q by the relevance $r_{i,j}$ of q_i and k_j and transform V to obtain a high-resolution feature representation T for feature transfer. Specifically, we first calculate a hard attention graph H where the i-th element h_i ($i \in [1, H_{LR} \times W_{LR}]$) is calculated from the relevance $r_{i,j}$:

$$h_i = \arg\max_j r_{i,j} \tag{2}$$

where the value of h_i can be considered as an index, which represents the most relevant position in the Ref to the i-th position in the LR image. Then we apply an index selection operation to the unfolded patches of V using the obtained hard-attention map as the index:

$$t_i = v_{h_i} \tag{3}$$

where t_i denotes the value of T in the i-th position, which is selected from the h_i-th position of V. Finally, we aggregate all the t_i ($i \in [1, H_{LR} \times W_{LR}]$) to get the HR feature T after feature matching.

Soft-Attention. We further apply different attention weights to different positions of the feature T by using the relevance $r_{i,j}$ of q_i and k_j. Specifically, we first calculate a soft attention graph S where the i-th element s_i ($i \in [1, H_{LR} \times W_{LR}]$) is calculated from $r_{i,j}$:

$$s_i = \max_j r_{i,j} \tag{4}$$

where s_i denotes the i-th position of the soft-attention map S. Then the soft-attention map S is applied to T to obtain the features F:

$$F = T \odot S \tag{5}$$

where $\odot$ denotes element-wise multiplication between feature maps.

2.2   Mixed Attention Module

The process of feature matching LR image with Ref was introduced in Sect. 2.1, and this part will introduce the process of implementing further processing on the obtained matched features. Inspired by ACTION-Net [4], we proposes a mixed attention module to do further feature screening on the transferred features extracted from the previous step.

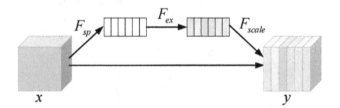

Fig. 2. Structure of channel attention module

Channel Attention. Inspired by RCAN [7], we first used a channel attention mechanism to direct the operational focus of deep neural networks to areas with more important information.

As shown in Fig. 2, for a feature mapping x of size $C \times H \times W$, the spatial information of the feature mapping is first aggregated using average pooling to generate a channel descriptor z_{avg} that can represent each channel, where the c-th element is computed from all elements of that channel:

$$z_{avg}^c = F_{sp}(x_c) = \frac{1}{H \times W} \sum_{i=1}^{H} \sum_{j=1}^{W} x_c(i,j) \tag{6}$$

where $x_c(i,j)$ is the value of the c-th channel feature at (i,j), $F_{sp}(\cdot)$ is the global average pooling function, and then this channel descriptor z_{avg} adaptively learns the weights through an activation network:

$$s = F_{ex}(z_{avg}) = \sigma(W_U \delta(W_D z_{avg}))$$ (7)

where $\sigma(\cdot)$ and $\delta(\cdot)$ denote the Sigmoid function and the ReLU function, respectively. W_D is the weight of the first layer of this activation network and W_U is the weight of the second layer of this activation network. The obtained weights of each channel further act on the input feature mapping x to achieve a different focus on each channel:

$$y_c = s_c \cdot x_c$$ (8)

where s_c and x_c are the weight and feature map in the c-th channel.

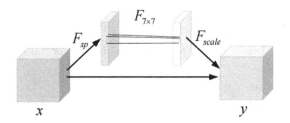

Fig. 3. Structure of spatial attention module

Spatial Attention. In order to further optimize the spatial information of features, we use a spatial attention as shown in Fig. 3. It uses the spatial relationship between features to generate a spatial attention map to act on the original features to achieve the deflation of location information. For a feature mapping x of size $C \times H \times W$, the position information of the feature mapping is first aggregated using average pooling to generate a position descriptor Z_{avg} that can represent each position, where the i-th element is computed from all the channel elements at that position:

$$Z_{avg}^i = F_{sp}(x_i) = \frac{1}{C} \sum_{k=1}^{C} x_i^k$$ (9)

where x_i^k is the value of the i-th position of the feature map x at the c-th channel and $F_{sp}(\cdot)$ is the global average pooling function. This position descriptor Z_{avg} is then passed through a convolution layer to generate a two-dimensional spatial attention map:

$$m = F_{7\times7}(Z_{avg}) = \sigma(W_c Z_{avg})$$ (10)

where $F_{7\times7}$ denotes the convolution operation with a filter size of 7×7, $\sigma(\cdot)$ denotes the Sigmoid function, and W_c is the weight of this convolution layer. The obtained spatial attention map is further acted on the input feature mapping x to realize different attentions for each position:

$$y = m \odot x \tag{11}$$

where $\odot$ denotes element-wise multiplication between feature maps.

As shown in Fig. 1(c), finally, we use the residual block to apply the mixed attention formed by combining channel attention and spatial attention to the feature F to obtain the final transferred features F_{out}:

$$F_{out} = F + SAM\left(CAM\left(W_2\delta\left(W_1F\right)\right)\right) \tag{12}$$

where W_1 and W_2 are the weights of the two convolutional layers in the residual block, CAM denotes the channel attention operation, and SAM denotes the spatial attention operation.

2.3   Loss Function

Reconstruction Loss. We adopt L1 loss as the reconstruction loss as:

$$\mathcal{L}_{rec} = \|I_{HR} - I_{SR}\|_1 \tag{13}$$

where I_{HR} and I_{SR} denote the ground truth image and the network output.

Perceptual Loss. We use the error between the predicted image and the target image on the feature space as the perceptual loss:

$$\mathcal{L}_{pec} = \|\phi_i^{vgg}\left(I_{HR}\right) - \phi_i^{vgg}\left(I_{SR}\right)\|_2 \tag{14}$$

where ϕ_i^{vgg} is the i-th layer feature map of VGG19, here we use $relu1\_1, relu2\_1,$ $relu3\_1$ and $conv5\_4$.

Adversarial Loss. We adopt the Relativistic GANs [6]:

$$\mathcal{L}_D = -E_{I_{HR}}\left[\log\left(D\left(I_{HR}, I_{SR}\right)\right)\right] = -E_{I_{SR}}\left[\log\left(1 - D\left(I_{SR}, I_{HR}\right)\right)\right] \tag{15}$$

$$\mathcal{L}_G = -E_{I_{HR}}\left[\log\left(1 - D\left(I_{HR}, I_{SR}\right)\right)\right] = -E_{I_{SR}}\left[\log\left(D\left(I_{SR}, I_{HR}\right)\right)\right] \tag{16}$$

The final loss is expressed as:

$$\mathcal{L} = \lambda_{rec}\mathcal{L}_{rec} + \lambda_{per}\mathcal{L}_{per} + \lambda_{adv}\mathcal{L}_{adv} \tag{17}$$

The weight coefficients λ_{rec}, λ_{per} and λ_{adv} are set to 1, 0.5, and 0.005, respectively.

3   Experiments

3.1   Datasets and Settings

In this paper, we mainly use CUFED5 as the training set in our experiments, it consists of a combination of 11871 training image pairs, each of which contains an original HR image and a reference image with 160×160 resolution. To validate the generalization capacity of our model, we test it on three popular benchmarks: CUFED5 testing set, Urban100 and Manga109. There are 126 test images in the CUFED5 test set, and each image corresponds to 4 reference images with different similarities. Urban100 contains 100 building images without references, and each image takes its LR image as a reference, and for Manga109, which also lacks reference images, the HR images in the randomly sampled dataset are used as reference images.

In the training process, we choose Adam as the algorithm for training optimization, and the parameters β_1 and β_2 are set to 0.9 and 0.999, respectively. The learning rate is set to 10^{-4} and the batch size is 9.

3.2   Evaluation

To evaluate the effectiveness of MAFT, we compare our model with other state-of-the-art SISR and RefSR methods. The SISR methods include EDSR [1], RDN [8], RCAN [7], ESRGAN [3], RSRGAN [6], The RefSR methods include TTSR [5], MASA [2]. All the models are trained on the CUFED5 training set, and tested on the CUFED5 testing set of Urban100 and Manga109. All experiments are performed with a scaling factor of 4× between LR and HR images.

Table 1. PSNR/SSIM comparison among different SR methods on 3 testing datasets. Methods are grouped by SISR (top) and RefSR (bottom).

Method	CUFED5	Urban100	Manga109
EDSR [1]	25.93/0.777	25.51/0.783	28.93/0.891
RDN [8]	25.95/0.769	25.38/0.768	29.24/0.894
RCAN [7]	26.06/0.769	25.42/0.768	29.38/0.895
ESRGAN [3]	21.90/0.633	20.91/0.620	23.53/0.797
RSRGAN [6]	22.31/0.635	21.47/0.624	25.04/0.803
TTSR [5]	25.53/0.765	24.62/0.747	28.70/0.886
TTSR-rec [5]	27.09/0.804	25.87/0.784	30.09/0.907
MASA [2]	24.92/0.729	23.78/0.712	27.23/0.845
MASA-rec [2]	27.54/0.814	26.09/0.786	30.18/0.908
MAFT	25.84/0.768	24.71/0.745	28.67/0.878
MAFT-rec	**27.72/0.825**	**26.11/0.788**	**30.52/0.910**

Quantitative Evaluations. MAFT denotes the model obtained by using total loss training, MAFT-rec denotes the model obtained by minimizing the reconstruction loss training only, and the same for the rest of RefSR methods. Table 1 shows the quantitative comparisons on PSNR and SSIM, where the best results are bolded to indicate. As shown in Table 1, our model outperforms state-of-the-art methods on all three testing sets.

Qualitative Evaluations. We show visual comparison between our model and other SISR and RefSR methods in Fig. 4. Our proposed MAFT outperforms other methods in terms of visual quality.

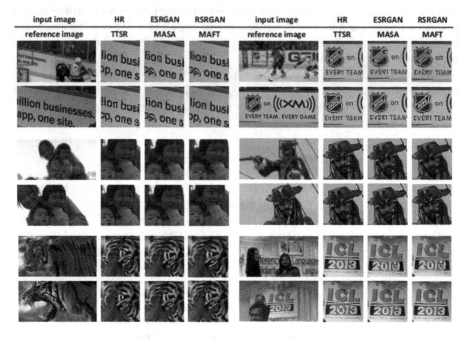

Fig. 4. Visual comparison among different SR methods on CUFED5, Urban100.

References

1. Lim, B., Son, S., Kim, H., Nah, S., Mu Lee, K.: Enhanced deep residual networks for single image super-resolution. In: Proceedings of the IEEE Conference On Computer Vision And Pattern Recognition Workshops, pp. 136–144 (2017)
2. Lu, L., Li, W., Tao, X., Lu, J., Jia, J.: Masa-sr: Matching acceleration and spatial adaptation for reference-based image super-resolution. In: Proceedings of the IEEE/CVF Conference on Computer Vision and Pattern Recognition, pp. 6368–6377 (2021)
3. Wang, X., et al.: ESRGAN: enhanced super-resolution generative adversarial networks. In: Leal-Taixé, L., Roth, S. (eds.) ECCV 2018. LNCS, vol. 11133, pp. 63–79. Springer, Cham (2019). https://doi.org/10.1007/978-3-030-11021-5_5
4. Wang, Z., She, Q., Smolic, A.: Action-net: Multipath excitation for action recognition. In: Proceedings of the IEEE/CVF Conference on Computer Vision and Pattern Recognition, pp. 13214–13223 (2021)
5. Yang, F., Yang, H., Fu, J., Lu, H., Guo, B.: Learning texture transformer network for image super-resolution. In: Proceedings of the IEEE/CVF Conference On Computer Vision And Pattern Recognition, pp. 5791–5800 (2020)

6. Zhang, W., Liu, Y., Dong, C., Qiao, Y.: Ranksrgan: Generative adversarial networks with ranker for image super-resolution. In: Proceedings of the IEEE/CVF International Conference on Computer Vision, pp. 3096–3105 (2019)
7. Zhang, Y., Li, K., Li, K., Wang, L., Zhong, B., Fu, Y.: Image super-resolution using very deep residual channel attention networks. In: Ferrari, V., Hebert, M., Sminchisescu, C., Weiss, Y. (eds.) ECCV 2018. LNCS, vol. 11211, pp. 294–310. Springer, Cham (2018). https://doi.org/10.1007/978-3-030-01234-2_18
8. Zhang, Y., Tian, Y., Kong, Y., Zhong, B., Fu, Y.: Residual dense network for image super-resolution. In: Proceedings of the IEEE Conference On Computer Vision And Pattern Recognition, pp. 2472–2481 (2018)

Automatic Report Generation Method based on Multiscale Feature Extraction and Word Attention Network

Xin Du, Haiwei Pan$^{(\boxtimes)}$, Kejia Zhang, Shuning He, Xiaofei Bian,
and Weipeng Chen

Harbin Engineering University, Harbin, People's Republic of China
panhaiwei@hrbeu.edu.cn

Abstract. A medical report is a textual description of the information presented in a medical image, which includes detailed information about different body organs and the radiologist's diagnosis from medical images. However, when summarizing the medical image content into a complete and accurate medical report, doctors usually face problems such as time-consuming and repetitive work. Although there are many studies in the field of automatic medical report generation, a lot of challenges still exist. First, when describing multiple organs and lesions presented in medical images, the generated report based on the single-scale feature extraction method is still inadequate and inaccurate. Second, when generating reports, most existing methods encounter problems such as duplicate words or lack of key descriptions. To solve the problems mentioned above, we propose Multiscale Feature Extraction and Word Attention Network (MFWAN) which is an automatic medical report generation model. The model contains three modules. In order to focus on abnormalities in different regions, the model includes the EPSA (Efficient Pyramid Split Attention) Multiscale Feature Extraction module which utilizes spatial information at different scales of medical images. After that, the visual features are classified by a Multi-Classification Context Generation Module to generate context messages. Then, by assigning different weights to the hidden layers of word LSTM, the Word-Attention-Based Report Generation module generates more accurate words with implicit disease critical information. Experimental results on benchmark datasets, IU X-Ray, show that our proposed MFWAN outperforms previous works and generates more accurate reports.

Keywords: Medical report generation · Word attention · Hierarchical LSTM · Multiscale

1 Introduction

Medical images, such as radiological and pathological images, are important for medical diagnosis and treatment. A medical report, on the other hand, is a

Supported by the National Natural Science Foundation of China under Grant No. 62072135 and No. 61672181.

B. Li et al. (Eds.): APWeb-WAIM 2022, LNCS 13422, pp. 520–528, 2023.
https://doi.org/10.1007/978-3-031-25198-6_40

textual description of the information presented in a medical image. Figure 1 is an example of a chest X-ray report. "FINDINGS" is a detailed description of each region of the image and "IMPRESSION" indicates the clinical diagnosis inferred by the radiologist through medical images and "FINDINGS". As the number of medical images increases, the workload for radiologists to write reports grows daily. The risk of miscalculations will increase under such a heavy workload, which is detrimental to medical diagnosis and treatment. Therefore, at this stage, there is an urgent need for methods that can generate high-quality medical reports in a short time to reduce the burden on doctors and patients.

FINDINGS: The heart, pulmonary XXXX and mediastinum are within normal limits. There is no pleural effusion or pneumothorax. There is no focal air space opacity to suggest a pneumonia. There is an interim XXXX cervical spinal fusion partly evaluated.
IMPRESSION: No acute cardiopulmonary disease.

Fig. 1. An example of a medical report.

With the development of deep learning, the encoder-decoder structure has achieved good results in image captioning tasks. For a long time, medical report generation models follow the encoder-decoder structure. For the encoder, most models directly utilize pre-trained Convolutional Neural Networks (CNN), however, abnormal lesion regions of medical images rely on local features, and different diseases exist in different regions [6]. In order to extract abnormal information from images more accurately, some models introduce attention mechanisms into the feature extraction stage [7]. However, using single-scale feature extraction cannot guarantee the models extract important spatial information. In addition, medical image reports consist of multiple sentences, so some models utilize a hierarchical LSTM [4] or Transformer [2] as the decoder. However, such approaches make redundant sentences or ignore key information. Recently some models [5] introduce external knowledge into the decoder, but such methods only generate words close to the real medical concepts and cannot extract key information from the images.

Our works focus on two main difficulties in medical report generation: (1) absence of important spatial information of multiple organs and lesions; (2) redundancy of words and lack of keywords. By simulating the practical process of radiologists' work, we propose the Multiscale Feature Extraction and Word Attention Network (MFWAN). Our model follows the standard encoder-decoder paradigm. First, the EPSA Multiscale Feature Extraction module utilizes spatial information at different scales to focus on abnormalities in different image regions, effectively solving the problem of missing important spatial information and establishing correlations between channels according to their importance. Second, the Multi-Classification Context Generation module classifies the extracted visual features and generates high-level semantic vectors with context.

Then, in order to improve the readability of reports, the Word Attention-Based Report Generation module generates words with key information by assigning weights to the hidden layer of word LSTM. Quantitative and qualitative experiment results on benchmark dataset IU X-Ray show that our model outperforms previous works.

2   Methodology

To address two problems of the existing medical report generation methods: (1)ignorance of important information based on single-scale feature extraction and (2)word redundancy, we propose an automatic medical report generation model MFWAN which follows the encoder-decoder structure. The framework of MFWAN is shown in Fig. 2, as the input of the encoder, the chest X-ray image is fed into the EPSA Multiscale Feature Extraction module. The output of this module is visual features containing spatial information. Then, the visual features are fed into a Multilabel Classification network (MLC) to predict labels noted as tags. The output of the encoder is a **ctx** vector with high-level semantic information. The decoder utilizes a hierarchical LSTM with **ctx** as input, generating semantic topics in stages and controlling the number of reported statements. The word LSTM takes the topic and the special word token START as input with a word attention mechanism. The generated words are concatenated to form the complete report. All the modules of the model are described as follows.

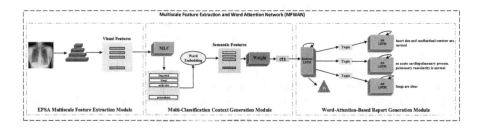

Fig. 2. The framework of proposed MFWAN.

2.1   EPSA Multiscale Feature Extraction Module

Most existing models utilize single-scale CNN for feature extraction of medical images. However, for chest X-ray images, there is more than one lesion area, such as the lung and heart, so the generated report should describe each area. For this reason, this module incorporates multiscale convolutional attention [8] mechanism based on ResNet152 to exploit spatial information at different scales, which helps the model better focus on different regions in the image and improves the accuracy of disease category prediction. Specifically, as shown in Fig. 3, the input feature $X \in R^{H \times W \times C}$ are divided into S groups at the channel level denoted as $[X_0, X_1, \ldots, X_{s-1}] \in R^{H \times W \times C'}$. For each X_i, the module utilizes different kernel sizes and groups, the generated and multiscale features are denoted as:

$$F_i = \text{Conv2d} \ (k_i \times k_i, G_i)(X_i) \quad i = 0, 1, \cdots, S - 1 \tag{1}$$

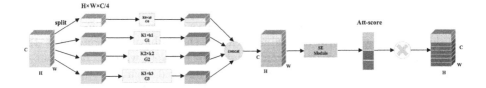

Fig. 3. The framework of EPSA Multiscale Feature Extraction Module.

where $F_i \in R^{H \times W \times C'}$. For each multiscale channel feature map F_i, we use the SEModule [3] to extract the channel attention weights. The weights are denoted as:

$$\text{Att}_i = \text{SEModule}\ (F_i)\quad i = 0, 1, \cdots, S-1 \tag{2}$$

where $Att_i \in R^{1 \times 1 \times C'}$, and each multiscale channel attention vector is concatenated to obtain $Att \in R^{1 \times 1 \times C}$. Then, we multiply the multiscale channel attention with the corresponding scale features to obtain a new feature map:

$$Y_i = F_i \times Att_i\quad i = 0, 1, \cdots, S-1 \tag{3}$$

where $Y_i \in R^{H \times W \times C'}$ pays more attention to the patient's abnormal region. Feature maps are concatenated to obtain $Y \in R^{H \times W \times C}$. As the output of this module, the feature $\mathbf{Y}$ contains not only the spatial information of different regions of the image, but also the association between each channel.

2.2   Multi-classification Context Generation Module

In order to predict the top n labels that are most suitable to represent the image, the visual feature $\mathbf{Y}$ is fed to a Multilabel Classification network (MLC). The classification result is recorded as tags to represent the whole image. Different from CoAtt [4] giving co-attention to visual and semantic features, we only distribute different attention weights α_n to the semantic features A_n, after that the pairs $\langle A_i, \alpha_i \rangle$ are multiplied and summed to obtain the context vector $\mathbf{ctx}$ as the input of the following modules.

2.3   Word-Attention-Based Report Generation Module

The medical report often contains a multi-sentence description with a global description. More detailed descriptions of each region are given with independent and nonrepetitive words. In order to generate multiple and long paragraph descriptions, most existing methods utilize a hierarchical LSTM. However, a single LSTM often suffers problems such as word redundancy and lack of keywords. To solve the problems mentioned above, this module distributes different attention to the hidden layer state of each word to generate more accurate words.

The use of sentence LSTM is the same as that of CoAtt [4]. As shown in Fig. 4, the word LSTM is a two-layer LSTM, which takes the topic obtained from sentence LSTM and a special word token **START** as inputs. The subsequent inputs at each step are the embedding vectors learned from the words. Considering that the words learned in each step should have different values, the hidden layer states generate different weights for different words. The output of the word LSTM is transformed by a linear layer, and then the importance of each word is calculated by a softmax layer. The result from the previous step is multiplied with the output of LSTM as the final word representation. This process can be formalized as:

$$Y_i = \text{softmax}\left(\tanh\left(W_w \cdot h_{it}\right)\right) \cdot LSTM(\text{topic}, START) \tag{4}$$

where $\text{softmax}\left(\tanh\left(W_w \cdot h_{it}\right)\right)$ denotes the weight corresponding to each word, Y_i denotes the last generated word, and W_w is the parameter matrix.

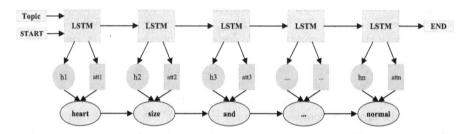

Fig. 4. The framework of Word-Attention-Based Report Generation Module.

2.4   Parameter Learning

For each training sample (X, T, Y), X is the chest X-Ray image, T is the classification label, and Y is the medical report, each report Y contains K sentences and each sentence contains L words. We train our model in an end-to-end manner, the loss function contains three parts, ℓ_{tag} is the Mean Square loss between the predicted tag $Pred_i$ and the ground-truth tag T_i, ℓ_p is the Cross-Entropy loss between the stop control distribution P_i' generated from sentence LSTM and the ground-truth stop control distribution P_i, ℓ_{word} is the Cross-Entropy loss between the generated word Y_i' and the real word Y_i in the report Y.

$$\text{loss} = \lambda_{\text{tag}}\, \ell_{\text{tag}}\left(Pred_i,\ T_i\right) + \lambda_p \sum_1^K \ell_p\left(P_i, P_i'\right) + \lambda_{\text{word}} \sum_1^K \sum_1^L \ell_{\text{word}}\left(Y_i, Y_i'\right) \tag{5}$$

where λ_{tag}, λ_p and λ_{word} are loss weights.

3   Experiments

3.1   Dataset and Experiment Details

We conduct our experiments on the Indiana University Chest X-Ray (IU X-Ray) which is a commonly used dataset for medical report generation tasks.

The dataset contains 7,470 images and 3,955 reports. For images in the dataset, we transform their size to 224 × 224. Tags in the dataset are automatically encoded using MeSH and MTI. We train our model with PyTorch 1.10.0 on a single NVIDIA GeForce RTX 3090 GPU. During the training process, the batch size is set to 16. We adopt the Adam optimizer in an end-to-end method with an initial learning rate of 1e-4, and the number of training epochs is 250. We adopt the ResNet152 and ESPA structure as encoder, where the group information S is set to 4 and the corresponding kernel sizes are 3, 5, 7, and 9. The number of classes in the MLC network is set to 210, and the 10 classes with the highest scores are selected for word embedding. The dimension of all hidden layer states and word embedding are set to 512. The threshold of stop control, the maximum number of sentences K_{max} and the maximum number of words L_{max} are set to 0.5, 6 and 30. λ_{tag}, λ_p and λ_{word} are set to 1, 1, and 0.5 respectively.

3.2   Quantitative Experiments

We compare our model (denoted as "Ours") with the following medical report generation methods: the encoder-decoder-based models CoAtt [4] and MVH [7], the Transformer-based models PPKED [5] and TransB [1], respectively. Note that we conduct the ablation experiments by setting the CoAtt [4] model as baseline. And we evaluate the models mentioned above and "Ours" with NLG metrics, which include BLEU, METEOR and ROUGE-L. According to Table 1, our model outperforms other models in most of the metrics. In the meanwhile, conventional encoder-decoder models perform better than Transformer-based models, which can be illustrated by the comparison between "Ours" and "TransB". The reason behind might be that most encoder-decoder models use cascade decoders, which generate more accurate words based on the semantics of each sentence.

Table 1. Comparison results with other models. The **best** results are highlighted.

Method	BLEU-1	BLEU-2	BLEU-3	BLEU-4	METEOR	ROUGE-L
CoAtt	0.502	0.357	0.295	0.267	0.241	0.430
Mul-Att	**0.529**	0.372	0.315	0.259	**0.343**	0.453
PPKED	0.483	0.315	0.224	0.168	–	0.376
TransB	0.479	0.359	0.219	0.160	0.205	0.380
MFWAN (Ours)	0.501	**0.437**	**0.381**	**0.340**	0.271	**0.473**

3.3   Qualitative Experiments

To further investigate the effectiveness of our model, we implement qualitative experiments on the IU X-Ray dataset. As shown in Fig. 5, we compare the generated reports from our model (denoted as "Generated") with CoAtt [4] and ground-truth, (a) and (b) are normal medical reports while (c) and (d) are abnormal medical reports. There are some findings from Fig. 5. First, the generated normal reports are very similar to the ground-truth. For images with abnormal regions, our model is able to predict keywords accurately and reduce repetitive statements, but the model also generates mismatched sentences. The reason behind this might be that we select the top ten highest-scoring classes in MLC. Second, according to the experimental results, the generated reports based on the normal images are of high quality, while the results of abnormal images are somewhat unsatisfactory. The reason might be that the number of normal medical reports in the dataset exceeds far more than the number of abnormal medical reports. Therefore, it is hard for the model to learn abnormal information.

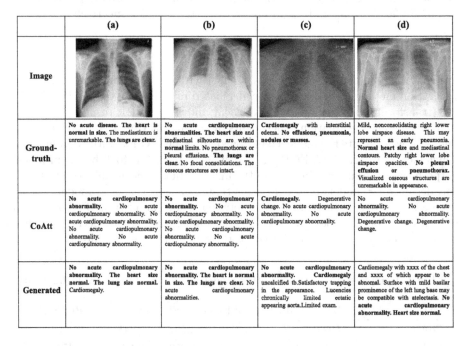

Fig. 5. Comparison results of the qualitative experiments.

3.4 Ablation Experiments

To illustrate the effectiveness of our proposed EPSA Multiscale Feature Extraction module and Word-Attention-Based Report Generation Module, we perform ablation studies with baseline model CoAtt [4] on the same dataset IU X-Ray. The NLG metrics of ablation studies are shown in Table 2. In addition, we conducted ablation experiments to verify how convolution kernel size in the EPSA Multiscale Feature Extraction module affects the model performance, the experimental results are shown in Table 3.

Table 2. The results of the ablation experiments on IU X-Ray dataset.

Method	BLEU-1	BLEU-2	BLEU-3	BLEU-4	METEOR	ROUGE-L
CoAtt	**0.502**	0.357	0.295	0.267	0.241	0.430
Ours (no-WordAtt)	0.445	0.373	0.302	0.271	0.241	0.434
Ours (no-ESPA)	0.487	0.399	0.346	0.318	0.252	0.363
MFWAN (Ours)	0.501	**0.437**	**0.381**	**0.340**	**0.271**	**0.473**

Table 3. Ablation experiments results of the EPSA module with different convolution kernel sizes.

Method	BLEU-1	BLEU-2	BLEU-3	BLEU-4	METEOR	ROUGE-L
k = 1,3,5,7	0.477	0.340	0.300	0.299	0.250	0.394
k = 3,5,7,9	**0.501**	**0.437**	**0.381**	**0.340**	**0.271**	**0.473**
k = 5,7,9,11	0.249	0.222	0.198	0.180	0.200	0.394

4   Conclusion

We propose a novel model MFWAN for medical report generation task. In our model, the EPSA Multiscale Feature Extraction module solves the problems of missing spatial information caused by single-scale feature extraction, hence improving the accuracy of disease prediction. The Multi-Classification Context Generation module predicts disease categories in medical images and generates high-level semantic features. The Word-Attention-Based Report Generation module improves the readability of the generated reports and solves the word redundancy. Experimental results on the IU X-Ray dataset demonstrate that our model outperforms previous works, results of ablation studies show the effectiveness of the three modules mentioned above. However, there is still room for improvement in generating high-quality abnormal medical reports. And we will evaluate our model on other datasets in the future.

Acknowledgements. The work was supported by the National Natural Science Foundation of China under (Grant No. 62072135 and No. 61672181).

References

1. Amjoud, A.B., Amrouch, M.: Automatic generation of chest x-ray reports using a transformer-based deep learning model. In: 2021 Fifth International Conference on Intelligent Computing in Data Sciences (ICDS), pp. 1–5. IEEE (2021)
2. Hou, B., Kaissis, G., Summers, R.M., Kainz, B.: RATCHET: medical transformer for Chest X-ray diagnosis and reporting. In: de Bruijne, M., de Bruijne, M., et al. (eds.) MICCAI 2021. LNCS, vol. 12907, pp. 293–303. Springer, Cham (2021). https://doi.org/10.1007/978-3-030-87234-2_28

3. Hu, J., Shen, L., Sun, G.: Squeeze-and-excitation networks. In: Proceedings of the IEEE Conference on Computer Vision and Pattern Recognition, pp. 7132–7141 (2018)
4. Jing, B., Xie, P., Xing, E.: On the automatic generation of medical imaging reports. arXiv preprint arXiv:1711.08195 (2017)
5. Liu, F., Wu, X., Ge, S., Fan, W., Zou, Y.: Exploring and distilling posterior and prior knowledge for radiology report generation. In: Proceedings of the IEEE/CVF Conference on Computer Vision and Pattern Recognition, pp. 13753–13762 (2021)
6. Sarki, R., Ahmed, K., Wang, H., Zhang, Y., Ma, J., Wang, K.: Image preprocessing in classification and identification of diabetic eye diseases. DSE 6(4), 455–471 (2021)
7. Xue, Y., et al.: Multimodal recurrent model with attention for automated radiology report generation. In: Frangi, A.F., Schnabel, J.A., Davatzikos, C., Alberola-López, C., Fichtinger, G. (eds.) MICCAI 2018. LNCS, vol. 11070, pp. 457–466. Springer, Cham (2018). https://doi.org/10.1007/978-3-030-00928-1_52
8. Zhang, H., Zu, K., Lu, J., Zou, Y.: Epsanet: An efficient pyramid squeeze attention block on convolutional neural network. arXiv preprint arXiv:2105.14447 (2021)

Dictionary-Induced Manifold Learning for Incomplete Multi-modal Fusion

Bingliang Xu[1], Haizhou Ye[1], Zheng Zhang[2], Daoqiang Zhang[1], and Qi Zhu[1(✉)]

[1] College of Computer Science and Technology, Nanjing University of Aeronautics and Astronautics, Nanjing 210016, China
zhuqi@nuaa.edu.cn
[2] Shenzhen Key Laboratory of Visual Object Detection and Recognition, Harbin Institute of Technology, Shenzhen 518055, China

Abstract. Data missing is a common problem in multi-modal fusion, and existing incomplete multi-modal methods usually only consider the case of two modalities and ignore the semantic information of samples during data recovery. In this paper, we propose dictionary-induced manifold incomplete multi-modal latent space representation, which reconstructs missing views with dictionary to assist consensus representation and captures the local manifold structure with reverse graph regularization. Specifically, we adopt dictionary learning to recover missing data with linear combinations of available samples for latent space alignment, and Laplacian matrix is utilized to embed the structural information of the high-dimensional space into the low-dimensional manifold latent space for optimizing the common representation. The proposed method can not only deal with multi-modal data fusion task, but also recovering missing data by effectively mining the structural information among different modalities. Experimental results demonstrate that our method performs better than other incomplete multi-modal fusion methods.

Keywords: Data recovery · Multi-modal data fusion · Dictionary learning · Manifold learning

1 Introduction

Multi-modal fusion has advantages in many fields with the complementarity of multi-modal data [1]. However, data missing limits the fusion performance due to uneven data quality. The current proposed incomplete multi-view methods can be divided into subspace-based methods and recovery-based methods.

Subspace-based methods focus on learning a common representation of multiple views through subspace mapping. Li et al. [2] proposed Partial Multi-view Clustering (PVC) algorithm based on Nonnegative Matrix Factorization (NMF) to learn the common representation of two views. Yin et al. [3] enhanced the clustering indication matrix through the similarity of samples within view and cross views. Wang et al. [4] integrated canonical correlation analysis and exclusive representation for incomplete cross-modal subspace clustering. However, the

© The Author(s), under exclusive license to Springer Nature Switzerland AG 2023
B. Li et al. (Eds.): APWeb-WAIM 2022, LNCS 13422, pp. 529–537, 2023.
https://doi.org/10.1007/978-3-031-25198-6_41

above methods ignore the effect of missing objects in the original space, and they cannot guarantee their effectiveness under high feature missing rates.

Recovery-based methods repair the incomplete samples through the spatial changes [5]. Pan et al. [6] proposed a feature-consistent generative adversarial network (FGAN), which shares network parameters to ensure the mutual generation of MR and PET images. Hu et al. [7] developed a disentangled multi-modal adversarial autoencoder (DMM-AAE) for missing modality by employing the shared information and specific information. However, the above reconstruction-based methods are prone to introduce noise through reconstruction, which may negatively affect the tasks with small amounts of complete data [8].

In addition, the above methods ignore semantic information, and only two views are taken into account in fusion. In this paper, we design a novel incomplete multi-modal method that adopts dictionary learning to recover the missing modalities for consensus representation and utilizes reverse graph regularization embedded with semantic information to capture the local manifold structure of the common representation.

In general, this paper has the following contributions: 1) The proposed incomplete multi-modal method can process multiple modalities for wider applicability. 2) We propose a unified framework with recovery and subspace representation, which adopts dictionary to restructure the incomplete modality with intra-modal similarity. 3) We guarantee the manifold structure of latent space through reverse graph regularization with considering the sample consistency cross modalities.

2   Method

2.1   Dictionary-Induced Manifold Incomplete Multi-modal Representation

The proposed method consists of following four sub-models, and the framework is shown in Fig. 1.

Consensus Representation Learning: Previous incomplete multi-modal methods focus on non-missing views to construct common representation, while ignoring the latent information of missing views. We introduce the multi-modal consensus representation [9], which exploits both non-missing and missing views:

$$\min_{E^{(v)}, U^{(v)}, P} \sum_{v=1}^{l} \left\| X^{(v)} + E^{(v)} A^{(v)} - U^{(v)} P \right\|_F^2, \quad s.t. \quad U^{(v)T} U^{(v)} = I \quad (1)$$

where $X^{(v)} \in \mathbb{R}^{m_v \times n}$ denotes the vth view composed of n samples with dimension m_v, and the missing samples in X are filled with 0. $E^{(v)} \in \mathbb{R}^{m_v \times n_v^m}$ is the error matrix used to model the missing samples n_v^m, while $U^{(v)} \in \mathbb{R}^{m_v \times d}$ is the basis matrix for mapping data into the latent space $P \in \mathbb{R}^{d \times n}$. And constraint $U^{(v)T} U^{(v)} = I$ is imposed for orthogonal basis matrices. $A^{(v)} \in \mathbb{R}^{n_v^m \times n}$ is the

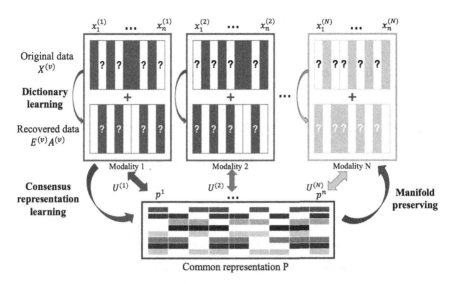

Fig. 1. The framework of the proposed dictionary-induced manifold incomplete multi-modal latent space representation.

index matrix of missing samples in each view:

$$A_{i,j}^{(v)} = \begin{cases} 1, & \text{if the j-th instance is the i-th missing instance in the vth view} \\ 0, & \text{otherwise} \end{cases} \tag{2}$$

It can be seen from Eqs. (1) and (2) that $X^{(v)} + E^{(v)}A^{(v)}$ to construct completed view and promote consensus representation.

Dictionary Learning Constraint: In order to better recover incomplete views with reducing the noise. We adopt dictionary learning to recover the missing sample with similar samples within modality, and the equation is as follows:

$$\min_{Z^{(v)}, E^{(v)}} \sum_{v=1}^{l} \left(\left\| E^{(v)}A^{(v)} - X^{(v)}Z^{(v)} \right\|_F^2 + \lambda \left\| Z \right\|_1 \right) \tag{3}$$

where the view $X^{(v)}$ is regarded as a dictionary, and the missing view $E^{(v)}A^{(v)}$ can be constructed by the coefficient matrix $Z^{(v)} \in \mathbb{R}^{n \times n}$ with L1 norm.

Reverse Graph Regularization: Considering the consistency of sample relations between modalities, the Laplacian matrix embedded with semantic information is utilized to capture common local manifold structure in low-dimensional latent space, and the representation is as follows:

$$\min_{U^{(v)}, P} \sum_{v=1}^{l} \sum_{j=1}^{n} \sum_{i=1}^{n} \left\| U^{(v)}P_{:,i} - U^{(v)}P_{:,j} \right\|_2^2 W_{i,j} \tag{4}$$

where the adjacency matrix $W \in \mathbb{R}^{n \times n}$ promotes similar samples to be closer in the latent space, and it is defined as:

$$W_{i,j} = \begin{cases} 1, & \text{If the labels of the j-th sample and the i-th sample are the same} \\ 0, & \text{otherwise} \end{cases}$$

(5)

According to the graph-based manifold learning,, we rewrite Eq. (4) as

$$\min_{U^{(v)}, P} \sum_{v=1}^{l} Tr(PL_w P^T)$$

(6)

where L_w is the Laplace matrix calculated by $L = D - W$ and $D_{i,i} = \sum_{j=1}^{n} W_{i,j}$. We constrain the low dimensional manifold of implicit space with $Tr(PL_w P^T)$ to retain the structural information of the original high-dimensional space.

Adaptive Multi-modal Weight: To differentiate the contribution of multiple modalities, we apply an adaptive weighting strategy [9] to balance the importance of different modalities:

$$\min_{\partial^{(v)}} \sum_{v=1}^{l} (\partial^{(v)})^r \Gamma^{(v)}, \quad s.t. \quad \sum_{v=1}^{l} \partial^{(v)} = 1, \partial^{(v)} \geq 0$$

(7)

where $\Gamma^{(v)}$ denotes the objective function of the v-th modality mentioned above. $\partial^{(v)}$ is a positive weight to balance the modal significance, and parameter $r > 1$.

Overall Objective Function: Finally, we integrate all sub-models by introducing regular parameters, i.e., $\lambda_1, \lambda_2, \lambda_3$, to form the overall objective function:

$$\min_{U^{(v)}, P, E^{(v)}, Z^{(v)}} \sum_{v=1}^{l} (\partial^{(v)})^r \left(\left\| X^{(v)} + E^{(v)} A^{(v)} - U^{(v)} P \right\|_F^2 \right.$$

$$+ \lambda_1 \left\| E^{(v)} A^{(v)} - X^{(v)} Z^{(v)} \right\|_F^2 + \lambda_2 Tr(PL_w P^T) + \lambda_3 \| Z^{(v)} \|_1)$$

$$s.t. \quad U^{(v)T} U^{(v)} = I, \sum_{v=1}^{l} \partial^{(v)} = 1, \partial^{(v)} \geq 0$$

(8)

2.2   Optimization Algorithm

We adopt the Alternating Direction Multiplier Method (ADMM) to solve the proposed objective function, which iteratively updates one variable with rest of the variables fixed until the function converges. We introduce a new auxiliary variable Z_1^v to further transform the problem (8) into the following form:

$$\min_{U^{(v)},P,C^{(v)},Z^{(v)}} \sum_{v=1}^{l} \partial^{(v)} (\left\| X^{(v)} + C^{(v)} - U^{(v)} P \right\|_F^2 + \lambda_1 \left\| C^{(v)} - X^{(v)} Z^{(v)} \right\|_F^2$$

$$+ \lambda_2 Tr(PL_w P^T) + \lambda_3 \| Z_1^{(v)} \|_1) \tag{9}$$

$$s.t. \quad C^{(v)} = E^{(v)} A^{(v)}, U^{(v)T} U^{(v)} = I, \sum_{v=1}^{l} \partial^{(v)} = 1, \partial^{(v)} \geq 0, Z_1^{(v)} = Z^{(v)}$$

The problem (9) can be optimized by Augmented Lagrange Multipliers (ALM):

$$L = \sum_{v=1}^{l} (\partial^{(v)})^r (\left\| X^{(v)} + E^{(v)} A^{(v)} - U^{(v)} P \right\|_F^2 + \lambda_1 \left\| E^{(v)} A^{(v)} - X^{(v)} Z^{(v)} \right\|_F^2$$

$$+ \lambda_2 Tr(PL_w P^T) + \lambda_3 \| Z_1 \|_1 + \left\langle Y_1, U^{(v)T} U^{(v)} - I \right\rangle + \left\langle Y_2, Z - Z_1 \right\rangle$$

$$+ \frac{\mu}{2} \left\| U^{(v)T} U^{(v)} - I \right\|_F^2 + \frac{\mu}{2} \| Z - Z_1 \|_F^2) \tag{10}$$

where Y_1 and Y_2 are Lagrange multipliers, $\mu > 0$ is the augmented Lagrange parameter. The problem (10) can be solved by the Inexact Augmented Lagrange Multiplier (IALM), when $\left\| U^{(v)T} U^{(v)} - I \right\|_\infty < \varepsilon$ and $\| \cdot \|_\infty$ is l_∞- norm.

The solution process of the algorithm is as follows:

Step 1 (Update P): P can be updated by solving:

$$P^* = \arg\min_P \lambda_1 \left\| E^{(v)} A^{(v)} - X^{(v)} Z^{(v)} \right\|_F^2 \tag{11}$$

The result of solving problem (11) is shown as:

$$P = (\sum_{v=1}^{l} (\partial^{(v)})^r U^{(v)T} (X^{(v)} + E^{(v)} A^{(v)}))(I + \lambda_2 L_w)^{-1} / \sum_{v=1}^{l} (\partial^{(v)})^r \tag{12}$$

Step 2 (Update Z): Z can be updated by solving the problem:

$$Z^{(v)*} = \arg\min_{Z^{(v)}} \sum_{v=1}^{l} (\partial^{(v)})^r (\left\| X^{(v)} + E^{(v)} A^{(v)} - U^{(v)} P \right\|_F^2 + \lambda_2 Tr(PL_w P^T) \tag{13}$$

The solution result is shown as:

$$Z^{(v)} = X^{(v)-1} E^{(v)} A^{(v)} \tag{14}$$

Step 3 (Update $Z_1^{(v)}$): We solve the problem (15) to update $Z_1^{(v)}$:

$$Z_1^* = \arg\min_{Z_1} \lambda_3 \| Z_1 \|_1 + \frac{\mu}{2} \left\| Z - Z_1 + \frac{Y_2}{\mu} \right\|_F^2 \tag{15}$$

The result is shown as:

$$Z_1 = shrink(Z + \frac{Y_2}{\mu}, \frac{\lambda_3}{\mu}) \qquad (16)$$

Step 4 (Update $U^{(v)}$): $U^{(v)}$ can be obtained by solving:

$$U^{(v)*} = \arg\min_{U^{(v)}} \left\| X^{(v)} + E^{(v)} A^{(v)} - U^{(v)} P \right\|_F^2 + \frac{u}{2} \left\| U^{(v)^T} U^{(v)} - I + \frac{Y_1}{\mu} \right\|_F^2 \quad (17)$$

The calculation result is:

$$U^{(v)} = BR^T \qquad (18)$$

where B and R^T are the left and right singular value matrices of $X^{(v)} + E^{(v)} A^{(v)} P^T$, respectively.

Step 5 (update $E^{(v)}$): $E^{(v)}$ can be updated by optimizing the problem (19):

$$E^{(v)*} = \arg\min_{E^{(v)}} \left\| X^{(v)} + E^{(v)} A^{(v)} - U^{(v)} P \right\|_F^2 + \lambda_1 \left\| E^{(v)} A^{(v)} - X^{(v)} Z^{(v)} \right\|_F^2$$
$$(19)$$

The result is shown as:

$$E^{(v)} = \frac{1}{1 + \lambda_1} (U^{(v)} P + \lambda_1 X^v Z_1^{(v)}) A^{(V)-1} \qquad (20)$$

Step 6 (update $\partial^{(v)}$): $\partial^{(v)}$ can be obtained by solving:

$$\arg\min_{\partial^{(v)} > 0, \sum_{v=1}^l \partial^{(v)} = 1} \sum_{v=1}^l (\partial^{(v)})^r d^{(v)} \qquad (21)$$

The calculation result is shown as:

$$\partial^{(v)} = (d^{(v)})^{1/(1-r)} / \sum_{v=1}^l (d^{(v)})^{1/(1-r)} \qquad (22)$$

Step 7 (update Y_1, Y_2 and μ): The Lagrange multipliers Y_1, Y_2 and iteration step μ can be update by:

$$W_{i,j} = \begin{cases} Y_1 = Y_1 + \mu(U^{(v)^T} U^{(v)} - I) \\ Y_2 = Y_2 + \mu(Z - Z_1) \\ \mu = \min(\rho U^{(v)}, \mu_{max}) \end{cases} \qquad (23)$$

3   Experiments

3.1   Dataset Introduction

In the paper, we adopt the ADNI dataset from the Alzheimer's Disease Neuroimaging Initiative to validate our proposed method. ADNI contains 202 subjects of three types, including 51 AD patients, 99 MCI patients and 52 normal controls. Refer to [10], we perform preprocessing work on the brain images of ADNI and obtain three modalities, i.e., MRI, PET and CSF, with 93 MRI features, 93 PET features and 3 CSF-related biomarkers.

3.2   Comparison with Baseline Completion Methods

In order to verify the effectiveness of our method, we choose four algorithms as comparison methods, including mean interpolation (Zero), k nearest neighbors (KNN), regular expectation maximization (RegEM) and singular value decomposition (SVD). Specifically, Zero fills the missing values with the average of all available samples in a certain modality, while KNN chooses the mean of the K nearest samples to replace the missing values. RegEM applies the regular estimation algorithm to the regression parameters of Gaussian data. For SVD, it is used to update the predicted missing values based on the KNN- completed data in our paper.

We perform experiments on AD vs. HC and MCI vs. HC tasks with ten-fold cross-validation. We randomly impose 10%, 30%, 50% deletions on the data and ensure that three modalities of the sample are not missing at the same time. All methods adopt support vector machines (SVM) for classification after data completion and connection of modalities, and grid search is used to select optimal parameters. The experimental results are shown in Table 1. It can be seen from the results that as the missing rate increases, accuracy drop due to the lack of information. However, compared with the baseline method, we improve the overall accuracy from 66.53% – 71.52% to 86.95% – 88.54% in AD vs. HC task, and from 63.09% – 65.88% to 72.33% – 73.66% in MCI vs. HC task. The results show that our method is effective in missing modality recovery.

Table 1. Comparison between our method and four baseline completion methods

Task	Missing rate	Zero	KNN	RegEM	SVD	**Ours**
AD vs. HC	10%	69.85	70.37	71.31	71.52	**88.54**
	30%	67.96	68.77	71.02	70.47	**88.17**
	50%	67.89	66.53	69.44	69.88	**86.95**
MCI vs. HC	10%	64.26	64.03	65.88	64.93	**73.66**
	30%	64.11	63.58	65.19	64.06	**72.91**
	50%	63.09	63.12	64.52	63.21	**72.33**

3.3   Comparison with Incomplete Multi-modal Methods

In this section, we compare the proposed method with three state-of-the-art incomplete multi-modal algorithms, i.e., Partial View Clustering (PVC), Partial View Clustering with Graph Regular Non-negative Matrix Factorization (GPVC), Incomplete Multi-Source Feature Learning (IMSF). Specifically, PVC works on the data of two modalities and embeds the structural information of the sample into the latent space. GPVC is an improved method based on PVC, which introduces graph Laplacian regularization and extends PVC to tasks with more than two modalities. IMSF is an ensemble-based method that tries to obtain a common representation of each sample and creates a classifier for each modality to ensemble classification.

Table 2. The comparison between our method and incomplete multi-modal method

Task		AD vs. HC				MCI vs. HC			
Missing rate	Method	ACC (%)	SEN(%)	SPE (%)	AUC (%)	ACC (%)	SEN (%)	SPE (%)	AUC (%)
10%	PVC	78.35	77.45	80.24	79.04	67.36	72.41	56.74	70.12
	GPVC	79.74	79.10	81.02	80.24	68.45	73.99	58.17	72.41
	IMSF	82.61	83.11	83.44	83.16	68.75	75.44	60.11	74.36
	Ours	**88.54**	**88.47**	**89.14**	**90.06**	**73.66**	**85.41**	**67.74**	**81.17**
30%	PVC	77.42	77.08	79.41	78.12	65.47	70.48	56.14	68.79
	GPVC	78.64	78.45	79.93	79.55	68.02	73.17	58.46	71.64
	IMSF	81.04	81.96	82.41	82.77	67.99	75.29	59.37	73.33
	Ours	**88.17**	**87.98**	**88.81**	**88.97**	**72.91**	**85.09**	**68.01**	**80.38**
50%	PVC	77.12	76.86	79.24	77.85	65.31	70.56	56.02	68.58
	GPVC	78.42	78.10	79.43	79.01	67.83	72.50	58.63	71.44
	IMSF	80.54	81.74	81.99	82.06	67.42	73.98	59.66	73.17
	Ours	**86.95**	**87.88**	**88.46**	**88.04**	**72.33**	**84.55**	**67.53**	**80.07**

Similar to Sect. 3.2, we conduct experiments under different missing rates in the two recognition tasks and adopt four indicators, i.e. accuracy (ACC), sensitivity (SEN), specificity (SPE), and area under the curve (AUC), to evaluate the diagnostic performance. The results are shown in Table 2. Specifically, most of the indicators decrease with the increase of the missing rate, which is consistent with our previous experimental results. Overall, our method achieves significant advantages in all missing rates and diagnosis tasks, which shows that our proposed method is effective in incomplete multi-modal data fusion problem.

4   Conclusion

In this work, we propose an dictionary-induced manifold incomplete multi-modal latent space representation. The algorithm recovers missing samples with available samples in the modality through dictionary learning. Furthermore, manifold learning incorporates semantic information for capturing local structural information in the learned common representation. Two experiments show that our method achieves convincing performance under different missing rates and tasks.

Acknowledgements . This work was supported in part by National Natural Science Foundation of China (Nos. 61732006, 62076129, 61501230, 62136004 and 61876082), National Science and Technology Major Project (No. 018ZX10201002), and the National Key R&D Program of China (Grant Nos.: 2018YFC2001600, 2018YFC2001602).

References

1. Zhu, Q., Yang, J., Wang, S., Zhang, D., Zhang, Z.: Multi-modal non-euclidean brain network analysis with community detection and convolutional autoencoder. IEEE Trans. Emerg. Topics Comput. Intell, 1–11 (2022)

2. Li, S.Y., Jiang, Y., Zhou, Z.H.: Partial multi-view clustering. In: Proceedings of the AAAI Conference On Artificial Intelligence, vol. 28 (2014)
3. Yin, Q., Wu, S., Wang, L.: Unified subspace learning for incomplete and unlabeled multi-view data. Pattern Recogn. **67**, 313–327 (2017)
4. Wang, Q., Lian, H., Sun, G., Gao, Q., Jiao, L.: icmsc: Incomplete cross-modal subspace clustering. IEEE Trans. Image Process. **30**, 305–317 (2021)
5. He, J., Han, X.: Efficient skyline computation on massive incomplete data. Data Sci. Eng. **7**(2), 102–119 (2022)
6. Pan, Y., Liu, M., Lian, C., Xia, Y., Shen, D.: Disease-image specific generative adversarial network for brain disease diagnosis with incomplete multi-modal neuroimages. In: Shen, D., et al. (eds.) MICCAI 2019. LNCS, vol. 11766, pp. 137–145. Springer, Cham (2019). https://doi.org/10.1007/978-3-030-32248-9_16
7. Hu, D., et al.: Disentangled-multimodal adversarial autoencoder: Application to infant age prediction with incomplete multimodal neuroimages. IEEE Trans. Med. Imaging **39**(12), 4137–4149 (2020)
8. Enders, C.K.: Applied missing data analysis. Guilford press (2010)
9. Wen, J., Zhang, Z., Xu, Y., Zhang, B., Fei, L., Liu, H.: Unified embedding alignment with missing views inferring for incomplete multi-view clustering. In: Proceedings of the AAAI Conference On Artificial Intelligence, vol. 33, pp. 5393–5400 (2019)
10. Zhang, D., Wang, Y., Zhou, L., Yuan, H., Shen, D., Initiative, A.D.N., et al.: Multimodal classification of alzheimer's disease and mild cognitive impairment. Neuroimage **55**(3), 856–867 (2011)

BSAM: A BERT-Based Model with Statistical Information for Personality Prediction

Bin Xu, Tongqing Wang$^{(\boxtimes)}$, Kening Gao, and Zhaowu Zhang

School of Computer Science and Engineering, Northeastern University, Shenyang, China
xubin@mail.neu.edu.cn, wangtongqing1997@163.com, gkn@cc.neu.edu.cn

Abstract. As a branch of psychology, personality plays an important role in distinguishing individuals in the society. The existing personality prediction models need to be further improved in precision and generalization. Recently, deep neural network (DNN) models are being applied to personality prediction tasks to obtain promising results. However, only extracting the semantic features of text through deep learning is very limited to improve the performance of the model. We propose a BERT-based Model for Personality Prediction named BSAM to extract semantic features and use the statistical information of corpus as external features. In this model, we concatenate the output of BERT with the statistical information and use bidirectional long short term memory networks (Bi-LSTM), bidirectional gated recurrent unit (Bi-GRU) and improved convolutional neural networks (CNN) to extract deep semantic features. We also compare the results with benchmark models on social media datasets and test the effectiveness of statistical features. The experimental results show that our model can effectively improve the classification performance of the five dimensions of the Big-Five personality.

Keywords: Personality prediction · BERT · Deep learning · Statistical information · Big-five

1 Introduction

As one of the branches of psychology, personality psychology mostly identifies people's internal personalities by their external behaviors and studies the relationship between them.

In order to unify the classification standard of personality, psychologists try to set some personality dimensions as the industry standard. But even professional researchers will follow their own rules when judging individual personality, which leads to the absence of an absolute standard personality classification

This work is supported by the Fundamental Research Funds for the Central Universities (N2116019) and the National Natural Science Foundation of China (U1811261).

B. Li et al. (Eds.): APWeb-WAIM 2022, LNCS 13422, pp. 538–545, 2023.
https://doi.org/10.1007/978-3-031-25198-6_42

model. In 1928, psychologists put forward the DISC self-assessment tool, which aims to help enterprises optimize personnel management. With the deepening of personality prediction research, Big-Five is the most widely used and recognized model in the field of personality prediction, which divides personality into five dimensions: Extraversion(EXT), Agreeableness(AGR), Conscientiousness(CON), Neuroticism(NEU) and Openness(OPN) [1].

The existing personality prediction methods mostly suffer from deficiencies in terms of precision and generalization. Therefore, we propose a new personality prediction model named BSAM to extract the semantic features of text, which are combined with the statistical information of the text.

2   Related Works

Similar to other text classification tasks, personality prediction based on social media data focuses on the semantic features of texts. Many researchers use deep learning models that have performed well in other fields. RNN and CNN were used to extract text features and outperformed benchmark models [2]. Majumder et al. [3] proposed a network based on CNN to predict users' personality and achieved a good result. Zhao et al. [4] proposed an attention-based LSTM network, which was extremely helpful to identify user personality. Thus, in our model BSAM, we integrate these deep learning networks to extract complex features of text.

Although the deep learning model can extract rich semantic features, it is limited to improving the performance of the model. In recent years, more and more studies have focused on external knowledge. Researchers have found many tools that can be applied in the field of personality prediction, such as LIWC [5], a tool for extracting statistical text features, and the Mairesse baseline set [6], a document-level benchmark feature set. Inspired by them, our model takes the correlation between labels and words into account by introducing the statistical feature of corpus: the word frequency information of labels.

In summary, the contributions are as follows:

- Based on deep learning, we propose a BERT-based model for personality prediction and merge statistical features into our model.
- We propose an adaptive adjustment strategy for balancing mean-pooling and max-pooling method.
- We conduct some experiments on benchmark datasets and the results indicate that our model outperforms all other models.

3   Proposed Model

3.1   Overview of the Model

The structure of BSAM model is shown in Fig. 1. We use BERT [7] and deep learning model to fulfill the feature extraction task. In order to rich text features, a statistical feature of text is introduced in our model.

Let $D = \{s_i\}_{i=1}^{N}$ be the set of sentences, which contains N sentences. Each sentence $s_i = \{w_{i1}, w_{i2}, ..., w_{im}\}$ contains m words. The goal of personality prediction is to get the most relevant labels in each dimension according to the input text representation.

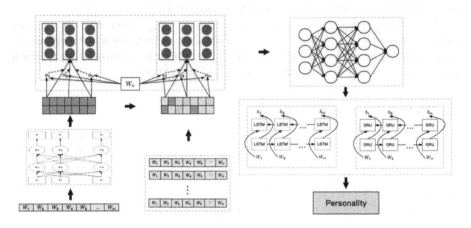

Fig. 1. The structure of proposed model BSAM

3.2   Statistical Features Extraction Network

For each sentence, we firstly use BERT to form all word vectors of the sentence into a $m \times d$ word embedding matrix $X = [X_1, X_2, ..., X_m]$. In order to use the feature map X efficiently, we use a dense layer for the output of BERT:

$$\widetilde{X} = Relu(W_X \cdot X + b_X) \tag{1}$$

Word frequency information is a static feature of the corpus and presnts the importance of a word to a label. Different words have different word frequencies on different labels. If a word frequency is higher in a label, it means that the word is more important to the label, and vice versa. As shown in Table 1, the label "yes" indicates that the person is of this kind of personality, and "no" indicates that he is not of this kind of personality. We can see in Extraversion personality, the word "happy" and "haha" appears 8327 times and 3286 times in label "yes", 1689 times and 108 times in label "no", but in Neuroticism personality, there is no significant difference in the word frequency of word "happy" and word "haha", which indicates that people whose posts contain the words "happy" or "haha" are more likely to be of extraversion personality.

Given a word w and a label with k classes in each personality dimension. For each word w, the word frequency is:

$$C^w = [C_{l_1}, C_{l_2}, ..., C_{l_k}] \tag{2}$$

Table 1. Word frequency information of sentences under different personality dimensions

Personality	Sentence						Label
	Traveling	Makes	Me	Very	Happy	Haha	
Extraversion	658	5073	15689	6734	8327	3286	Yes
	246	4968	12487	7835	1689	108	no
Neuroticism	364	3278	16992	7823	2551	81	Yes
	337	4116	12487	5456	4877	177	No

where C_{l_i} represents the occurrences of the word w in label l_i. For a sentence, the word frequency matrix is:

$$C^s = [C_1^w, C_2^w, ..., C_m^w] \tag{3}$$

where C_i^w is the word frequency of the i-th word of a sentence. We use a dense layer for the statistic information C:

$$\tilde{C} = Relu(W_C \cdot C + b_C) \tag{4}$$

3.3   Deep Features Extraction Network

In order to fully utilize the personality information of social media text dataset, we improve CNN to extract different local features. CNN uses convolutional filters to extract local semantic traits in text, then sentence features are connected by local features of different lengths. At first, we fuse $\tilde{X}$ and $\tilde{C}$ to get a new latent feature:

$$H = \tilde{X} \odot \tilde{C} \tag{5}$$

where $\odot$ denotes an element-wise product. Then we use attention mechanism to combine latent feature map H and word embedding matrix X:

$$\tilde{H} = attention(H, X) \tag{6}$$

In the improved-convolution network, we use 2,3,4-gram kernels to extract the local sentence information. The convolutional layer performs an $e \times d$ convolution operation on the $m \times d$ word embedding matrix X, the convolution kernel $W \in R^{m \times d}$. The local semantic sentence features are computed as:

$$C_i = f(W_{\tilde{H}} \times \tilde{H}_{i:i+k-1} + b_{\tilde{H}}) \tag{7}$$

where C_i is local semantic features, $\tilde{H}_{i:i+k-1}$ is the word vector matrix from word i to word $i+k-1$, b is a bias term, $W_{\tilde{H}}$ is the filter and f is the activation function.

In addition, there are two common methods to reduce parameters while retaining the local optimal features: max-pooling and mean-pooling. These two pooling methods have their own advantages in different situations. We propose a mixed-pooling method suitable for most situations and realize adaptive selection of pooling method by learning a weight parameter α. The equation is calculated as follows:

$$f_{mixed}(C_i) = \alpha f_{mean}(C_i) + \frac{1}{\alpha} f_{max}(C_i) \tag{8}$$

Because our dataset comes from social media posts which are of vast quantities and in various forms. There are contextual dependencies in the structure of long text, which can extract deeper textual semantic information if the connection between contexts is considered. In our experiments, we apply Bi-LSTM and Bi-GRU to the output of the convolutional layer H^C in parallel, which can analyze complex sentence structure and extract forward and backword dependencies in the anterior-posterior direction:

$$H^T = Tanh(W_T(H^L \odot H^G) + b^T) \tag{9}$$

where H^L and H^G are the outputs of Bi-LSTM and Bi-GRU. The output of this layer is fed to sigmoid layer for classification.

4   Experiment and Analysis

4.1   Experimental Dataset

We conducted experiments on two classical datasets. The first dataset Stream-of-Consciousness Essays dataset [8] is authoritative in personality prediction. It contains 2467 anonymous articles that label the personality of authors. The mean length and max length of sentences are 662 and 3836.

The second dataset MyPersonality contains 250 users from Facebook and 9917 statuses labeled on the Big-Five personality. The mean length and max length of sentences are 163 and 687.

4.2   Experiment Results and Analysis

In this section, we choose several classical models in personality prediction as our contrast models: 2CLSTM [9], Deep-CNN [3], Text-CNN [10]. To measure the effectiveness of our model in personality prediction, we use precision as the evaluation index, the results of contrast models and our model are shown as Table 2 and Table 3. The bold data represent the best experimental results for each dimension.

On the dataset Essays, our model BSAM has achieved the highest precision in five dimensions. On the dataset MyPersonality, BSAM outperform other models in four dimensions. This is because BSAM has good performance in processing long text and short text and statistical features can capture words that are important to the label. Furthermore, our model solves the problem of poor generalization of the model in previous studies. Almost in every dimension, the precision is more than 70% on both two datasets, and there is no situation where the experimental results of two dimensions differ significantly.

Table 2. The precision under different classification methods by using Essays dataset

Models	EXT	NEU	AGR	CON	OPN
2CLSTM	58.87%	55.64%	53.52%	56.77%	54.19%
Deep-CNN	50.55%	50.20%	48.10%	50.47%	50.60%
Text-CNN	50.29%	53.21%	49.23%	48.59%	47.73%
BSAM	**70.73%**	**69.58%**	**70.83%**	**71.34%**	**70.03%**

Table 3. The precision under different classification methods by using MyPersonality dataset

Models	EXT	NEU	AGR	CON	OPN
2CLSTM	62.58%	65.46%	60.89%	62.83%	72.48%
Deep-CNN	62.48%	60.67%	65.39%	52.01%	**78.69%**
Text-CNN	59.47%	58.21%	58.46%	57.66%	70.23%
BSAM	**72.36%**	**70.59%**	**71.79%**	**67.82%**	76.22%

4.3 Ablation Experiment

To further test the performance of the statistical features, we conducted ablation experiments and proposed BAM model. Compared with BSAM, this model has the same parameters and removes word frequency matrix.

Table 4. Ablation experiment results on Essays and MyPersonality datasets

Models	EXT	NEU	AGR	CON	OPN
Essays					
BAM	62.63%	60.83%	61.57%	58.16%	64.40%
BSAM	70.73%	69.58%	70.83%	71.34%	70.03%
MyPersonality					
BAM	68.33%	64.74%	66.98%	64.02%	72.56%
BSAM	72.36%	70.59%	71.79%	67.82%	76.22%

Table 4 demonstrates the experimental results. Even without introducing the statistical features of the text, the performance of BAM is better than previous model in three dimensions on the dataset Essays and four dimensions on the dataset MyPersonality, which also proves that our model has strong feature extraction ability and can mine the deep semantic features of the text. After introducing word frequency information, the precision of BAM is significantly better than other models. This also verifies the effectiveness of introducing word frequency matrix into our experiment.

4.4   Case Analysis of Attention

In the study of personality prediction, each user document corresponds to five personality dimensions. As the attention mechanism is used in predicting different labels of the same document, and each word and phrase in the document has different attention weight to the label. We cite a case to visualize the attention weight corresponding to the label in Fig. 2, and the highlighted words represent the information that the current tag pays more attention to.

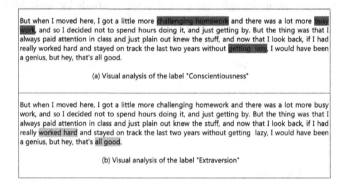

Fig. 2. Important words captured by different labels

We can see that the label "Conscientiousness" is more closely related to the words "challenging", "busy" and phrase like "without getting lazy". Sentence contains phrase "worked hard" and "all good" carry high "Extraversion" personality information. Moreover, our personality prediction model can't only automatically select informative words and assign larger weight according to labels, but also analyze the structural information of complex sentences. For example, note sentence "and stayed on track the last two years without getting lazy" in Fig. 2(a), this sentence can easily be mistaken for not belonging to "Conscientiousness", but our model takes the word "without" in the previous content into account and correctly predicts the personality. This further proves that the introduction of attention mechanism can more accurately predict the user's personality.

5   Conclusion

Personality prediction, as an emerging research area, has been increasingly noticed by researchers and the public with the rapid development of social networks, including automatic inference of users' personality based on social network. We propose a deep learning-based model and introduce statistical features of text. After evaluating the performance of our model on the benchmark dataset, the results indicate our model has significant improvement than contrast models and has good generalization. We also conduct further experiments to prove the effectiveness of using BERT and word embedding matrix. In the following research, we will shift our focus from single personality prediction task to multiple personality prediction problem.

References

1. Digman, J.M.: Personality structure: Emergence of the five-factor model. Annu. Rev. Psychol. **41**(1), 417–440 (1990)
2. Xue, D., et al.: Deep learning-based personality recognition from text posts of online social networks. Appl. Intell. **48**(11), 4232–4246 (2018). https://doi.org/10.1007/s10489-018-1212-4
3. Majumder, N., Poria, S., Gelbukh, A., Cambria, E.: Deep learning-based document modeling for personality detection from text. IEEE Intell. Syst. **32**(2), 74–79 (2017)
4. Zhao, J., Zeng, D., Xiao, Y., Che, L., Wang, M.: User personality prediction based on topic preference and sentiment analysis using lstm model. Pattern Recogn. Lett. **138**, 397–402 (2020)
5. Safitri, G., Setiawan, E.B., et al.: Optimization prediction of big five personality in twitter users. Jurnal RESTI (Rekayasa Sistem Dan Teknologi Informasi) **6**(1), 85–91 (2022)
6. Mairesse, F., Walker, M.A., Mehl, M.R., Moore, R.K.: Using linguistic cues for the automatic recognition of personality in conversation and text. J. Artifi. Intell. Res. **30**, 457–500 (2007)
7. Devlin, J., Chang, M.W., Lee, K., Toutanova, K.: Bert: Pre-training of deep bidirectional transformers for language understanding. arXiv preprint arXiv:1810.04805 (2018)
8. Pennebaker, J.W., King, L.A.: Linguistic styles: language use as an individual difference. J. Pers. Soc. Psychol. **77**(6), 1296–1312 (1999)
9. Sun, X., Liu, B., Cao, J., Luo, J., Shen, X.: Who am i? personality detection based on deep learning for texts. In: 2018 IEEE International Conference on Communications (ICC), pp. 1–6. IEEE (2018)
10. Wei, H., et al.: Beyond the words: ACM Predicting User Personality from Heterogeneous Information/Proceedings of The Tenth International Conference on Web Search and Data Mining (2018)

A Combined Model Based on GRU with Mahalanobis Distance for Oil Price Prediction

Shichen Zhai and Zongmin Ma[✉]

College of Computer Science and Technology, Nanjing University of Aeronautics
and Astronautics, Nanjing 211106, China
zongminma@nuaa.edu.cn

Abstract. Oil plays a key role in economic development. It could help people to make plans and decisions. Because oil price is affected by some factors, it is difficult for people to predict it accurately with current existing models. With the development of deep learning, it is often used to solve multivariate nonlinear problems. In this paper, a combined model based on gated recurrent units is proposed, Mahalanobis distance is used to eliminate outliers and the multivariate nonlinear regression model is constructed, Spearman correlation coefficient is used as feather selection and evaluation metric. Experiment shows that the proposed combined method performs better and could predict the price of crude oil effectively.

Keyword: Time series · Mahalanobis distance · Outlier detection · GRU

1 Introduction

Oil is a major source of energy. Fluctuations in the price of crude oil have a huge impact on a country's macro-economy. The factors affecting crude oil price include origin, trade, interest rate, income and so on, which makes it a difficult task to predict WTI crude oil price effectively. There are complex nonlinear and non-stationary characteristics, which makes the oil prediction more difficult. Especially due to the recent epidemic outbreak, we have observed large fluctuations in crude oil prices. Therefore, it is necessary to build an efficient and intelligent mathematical model for the prediction of industrial crude oil prices.

In recent years, some studies have shown that hybrid models can improve the accuracy of prediction. Li et al. [1] developed a text-based crude oil price prediction model using deep learning techniques and sentiment analysis. They observed that news text can add useful information to crude oil predictions. Chen et al. [2] used deep learning to predict crude oil prices. They constructed a hybrid model combining ARMA and deep learning-based prediction. They observed that the deep learning model for crude oil price prediction improved the accuracy of the prediction value. Chen et al. [3] proposed a grey-wave prediction technique for predicting daily crude oil prices early. Gumus and Kiran [4] used the machine learning algorithm XGBoost to predict oil prices. Luo et al.

[5] proposed a CNN-based prediction model for predicting short-term crude oil futures prices, in which they used matrix inputs and claimed to be more powerful than the benchmark model. Salvi et al. [6] highlighted the advantages of the LSTM in predicting oil prices.

In other area, Mahajan R et al. [7] proposed to predic geolocation of tweets by using CNN and LSTM. Sugiartawan et al. [7] developed a hybrid model with wavelet transform and LSTM. Cen and Wang et al. [9] applied the LSTM model to predict the volatility of crude oil prices. They set up a new data transmission mode to improve the predictability of global crude oil price fluctuations. Siddhaling et al. [10] designed several variants of the multivariate model based on the combination of feature transformation and LSTM to predict crude oil prices.

The recurrent network has derived many variant models such as LSTM (Long Short-Term Memory) and GRU (Gated Recurrent Unit). LSTM adds the gate mechanism to control the circulation and loss of features. GRU is also proposed to solve the problems of long-term memory and gradient in back propagation. It is a new neural network deep learning algorithm suitable for solving nonlinear and sequential problems, and has one less gate mechanism than LSTM. Moreover, the overall training speed of GRU is faster, which can greatly improve the training efficiency. Therefore, GRU will be preferred in many cases.

This paper proposes a method based on outlier processing and multivariate GRU model to predict crude oil prices with Mahalanobis distance to eliminate outliers, select the most effective features for identification with Spearman correlation coefficient. Then, we construct and train the hybrid model MGRU-3F-RM with three features and Mahalanobis distance, and finally compare it with other models, the experiments show that our proposed method performs better and has higher accuracy. The main innovations and contributions are as follows:

(1) A novel method is proposed to predict oil price for multiple nonlinear regression models with deep learning model.
(2) The method uses Mahalanobis distance transformation to eliminate outliers during time series data from the statistical view.
(3) The method uses Spearman correlation coefficient not only for feature extraction but also for model evaluation metrics.
(4) The method shows better sensitivity and consistent behaviors under model evaluation metrics than the previous approaches.

The rest of this paper is arranged as follows. In the second part, we introduce Mahalanobis distance and our proposed MGRU model. In the third part, we introduce the specific experiments, including data preprocessing, feature engineering, model training and so on. In the fourth part, the combined model we constructed is compared with other models based on the model evaluation metrics. Finally, we make a summary and outlook of our research work.

2 Methodology

2.1 Mahalanobis Distance

Mahalanobis distance was proposed by Indian statistician P. C. Mahalanobis, which represents the distance between a point and a distribution. It is an effective method to calculate the similarity of two unknown sample sets, and it can effectively represent the multivariate distance measure, which can measure the distance between points and distributions. Mahalanobis distance can also eliminate the interference of correlation between variables. It can also be defined as the degree of difference between two random variables that follow the same distribution. If the covariance matrix is an identity matrix, then the Mahalanobis distance is reduced to the Euclidean distance and hyperbolic distance described by [11]. it can also be used for outlier detection as described by Titouna [12]. David et al. [13] illustrate the application of the Mahalanobis distance and point to their observation that for multivariate non-Gaussian data, it's used to detect and eliminate outliers.

For a multivariate X with mean is μ and covariance matrix C, the Mahalanobis distance is as shown in formula (1):

$$D = \sqrt{(x - \mu)^T \cdot C^{-1} \cdot (x - \mu)} \tag{1}$$

If the covariance matrix is a diagonal matrix, then it can also be called the normalized Euclidean distance. Mahalanobis and Euclidean distance relationships are shown in Fig. 1.

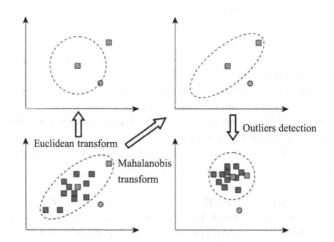

Fig. 1. Outlier detection with Mahalanobis distance transformation

As shown in the above figure, Mahalanobis distance effectively solves the problem of large-scale outliers and the correlation of variables after the rotation transformation scaling. The geometric meaning of the Mahalanobis distance is that outliers are successfully separated.

2.2 Our Model

GRU is a kind of RNN, which has been widely used in complex prediction problems, such as time series prediction. In GRU, it has a special input, as well as an oblivion gate to better control the gradient by deciding what information to remember. This helps to maintain a long range of output by solving common RNN problems. Compared with traditional neural networks, GRU can remember a series of data points for a long time and overcome the limitation of gradient disappearance. GRU is a variant of LSTM, combines the forget gate and the input gate into a single update gate. It also mixes the cellular state with the hidden state, with a few other changes. The resulting model is simpler than the standard LSTM model and is a very popular variant.

Our proposed the combined framework of oil price prediction model using GRU and Mahalanobis distance is shown in Fig. 2. The time series data and other financial characteristics are derived from internet sources, which need to be cleaned and preprocessed with visual analysis, correlation coefficient analysis, feature selection, outlier elimination, missing value filling, data standardization and splitting data. Feature selection is carried out through the relationship between feature variables and target variables by the Spearman correlation coefficient. Mahalanobis transformation makes the transformed data have the properties of characteristic uncorrelation and standard deviation, these transforms can effectively for outlier detection and remove some outliers, then we build and train the Multivariate GRU model and tune the parameters optimization, which was helpful for improving the accuracy of the model. After completing evaluation and comparation of model, the innovative model is trained finally, and then we will apply it to predict oil price in practice.

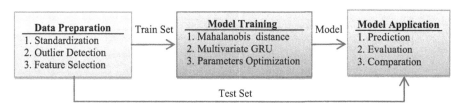

Fig. 2. Work flow of MGRU model

3 Experiment

3.1 Feature Engineering

We collect historical time series data sets of West Texas crude oil prices and influencing factors for experiments, It's from websites (https://in.investing.com/) and (https://finance.yahoo.com/). This study mainly considers five factors affecting oil prices for time series regression analysis. WTI is the target variable and GOLD, SP500, USDINDEX, US10B, DJU are the features. We will preprocess the data, including that we fill the missing value of the attribute by taking the value of the previous day, normalize all data for preprocessing, lag the WTI and then, according to the Spearman correlation coefficient, we eliminate features under coefficient less than 0.4 shown with the heatmap in Fig. 3(a). we could see that the annual price of WTI crude oil fluctuates greatly. It's shown by the boxplots in Fig. 3(b). There are some outliers in 2002, 2003, 2008, 2014 and 2018 years.

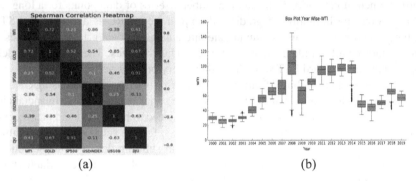

(a)   (b)

Fig. 3. Spearman correlation coefficient heat map and box chart of WTI

We use the distribution of z-score and Mahalanobis distance to eliminate outliers respectively. we consider all the z-score points that are more than 2.5 times away from the mean as outliers shown in Fig. 4(a). Another method is Mahalanobis distance, which can be effectively used to eliminate outliers. We consider points with a Mahalanobis distance greater than 10 to be outliers. It's shown in Fig. 4(b).

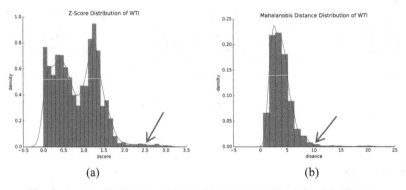

(a)   (b)

Fig. 4. Density distribution of WTI with Z-score and Mahalanobis distance

3.2 Model Training

We will split the dataset to train and test parts, then develop four MGRU models: MGRU-5F, MGRU-3F, MGRU-3F-RZ and MGRU-3F-RM to predict WTI oil prices, which represent five features, three features, three features with Z-score outlier elimination, and three features with Mahalanobis distance. All four models were trained over 50 epoch cycles with a batch size of 32. Sixty times steps were used for all MGRU models, 60 previous day records were considered to calculate day 61, It can be seen that with the extension of the training period, the error and loss values of the Basic MGRU model gradually decrease and tend to steady, and the curves of model training are shown in Fig. 5.

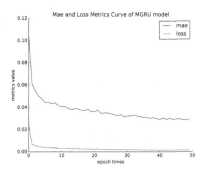

Fig. 5. Basic MGRU model training loss curve

A plot of the predicted and actual WTI values for all these models on the test data is shown in Fig. 6. It shows the predicted WTI values based on the models MGRU-3F-RM and MGRU-3F-RZ respectively. Obviously, we could see the predicted value is much closer to the actual WTI value and it can be observed in Fig. 6(a).

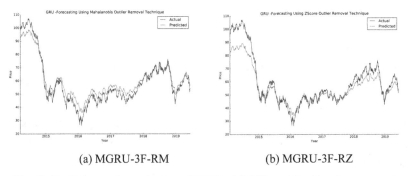

(a) MGRU-3F-RM   (b) MGRU-3F-RZ

Fig. 6. Prediction and actual value of WTI by MGRU model with outers removal

4 Discussion

4.1 Model Evaluation

In order to further study the predictive ability of the model, we use some regression model evaluation indicators, They are Mean absolute error (*MAE*), Root mean square error (*RMSE*), Coefficient of Determination (R^2). The specific formula is as follows from formula (2) to (5), where y_i is the true value, $f(x_i)$ is the predicted value, $\bar{y}$ represents the average value of the sample, m is the number of samples and where d_i represents the difference between the ranks of the corresponding variables and n is the number of observations.

$$SCC = 1 - \frac{6\sum_{i=1}^{n} d_i^2}{n(n^2 - 1)} \tag{2}$$

$$MAE = \frac{1}{m}\sum_{i=1}^{m} (|y_i - f(x_i)|) \tag{3}$$

$$RMSE = \sqrt{\frac{1}{m}\sum_{i=1}^{m} (y_i - f(x_i))^2} \tag{4}$$

$$R^2 = 1 - \frac{MSE(y_i, f(x_i))}{\frac{1}{m}\sum_{i=1}^{m} (y_i - \bar{y})^2} \tag{5}$$

4.2   Model Comparisons

In this study, we compare the proposed method with the common Linear Regression, Polynomial Regression, SVR, Random Forest, XGBoost, deep learning LSTM, GRU and other models in order to further verify the sensitivity of our model. The comparison results are shown in Table 1.

Table 1. Comparison of prediction effects of different models.

Models	SCC	MAE	RMSE	R^2
Linear regression	0.893	10.701	13.115	0.747
Polynomial regression	0.907	8.229	10.651	0.879
SVR	0.852	12.592	16.728	0.695
Random forest regressor	0.956	2.251	3.766	0.956
XGBoost	0.953	4.612	6.511	0.968
LSTM	0.976	2.793	3.212	0.974
MGRU-5F	0.971	0.483	0.556	0.690
MGRU-3F	0.964	0.378	0.522	0.727
MGRU-3F-RZ	0.984	0.264	0.318	0.890
MGRU-3F-RM	**0.986**	**0.163**	2.212	**0.986**

It can be seen that the proposed model MGRU-3F-RM performs well in all performance indicators, although it has not the lowest RMSE. In addition, it can be seen that although the R^2 value of LSTM and XGBoost is similar, the RMSE has still a certain gap. With the increase of data volume, the deep learning model may be better.

5 Conclusion

In this study, a framework for oil price forecasting was developed based GRU using Mahalanobis distance transform to remove outliers in combination with MGRU. We use Spearman correlation coefficient for feature selection and eliminate outliers by Mahalanobis distance, we construct MGRU-3F-RM model based on data-driven. Four metrics were used to measure the performance of these models. Our model MGRU-3F-RM achieves a better performance level. However, the oil price may be influenced by other events, it's still very challenging to predict the price precisely, so we will focus on exploring other new intelligent models combined with specific real business to predict oil price well in the future.

References

1. Li, X., Shang, W., Wang, S.: Text-based crude oil price forecasting, a deep learning approach. Int. J. Forecast. **35**, 1548–1560 (2019)
2. Chen, Y., He, K., Tso, G.K.F.: Forecasting crude oil prices: a deep learning based on model. Procedia Comput. Sci. **122**, 588–595 (2017)
3. Chen, Y., Zou, Y., Zhou, Y., Zhang, C.: Multi-step-ahead crude oil price forecasting based on grey wave forecasting method. Procedia Comput. Sci. **91**, 1050–1056 (2016)
4. Gumus, M., Kiran, M.S.: Crude oil price forecasting using XGBoost. In: International Conference on Computer Science and Engineering, pp.1100–1103. IEEE, Antalya (2017)
5. Luo, Z., Cai, X., Tanaka, K., Takiguchi, T., Kinkyo, T., Hamori, S.: Can we forecast daily oil futures prices? Experimental evidence from convolutional neural networks. Risk Finan. Manage. **2**(12), 519–530 (2019)
6. Salvi, H., Shah, A., Mehta, M., Correia, S.: Long Short-term model for Brent oil price forecasting. Eng. Tech. **5**(7), 315–319 (2019)
7. Mahajan, R., Mansotra, V.: Predicting geolocation of tweets: using combination of CNN and BiLSTM. Data Sci. Eng. **6**(4), 402–410 (2021)
8. Sugiartawan, P., Pulungan, R., Kartika Sari, A.: Prediction by a hybrid of wavelet transform and long-short-term-memory neural network. Int. J. Adv. Comput. Sci. Appl. **8**(2), 287–296 (2017)
9. Cen, Z., Wang, J.: Crude oil price prediction model with long short term memory deep learning based on prior knowledge data transfer. Energy **169**, 160–171 (2019)
10. Urolagin, S., Sharma, N.: Tapan Kumar Datta: A combined architecture of multivariate LSTM with Mahalanobis and Z-Score transformations for oil price forecasting. Energy **231**, 975–987 (2021)
11. Wu, J., Li, B., Ji, Ye., Tian, J., Xiang, Y.: Text-enhanced knowledge graph representation model in hyperbolic space. In: Li, B., Yue, L., Jiang, J., Chen, W., Li, X., Long, G., Fang, F., Yu, H. (eds.) ADMA 2022. LNCS (LNAI), vol. 13088, pp. 137–149. Springer, Cham (2022). https://doi.org/10.1007/978-3-030-95408-6_11
12. Titouna, C., Titouna, F.: Outlier detection algorithm based on mahalanobis distance for wireless sensor networks. In: Proceedings of the 2019 International Conference on Computer Communication and Informatics, pp.1567–1576. IEEE, Coimbatore (2019)
13. Drumond, D.A., Rolo, R.M., Costa, J.F.C.L.: Using Mahalanobis distance to detect and remove outliers in experimental covariograms. Nat. Resour. Res. **28**, 1056–1067 (2018)

Author Index

Printed in the United States
by Baker & Taylor Publisher Services